10 / 15

To Richard:
With my compliments!

George Fisher

Springer Series in Operations Research

Editors:
Peter Glynn   Stephen M. Robinson

Springer
New York
Berlin
Heidelberg
Barcelona
Hong Kong
London
Milan
Paris
Singapore
Tokyo

Springer Series in Operations Research

George S. Fishman

Discrete-Event Simulation
Modeling, Programming, and Analysis

With 83 Illustrations

Springer

George S. Fishman
Department of Operations Research
University of North Carolina at Chapel Hill
CB#3180
210 Smith Building
Chapel Hill, NC 27599-3180
USA
gfish@email.unc.edu

Series Editors:
Peter Glynn
Department of Operations Research
Stanford University
Stanford, CA 94305
USA

Stephen M. Robinson
Department of Industrial Engineering
University of Wisconsin–Madison
1513 University Avenue
Madison, WI 53706-1572
USA

Library of Congress Cataloging-in-Publication Data
Fishman, George S.
    Discrete-event simulation: modeling, programming, and analysis / George S. Fishman
        p.   cm. — (Springer series in operations research)
    Includes bibliographical references and index.
    ISBN 0-387-95160-1
    1. Simulation methods.   2. Discrete-time systems.   I. Title.   II. Series.
    T57.62.F58   2001
    658.4$'$0352—dc21   00–052608

Printed on acid-free paper.

Production managed by Lesley Poliner; manufacturing supervised by Erica Bresler.
Typeset by The Bartlett Press, Inc., Marietta, GA.
Printed and bound by Hamilton Printing Co., Rensselaer, NY.
Printed in the United States of America.

9 8 7 6 5 4 3 2 1

ISBN 0-387-95160-1   SPIN 10785084

Springer-Verlag   New York Berlin Heidelberg

A member of BertelsmannSpringer Science+Business Media GmbH

To Paula Sue,

my wife and best friend

Preface

As tools for scientific inquiry, the methodologies of discrete-event simulation have been with us for almost 50 years. They enable us to study discrete-event dynamical systems in which delay is an intrinsic feature. Examples easily come to mind in manufacturing, communications networks, transportation, health-care delivery, and public-sector services.

The methodologies embrace modeling, programming, input-data analysis, pseudorandom variate generation, output-data analysis, and presentation of results. These 50 years have seen these topics evolve in content, sophistication, and, most compellingly, accessibility. In this last regard, the mushrooming PC environment of the past 20+ years has facilitated a relatively rapid shift of many simulation techniques from concept to practical use. Witness the many discrete-event simulation software systems available to the would-be simulationist.

Time was when a distinct demarcation existed between modeling and programming. One laid out the dynamic logic of a system, converted it to executable computer code, and executed the resulting simulation program. Today, these distinctions are less apparent. Taking advantage of a highly interactive PC environment available to many, vendors market simulation software that offers macrostatements or modules that correspond to frequently encountered modeling constructs, These macrostatements are collections of more elementary statements in an underlying computer language. Some apply directly to delay modeling;

others apply to sampling on a computer; still others streamline data collection and report preparation. Using these macrostatements effectively merges modeling and programming activities into a single effort.

In many respects, these developments are welcome. They eliminate much of the drudgery and delay in getting a simulation program to execute. However, they have drawbacks. The macrostatements cover only a subset of all the dynamic logic that may be needed to model a system. When no macrostatement represents the logic of a particular problem, a simulationist either has to resort to programming in the underlying language to create a correct model, or has to create a *kluge* from existing macrostatements, sometimes awkwardly, to create an equivalence. Most regrettably, a simulationist sometimes merely changes the model to accommodate the available macrostatements.

Virtually all simulation software systems allow a simulationist to select sampling distributions from libraries or menus. When the desired distribution is not offered, the simulationist's only recourse may be to a general-purpose programming language like C. Alternatively, he/she may choose to use whatever the simulation software offers and ignore the distribution(s) that the problem setting dictates. While offering options for customizing reports, automatic assessment of the statistical accuracy of a simulation's output is available in few simulation software systems.

Since students often learn about discrete-event simulation through a particular software's modeling, sampling, analysis, and reporting capabilities, their skills for overcoming the impediments mentioned above may be severely limited unless they are well versed in the basic concepts of discrete-event simulation. This book supplies a means for providing this background. It gives a relatively detailed account of the principal methodologies of discrete-event simulation. Modeling (Chapter 2), programming (Chapter 4), output analysis (Chapter 6), making sense of the output (Chapter 7), pseudorandom variate generation (Chapters 8 and 9), and input-data analysis (Chapter 10) are all treated in depth sufficient to acquaint a would-be simulationist with the central issues.

The book also offers an excursion into the underlying models of data collection (Chapter 3), explaining why using these models leads to sample averages that approximate the values of unknown system parameters with errors that diminish as the simulation runs for longer and longer amounts of computing time. It also characterizes these errors when a simulation is run, more realistically, for a finite computing time and, therefore, a finite sample-path length. By sensitizing the reader to the presence of error prior to running a simulation program, we hope to give her/him a better understanding of experimental design considerations, which are so important for carrying out simulation experiments in a knowledgeable way, and on which Chapters 4 and 6 touch.

Chapter 5 offers another excursion, this time into the topic of computing time versus size of problem. To obtain a specified level of statistical accuracy in output sample averages, discrete-event simulation requires computing time that tends to grow with the size of the system under study, sometimes superlinearly. If a simulationist is to avoid an excessive amount of computing, he/she needs to be acquainted with the principal determinants of this growth. Chapter 5 describes how this growth occurs for several delay models.

The book is suitable for advanced undergraduates and graduate students in the management, mathematical, and physical sciences and in the engineering sciences. A background in calculus, probability theory, and intermediate statistics is necessary. Familiarity with stochastic processes considerably eases the study of Chapter 6. Knowledge of a programming language significantly reduces the setup cost in mastering Chapter 4. The book also serves as a ready reference for professionals who want to acquaint themselves with some of the deeper issues of discrete-event simulation, such as estimating the accuracy of results (Chapter 6) and the comparison of results for different scenarios (Chapter 7).

My students at the University of North Carolina at Chapel Hill have served as the test bed for most of the material in this book. The course assigns Chapter 1 for background reading on delay systems and as an introduction to discrete-event simulation. Detailed lectures on Chapter 2, followed by selective lectures on sections in Chapter 3, and by rather complete lectures on Chapter 4, work well to prepare students to do the exercises at the ends of Chapters 2 and 4. In particular, Chapter 4 uses SIMSCRIPT II.5 and Arena to illustrate programming concepts. SIMSCRIPT II.5 is a full-service programming language (like C) with a comprehensive collection of macrostatements for discrete-event simulation. Arena is a highly integrated simulation software system that in addition to providing macrostatements in the SIMAN simulation language allows a user to engage menu-driven software that performs many ancillary tasks that accompany simulation experimentation. These include an input-data analyzer, an output-data analyzer, customized report generation, and output graphical display. Instructors who choose simulation software other than SIMSCRIPT II.5 or Arena can easily substitute lectures (for those on Chapter 4) on their preferred software. However, students find intriguing a lecture on the variance-reducing technique known as *common pseudorandom numbers* (Section 4.22).

Selective lecturing on Chapter 5 is an eye opener for most students. I customarily focus on Sections 5.1 through 5.4.5. Lecturing on Chapter 6, especially on Sections 6.1 through 6.8, with the aid of the LABATCH.2 software, available on the Internet, significantly heightens the student's ability to assess the accuracy of results. Selective lecturing based on Sections 7.1 through 7.6.1 and, if time permits, on Sections 7.7 through 7.11, creates an awareness among students that there is much more to output analysis and presentation that looking at a general-purpose automatically generated report.

Lectures on Sections 8.1 through 8.6 provide the basics of random variate generation. Examples taken from Sections 8.9 through 8.23 illustrate the concepts. For instance, Section 8.13 offers a good example of a sophisticated algorithm that has bounded expected computing time for Gamma sampling. I also lecture on Chapter 9, which addresses pseudorandom number generation. I like to include Section 9.7, which describes a relatively new type of pseudorandom number generator that ameliorates many of the limitations of the commonly employed linear congruential generators. Although this and like generators are not now in common use, they are a preview of things to come. Chapter 10 describes techniques for input-data analysis. These can be covered selectively, depending on the types of data available for classroom analysis.

Four problems in the modeling exercises at the end of Chapter 2 are carried forward through programming in Chapter 4, output analysis in Chapter 6, and comparison of results in Chapter 7. Some problems demand more modeling and programming effort than others, which require more effort in estimation and interpretation of results. To give students a sense of satisfaction in solving a simulation problem from beginning to end, I customarily choose one of the problems in Chapter 2 and assign its corresponding exercises there and in Chapters 4, 6, and 7. Some exercises in Chapter 8 allow students to gain perspective on the wide applicability of the methods of Sections 8.1 through 8.6. Others offer hands-on experience in programming and executing algorithms. Those at the end of Chapter 9 offer students an opportunity to hone their skills in analyzing pseudorandom number generators.

When I began writing the book in 1990, little did I know that such massive changes would take place in simulation software during the following decade. I describe many of these developments while retaining much of the well-established methodologies that have so well served simulationists in the past. However, space considerations have inevitably led to a selectivity that reflects my best judgment as to what a simulationist should know. The book limits its discussion of variance-reducing techniques to those that I have found most useful in the context of discrete-event simulation. While it addresses various aspects of experimental design, more on this topic would be beneficial. Although important, the topics of verification and validation have not made the cut, the rationale being that these are considerations that apply to virtually every aspect of scientific modeling, not just discrete-event simulation. More specialized topics, including regenerative simulation and infinitesimal perturbation analysis, are also omitted. While of considerable conceptual interest, neither of these two topics can claim to be an intrinsic part of discrete-event simulation as practiced today.

I am grateful to Christos Alexopoulos of Georgia Tech and Jim Henriksen of Wolverine Software for their helpful comments on parts of the manuscript, and to Russell Cheng, Daryl Daley, Debasis Mitra, and Les Servi for their direct or indirect help with several of the exercises. Thanks also go to Pierre L'Ecuyer for his assistance in obtaining and operationalizing software for the combined pseudorandom number generator described in Section 9.7. My colleagues Vidyadhar Kulkarni, Scott Provan, and Sandy Stidham have all been generous with their time and counsel, often steering me onto a much more constructive path than I might have otherwise taken. I am most grateful to them. Thanks go to all my students of the past decade who have subtly and, occasionally, overtly offered their comments as to what works and what does not. Kathleen Dulaney, Betty Richmond, and Barbara Meadows typed sections of this manuscript over the past decade. For their patience with my repeated modifications and corrections I am again most grateful.

Errata for this book are available at http://www.or.unc.edu/~gfish .

George S. Fishman
Chapel Hill, North Carolina
August 2000

Contents

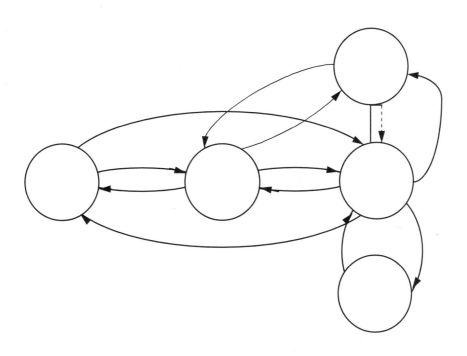

CHAPTER 1

Simulation in Perspective

Discrete-event simulation consists of a collection of techniques that, when applied to the study of a discrete-event dynamical system, generates sequences called *sample paths* that characterize its behavior. The collection includes:

- Modeling concepts for abstracting the essential features of a system into a coherent set of precedence and mathematical relationships among its elements

- Specially designed software for converting these relationships into computer-executable code capable of generating the requisite sample-path data

- Procedures for converting these data into estimates of system performance

- Methods for assessing how well these estimates approximate true, but unknown, system behavior.

Modeling complex systems has become a way of life in many fields, most especially in the engineering, health, management, mathematical, military, social, telecommunications, and transportation sciences. It provides a relatively low-cost way of gathering information for decision making. Since the size and complexity of real systems in these areas rarely allow for analytical solutions to provide the information, discrete-event simulation executed on a computer has become the method of choice. As with all tools of scientific inquiry, success in applying discrete-event simulation to a particular problem depends on the depth and breadth of its underlying model as an approximation of the system, and on the investigator's or *simulationist's* skill in applying her/his collection of techniques to penetrate the mysteries of the problem.

As a basis for acquiring these skills, this book describes concepts and methods and provides guidance for:

- Modeling the behavior of a discrete-event system (Chapter 2)

- Translating the model into code executable on a computer (Chapters 3 and 4)

- Executing the code and generating output in response to one or more input scenarios, some assigning different numerical values to system parameters and others based on different strategies for dynamically operating the system (Chapter 4)

- Analyzing each output to infer the system behavior that each scenario induces. This is done by assessing, by means of statistical methods, how well estimates of system performance measures approximate their true, but unknown, values (Chapter 6)

- Comparing these inferred behaviorial patterns for the different scenarios (Chapter 7).

Knowledge of these topics is essential for an analyst who plans a discrete-event simulation experiment, executes it, and wants to draw conclusions from its experimental results that can withstand scientific scrutiny. In today's world of readily available software package that virtually automates simulation modeling and execution, the newcomer to the methodology of discrete-event simulation may easily conclude that there is only one way to model a system and one way to encode it for execution. Indeed, many courses on discrete-event simulation focus on the features of a particular software package as the critical issues. While this style of pedagogy may suffice for teaching students how to simulate relatively small simple problems, it denies them the broader perspective about modeling alternatives and an awareness of all options available to them. Chapters 2 through 4 offer the reader a broader view of what is possible.

No methodology within discrete-event simulation has been as neglected as that for assessing how well sample averages approximate true, but unknown, long-run time averages that characterize system performance. Although some simulation software does provide a *batch-means* capability, the burden is on the user to figure out how to employ it in a way consistent with good scientific practice. This is not easy. To reduce this burden, Chapter 6 decribes implementable software, LABATCH.2, based on the batch-means method that with minimal user effort can be applied to simulation output, either during or after execution, to make this assessment. The software is available on the Internet at http://www.or.unc.edu/~gfish/labatch.2.html.

The book also addresses:

- Computational complexity as a function of problem size (Chapter 5)

- Sampling from diverse distributions on a computer (Chapter 8)

- Pseudorandom number generation (Chapter 9)

- Using empirical data to estimate the values of the input parameters of a simulation model (Chapter 10).

In practice, dramatic increases in computing time and memory requirements frequently occur when a simulation user increases the size of the system being modeled. This consequence often puzzles users, since the cause of these increases is not apparent to them. Our account of *computational complexity* provides a reader with an understanding of how these increases can arise, thus encouraging the choice of modeling techniques that do most to control the inevitable increase in computing time and memory.

Virtually all simulation software provides for sampling from a finite collection of continuous and discrete probability distributions. Chapter 8 expands the options open to the simulationist by describing methods for sampling from selected theoretical and tabled empirical distributions. It does so with an emphasis on sampling algorithms characterized by *bounded-mean computing times*, regardless of the values assigned to the parameters of the distribution. It also describes algorithms for a wider range of theoretical distributions than most simulation software offers.

All simulation sampling on computers relies on a source of *pseudorandom numbers*. Chapter 9 gives a concise description of the method of generating these numbers found in most simulation software, the properties of these numbers, and procedures for generating pseudorandom numbers with improved properties.

Before a simulation experiment can be executed, numerical values must be assigned to its input parameters. These values come from either expert judgment or the analysis of empirical data. Chapter 10 provides a concise account of the most essential details of this data analysis for several environments that arise in simulation modeling.

1.1 DISCRETE-EVENT SYSTEMS

In a discrete-event system, one or more phenomena of interest change value or *state* at discrete points in time, rather than continuously with time. For example, Figure 1.1 depicts a bus route partially characterized by:

- Number of passengers on the bus

- Number of individuals waiting at each of the five stops

- Location of the bus.

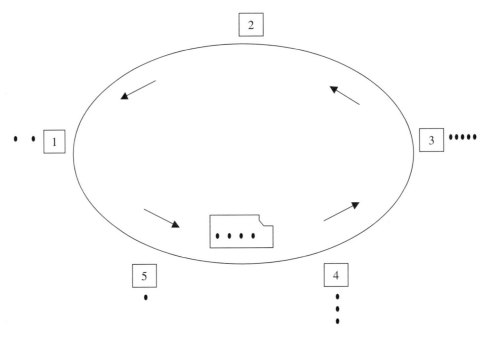

Figure 1.1 Bus model

The first two descriptors are discrete, whereas bus location is continuous on the circuit shown in Figure 1.1. However, the number of passengers on the bus can change only when the bus location assumes one of the five values that uniquely identify the stops. Moreover, the number waiting at a stop can change only when the bus arrives there or when a new individual comes to the stop.

Clearly, the sequences of bus arrival times and individual arrival times, together with bus capacity, determine how many passengers wait at each stop and bus occupancy at each moment in time. They also determine how long each individual waits for the bus.

Manufacturing plants, inventory systems, distribution systems, communications networks, transportation networks, health-care delivery systems, as well as many other environments are amenable to modeling as discrete-event systems. Virtually all measure their performance in terms of delay, number waiting, throughput, and resource utilization.

Delay denotes time spent waiting for resources. *Buffer occupancy* denotes number of items, jobs, or individuals waiting for resources. Number of finished units emerging from the system per unit time characterizes *throughput*, and proportion of time that resources are busy relative to total time describes *resource utilization.*

As examples, throughput in a manufacturing setting denotes the number of finished goods that a factory produces per unit of time. In a telecommunication network, the rate at which messages pass from sender to receiver constitutes a measure of throughout. The number of vehicles that cross a bridge per unit time characterizes throughput in a particular type of transportation system.

Buffer occupancy denotes the number of items in the system awaiting processing or service and is sometimes called *queue length*. Delay denotes the time that these items wait and alternatively is called *waiting time*.

Occasionally, *system loss rate* is an additional performance descriptor. It measures the rate at which objects leave a system, without successfully completing their purposes, when they encounter delays exceeding specified tolerances. For example, work-in-process within a manufacturing system may lose value unless delays between successive steps in the production process are kept within specific limits. This deterioration, which often arises because of the properties of materials and the production process, occurs in some forms of semiconductor wafer fabrication.

In the bus example, passenger waiting times describe delay, number waiting at each station describes queue length, number of completed passenger trips per unit time describes throughput, and the ratio of time-average seat occupancy and seat capacity describes resource utilization. In this case, the buffer or waiting room at each stop has infinite capacity. However, in many systems the buffer is finite, thus requiring the modeler to specify explicit rules for what to do when an arrival occurs to an already-full buffer. More generically, every discrete-event system embodies at least seven concepts:

- Work

- Resources

- Routing

- Buffers

- Scheduling

- Sequencing

- Performance.

Work denotes the items, jobs, customers, etc. that enter the system seeking service. Would-be passengers are the work in the bus example. *Resources* include equipment, conveyances, and manpower that can provide the services, for example, the bus in Figure 1.1. Associated with each unit or batch of work is a *route* delineating the collection of required services, the resources that are to provide them, and the order in which the services are to be performed. For the bus illustration, each passenger uses the bus to move from the stop at which the passenger gets on to the stop at which he/she gets off.

Buffers are waiting rooms that hold work awaiting service. They may have infinite capacity, as in the bus example, or may have finite capacity, as occurs at the retransmission nodes (resources) in a telecommunication network that transmits packets of information.

When buffers have finite capacity, explicit rules must be adopted that account for what happens to arriving work that finds a buffer full.

Scheduling denotes the pattern of availability of resources. For example, suppose there are two buses in Figure 1.1 and operating costs suggest that both buses be used only during daily periods of high demand. A schedule would consist of the times at which each bus is to provide service, taking into consideration time out for maintenance and cleaning.

Sequencing denotes the order in which resources provide services to their waiting work. It may be in first-come-first-served order; it may be dictated by the amount of work awaiting the resource or other resources from which it receives or transmits work. Sometimes the rule for sequencing is called the *queueing discipline*.

1.2 OPEN- AND CLOSED-LOOP SYSTEMS

Collectively, the concepts of time delay, number waiting, resource utilization, and through-put characterize performance measures in a *queueing model*. Originally formulated to study problems in telephony, queueing models, in varying degrees of detail, are today the intrinsic structures for studying many discrete-event systems in which the routing of items, the se-quencing of jobs for service to be performed on the items at each station, and the scheduling of resources to perform these services are principal determinants of delay, number of items in system, resource utilization, and throughput.

Most queueing systems are either *open-loop* or *closed-loop systems*. Work arrives to an open-loop system from outside the system at a rate independent of the state of the system and beyond the control of system managers. Figure 1.2 depicts five examples of open-loop systems. Analysts study systems like these for the purpose of finding work-routing, work-selection (sequencing), and resource-scheduling policies that balance the competing needs to keep queue length and waiting time low and resource utilization high.

The five formulations in Figure 1.2 reveal a small portion of the wide variety of behavior that open-loop queueing systems can represent. Figure 1.2a illustrates the most elementary model. A customer arrives at a single-server facility and waits in the buffer until the server becomes available and selects her/him for processing, according to the queueing discipline. Once processing is complete, the customer departs and, if any customers remain in the buffer, the server selects one for processing. A barber shop with one chair, a medical practice with a single physician, and a shoe-repair service with a single cobbler fit this model.

Figure 1.2b extends this concept to a multiple-server environment in which any one of r servers can process any customer. Figure 1.2c extends the concept of Figure 1.2a to one in which a customer requires k distinct types of service in series. A generalization to multiple servers at each station is easily made.

Figure 1.2d introduces the concept of *quality control* in which a *tester* determines whether or not the work performed on items is satisfactory. When it is not, the item must be reprocessed by the server. In this example, p denotes the proportion of tests that are

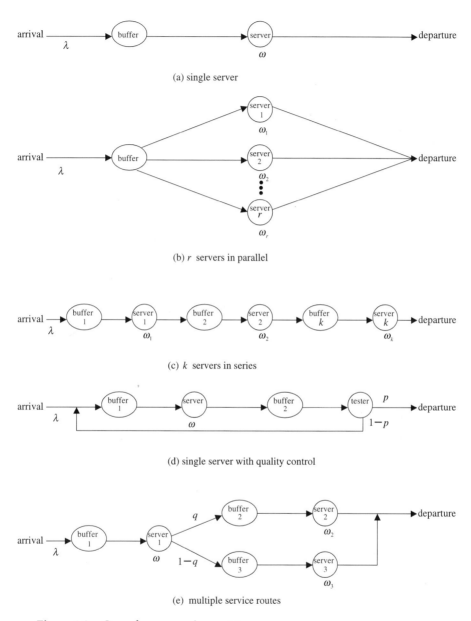

(a) single server

(b) r servers in parallel

(c) k servers in series

(d) single server with quality control

(e) multiple service routes

Figure 1.2 Open-loop queueing systems

successful. Figure 1.2e exemplifies a system in which the a proportion q of the arrivals requires service at stations 1 and 2 and the remaining proportion $1 - q$ requires service at stations 1 and 3.

This is merely a microcosm of the many variations that can arise in a service environment in which each arrival comes equipped with a route through the system that indicates the services it requires, each server selects items for service according to its queueing discipline, and there exists a schedule giving the time intervals during which each server is available to do work.

Control over work arrival times distinguishes closed-loop from open-loop systems. Figure 1.3 illustrates the closed-loop concept. A *new* item arrives for service at the moment the server completes processing its *old* work. By enforcing this policy, the system manager keeps throughput and resource utilization high and queue length and waiting time low. In particular, queue length and waiting time are necessarily zero in this illustration.

Closed-loop systems characterize many manufacturing, transportation, and telecommunications systems with controlled access. In addition to choosing routing, sequencing, and scheduling policies, the ability to control admission based on the state of the system offers system managers a valuable tool for high throughput and resource utilization and low system occupancy and delay. This motivates the study of admission policies presumably designed to accomplish these objectives. From time to time, we return to this distinction between open-loop and closed-loop systems to emphasize the implications of each for discrete-event simulation.

1.3 TIME AVERAGES

As mentioned earlier, queue length, waiting time, throughput, and resource utilization all relate to system performance. To be useful concepts for decision making, we need a way of summarizing what the output of a simulation run tells us about these phenomena. Time averages provide a basis for doing this, as we illustrate for the single-server model in Figure 1.2a.

Assume that the buffer in Figure 1.2a has infinite capacity and arriving items have infinite patience. For $t > s \geq 0$ let

$$A(s, t) := \text{number of arrivals in time interval } (s, t],$$

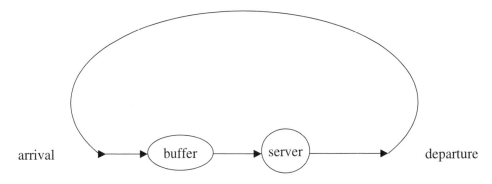

Figure 1.3   Closed-loop single-server system

$$N(s, t) := \text{number of completions in time interval } (s, t],$$

$$Q(t) := \text{queue length at time } t.$$

$$B(t) := \begin{cases} 1 & \text{if the server is busy at time } t, \\ 0 & \text{otherwise,} \end{cases} \tag{1.1}$$

$$W_i := \text{waiting time of departure } i,$$

and note the accounting identities

$$A(s, t) = A(0, t) - A(0, s),$$
$$N(s, t) = N(0, t) - N(0, s),$$

and

$$A(0, t) - N(0, t) = Q(t) - B(t).$$

Suppose that a simulation is run over a time interval $(0, t]$ and that these phenomena are observed over the interval $(s, t]$, s being a *warm-up interval* that reduces the influence of the starting conditions on the behavior of the phenomena over $(s, t]$ (Section 6.1) Then the system has sample average arrival rate

$$\bar{A}(s, t) := \frac{A(s, t)}{t - s},$$

average throughput

$$\bar{N}(s, t) := \frac{N(s, t)}{t - s},$$

average number waiting or queue length

$$\bar{Q}(s, t) := \frac{1}{t - s} \int_s^t Q(u) du, \tag{1.2}$$

average server utilization

$$\bar{B}(s, t) := \frac{1}{t - s} \int_s^t B(u) du, \tag{1.3}$$

and average waiting time

$$\bar{W}(s, t) = \frac{1}{N(s, t)} \sum_{i=N(0,s)+1}^{N(0,t)} W_i. \tag{1.4}$$

If the pattern of arrivals is deterministic at rate λ and each item has processing time $1/\omega$, then the system in Figure 1.2a is said to be *deterministic* and $\bar{A}(s,t) \rightarrow \lambda$, where notation "$a(t) \rightarrow b$ as $t \rightarrow \infty$" means that "$a(t)$ converges to b as t grows without bound." Moreover, the limiting behaviors of $\bar{Q}(s,t)$, $\bar{B}(s,t)$, and $\bar{W}(s,t)$ depend on λ and ω. In particular, as $t \rightarrow \infty$

$$\left.\begin{array}{ccc} \bar{N}(s,t) & \rightarrow & \lambda \\ \bar{Q}(s,t) & \rightarrow & 0 \\ \bar{B}(s,t) & \rightarrow & \lambda/\omega \\ \bar{W}(s,t) & \rightarrow & 0 \end{array}\right\} \quad \text{if } \lambda < \omega$$

and   (1.5)

$$\left.\begin{array}{ccc} N(s,t) & \rightarrow & \omega \\ \bar{Q}(s,t) & \rightarrow & \infty \\ \bar{B}(s,t) & \rightarrow & 1 \\ \bar{W}(s,t) & \rightarrow & \infty \end{array}\right\} \quad \text{if } \lambda > \omega.$$

If interarrival times, service times, or both are random and have finite first and second moments, then the limits (1.5) continue to hold as $t \rightarrow \infty$ *with probability one*, which we denote by w.p.1. This characterization merely emphasizes that the limit is attained in accordance with rules that acknowledge the random behavior inherent in the system.

The condition $\lambda < \omega$ is necessary for *stability*, and every queueing system must meet an analogous condition to keep queue length and waiting time from growing without bound as time elapses. For example, the multiserver system in Figure 1.2b requires $\lambda < \omega_1 + \cdots + \omega_r$, whereas the k-station system of Figure 1.2c requires $\lambda < \min(\omega_1, \ldots, \omega_k)$. Since the tester in Figure 1.2d induces total mean service time $1/\omega + (1-p)/\omega + (1-p)^2/\omega + \cdots$, the stability requirement is $\lambda < \omega p$, whereas the dual route environment in Figure 1.2e requires $\lambda < \omega_1, q\lambda < \omega_2$, and $(1-q)\lambda < \omega_3$; or equivalently $\lambda < \min[\omega_1, \omega_2/q, \omega_3/(1-q)]$.

A more general characterization of the systems in Figures 1.2a and 1.2c reveals the central role that stability plays in performance. Consider a k-station series system with m_i servers each operating at rate ω_i at station i for $i = 1, \ldots, k$, and arrival rate λ. Let

$$\rho := \frac{\lambda}{\min_{1 \leq i \leq k}(m_i\omega_i)}$$

and assume random interarrival and service times. Then under relatively mild conditions as $t \rightarrow \infty$, throughput behaves as

$$\bar{N}(s,t) \rightarrow \lambda \min(1, 1/\rho) \qquad \text{w.p.1,}$$

whereas sample average queue length (number in system waiting for service) and waiting time (system delay time) behave as

$$\bar{Q}(s,t) \rightarrow \begin{cases} O(\dfrac{1}{1-\rho}) & \text{if } \rho < 1, \\ \\ \infty & \text{if } \rho > 1, \end{cases} \quad \text{w.p.1,}$$

and

$$\bar{W}(s,t) \rightarrow \begin{cases} O(\dfrac{1}{1-\rho}) & \text{if } \rho < 1, \\ \\ \infty & \text{if } \rho > 1, \end{cases} \quad \text{w.p.1,}$$

where the notation $O(u(z))$ denotes a function $\{v(z)\}$ for which there exist constants $c > 0$ and z_0 such that $v(z) \le cu(z)$ for all $z \ge z_0$.

Recall that in an open-loop system items arrive at a rate λ (> 0) independent of the number of items in the system awaiting or receiving service and independent of the status of all resources. The arrivals may be deterministic; for example, occurring in batches of fixed size n at times $1/\lambda, 2/\lambda, 3/\lambda, \ldots$. Alternatively, they may occur randomly in time with mean interarrival time $1/\lambda$ and random batch size. By contrast, the workload and its distribution among buffers and servers in a closed-loop system feed into an action space that determines the arrival pattern.

1.3.1 EXCEEDANCE PROBABILITIES

In addition to means, performance is sometimes viewed in terms of the *exceedance probabilities*:

- $\text{pr}[Q(t) > q] = 1 - F_Q(q), \qquad q \ge 0,$
- $\text{pr}[W_i) > w] = 1 - F_W(q), \qquad w \ge 0,$

where F_Q and F_W are distribution functions (d.f.s). For example, $\text{pr}[Q(t) > 0]$ denotes the probability that at an arbitrarily selected time t at least one item is waiting for service, and $\text{pr}(W_i > 0)$ denotes the probability that an arbitrarily chosen customer i must wait. Whereas means characterize central tendency, exceedance probabilities shed light on extreme behavior, especially as q and w increase.

Suppose that a simulation is run for the time interval $(0, t]$. Then the sample time averages

$$P_Q(q,s,t) := \frac{1}{t-s} \int_s^t I_{(q,\infty)}(Q(u))\mathrm{d}u, \quad q > 0, \tag{1.6}$$

and

$$P_W(w, s, t) := \frac{1}{N(s, t)} \sum_{i=N(0,s)+1}^{N(0,t)} I_{(w,\infty)}(W_i), \quad w > 0, \tag{1.7}$$

denote the proportion of time $(s, t]$ that the queue has more than q items and the proportion of customers who begin service in $(s, t]$ that wait more than w time units.

Under relatively widely applicable conditions, for fixed s

$$P_Q(q, s, t) \to 1 - F_Q(q) \text{ w.p.1 as } t \to \infty$$

for each $q \geq 0$. Also,

$$P_W(w, s, t) \to 1 - F_W(q) \text{ w.p.1 as } t \to \infty$$

for each $w \geq 0$. Sections 3.4 and 3.5 discuss these limits in more detail and consider them when a simulation is run until a fixed number of items enter service. The photocopier problem in the next section illustrates the use of means and exceedance probabilities.

We next describe several contexts in which simulation is an invaluable aid to discovery, thus allowing the reader to study the methods in this book with a heightened awareness of their potential value.

1.4   EXAMPLE: REDUCING WAITING TIME FOR PHOTOCOPYING

Photocopying in a library offers a good example of an open-loop queueing system. The Health Sciences Library (HSL) at the University of North Carolina at Chapel Hill has photocopying machines available for university-affiliated patrons, including students, faculty, and staff, and for commercial customers. Figure 1.4 roughly illustrates work-flow in a two-floor version of the problem. For a six-floor library, each buffer would have five flows from and five flows to other floors, for a total of 30 dashed lines, in a corresponding flowchart.

The HSL has 11 machines; 1 in the basement and 1, 2, 3, 3, 1, located on floors 1 through 5, respectively. Customers usually make copies on the floor on which they find their material. If all operational machines on a floor are busy, an arriving customer may choose to go to another floor. Some patrons for go copying when the delay becomes too long.

Photocopying machines are subject to paper jams that make them inoperative. The HSL employs a single *dejammer* to remedy these problems. When a machine becomes inoperative, the customer calls the front desk on the first floor to report the jam. The deskperson receiving the call then signals the dejammer by beeper, indicating the floor on which the jam has occurred. The dejammer services inoperative machines in the order in which he/she receives notification of the jams.

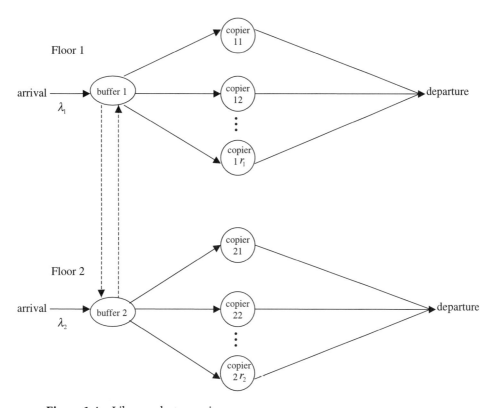

Figure 1.4   Library photocopying

Copiers are also subject to breakdowns requiring repair. When this occurs, the HSL calls a service that dispatches a repairperson. The concomitant loss of operating time is known to be a principal contributor to customer delay.

The purpose of studying this system was to evaluate the extent to which modifications to the operation of this photocopying system could reduce waiting time. A discrete-event simulation was developed to study the problem (Tzenova 1999). Simulation experiments were run under several different scenarios, including:

- Replacing old copiers that require frequent repair with new copiers
- Adding new copiers to selected floors
- Moving copiers among floors to be more responsive to the demand pattern
- Each of the above with a 50 percent increase in customer arrivals.

Statistical tests applied to library data revealed significant differences at the 5% level in arrival rates by floor, time of day, and day of week. Table 1.1 shows a partition into six groups that captured these differences. Since data limitations prevented a detailed analysis for the basement and floors 1, 2, and 5, they were treated individually as groups 7 through

Table 1.1 Partition of arrivals

Group	Floor	Peak hours 11 a.m.–4 p.m.	Off-peak hours 10–11 a.m., 4–5 p.m.	Day	Mean interarrival time (minutes)
1	3	x		All	4.04
2	3		x	All	4.53
3	4	x		Tuesday	3.38
4	4		x	Tuesday	3.77
5	4	x		All but Tuesday	4.28
6	4		x	All but Tuesday	4.30
7	Basement				16.14
8	1				10.15
9	2				5.40
10	5				23.87

10. Since these arrival rates were notably smaller than those for groups 1 through 6, this aggregation had no more than a negligible effect on simulation output.

For each scenario, differences in arrival rates by time of day and day of week necessitated four sets of runs to be executed; namely (off-peak hours, Tuesday), (peak hours, Tuesday), (off-peak hours, all but Tuesday), and (peak hours, all but Tuesday). For each, sample path data were collected for 1000 simulated five-day weeks. Omitting data collection during the warm-up interval reduced the influence of the arbitrarily selected initial conditions in each simulation run (Section 6.1).

The data were used to estimate:

- Mean waiting time

- Probability of waiting

- Probability of waiting more than five minutes

- Probability of an arriving customer not waiting for service when copiers are busy

plus additional performance measures. The three exceedance probabilities provide a picture of extreme behavior, whereas the mean reveals central tendency. Figure 1.5a shows point estimates and approximate 99% confidence intervals for the mean for each group for the library's current operating policy. The study used the ratio-estimation methodology of Section 6.11 with the LABATCH.2 statistical package in Section 6.7 to derive the confidence intervals. This ratio approach was necessitated by the need to account for random variation in the number of arrivals in a simulation run for a fixed number of days (Section 3.4). Figure 1.5b displays the corresponding point estimates of the exceedance probabilities.

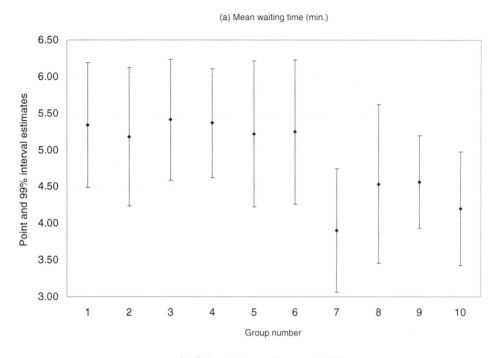

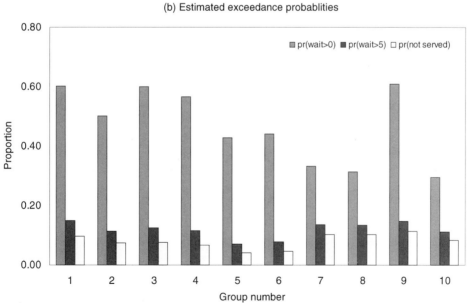

Figure 1.5 Photocopier waiting time analysis: Scenario 1

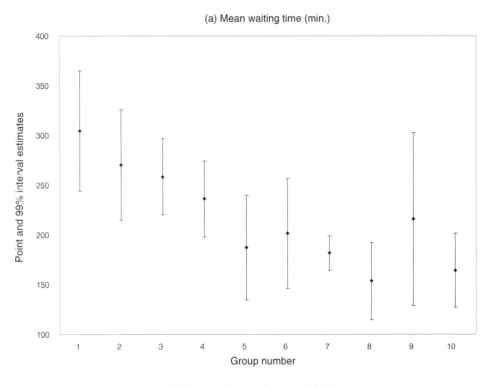

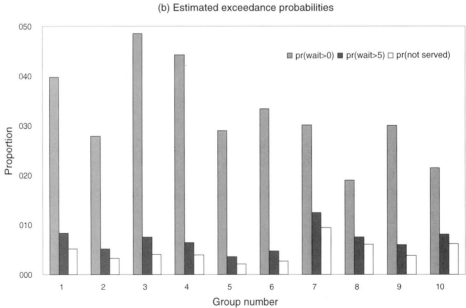

Figure 1.6 Photocopier waiting time analysis: Scenario 11

Among all scenarios considered, scenario 11, which replaced the four photocopiers with the highest failure rate, led to the most substantial reductions in waiting time. These were on the busiest floors, 3 and 4, which comprise Groups 1 through 6. Figure 1.6a shows the means, and Figure 1.6b, the exceedance probabilities. In particular, note that for a given group the corresponding confidence intervals in Figures 1.5a and 1.6a do not overlap, offering support to the contention that the difference in means is real. Also, note the reduction in exceedance probabilities, especially for the two busiest floors 3 and 4. While the differences are to be expected when new equipment replaces old, the real improvement on floors other than 3 and 4 in Groups 7 through 11 shows the benefit to customers on these floors as well.

These and other alternatives need to be evaluated with regard to the benefit each offers for the cost of the additional resources they require. Sections 6.11 and 7.6 return to this example.

1.5   EXAMPLE: INCREASING THROUGHPUT AT AN INCINERATOR

The process by which companies dispose of harmful liquid waste epitomizes a closed-loop system. Feel Good Incorporated is a large pharmaceutical company with extensive research and manufacturing facilities on four geographically close campuses or sites. Its research and development activities generate liquid waste, which the company disposes of in its own incinerator, located at its Environmental Safety Facility (ESF), and by arrangement with independent contractors. Since local disposal is considerably less costly per unit than contract disposal, the company would like to increase the throughput of its own incinerator. Current operating policies and storage capacity limit its ability to do so.

Each site generates liquid waste that is stored in 55-gallon metal drums on site, if space is available, until they can be transferred to the ESF, if it has available storage space. After being brought to the ESF, the waste is eventually pumped out of the drums into one of three vats. Two hold 500 gallons each and one holds 1000 gallons. After the emptied drums undergo a time-consuming triple rinsing process, they are returned to the holding areas at the manufacturing sites. The total number of 55-gallon drums is fixed.

Once a vat is filled, a sample of its liquid waste is sent to an independent laboratory for assay. During this period, which can be as long as a week, the waste remains in the vat. The assay determines the composition of the waste and specifies the temperature and time duration for incineration of the vat's contents. The incinerator burns waste from one vat at a time. After incineration, the ash residue is put in plastic drums and removed from the incinerator complex.

Under this policy, more drums of waste are occasionally generated than can be stored until transfer to the incinerator complex. Then Feel Good employs an outside contractor to remove the excess drums and dispose of the waste. Therefore, we can think of Feel Good's waste incinerating process as a closed-loop system to which admission is controlled by the availability of storage space at the four sites. Collectively, the four sites have space to store up to 300 drums, and the ESF has space for 11 drums. As a consequence, waste can be stored on the campuses only if space is available. Hereafter, we refer to a drum handled by an outside contractor as *lost*.

Disposing of waste by outside contractor is considerably more expensive per drum than disposal at the incinerator complex. Although the need for outside contracting rarely occurs at the current waste generation rate, this rate is expected to increase in the future. In anticipation of this growth, the company requested a study to:

- Estimate the volume of outside contracting to expect as the amount of waste generation grows in the future

- To identify which of the several time-consuming steps of transporting dums, pumping their contents into vats, assay, rinsing drums, and incineration is the principal retardant of a higher incinerating rate. This was to be studied for a fixed number of drums and storage space at the four campuses and at the ESF.

A discrete-event simulation experiment (Ma 1999) was developed and executed for selected annual waste-generation growth rates. Of particular interest for each growth rate was $\mu :=$ mean annual number of lost drums. In the notation of Section 1.3, let

$A_i(0, t) :=$ number of generated drums at site i in the time interval $(s, t]$,

$Q_i(t) :=$ number of stored drums at time t at site i,

$Q'(t) :=$ number of drums at time t at the incinerator being stored or processed,

$N(s, t) :=$ number of drums processed at the incinerator during the time interval $(s, t]$.

Then the number of lost drums in $(s, t]$ is

$$D(s, t) := \sum_{i=1}^{4} A_i(s, t) - \sum_{i=1}^{4} Q_i(t) - Q'(t) - N(s, t),$$

and the time-average annual number of lost drums over τ years is

$$\hat{\mu}(\tau) := \frac{1}{\tau} D(s, s + \tau) = \frac{1}{\tau} \sum_{j=1}^{\tau} D(s + j - 1, s + j),$$

where data collection begins at year s, thereby diluting the influence of initial conditions in the simulation (Section 6.1) on the sample-path data. The summation partitions the sample-path data into increments that serve as the basis for a statistical analysis to assess how well $\hat{\mu}(\tau)$ approximates the unknown μ (Section 6.6).

Table 1.2 shows $\hat{\mu}(\tau)$ for selected growth rates α. To assess the statistical error in approximating $\mu(\tau)$ by $\hat{\mu}(\tau)$ due to randomness in drum generation in the simulation, a

Table 1.2   Number of drums handled by outside contractor[a] (current
waste generation = 380 drums/year)

Growth rate α	Sample mean $\hat{\mu}(\tau)$	99% Confidence interval[b]		
		L	U	$\frac{U-L}{\hat{\mu}(\tau)}$
$0 < \ \leq .05$	0	0	0	–
.06	.1065	0	.2562	2.41
.07	.9268	.6061	1.247	.69
.08	3.269	2.743	3.795	.32
.09	5.845	5.338	6.352	.17
.10	8.552	8.016	9.088	.13
.11	11.22	10.63	11.80	.10
.12	14.04	13.50	14.58	.08
.13	16.83	16.29	17.37	.06
.14	19.58	19.06	20.10	.05
.15	22.34	21.76	22.92	.05
.16	25.19	24.64	25.73	.04
.17	28.15	27.55	28.74	.04
.18	30.85	30.22	31.47	.04
.19	33.67	33.04	34.30	.04
.20	36.44	35.87	37.02	.03
.21	39.50	37.65	41.34	.09
.33	72.78	70.85	74.71	.05
.46	110.1	108.1	112.2	.04
.61	150.8	149.0	152.5	.02

[a] For $0 < \alpha \leq .20$, the simulations were run for $\tau = 10,000$ years.
For $\alpha \geq .21$, they were run for $\tau = 1,000$ years.
[b] $L :=$ lower limit; $U :=$ upper limit.

statistical analysis using LABATCH.2 (Section 6.7) gave the approximating .99 confidence
intervals for μ. That is, $\hat{\mu}(\tau)$ is a point estimate for μ, and $[L, U]$ covers μ with approximate
probability .99. The difference $U - L$ measures the absolute accuracy of $\hat{\mu}(\tau)$, and $(U - L)/\hat{\mu}(\tau)$ its relative accuracy. For example, $\alpha = .14$ has $(U - L)/\hat{\mu}(\tau) = .05$, indicating that
the length of the interval is 5% of the sample mean. This suggests a relatively high level of
accuracy. Only $\alpha = .06$ has a relatively inaccurate point estimate, but the absolute accuracy
measure $U - L$ is tolerable for present purposes.

Figure 1.7 plots the sample mean $\hat{\mu} := \hat{\mu}(\tau)$ versus α. The graph shows a straight line

$$\hat{\mu} = -19.46 + 279.9\,\alpha, \qquad \alpha > .07, \tag{1.8}$$

as computed by least-squares regression analysis. If 50 lost drums per year were a crucial
number, then the regression line reveals that an annual growth rate of

$$\alpha = \frac{50 + 19.46}{279.9} = 0.248$$

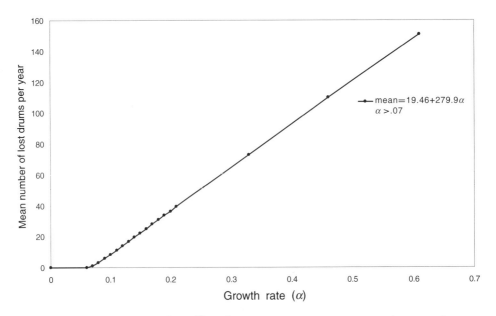

Figure 1.7   Mean number of lost drums per year vs. waste-generation growth rate

would generate that amount on average in one year. More generally, let

$$\mu(\alpha, t) := \text{mean number of lost drums } t$$
$$\text{years in the future with annual}$$
$$\text{growth rate } \alpha \text{ for waste generation.}$$

Then the regression line (1.8) gives

$$\mu(\alpha, t) = -19.46 + 279.9[(1 + \alpha)^t - 1], \qquad t > \frac{\ln 1.07}{\ln(1 + \alpha)}, \quad \alpha > 0, \qquad (1.9)$$

enabling us to make projections as in Table 1.3.

As illustration, the table reveals that the mean annual number of lost drums exceeds 50 in years 5, 4, and 3 for annual growth rates $\alpha = .05, .06,$ and $.08$, respectively. More generally, the table provides a basis for estimating growth in the number of lost drums.

The study also made simulation runs to evaluate five changes in operating policy, each intended to reduce the annual number of lost drums. One of these moved the taking of samples for testing content and the concomitant time delay in waiting for the resulting assay from the ESF to the four drum-originating campuses. Under this policy, the simulation results revealed zero lost drums for annual growth rates $0 < \alpha \leq .10$ over a five-year period. None of the remaining four scenarios achieved this zero loss.

Table 1.3   Mean annual number of lost drums[a]

| | Annual waste-generation growth rate α | | | | | | |
Year t	0.05	0.06	0.07	0.08	0.09	0.10	0.15
1	0	0	0.13	2.9	5.7	8.5	22.5
2	9.2	15.1	21.1	27.1	33.2	39.3	70.8
3	24.7	34.0	43.5	53.2	63.1	73.2	126.3
4	40.9	54.0	67.5	81.4	95.7	110.4	190.1
5	57.9	75.2	93.2	111.9	131.3	151.4	263.5
10	156.6	201.9	251.9	304.9	363.2	426.6	832.7

[a] Based on expression (1.9).

To effect this policy, the samples had to be taken from the individual drums at the sites rather than from a vat at the incinerator after the contents of the drums had been pumped in. This was deemed worth the effort to eliminate the time delay at the ESF due to waste remaining in a vat until the results of its assay were available.

1.6   WHY DISCRETE-EVENT SIMULATION

To put discrete-event simulation into context as a tool of analysis, we next describe the way in which it relates to modeling in general. The first step in studying a system is using accumulated knowledge to build a model. A model can be a formal representation based on theory or a detailed account based on empirical observation. Usually, it combines both. A model serves many purposes. In particular, it:

1. Enables an investigator to organize her/his theoretical beliefs and empirical observations about a system and to deduce the logical implications of this organization

2. Leads to improved system understanding

3. Brings into perspective the need for detail and relevance

4. Expedites the speed with which an analysis can be accomplished

5. Provides a framework for testing the desirability of system modifications

6. Is easier to manipulate than the system

7. Permits control over more sources of variation than direct study of a system allows

8. Is generally less costly than direct study of the system.

Except in relatively simple cases, the values of system performance descriptors are unknown, and determining then remains beyond the capacity of the analytical tools of queueing theory. Discrete-event simulation offers techniques that can approximate the values of these descriptors to within a remarkably small error. The approximations come from analyses of data observed on *sample paths* generated during execution of a simulation program corresponding to the model of interest, usually on a queue-length process $\{Q(z), s < z < t\}$ and on a waiting-time process $\{W_{N(0,s)+1}, \ldots, W_{N(0,t)+1}\}$ (Section 1.3). This book describes many of these methods in detail.

Often one creates and executes a simulation of a complex system simply to study alternatives. How does the system respond to a change in input parameter values? How does it respond to a change in routing, sequencing, or scheduling rules? Simulation-generated data often can provide sufficiently accurate answers to these questions, thereby enabling an investigator to sharpen her/his understanding of the system. Occasionally, interest focuses on the selection of one of several system configurations that leads to the *best* performance with regard to explicitly stated criteria. Carefully constructed and executed simulations of each system alternative can provide the sample data on which one can apply well-established statistical tests to detect superior performance (Sections 7.2, 7.5, and 7.6).

Another context in which the benefits of simulation are considerable is in the development of optimal or near-optimal policies for system management. This area involves a partnership between simulation and analytical methods. In many situations one constructs an analytical model of a system in order to determine an optimal policy, with regard to a specified criterion, for managing the system. However, the need to simplify the representation to obtain an analytical solution often removes from the model critical characteristics of the system that may affect performance. Here, the analyst who has found an optimal policy for the simplified analytical model ponders the extent to which the policy remains optimal in the presence of the features suppressed for analytical convenience. Discrete-event simulation can play a vital role in resolving this issue. It enables one to build a (more realistic) simulation model containing the features in question, first execute it using the analytically based optimal policy, and then execute it using other management policies that institutional considerations may suggest. Comparing the simulation results for the different policies enables the simulationist to determine the degree to which the optimal policy prevails over the institutionally based policies in the more realistic simulation. This joint use of analytical and simulation methods considerably enhances the value of both methods.

1.6.1 DETAIL

As mentioned earlier, discrete-event simulation becomes a legitimate research tool when known analytical methods cannot supply a solution to a problem. Once simulation is adopted, certain admonitions made earlier in regard to modeling must be reexamined. Principal among these considerations is the issue of detail.

When building a model, an investigator repeatedly faces the problem of balancing the need for structural detail with the need to make the model amenable to problem-solving techniques. Being a formalism, a model is necessarily an abstraction. However, the more explicit detail a model includes, presumably the better it resembles reality. Also, including detail offers greater opportunities for studying system response when structural relationships within the model are altered. First, more combinations of structural change can be considered, and second, more aspects of the response can be studied.

On the other hand, detail generally impedes problem solution. Added detail often shifts the method for solving a problem from an analytical to a numerical one, thereby giving up the generality that an analytical solution offers. Detail also increases the cost of solution. However, the most limiting factor in the use of detail is that we rarely know enough about the system under study to specify more than a modicum of its characteristics. Every model must limit detail in some respect. To fill in a system description in place of detail, one makes *assumptions* about system behavior. Since one does not want these assumptions to conflict with the observable behavior of the system, testing them against observation is desirable whenever possible.

Our earlier remarks stressed the fact that the amount of detail in a model is generally inversely related to our ability to obtain an analytical solution. However, if the *minimal model*, which contains the least detail needed for useful study, is not amenable to analytical solution and we adopt an approach based on discrete-event simulation, we have the prerogative of building as much detail into our model as we like without concerning ourselves about the absence of an analytical solution. It is precisely this descriptive ability that holds the most attraction for modelers, for the greater the amount of detail, the more realistic the model is and consequently the closer they expect the results of a simulation run to conform to reality. However, practice indicates that judicious restraint in regard to detail is often the better policy. At least three reasons support this assertion:

- To include detail we must devote effort and time to preliminary observation of the individual characteristics of the system under scrutiny. These actions induce a cost, the justification for which a modeler has to weigh with respect to her/his objective. This issue arises in most simulations, and no universal course of action fits all contingencies.

- Additional detail requires more programming, especially for the contingencies that detail specifies. Moreover, the inclusion of great detail at the outset of a modeling effort makes the job of locating the sources of error in the resulting computer program especially difficult because of the many potential trouble spots.

- Because of the increased number of calculations that a computer program based on a detailed model must perform, we expect a concomitant increase in execution time and need for computer memory. These increase cost. This third reason for restraint is not always apparent until actual simulation execution begins and we see computing cost quickly mounting (Chapter 5).

These sobering remarks provide a useful perspective for evaluating the degree of detail worth having. Often, a simulationist gets a limited model running first and then introduces detail where this model provides inadequate answers. This bootstrapping approach makes for a more judicious allocation of a modeler's time and effort and reduces debugging and computing times and memory requirements.

1.6.2 MODELING DANGERS

Although modeling clarifies many relationships in a system, there are at least two warnings that an investigator should always keep in mind. First, no guarantee exists that the time and effort devoted to modeling will return a useful result. Occasionally, failure occurs because the level of resources is too low. More often, however, the simulationist has relied more on the simulation language and its tools and not enough on ingenuity when the proper balance between both is the recipe that enhances the probability of success.

The second warning concerns the use of the model to predict beyond its range of applicability without proper qualification. Being an abstraction, a model inevitably produces numbers that serve as approximations to the true values of unknown performance measures. Do these numbers have a tendency to overstate or understate the true values? How close are the generated values to the true values? Good modeling analysis demands answers to both these questions. For example, the .99 confidence intervals in Table 1.2 attest to the sample means (point estimates) being good approximations to the unknown mean number of lost drums. Most importantly, the answers demarcate the range of applicability of model output. Neglecting to provide this qualification constitutes one major cause of model misuse in practice. Chapter 6 addresses the issue of statistical accuracy.

1.7 TECHNICAL ATTRACTIONS OF SIMULATION

Discrete-event simulation on a digital computer, or *computer simulation*, offers many conveniences that make it an attractive tool for analysis.[1] It *compresses* time so that years of activity can be simulated in minutes or, in some cases, seconds. This ability enables an investigator to run through several competing operational designs in a very small fraction of the time required to try each on the real system.

Computer simulation can also *expand* time. By arranging for statistics of interest to be produced over small intervals of simulated time, an investigator can study by computer simulation the detailed structure of change in a real system which cannot be observed in real time. This figurative *time dilation* is especially helpful when few data are available on change in the real system. For example, the "list" statement in the SIMSCRIPT II.5 programming language offers an example of this capability.

One important consideration in any experiment is the ability to identify and control *sources of variation*. This is especially important when a statistical analysis of the relationship

---

[1]The term computer simulation should not be confused with a simulation study of a computer system. Such studies have become major users of computer simulation.

between independent (input) and dependent (output) factors in an experiment is to be performed. In the real world, the ability is to a major extent a function of the system under study. In a computer simulation, a simulationist, of necessity, must explicitly specify the sources of variation and the degree of variation due to each source in order to make a simulation run. This requirement enables her/him to eliminate unwanted sources of variation simply by omitting them from the simulation. However, this ability also requires the simulationist to devote sufficient attention to a system in order to derive an adequate understanding of how to describe quantitatively the sources of input variation that he/she wishes to include. Sections 4.7 and 4.22 illustate how this control of sampling variation can improve statistical efficiency.

In a field experiment, *errors of measurement* inevitably arise in recording data. These occur because no perfect measuring device exists for suppressing all external influences. By contrast, only round-off errors occur in a computer simulation, since the programmed simulation produces numbers free of any superimposed variation due to external and uncontrollable sources. The round-off errors are due to finite word length in a computer. Using extended-precision computation, as needed, can make these errors relatively negligible.

During execution, it is occasionally desirable to stop the experiment and review results to date. This means that all phenomena associated with the experiment must retain their current states until the experiment resumes execution. In field experiments a complete halt of all processes is seldom possible. However, computer simulation offers this convenience, provided that the termination part of the program contains instructions for recording the states of all relevant variables. When the experiment resumes, the terminal states become the initial states for those variables so that no loss of continuity occurs. Most simulation languages offer an option for programming this feature.

The ability to restore the state of the simulation provides another benefit. At the end of a computer simulation run, analysis of the results may reveal a situation explainable by data that were not collected. Here, the simulationist can reprogram the data-collecting subprogram to record the desired data and can run the simulation again with the *same* initial conditions. The simulation runs identically as before with the additional data being made available.

Simulation also enables us to *replicate* an experiment. To replicate means to rerun an experiment with selected changes in parameter values or operating conditions made by the investigator. For example, an independent replication is one in which the simulation model is unchanged but the sequence of pseudorandom numbers used to drive the model in the replication is independent of the sequence employed in all previous runs.

1.8 MEDIUM FOR MODELING

Modeling a system inevitably relies upon some *medium* to connect the realities of the system into a collection of abstractions amenable to manipulation for purposes of solution. In physics, partial differential equations are the abstractions that result from applying the infinitesimal calculus as the medium. Difference equations are often the abstractions that result from employing the calculus of finite differences to model a dynamical economic system.

Discrete-event delay systems are also amenable to abstraction using these two media, and indeed, the analytical derivation of queue-length, waiting time, and resource-utilization distributions relies crucially on them. As illustration, consider a single-server queueing system, as in Figure 1.2a, with interarrival times $A_1, A_2, \ldots$ and service times $B_1, B_2, \ldots$. Let W_i denote the waiting time in queue for arrival i and suppose the first arrival finds the queue empty and the server idle. Then it is easily seen that for a first-come-first-served queueing discipline,

$$
\begin{aligned}
W_1 &= 0, \\
W_2 &= (B_1 - A_2)^+, \\
W_3 &= (W_2 + B_2 - A_3)^+, \\
&\ \vdots \quad\ \vdots \quad\ \vdots \quad\ \vdots \\
W_{i+1} &= (W_i + B_i - A_{i+1})^+,
\end{aligned}
$$

where

$$
(x)^+ := \begin{cases} x & \text{if } x \geq 0, \\ 0 & \text{otherwise.} \end{cases}
$$

Known as *Lindley's formula*, this notation incorporates the calculus of finite differences with *one boundary constraint*. In principle, this calculus can be used to model a system with m servers in parallel at a single station but with multiple boundary constraints that grow exponentially in number with m. Moreover, other queueing disciplines demand more complicated notation.

Recognizing the limitations of this approach to modeling discrete-event dynamical systems and aware of the need to adopt modeling conventions that could be easily converted into executable computer code, early simulation methodologists devised two paradigms for abstracting these systems. The *event-scheduling approach* focuses on the moments in time when state changes occur (Section 2.3). The *process-interaction approach* focuses on the flow of each entity through the system (Section 2.4). Both allow for variable updating and flow-branching, characteristics that epitomize discrete-event delay systems. Both allow a modeler to create a collection of flowcharts that incorporate all system dynamics needed to write a simulation program conveniently.

In practice, experienced simulation users rarely turn to explicit flowcharting when modeling a system. More often, they rely on the constructs available to them within the simulation programming packages they employ. These constructs implicitly embrace either the event-scheduling or process-interaction approach, more often the latter than the former. Although skipping the flowcharting step saves considerable time for a modeler, the newcomer to discrete-event simulation cannot possibly develop an appreciation of the options available to her/him for modeling delay systems by totally omitting a study of the principles of the event-scheduling and process-interaction approaches. Familiarity with these options is critical in modeling large complex systems. Indeed, total reliance on language constructs can occasionally keep the modeler from seeing the simplest way to account for behavior. Chapter 2 describes these principles in detail.

1.9 DISCRETE-EVENT SIMULATION PROGRAMMING LANGUAGES

Virtually all discrete-event simulation programming languages contain capabilities for:

- Maintaining a list of events scheduled to be executed as time evolves

- Maintaining other dynamical data structures

- Generating pseudorandom numbers

- Generating samples from a variety of distributions

- Computing summary statistics including time averages, variances, and histograms

- Generating reports

- Detecting errors peculiar to simulation programs.

Also, having the capability to:

- Perform conventional arithmetic operations and to evaluate functions such as $\max(x, y)$, $\log x$, $\sin x$

adds to a language's appeal, although some languages that provide less than this capability have managed to survive and prosper.

In today's PC environment, many simulation languages have become embedded in larger software systems with enhanced features that include the capabilities to:

- Operate in Microsoft Windows 98, Windows NT, and Unix environments

- Display data graphically

- Display simulation animation.

Some also have capacities to:

- Import CAD (computer-assisted design) drawings to provide icons for animation

- Draw icons including all standard CAD objects such as rectangles and ellipses

- Allow a user to incorporate other computer applications such as Excel spreadsheets, Microsoft Word files, clipart, and Microsoft PowerPoint presentations

- Integrate database systems that comply with the Microsoft ODBC standard such as Dbase, Access, Fox Pro, and Excel

- Allow user-coded inserts in other computer languages such as Visual Basic, C, and C++.

The principal general-purpose discrete-event simulation languages include GPSS/H (Wolverine Software), which is now a subset of the more general SLX simulation programming language; SIMAN (Systems Modeling Corporation), contained within the Arena general-purpose simulation system; SIMSCRIPT II.5 (CACI); and SLAM, which is embedded in the AweSim general-purpose simulation system (Symix Systems, Inc./Pritsker Division). Because of frequent updating of these software products, we merely include as references the name of the vendor identified with each. More current information is obtainable at their sites on the Internet. For a listing of other simulation software, see Swain (1999).

In principle, two options exist for constructing a discrete-event sinulation program. One allows a simulationist to enter statements via a command line. Sections 4.2 through 4.21 demonstrate this option using SIMSCRIPT II.5. The other displays icons corresponding to the language statements, or aggregates thereof, on the screen and allows a simulationist to construct a program by pointing, clicking, dragging, and dropping these into a "model window." Sections 4.23 through 4.27 use SIMAN to illustrate this option. For several reasons, the second option has come to prevail in practice. In a PC environment, working with a collection of screen-based icons, one for each of the powerful statements in the language, allows a user quickly to:

- Construct a block-diagram flowchart of a program

- Use menus associated with each icon to assign numerical values to parameters

- Specify operating statistics to be calculated during a simulation experiment

- Have the block diagram converted into executable code

- Execute the simulation experiment

- Display the final operating statistics.

The availability of PC-based simulation systems that provide all or most of these conveniences has significantly reduced the time and effort for the newcomer to discrete-event simulation to create an executing program.

1.9.1   SPECIAL FEATURES

Within specialized areas, additional reductions in programming effort are possible. As illustration, many manufacturing and materials-handling systems use *conveyors* and *automatic guided vehicles* (AGVs) to move items within factories and warehouses. Some also rely on automated storage and retrieval systems (ASRS), robots, and bridge cranes. Modeling these objects and the dynamics of their interactions with other parts of the system presents a formidable challenge to the average simulationist. However, manufacturing and materials-handling simulation has grown into an industry of its own, prompting software vendors to develop and offer precoded modules that simulate these special behaviors.

Among these offerings are the advanced manufacturing template in Arena (Systems Modeling Corporation), AutoMod (AutoSimulations, Inc), FACTOR/AIM (Symix Systems, Inc./Pritsker Division), ProModel (PROMODEL Corporation), SIMFACTORY II.5 (CACI), Taylor II (F & H Simulation, Inc.), and WITNESS (Lanner, Inc.). These concepts are usually iconized on the monitor screen and have menus that allow a user to customize features to the particular problem at hand. Sections 4.23 and 4.24 discuss limitations of this approach.

1.9.2   OBJECT-ORIENTED PROGRAMMING

In practice, code reuse plays a central role in establishing a workable balance between programming effort and computational effciency. If a large-scale simulation program is to be developed and then executed many times, merely with different numerical input, the investment in programming effort to achieve highly efficient code can be negligible when amortized over the useful lifetime of the program. By contrast, a study that requires a sequence of simulation programs that successively differ in logic in relatively small ways may call for so much programming effort at each stage using a conventional simulation language as to discourage one from carrying out the full investigation. In this environment, keeping programming effort within an acceptable bound becomes the overriding objective. In the current world of highly interactive computing, this second modality has grown in relative importance and led to an emphasis on point-click-drag-drop programming in conjunction with *object-oriented programming*. Indeed, this concept underlies the icon/menu-driven environment so commonplace on PCs and workstations.

An *object* essentially denotes a collection of data structures for storing critical data. For example, in a network model a message is an object with length, priority, origin, and destination. In manufacturing, a part is an object whose data include a route identifying the sequence of processing stations it visits on the way to completion. The data also include processing times and priorities for service at the stations. Each *class* of objects is associated with *functions* or algorithms that contain recipes for operating on the data stored in *instances* of the objects. An airport simulation clarifies the meanings of class, object, instance, and function.

Aircraft form a class, and particular types of aircraft are objects. An individual aircraft denotes an instance of a particular type (object) of aircraft within the class of aircraft. Associated with these are the functions of landing an aircraft, taxiing on a runway, loading an aircraft with cargo, unloading cargo from an aircraft, loading passengers onto an aircraft, unloading passengers from an aircraft, refueling an aircraft, and aircraft takeoff. In principle, programs created to represent the class called aircraft and to compute the functions of landing, loading, unloading, refueling, and takeoff can be used in many different scenarios of airport simulation. For example, one simulation may study congestion at an airport dedicated exclusively to passenger traffic, while another may study capacity limitations at an air-cargo terminal. Both simulations would use the aircraft class but with different mixes of types and different utilization of functions.

Ease of programming and program reuse provide the principal motivations for adopting an object-oriented approach to discrete-event simulation. This is especially true in an environment where an iconized collection of preprogrammed modules allows a simulationist to construct a simulation model on the monitor's screen with relatively little knowledge of the components of each module. If results obtained from executing the simulation encourage the analyst to consider an amended simulation program, the point, click, drag, and drop capacities facilitate this modification by assembling a new program on-screen made up of parts of the old program and new modules taken from the collection of preprogrammed modules in the displayed collection of icons.

As examples, AutoMod, ProMod, and WITNESS employ objected-oriented programming within their modules for manufacturing and materials handling. MedModel (PROMODEL Corporation) uses object-oriented programming to create its modules within a health-care delivery system. Other simulation software products use a semblance of the object-oriented programming approach when constructing conveyors, etc.; but in reality they do not entirely adhere to the formal rules for this approach to programming. For business applications, Technology Economics, Incorporated offers FlowModel, which contains a preprogrammed collection of business-oriented objects and functions.

This capacity for program reuse and reduced programming effort has limitations. If the menu of preprogrammed objects and functions does not provide for a particular application, then either simulating the application cannot be done using this collection, or else programming is required to effect the simulation. Sometimes this last option results in relatively awkward use of objects and functions for which they were not originally intended. To provide a more global setting, CACI offers an object-oriented simulation language called MODSIM III that allows a user to program her/his own objects and functions.

To regard the term "object" as merely data and a function as merely another word for subroutine would be an oversimplification. Whereas a subroutine may use global as well as local data structures for computation, an object must have all its required data stored within its own local data structure, and a function in an object-oriented environment operates on these local data. This distinction clearly makes an object and a function less dependent on the overall problem than a subroutine would be, and this offers a considerable advantage in making changes.

Customarily, a program consists of objects linked together by well-defined connections (which are themselves objects). Each object is compiled separately, with the underlying simulation language lurking in the background and usually out of the simulator's consciousness. When a program requires change, only those objects and connections affected by the change need to be recompiled. If one thinks of compilation as a process of optimization with regard to computational efficiency, then this procedure of selective recompilation (suboptimization) inevitably results in reduced run-time effciency when compared to a globally compiled (optimized) program. Also, a program consisting of objects requires more space for data structures than a more traditionally written program would. Nevertheless, proponents of the object-oriented approach to simulation regard the programming convenience that it offers as outweighing the penalties incurred by increased space requirements and reduced computational efficiency.

The concept of object-oriented programming is not new to simulation. SIMULA (Birtwistle et al. 1973), one of the first simulation programming languages, used the concept. Moreover, the property of *inheritability* of attributes, an intrinsic feature of object-oriented programming, has been a feature of GPSS from its beginning as an IBM programming product. In particular (e.g., Schriber 1991, p. 368), the SPLIT statement in GPSS creates a specified number of copies of a process and endows each with the data structures and functions in the original process. However, each copy executes independently of all remaining copies. In this way a batch of identical parts entering a job shop can be SPLIT for processing and later reassembled into a batch using the ASSEMBLE statement of GPSS. The SPLIT statement allows the spawned instances of the object to be numbered sequentially, if desired, for easy identification.

Virtually all object-oriented simulation programming languages translate their statements, into C^{++} statements, which are then compiled into executable code. C^{++} is an enhancement of the C programming language endowed with a syntax for object-oriented programming.

At present, every simulation software product is of one of two forms. Either it embraces the traditional sequential programming approach reminiscent of FORTRAN and C or it embraces the object-oriented paradigm as in C^{++} and JAVA. As to whether this coexistence will continue or one of these formats will come to dominate remains for the future to reveal.

1.9.3 WEB-BASED DISCRETE-EVENT SIMULATION

The Internet has created opportunities for new modes of modeling, programming, and executing discrete-event simulation. Modules created in the JAVA programming language, using its object-oriented programming features, reside at accessible websites. Modelers browse these sites and retrieve off-the-shelf modules that appear to fit the problem at hand. All this presumes that the browser is using a JAVA-based discrete-event simulation language possessing all the features delineated in Section 1.9 and that the off-the-shelf modules meet a universally accepted standard that allows their incorporation, without incident, into the browser's program.

More ambitious modes of web-based simulation are also possible. For example, under certain conditions, a simulationist could run a web-based simulation program t times in parallel to generate sample paths for independent replications by distributing the workload over t distinct computing platforms. Conceptually, this approach reduces elapsed clock time to obtain results by a factor of t. Alternatively, a web-based simulation program can be used simultaneously by many users. See Whitman et al. (1998).

1.9.4 NO PROGRAMMING

To make their products less intimidating and thus more appealing to a wider range of potential users, some vendors have introduced simulation software that attempts to minimize or, if possible, to eliminate user-originated programming. The products encourage a simulationist to point and click on icons, corresponding to precoded constructs available in an on-screen template and to drag and drop these constructs into the simulation program being created, thus reducing the user's time and effort spent programming. The modules discussed in the last several sections exemplify these precoded blocks.

While noble in concept, the practice has limitations. Most notably, prebuilt constructs often represent specific environments that are difficult to modify in order to accommodate the problem at hand. Indeed, it may not be apparent to the simulationist that a misrepresentation of problem logic is about to occur when inserting a prebuilt module into her/his program. For example, see Section 4.28.

1.9.5 TEACHING DISCRETE-EVENT SIMULATION

Every introductory course in discrete-event simulation seeks to turn out students capable of performing discrete-event simulation experiments. Inevitably, this must include the capability to create and execute programs. However, the programming skill level is problematical. Catering to the little-or-no programming philosophy tends to produce students whose knowledge is restricted to the software to which they have been exposed and possibly to a narrow subset of problems encountered within discrete-event simulation.

A considerably more desirable outcome for the course is to equip each student with the knowledge of how to model a system and translate the model into executable code in at least one language and, at the same time, to familiarize them with the breadth and depth of concepts that they are likely to encounter when asked to work in a language other that those taught in the course. This book embraces this strategy. While it does offer examples in particular languages in Chapter 4, our account works best when the instructor uses it in parallel with an adopted relatively comprehensive discrete-event simulation language such as GPSS/H, SIMAN, SIMSCRIPT II.5, SLAM, or SLX. It works less well when used with special-purpose simulation languages that encourage minimal programming.

What about object-oriented programming? Since the purpose of this book is to acquaint the reader with the most basic concepts of discrete-event simulation, we have chosen to forego object-oriented programming as a pedagogic medium in favor of an account that exposes the reader to the conceptual formalisms of the event-scheduling and process-interaction approaches and with their conversion to executable code in two major discrete-event simulation languages, SIMSCRIPT II.5 and SIMAN. Their syntactical comprehensiveness motivated this choice. After having studied the present account, a reader will be well prepared to work with any general-purpose discrete-event simulation language or any object-oriented discrete-event simulation language, subject to familiarizing herself/himself with its particular syntax.

1.10 REFERENCES

Birtwistle, G., O.J. Dahl, B. Myhraug, and K. Nygaard (1973) *SIMULA BEGIN*, Auerbach, Philadelphia.

Ma, Xudong (1999). Managing the hazardous liquid waste input to Glaxo Wellcome's environmental safety facility incinerator, OR350 project, Operations Research Department, University of North Carolina, Chapel Hill.

Schriber, T.J. (1991). *An Introduction To Simulation Using GPSS/H*, Wiley, New York.

Swain, J.J. (1999). 1999 Simulation software survey, *ORMS Today*, **26**, 38–51

Tzenova, E. (1999). An analysis of operations for the health sciences library photocopier service, OR350 project, Operations Research Department, University of North Carolina, Chapel Hill.

Whitman, L., B. Huff, and S. Palaniswamy (1998). Commercial simulation over the web, *1998 Winter Simulation Conference*, D.J. Medeiros, E.F. Watson, J.S. Carson, and M.S. Manivannan, editors, Association for Computing Machinery, New York, 335–341.

CHAPTER 2

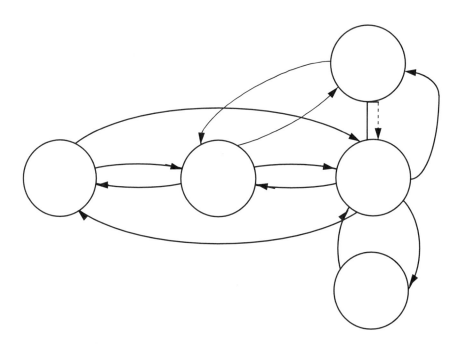

Modeling
Concepts

Every simulation study consists of at least 10 steps:

1. Translating the qualitative description of a system into a formal abstraction that explicitly accounts for all logical and mathematical relationships

2. Identifying all parameters of the abstraction that require numerical values as input

3. Identifying all measures of performance whose values require estimation

4. Estimating the values of all unknown input parameters from available data, expert opinion, etc.

5. Designing the set of sampling experiments, at least one for each distinct set of values for the vector of input parameters

6. Converting the logical abstraction to executable code in a simulation programming language

7. Incorporating into the code statements for computing time averages of performance measures during program execution

8. Performing the set of sampling experiments by executing the code repeatedly, at least once for each vector of input-parameter values

9. For each measure of performance, evaluating how well its time average approximates its unknown long-run average

10. Comparing corresponding sample time averages for each performance measure across experiments

While this numbering allows for orderly progress from beginning to end, actual practice may turn out to blur the distinction between steps, rearrange their order, or in some cases, neglect or pay too little attention to them. For example, the availability of point-click-drag-drop simulation software entices a simulationist to merge steps 1 and 6, possibly making her or him oblivious to a wider range of modeling concepts than the selected language implementation may accommodate. Frequently, the choice of long-run time averages to estimate in step 3 depends solely on what the programming language automatically makes available in converting the model to code in step 6. Too often, step 9 is ignored, making any conclusions based on comparisons across experiments difficult to justify statistically.

This divergence between our idealized list of steps and practice stems from the premium that simulationists often put on their time. There is no doubt that the point-click-drag-drop option saves time in model development and coding, and produces models and, subsequently, code that well approximates many a system. However, relying exclusively on this option limits the diversity of models that can be formulated, an issue of considerable relevance when the system under study calls for more structure than direct application of this software can accommodate. To remove a potential impediment, this chapter focuses on step 1. In particular, it presents concepts germane to the construction of a discrete-event simulation model. Most importantly, it describes two alternative, but logically equivalent, ways of representing system behavior. Section 2.3 describes the *event-scheduling approach* and Section 2.4, the *process-interaction approach*. Each provides directions for writing a computer program whose execution simulates system behavior.

Discrete-event simulation has a special vocabulary that we use throughout this book. Its principal terms include:

System. A collection of objects that interact through time according to specified rules.

Model. An abstract logical and mathematical representation of a system that describes the relationship among objects in a system.

Entity. An object in a system that requires explicit representation within a model of the system. For example, servers, customers, and machines all denote objects in a service delay system.

Attribute. A descriptor of a property of an entity. For example, a server may have a skill attribute that identifies its skill level and a status attribute that identifies whether it is idle or busy.

Linked List.	A collection of records each identified with an entity and chained together in an order according to values assumed by a particular attribute of each entity. For example, arriving jobs in a system are entities. If they have to await service, they may be filed in a list ordered by the value of their arrival-time attribute.
Event.	A change in state of a system.
Event notice.	A record describing when an event is to be executed.
Process.	A collection of events ordered in time. Time elapses in a process.
Reactivation point.	An attribute of an event notice associated with a process. It specifies the location within the process at which execution is to resume.
Future-event set.	A linked list containing event notices ordered by desired execution time.
Timing routine.	Procedure for maintaining the future-event list set and advancing simulated time.
Activity.	A pair of events, one initiating and the other completing an operation that transforms the state of an entity. Time elapses in an activity.

The reader has already encountered several of these terms; for example, system and model. Others, such as the concept of a linked list, are essential for an understanding of how the computer code for a discrete-event simulation carries out its tasks. Briefly, each record in a singly linked list has an attribute, called the *successor address*, that points to the record that immediately follows it in the list. This facilitates search starting from the first record in the list, called the *header*. In addition to a successor address, each record in a doubly linked list has an attribute, called the *predecessor address*, that points to the record that immediately precedes it in the list. Collectively, predecessor and successor addresses allow for search beginning either at the header or the *trailer*, the last record in the list. Section 5.2 discusses the concept of list processing in more detail. Unless clarity demands otherwise, we hereafter refer to a doubly linked list simply as a list.

2.1 NEXT-EVENT TIME ADVANCE

Discrete-event simulation concerns the modeling on a digital computer of a system in which state changes can be represented by a collection of discrete events. Here the appropriate technique for modeling a particular system depends on the nature of the *interevent intervals*. These time intervals may be random or deterministic. If they are random, the modeling technique must allow for intervals of varying lengths. If the intervals are deterministic, they may vary according to a plan, or they may all be of equal length. When they vary, the modeling techniques must again acknowledge the nonuniform nature of the interevent intervals. When they are constant we can employ a considerably simpler modeling technique to structure change in a model. For present purposes, we assume that they vary.

In a discrete-event system, a state change implies that an event occurs. Since the states of entities remain constant between events, there is no need to account for this inactive time in our model. Accordingly, virtually all computer simulation programming languages use the *next-event approach* to time advance. After all state changes have been made at the time corresponding to a particular event, simulated time is advanced to the time of the next event, and that event is executed. Then simulated time is again advanced to the scheduled time of the next event, and the procedure is repeated. Most importantly, the procedure enables a simulation to skip over inactive time whose passage in the real world we are forced to endure. Hereafter, we sometimes refer to simulated time, more concisely, as time.

Although the next-event approach is the most commonly used, the reader should note from the outset that other methods exist for advancing time and processing events. (e.g., Fishman 1973, pp. 38-40 and Pritsker et al. 1997, p. 25) Unfortunately, use of an alternative necessitates programming it, hence the appeal of the next-event approach, which is a standard feature of most discrete-event simulation programming languages.

2.2   ARITHMETIC AND LOGICAL RELATIONSHIPS

In modeling a system for eventual computer simulation, two different structures play significant roles. One includes the mathematical relationships that exist between variables (attributes) associated with the entities. For example, if Q is the number of patients waiting in an outpatient clinic, then we set $Q \leftarrow Q + 1$ when a patient arrives and $Q \leftarrow Q - 1$ when a patient begins treatment. Sometimes, the specification of the mathematical relationships for a system serves to describe completely the way in which state changes occur, as, for example, in a model of a national economic system.

Logical relationships constitute the other set of structures that are used to describe a system. In a logical relationship, we check to see whether a condition is true or false. If true, we take a certain action. If false, we take an alternative action. For example, when a server becomes available in a maintenance shop, he or she checks to determine whether a job is waiting for service. If one is, service begins. If no job is waiting, then the server remains idle. If several jobs are waiting, the server selects one according to an established *logical-operating* or *job-selection rule*. A logical-operating rule describes a set of mutually exclusive conditions with corresponding actions. For example, if several jobs are waiting, the rule may be to select the job that has been waiting for the longest time. If there are n jobs waiting, this rule is formally, For $i = 1, \ldots, n$, if job i has been waiting for the longest time, select job i for service.

Our purpose in discussing time passage and relational questions in this introduction is to establish their importance in modeling a discrete-event system. The remainder of the chapter describes modeling methods that account for these concepts in different ways.

The concepts of *event, process*, and *activity* are especially important in modeling a system. To illustrate the relationships among them, we consider a job arriving at a maintenance shop with two tasks to be performed. Figure 2.1 shows the arrival and the eventual service that each task receives. It also shows the relationships that exist among events, processes, and activities. Collectively, these concepts give rise to three alternative ways of building discrete-event models. The *event-scheduling approach* emphasizes a detailed description of the steps that occur when an individual event takes place. Each type of event naturally has a distinct set of steps associated with it. The *activity-scanning approach* emphasizes a review of all activities in a simulation to determine which can be begun or terminated at each advance of the clock. The *process-interaction approach* emphasizes the progress of an entity through a system from its arrival event to its departure event.

The development of the three concepts is related to the world views adopted by simulation programming languages. Among the major ones, SIMAN, SIMSCRIPT II.5, and SLAM offer a user the choice of event scheduling or process interaction, whereas GPSS/H exclusively offers process interaction. While activity scanning is a logically equivalent approach, it is computationally less efficient for most discrete-event scenarios. Accordingly, we omit further mention of it.

2.3 EVENT-SCHEDULING APPROACH

Every discrete-event system has a collection of *state variables* that change values as time elapses. A change in a state variable is called an *event*, and this concept is the basic building block of every discrete-event simulation model based on the event-scheduling approach. In

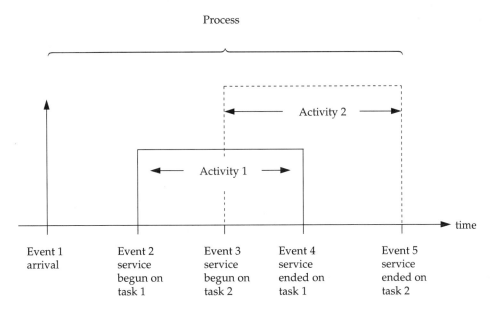

Figure 2.1 Event, process, and activity

a queueing system, an arrival constitutes an event because it increases the number of items in the system by one. A departure is also an event because it decreases the number of items in the system by one. Beginning service and completing service are also events, for each signals a change in state for one or more resources of the system and a change in status for a particular item. To illustrate the significance of the event-scheduling approach, we describe an airline reservation system. We encourage the reader to recognize that the concepts described here are generic to all simulation modeling based on this approach.

The principal features of the airline reservation model are:

a. All customers call on the telephone.

b. Customers are willing to wait indefinitely for service.

c. No restriction exists on the number of customers that can wait at one time; i.e., an infinite number of telephone lines exists.

d. Each customer wants a reservation for either a single-destination or a multiple-destination trip.

e. If all reservationists are busy when a call occurs, a 9-second recorded message asks that customer to wait.

f. The airline office has m reservationists. Calls occur at rate λ, and intercall times are i.i.d. random variables that follow the exponential distribution

$$g(t) = \lambda e^{-\lambda t}, \qquad \lambda > 0, \quad t \geq 0,$$

denoted by $\mathcal{E}(1/\lambda)$. The proportion of customers who want a multiple-destination reservation is p, whereas the proportion that desire a single-destination reservation is $1 - p$. Single-destination service times are i.i.d. random variables that follow the exponential distribution

$$f_1(t) = \omega e^{-\omega t}, \qquad \omega > 0, \quad t \geq 0, \tag{2.1}$$

denoted by $\mathcal{E}(1/\omega)$. Multiple-destination service times are i.i.d. random variables that follow the Erlang distribution

$$f_2(t) = \omega^2 t e^{-\omega t}, \qquad \omega > 0, \quad t \geq 0, \tag{2.2}$$

denoted by $\mathcal{E}_2(2/\omega)$. The service time of an arbitrarily selected caller is then a random variable with probability density function (p.d.f.)

$$f(t) = (1 - p)f_1(t) + pf_2(t) \tag{2.3}$$
$$= \omega(1 - p + p\,\omega t)e^{-\omega t}, \qquad 0 \leq p \leq 1, \quad \omega > 0, t \geq 0.$$

Note that a random variable from $\mathcal{E}_2(2, 1/\omega)$ is equivalent to the sum of two independent random variables, each from $\mathcal{E}(1/\omega)$.

g. The purpose for simulating this system is to compare its performance as the number of reservationists m grows while λ, ω, and p remain fixed. In support of this objective it is of interest to estimate the long-run averages and, more generally, the distributions of waiting time, number of customers awaiting service, and number of busy reservationists as the number of reservationists m varies.

In terms of the 10 major steps in a simulation study described in the introduction, the present description accomplishes steps 2 and 3. To effect step 1, we first need to identify the entities in the system and their attributes. For this problem we have two classes of entities:

Entity	**Attribute**	**Possible attribute assignments**
CUSTOMER	TYPE	ONE_DESTINATION / MULTIPLE_DESTINATIONS
RESERVATIONIST	STATUS	IDLE / BUSY

In the event-scheduling approach to modeling, a single flowchart usually incorporates all state changes that occur at the same time. Recall that a state change is called an event. In the airline reservation problem we can identify 10 types of state change. These *elementary events* are:

Customer	**Reservationist**
1. Call	8. Remove from waiting line
2. Hang up	9. Become busy
3. Begin service	10. Become idle
4. End service	
5. Begin recorded message	
6. End recorded message	
7. File in waiting line	

Flowcharts provide a convenient way of describing behavior that is eventually to be transformed into a computer program. Moreover, alternative flowchart representations are possible for the same problem, the choice of which to use being a function of user familiarity with different languages and the degree of insight that each representation offers. We mention familiarity before insight because experience regrettably shows this to be an overriding consideration in practice.

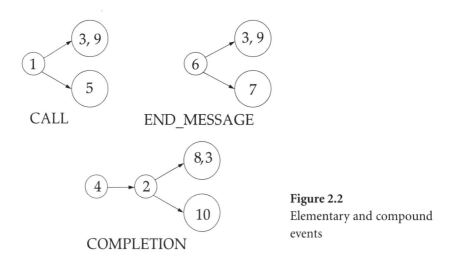

Figure 2.2
Elementary and compound events

Figure 2.2 shows the logic flow that enables us to incorporate these elementary events into three *compound events* entitled CALL, END_MESSAGE, and COMPLETION. From the original description of the problem we see that whenever a call occurs (1), the caller either begins service (3) if a reservationist is idle, or otherwise begins listening to a 9-second recorded message (5). If service begins (3), then the reservationist who begins service also becomes busy (9). While all four of these state changes are elementary events, they differ in at least one important way: (1) is an *unconditional event*, whereas (3), (5), and (9) are *conditional events*. An unconditional event occurs independently of the current state of the system, whereas a conditional event occurs only when another event puts the system in a prescribed state. In particular, (3) and (9) occur at the same time as (1) if and only if a reservationist is idle; but (5) occurs simultaneously with (1) if and only if all reservationists are busy. Collectively, the CALL compound event summarizes this behavior.

In a like manner, the END_MESSAGE compound event summarizes state-change possibilities whenever a recorded message ends (6), and the COMPLETION compound event summarizes the elementary events that can occur when a caller completes service (4). Since the beginning of service (3) coincides with a customer call (1), an end of the recorded message (6), or a service completion (4), all three CALL, END_MESSAGE, and COMPLETION events incorporate the beginning of service representation. Because a customer hangs up (2) upon service completion (4), this event (2) is modeled as part of the COMPLETION compound event. Lastly, the idling of a reservationist (10) occurs when either no customers await service or those that are waiting are listening to the recorded message. Consequently, it is also modeled as part of the COMPLETION event.

Figure 2.3 defines the symbols used for flowcharting. We usually refer to the symbols as *blocks*. The reader should focus attention on the "scheduling" and the "next-event selection" blocks, since all simulation programming languages contain corresponding special subprograms. If one chooses to program a simulation model in a nonsimulation language, he/she must write code for both of these subprograms. As the description of the airline reservation problem evolves, the convenience of having these subprograms available in a simulation programming language becomes apparent.

At each moment in time, the simulated system is in a particular state. For example, the system state for the airline reservation problem has entries for:

- Number of busy reservationists (i)

- Number of customers listening to recorded message (j)

- Number of customers waiting for service after having listened to the recorded message (k)

- Remaining time until next call (S)

- For each caller l listening to the recorded message:

  Remaining time until completion of message (T_l)

- For each busy reservationist l:

  Remaining time to complete service (T_l').

Note that the first three entries can assume only nonnegative *integer* values, whereas the last three assume nonnegative values. That is, the first three are discrete-state variables, whereas the last three are continuous-state variables. More generally, every simulation has a state vector some of whose entries are discrete and others of which are continuous.

Selecting the next event for execution and its timing depend crucially upon the entries in this state vector. As illustration, the remaining time until the next event in the airline reservation problem is

$$V = \min(S, T_1, \ldots, T_i, T_1', \ldots, T_j').$$

Moreover, this event is a new call if $S = V$, the end of a recorded message if $T_l = V$ for some $l = 1, \ldots, i$, or a completion of service if $T_l' = V$ for some $l = 1, \ldots, j$. As we presently show, one rarely has to concern oneself with maintaining this vector explicitly.

Although brief, this description of the airline reservation problem illustrates several important properties generic to the event-scheduling approach:

- An event is executed if and only if a state change occurs.

- Simulated time remains constant while an event is being executed.

- A compound event is a recipe for executing a sequence of actions all at the same time.

- A discrete-event simulation consists of a sequence of instances of compound events of possibly different types ordered according to their scheduled execution times, which may be randomly generated.

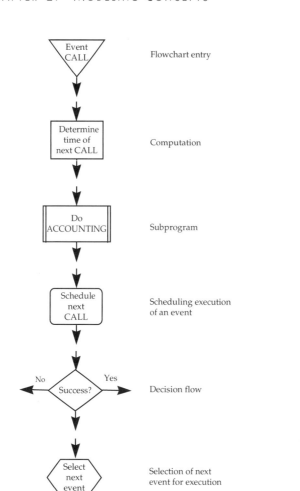

Flowchart entry

Computation

Subprogram

Scheduling execution
of an event

Decision flow

Selection of next
event for execution

Figure 2.3
Flowchart conventions

Transfer

This last property immediately raises the following issue:

- How does one determine the order of execution by type of compound event and by time, and how does one arrange for this schedule of executions in practice?

Every discrete-event simulation programming language provides a capability for scheduling events and for executing them. This distinction between scheduling and execution is an important one. In general-purpose programming languages, code within a program is executed sequentially. In a simulation language, a time delay may occur. The act of scheduling creates a record that identifies the event that one wants to execute and that contains the simulated time at which this execution is desired. The time may be any time no less than the current simulated time. This record is then added to a list, often called the *future-*

event set (FES), using a filing protocol that allows retrieval of the records in the order of desired execution time. The form of this protocol differs among simulation languages (Section 5.4).

A simulation programming language also provides a *timing routine* that operates on the FES to make event execution occur in the desired order and at the desired simulated times. Figure 2.4 illustrates the concept. The dashed line in the User's Program denotes elapsed simulated time. To prime the future-event set, the User's Program first schedules at least one event. Then it transfers control to the timing routine, which selects the event with the smallest execution time for execution (block 1). By executing an event in block 3 we mean passing control to a block of programmed logic corresponding to the changes in state that this particular event describes, executing the block, and returning control to the timing routine in block 4. Note that the event corresponds to a particular instance of this programmed logic. The act of destroying the event notice (block 4) makes the space it occupies available for other uses. Rather than destroying the event notice, some simulations selectively choose to save and reuse the space for a newly created event of the same class. Doing so eliminates the need to find space for the new event. Section 5.3 addresses this issue.

Note that once the user's program passes control to the timing routine, it remains there, or at the event it executes (block 3), until the timing routine finds the FES empty (block 5). Then it terminates the simulation by transferring control back to the user's program (block 6). This is one of several ways of ending a simulation.

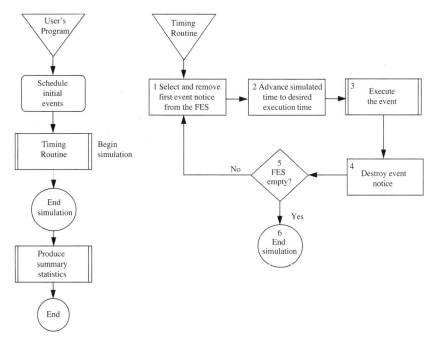

Figure 2.4 Prototypical timing routine, event-scheduling approach

While Figure 2.4 shows how to *prime* the FES with the initial events that start the simulation, it provides no explicit account of how successive events are scheduled as simulated time elapses. This scheduling actually occurs within executing events, as we now show. Figure 2.5 shows the steps executed each time a CALL event occurs. Here words in capital letters denote terms that have meanings particular to our description of the airline reservation problem. Block 1 denotes the creation of a record of an entity called CUSTOMER. The record presumably contains space for predefined *attributes* of the CUSTOMER. In block 2, TYPE denotes an attribute of CUSTOMER that contains information about whether the CUSTOMER wants a single- or multiple-destination reservation. This record is a block of contiguous words in memory that conceptually has the form

$$\text{entity record:} \left(\text{identification, attribute 1,...,attribute } r, \quad \begin{matrix} \text{predeccessor} \\ \text{address,} \end{matrix} \quad \begin{matrix} \text{successor} \\ \text{address} \end{matrix} \right),$$

$$(2.4)$$

providing space for the values of r attributes and two pointers. Since CUSTOMER in the present problem has one attribute TYPE (block 2), $r = 1$. The fields entitled predecessor address and successor address are included only if the record is to be chained to a doubly linked list (Section 5.2). For a singly linked list, only the successor (predecessor) address is included if the search is to begin from the front (back) of the list. No predecessor and successor address fields appear if the entity record is not to be linked to a list. For the present problem, the END_MESSAGE (Figure 2.6) and COMPLETION (Figure 2.7) events make use of a list called QUEUE that contains records of CUSTOMERs.

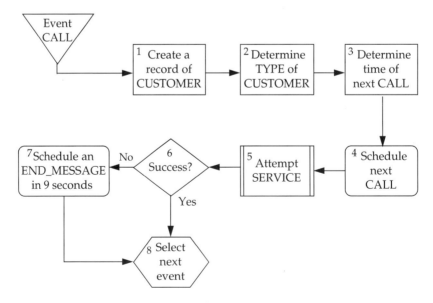

Figure 2.5 CALL event: event-scheduling approach

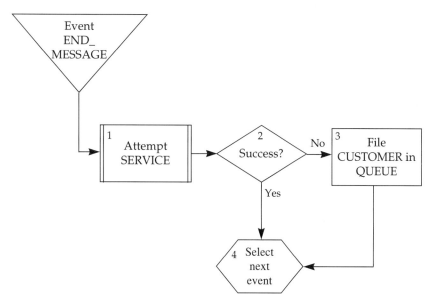

Figure 2.6   END_MESSAGE event: event-scheduling approach

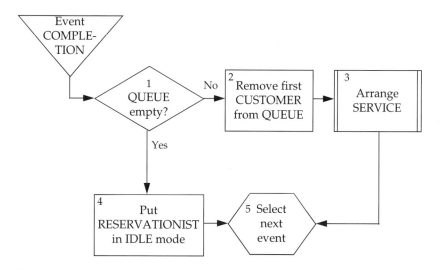

Figure 2.7   COMPLETION event: event-scheduling approach

Blocks 3 and 4 in Figure 2.5 provide a mechanism for executing the *next* CALL event. In particular, block 3 determines when the event occurs, and block 4 creates an *event notice* corresponding to the event, assigns the type of event to its *class* attribute and its desired execution time to its *clock* attribute of the event notice, and files this event notice in the future-event set. In the present case, the class is CALL.

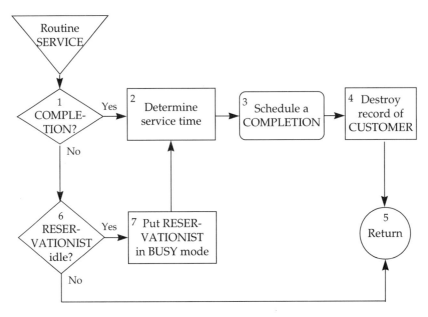

Figure 2.8   SERVICE routine: event-scheduling approach

Conceptually, an event notice takes the form

event notice:

$$\left(\text{identification, class, clock, attribute 2, \ldots, attribute } r, \quad \begin{matrix} \text{predeccessor} \\ \text{address,} \end{matrix} \quad \begin{matrix} \text{successor} \\ \text{address} \end{matrix} \right),$$

(2.5)

where the predecessor and successor address fields allow for chaining the record to the FES and for backward as well as forward searching of the FES.

The steps in blocks 3 and 4 represent bookkeeping procedures essential to all simulation modeling based on the *next-event* approach. A study of block 8 indicates why this is so. When entry into this block occurs, program control transfers to the timing routine. As Figure 2.4 shows, this routine searches the future-event set, selects the first event notice, removes it from the set, *advances* time to the desired execution time, and transfers control to the subprogram corresponding to the event name in the event notice. Execution of the event then occurs. This *variable-time advance* procedure skips all the intervening time between events and concentrates on the next event. Doing so contrasts with *fixed-time advance* procedures that increase time in small increments and, at each advance, execute all events that call for execution at that time.

Returning to Figure 2.5, we observe that the subprogram in block 5 attempts to provide service to the CUSTOMER. Since attempts at service also occur in other flowcharts, describing these steps separately, as in Figure 2.8, avoids redundancy. Note the use there of another entity called RESERVATIONIST. The decision flow in block 6 of Figure 2.5 indicates that if no RESERVATIONIST is free, a message indicating a wait is to occur. Notice that the scheduling in block 7 identifies the time that the message ends as the event execution time. Doing so enables the END_MESSAGE event (Figure 2.6) to check to see whether a RESERVATIONIST has become available during the nine seconds. Here scheduling creates an event notice for an END_MESSAGE and files it in the future event set. Also notice that entry into block 8 occurs when the CALL event is complete.

Figure 2.6 illustrates the END_MESSAGE event. Block 3 describes a new feature wherein the record of the CUSTOMER is filed in a list called QUEUE. Although filing can occur based on any attribute of the CUSTOMER, it is customary to order this list so that the next available RESERVATIONIST can find the record of the next CUSTOMER for service first in the list. Often, time of filing is the ordering attribute; however, another attribute may be the basis for ordering.

The COMPLETION event in Figure 2.7 describes actions that occur whenever a RESERVATIONIST completes service to a CUSTOMER. The RESERVATIONIST checks to see whether any CUSTOMERs await service (block 1). Each RESERVATIONIST is an entity with an attribute STATUS whose value denotes BUSY or IDLE status. When a COMPLETION occurs this attribute of the corresponding RESERVATIONIST indicates the BUSY mode. If no one is waiting for service, then block 4 indicates that STATUS should be changed to IDLE, and simulation continues via selection in block 5.

If at least one CUSTOMER is waiting in the QUEUE, the RESERVATIONIST removes the record of the first CUSTOMER from the QUEUE (block 2) and service is arranged via the SERVICE routine (block 3). Whereas the transfers to this routine in the CALL and END_MESSAGE events are attempts to get service, here service is assured because of the availability of the RESERVATIONIST. Also, note that the design of this event eliminates the need to update STATUS when customers are waiting.

The decision flow in block 1 of the SERVICE routine (Figure 2.8) allows for the contingency that a RESERVATIONIST is definitely available. Block 2 determines service time. The scheduling of a COMPLETION in block 3 establishes the time at which a corresponding COMPLETION event in Figure 2.7 for this RESERVATIONIST is to be executed. Here again an event notice is created and filed in the future-event set. We now begin to see how crucial a role this set plays in a simulation.

Block 4 destroys the record of the CUSTOMER entity about to begin service. The action indicates that each CUSTOMER is a *temporary entity*, since it can be destroyed (Figure 2.8) as well as created (Figure 2.5). This contrasts with each RESERVATIONIST, which is a *permanent entity* for the life of the simulation. The distinction is important, for as Section 4.2 shows, it leads to considerably different programming procedures.

A little thought makes the convenience of using temporary entities apparent. Unless we destroy CUSTOMER records as they receive service, the number of such records grows as simulated time elapses. Since each record occupies space in the computer, this growth can conceivably consume all available space before desired termination of the simulation occurs. Destroying records that are no longer needed reduces the probability of this occurrence.

Careful space management can also lead to a reduction in computing time. For simulation languages that rely on a *dynamic space allocator* to assign space to records as the simulation program executes, destroying unneeded records makes the program run more rapidly. That is, the more space that is available, the faster the space can be found and allocated. Conversely, an accumulation of unpurged, unnecessary records reduces the amount of space available to the dynamic space allocator and forces it to spend extra computing time searching for the remaining unallocated space. Section 5.3 returns to this topic.

Note that Figure 2.8 ends with a transfer of control back to the calling event. Although we may also regard events as subprograms, they have the distinctive feature of always returning control to the timing routine. Throughout this book we maintain this distinction between a routine and an event.

2.3.1 PRIORITY ORDER

The airline reservation problem assumes that intercall and service times are continuous random variables. This implies that if T_1 and T_2 are the random times at which two distinct events occur, then $\mathrm{pr}(T_1 = T_2) = 0$. However, all computers have finite word size. If T_{11} and T_{22} are the finite-digit representations of T_1 and T_2, respectively, in a computer executing a simulation with these two events, then it is possible for $T_{11} = T_{22}$. That is, $\mathrm{pr}(T_{11} = T_{12}) > 0$.

If an executing simulation encounters this situation, it needs guidance as to the order in which it should execute the two or more events scheduled for execution at the same time. To resolve this issue, we assign a *priority* to each event class. If event class CALL has a higher priority than event class COMPLETION, then, when a CALL event has the identical execution time as a COMPLETION event, the CALL event is executed first. Hereafter, we let the notation CALL $\succ$ COMPLETION denote this priority order. For the airline reservation problem, we assume CALL $\succ$ COMPLETION $\succ$ END_MESSAGE. All simulation languages provide an explicit means for making these declarations.

It is possible that two or more events within the same event class have identical scheduled execution times. When this happens, most simulation programming languages break the ties by executing events in the order in which they are scheduled. In summary, ties are broken first by class and then by time scheduled within class.

2.4 PROCESS-INTERACTION APPROACH

The *process-interaction approach* to modeling differs substantially from the event-scheduling approach. It caters to the flow environment that characterizes queueing systems and produces results identical to those for the event-scheduling approach when the same logical relationships, input, sampling methods, and output analyses are employed. The four principal discrete-event simulation languages, GPSS/H, SIMAN, SIMSCRIPT II.5, and SLAM, all allow model representation based on this concept of flow logic.

Before beginning our description, we offer one viewpoint that the reader will find helpful when weighing the benefits of the two alternative modeling schemes. Whereas the process-interaction method often appears conceptually simpler to a model builder, this simplicity usually comes at the expense of less programming control and flexibility than the event-scheduling approach offers. The truth of this assertion becomes apparent as this section evolves. For each simulation problem, a model builder who plans to use a general-purpose simulation language must decide whether or not to opt for conceptual simplicity or programming control. Considerations that enter this decision include familiarity with alternative languages, the availability of languages at a user's computing facility, and the extent to which a particular simulation program product is to be made available to other potential users with different backgrounds at other computing facilities. Alternatively, suppose that a model builder employs a special-purpose simulation package, as often occurs in telecommunication and manufacturing simulation. Since most of these special-purpose packages use the process-interaction approach, the model builder gives up the ability to choose, presumably in return for the benefit of reduced programming effort.

Process interaction relies on a conceptually different representation of behavior than event scheduling does. In the event-scheduling approach, an event contains all decision flows and updating that relate to the change in state that execution of the event requires. Since time elapses between events, one can regard a simulation as the execution of a sequence of events ordered chronologically on desired execution times. However, no time elapses within an event. By contrast, the process-interaction approach, either implicitly or explicitly, provides a *process* for each entity in a system. To illustrate this concept, we return to the airline reservation problem of Section 2.3. Figure 2.9 organizes the elementary events of the problem into two sequences, one corresponding to the temporary entity CUSTOMER and the other to the permanent entity RESERVATIONIST. Each generic sequence is called a process, and dashed lines denote simulated time flow. For example, elementary event 6 (finish nine-second message) occurs nine seconds after elementary event 5 (begin nine-second message). Each temporary entity moves through the system and consequently through time. Occasionally, a temporary entity encounters an impediment to progress and must *wait*. For example, if all RESERVATIONISTs (permanent entities) are answering calls when a new CUSTOMER (temporary entity) materializes, this CUSTOMER must wait for service. The inclusion of time flow in a process distinguishes this approach from event scheduling. Moreover, it calls for a more indirect form of computer programming when moving from logic flow to executable program.

Figures 2.10 and 2.11 depict prototypical representations of these processes. Each CUSTOMER has an attribute TYPE that identifies whether a single-destination or multiple-destination reservation is desired. TYPE plays the same role here as in the event-scheduling approach in Figures 2.4 and 2.7. Note in block 4 of the CUSTOMER procedure that an entry into the procedure at block 1 is scheduled at a future simulated time. This scheduling sets the stage for the creation of the next CUSTOMER. This form of scheduling coincides with that of the event-scheduling approach, which always specifies entry to the first block of a flowchart. Consequently, block 4 creates an event notice for this entry and files it in the

CUSTOMER   RESERVATIONIST

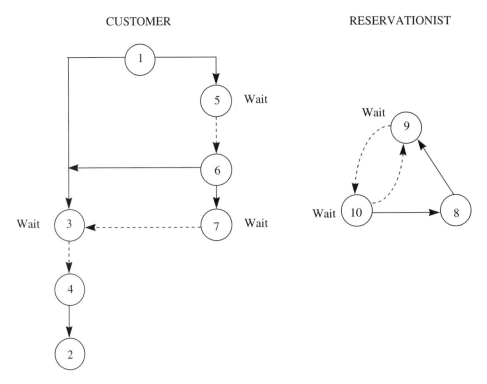

Figure 2.9   Processes

future event set for later execution. Block 6 also schedules an event, namely, a completion of the busy message. Flow then proceeds to block 7, where the CUSTOMER *waits* until the message is complete. After the nine seconds for the message elapse, flow continues to block 8.

The flow through blocks 6 and 7 represents one distinguishing feature of the process-interaction approach. When the simulation encounters block 6, it creates an event notice and assigns current simulated time plus nine seconds to its clock attribute as the desired execution time. This event notice also has a *reactivation point* attribute to which the scheduling action assigns a pointer to block 8. Then it files the notice in the future-event set. Flow continues to block 7, where control returns to the timing routine, which then selects the next event for execution. Here, block 7 represents a *deactivation point*. In contrast to the event-scheduling approach, here entry into a process for execution does not necessarily start at the first block. Also, time elapses within the CUSTOMER process. The dashed line between blocks 7 and 8 indicates that execution of block 8 occurs at a later time than execution of block 7 does. This time evolution within a process distinguishes it in a major way from an event where all computation takes place at the same moment in time.

Since reactivation occurs in a specified number of seconds, the wait in block 7 is *unconditional*. However, the wait in block 10 is a *conditional* one. Here, reactivation at block

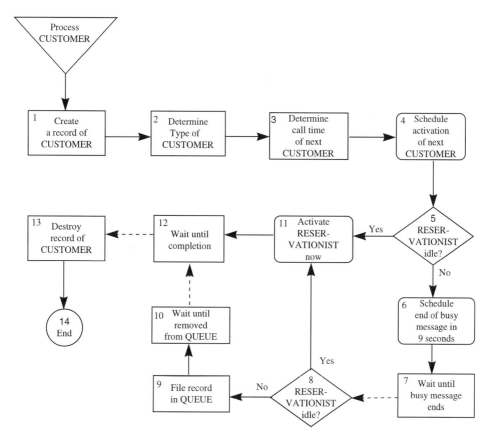

Figure 2.10 CUSTOMER process: process-interaction approach

12 occurs once a RESERVATIONIST in Figure 2.11 selects the CUSTOMER for service. At least two alternative modeling representations of a conditional wait occur in practice. First, a notice that a CUSTOMER is waiting for a RESERVATIONIST may be filed in a list associated with the RESERVATIONIST process. In Figure 2.10 this list is called QUEUE. Then, whenever a RESERVATIONIST in Figure 2.11 completes service at block 7, a scan of the relevant list results in the selection of the next CUSTOMER to receive service. The notice corresponding to this CUSTOMER is then removed from the list in block 4. If no CUSTOMER is waiting, the RESERVATIONIST becomes idle in block 3.

If several alternative types of servers exist in a simulation problem, each would have a distinct representation as a process and each would have a corresponding list on which to store notices of demands for service. By contrast, a second approach to modeling a conditional wait has a single list in which to store notices of all service demands, regardless of the type of server demanded. By aggregating all demands, this alternative reduces the number of lists to one, thereby reducing space requirements within a program. However, it requires additional search time on the part of each type of server, who now has to search through a longer list of

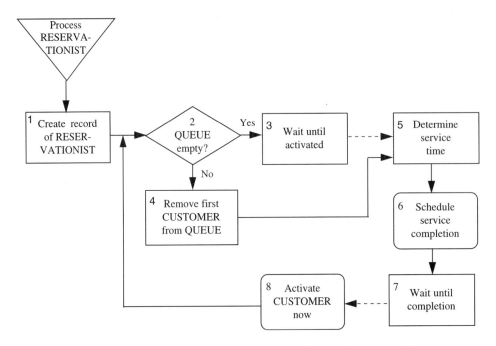

Figure 2.11   RESERVATIONIST process: process-interaction approach

notices. Both options are available in most simulation languages, with multiple lists being the more common approach. Regardless of which option is chosen, we hereafter refer to these lists as the *conditional-event* set (CES).

Time also elapses in the RESERVATIONIST process in Figure 2.11. Unlike every instance of the CUSTOMER process, however, each instance of the RESERVATIONIST process remains in existence during the life of a simulation run. Each RESERVATIONIST has an unconditional delay at block 7 and a conditional delay at block 3. Collectively, these reveal the interaction that occurs between processes. In particular, block 11 in Figure 2.10 signals the idle RESERVATIONIST in block 3 of Figure 2.3 that service is required, whereas the removal of a CUSTOMER from the QUEUE in block 4 of Figure 2.11 signals the waiting CUSTOMER in block 10 of Figure 2.10 that service is about to begin.

In a simulation based on the process-interaction approach, the timing routine does more work than in one based exclusively on event scheduling. As Figure 2.12 shows, this extra work arises from the need to manage the conditional event set as well as the future events list. For conceptual simplicity, this depiction relies on a single CES. Here, block 12 is equivalent to a scheduling statement, whereas reactivating a process in block 5 is equivalent to transferring control to a particular instance of a process. Also, after selecting a scheduled process from the FES in block 3 and executing it in block 5, the timing routine scans the CES in block 1 to identify any processes whose conditions for continued execution are now met. If found, it removes these processes, one at a time, from the CES in block 11 and files them

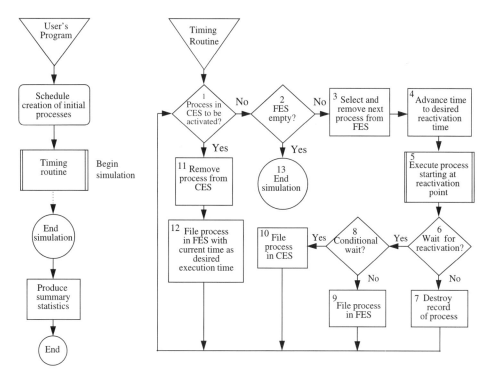

Figure 2.12   Prototypical timing routine, process-interaction approach

in the FES with current simulated time as the desired execution time (block 12). As with Figure 2.3, we emphasize that Figure 2.12 describes a prototypical timing routine. Other designs may exist that manage processes in a different, but logically equivalent, manner.

2.5   ACTIVE PROCESSES AND PASSIVE RESOURCES

Perusal of Figures 2.10 and 2.11 reveals that the CUSTOMER process provides the stimulus to action, whereas the RESERVATIONIST process merely responds to the demand of a newly arriving CUSTOMER (block 11 of Figure 2.10) or to the status of the QUEUE (block 2 of Figure 2.11). More generally, the demand for resources in the process-interaction approach induces a sequence of steps very much akin to those that Figure 2.11 depicts.

Recognition of the commonality of these resource-related steps across simulations based on the process-interaction approach has led to simulation software with macrostatements that account for this behavior. For example, most simulation programming languages have SEIZE or REQUEST commands that "activate" a resource and RELEASE commands that indicate the completion of service and provide a signal to resources to seek out other work to do. In particular, these and like simulation statements do away with the need to model the RESERVATIONIST process explicitly as in Figure 2.11.

The concept of active processes and passive resources originated with GPSS (Gordon 1969). Its appeal has been so great that virtually every simulation programming language based on the process-interaction approach embraces it. Section 4.21 illustrates the concept for SIMSCRIPT II.5, and Sections 4.24 through 4.25 for SIMAN.

If its appeal is so great, why do we bother to describe the option in which both CUSTOMER and RESERVATIONIST are active processes? Flexibility is the answer. Macrostatements such as SEIZE and RELEASE correspond to recipes for performing a multiplicity of steps in sequence. They serve a constructive purpose, provided that the problem at hand has a logic flow that corresponds to that in the statements. If it does not, the simulationist has two options, to ignore the difference and take advantage of the macrostatements, or to write her/his own sequence of statements in the selected language to perform the steps required in the problem at hand. Our discussion provides this guidance for effecting the latter option.

2.6 TELECOMMUNICATIONS NETWORK MODELING

Previous sections concentrate on the flow of discrete objects through a delay system. Each object is identified and its path tracked from entry to exit. In principle, telecommunications systems also fit this characterization. A packet (object) is transmitted (flows) from an origin node (entry) to a destination node (exit). En route it visits a sequence of nodes (stations), collectively called a route. Each channel or link connecting a pair of nodes has an upper limit on bandwidth (service rate), and each node has a buffer with finite capacity where en-route messages wait until selected for transmission.

Discrete-event simulation is often used to study how to set buffer sizes and node retransmission rates in order to control message loss, due to encountering a buffer with no waiting room, and to evaluate buffer utilization. However, major differences exist between these simulation models and those that arise in the job-shop settings described earlier. In particular, the volume of messages is so large that any attempt to identify and track each within an executing simulation generally meets with failure. To resolve this problem, messages are viewed as part of a *fluid* that flows from node to node. While the concept of discrete events is preserved, the fluid representation considerably reduces the computational burden, thus allowing a simulationist to study aggregate behavior without tracking its atomistic parts. This section describes a prototypical telecommunications network simulation model. It shows how the concepts of discrete-event modeling apply and notes those features generic to most telecommunications networks encountered in practice. Section 10.1 returns to the topic of fluid-flow arrivals.

Consider a system of k buffers in series as in Figure 2.13 with r exogenous sources putting fluid into buffer 1 at rates $\alpha_1(\tau), \ldots, \alpha_r(\tau)$ at simulated time τ. Each source j

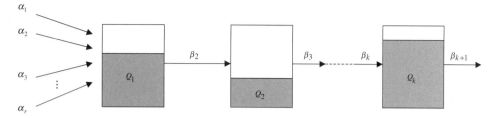

Figure 2.13   Fluid flow network

changes rate at discrete points in time. The times between changes for a particular source may be i.i.d. random variables, in which case they are sampled from a probability distribution. Alternatively, they may be deterministic with possibly constant increments. The simplest rate change structure has $\alpha_j(\tau)$ alternating between a low value α_{j0} and a high value α_{j1}; but other behavior is possible.

For each buffer $i = 1, \dots, k$, let

$$c_i := \text{capacity},$$
$$Q_i(\tau) := \text{volume occupied at time } \tau,$$
$$\tau_{i,\text{next}} := \text{time of next input rate change},$$
$$\gamma_{i-1} := \text{maximal allowable outflow rate from buffer } i - 1 \text{ into buffer } i.$$

Buffer 0 is taken to be the exogenous source.

The aggregate exogenous input rate to buffer 1 is

$$\beta_1(\tau) = \min\left[\gamma_0, \alpha_1(\tau) + \cdots + \alpha_r(\tau)\right].$$

Fluid flows out of buffer 1 at rate

$$\beta_2(\tau) = \begin{cases} \min\left[\beta_1(\tau), \gamma_1\right] & \text{if } Q_1(\tau) = 0, \\ \gamma_1 & \text{if } Q_1(\tau) > 0, \end{cases}$$

and this is the inflow rate to buffer 2. More generally, the outflow rate from buffer i and the inflow rate to buffer $i + 1$ for $i = 1, \dots, k$ is

$$\beta_{i+1}(\tau) = \begin{cases} \min\left[\beta_i(\tau), \gamma_i\right] & \text{if } Q_i(\tau) = 0, \\ \gamma_i & \text{if } Q_i(\tau) > 0. \end{cases}$$

Note that $\beta_{k+1}(\tau)$ is the outflow rate of the network at time τ.

If $\tau < \tau_{i,\text{next}}$ and $\beta_i(\tau) > \beta_{i+1}(\tau)$, then

$$\left[\beta_i(\tau) - \beta_{i+1}(\tau)\right]\left(\tau_{i,\text{next}} - \tau\right) + Q_i(\tau) > c_i,$$

and buffer i overflows at time

$$\tau_f = \tau + \frac{c_i - Q_i(\tau)}{\beta_i(\tau) - \beta_{i+1}(\tau)} \quad < \tau_{i,\text{next}}$$

and fluid is lost to the system. If $\tau < \tau_{i,\text{next}}$ and $\beta_{i+1}(\tau) > \beta_i(\tau)$, then buffer i empties at time

$$\tau_e = \tau + \frac{Q_i(\tau)}{\beta_{i+1}(\tau) - \beta_i(\tau)},$$

so that $Q_i(\tau_e) = 0$.

In this system the design parameters are the buffer capacities $c_1, \ldots, c_k$ and possibly the maximal outflow rates $\gamma_1, \ldots, \gamma_k$. Over a simulated time interval $[s, t)$, the sample average unused capacity

$$\bar{B}(s, t) = \frac{1}{t - s} \int_s^t \sum_{i=1}^k [c_i - Q_i(\tau)] \, d\tau$$

and the sample loss rate

$$L(s, t) = \frac{1}{t - s} \int_s^t [\beta_{k+1}(\tau) - \beta_1(\tau)] \, d\tau$$

estimate performance. For each buffer i the sample average unused capacity

$$\bar{B}_i(s, t) = \frac{1}{t - s} \int_s^t [c_i - Q_i(\tau)] \, d\tau$$

and sample average unused bandwidth

$$\bar{R}_i(s, t) = \frac{1}{t - s} \int_s^t [\gamma_i - \beta_{i+1}(\tau)] \, d\tau$$

may also be of interest.

More complex models exist, some with thresholds and volumes for each buffer that indicate when to reduce inflow rates to limit the frequency of overflow or the buffer becoming empty. These are more in the spirit of dynamic control problems in which interest focuses on how the choice of threshold affects unused capacity and loss rate. Other models consider the flow of several different fluids. For example, see Kumaran and Mitra (1998).

Four elementary events characterize our model:

1. Change in buffer inflow rate

2. Change in buffer outflow rate

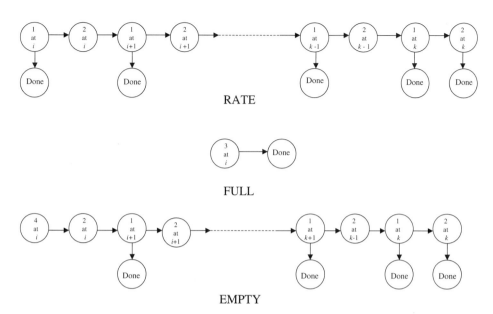

Figure 2.14 Fluid-flow compound events

3. Buffer becomes full

4. Buffer becomes empty.

Note that at time τ, $\beta_1(\tau)$ denotes the exogenous rate to buffer 1; for $i = 2, \ldots, k$, $\beta_i(\tau)$ denotes the inflow rate to buffer i and the outflow rate for buffer $i - 1$; and $\beta_{k+1}(\tau)$ denotes the output rate from the network. If, as in some models, controls were applied to outflow rates, then the twofold definitions of $\beta_2(\tau), \ldots, \beta_k(\tau)$ would not apply.

The elementary events form the compound events in Figure 2.14. These compound events reveal how the event-scheduling approach can conveniently represent the iterative updating of inflow-rate changes that potentially can occur at all buffers $j \geq 1$ when the exogenous flow rate changes, or at all buffers $j \geq i$ when buffer i becomes empty. Notice that executing an overflow event allows for recording when overflow begins.

Figure 2.15 provides more details. Because fluid flows continuously, it is necessary to schedule events to handle fluid exhaustion and overflow. The EMPTY event updates a change in a buffer's outflow rate to reflect the fact that fluid can exit an empty buffer at a rate no greater than the rate at which it enters it. In the current setting, the FULL event merely acknowledges that a buffer has become full and fluid (messages) will overflow and be lost. In other operating environments, a decision may be made to reduce the inflow rate prior to or at overflow. This action calls for rate changes in previous as well as successive buffers. The FULL event here allows changes only in successive buffers.

The RATE event executes every time an exogenous or endogenous rate change occurs. It introduces the reader to a new concept, namely, a *cancellation*. When $\beta_i(\tau)$ changes at time

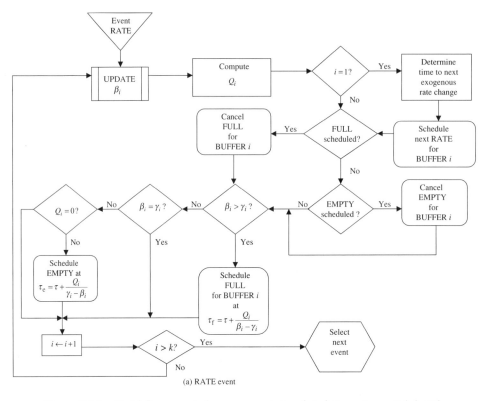

Figure 2.15 Fluid flow events ($\tau :=$ current simulated time, $\beta_i := \beta_i(\tau) \; \forall i$)

τ, the scheduled execution for an already scheduled FULL or EMPTY event may change. Canceling an event causes the simulation to remove the corresponding event notice from its FES and, possibly, to schedule a new FULL or EMPTY event. Indeed, this can happen as many as $k - i + 1$ times in a RATE event. Routine UPDATE describes the three alternative ways that a change in flow rate can occur.

2.7 LESSONS LEARNED

- An *event* denotes a change in state.

- Every discrete-event simulation model is composed of *elementary events*. These can be combined to form *compound events*, leading to the *event-scheduling approach*. Alternatively, they can be combined into *processes*, leading to the *process-interaction approach*.

- For a particular problem, modeling with the event-scheduling and process-interaction approaches both lead to identical output for identical input.

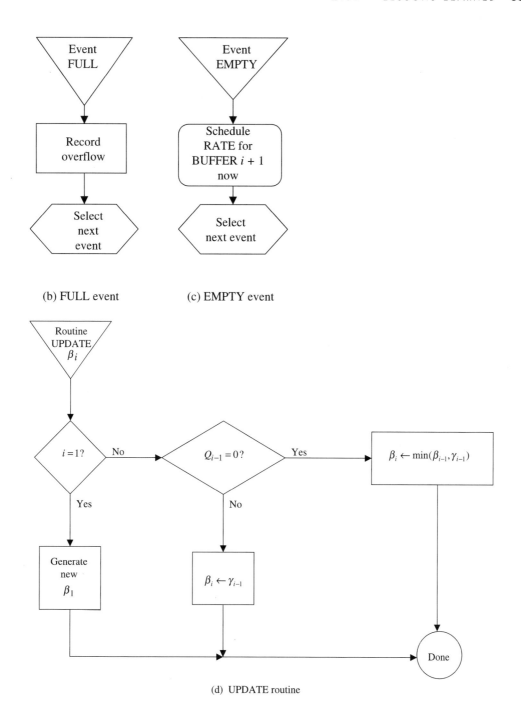

(b) FULL event

(c) EMPTY event

(d) UPDATE routine

Figure 2.15

- During execution of program code for an event, simulated time remains fixed.

- During execution of program code for a process, simulated time may advance. Indeed, execution may terminate within the code, and after an elapsed simulated time execution may begin again at a *reactivation point* in the code.

- An *event notice* with an *attribute* for desired execution time is used to schedule execution.

- An event or process notice with attributes for desired execution time and reactivation point is used to schedule execution.

- Every discrete-event simulation language keeps its event notices in a *future-event set* (FES).

- Every discrete-event simulation language has a *timing routine* that operates on the FES to cause the simulation to execute events in the desired order.

- The process-interaction approach also has some form of a *conditional-events set* that keeps track of when conditions are met that allow execution to continue for each process.

2.8 EXERCISES

GENERAL INSTRUCTIONS

For each exercise modeled with the event-scheduling approach, define all entities, attributes, lists, and elementary events, as well the compound events they form. For each exercise modeled with the process-interaction approach, define all entities, attributes, lists, elementary events, and the processes formed from them.

QUARRY-STONE CRUSHING

✐ **2.1** Hard Rock Corporation owns a quarry and three power shovels that load its truck with stone to be delivered to its crusher. The distances between each shovel and the crusher vary. Two 20-ton and one 50-ton trucks are assigned to each shovel. A shovel loads its trucks in first-come-first-served order. However, unloading at the crusher occurs first based on truck size, with largest going first and then on a first-come-first-served basis. Table 2.1 gives loading, dumping, and travel times, where $a + \mathcal{E}(b)$ denotes the sum of a constant a and an exponentially distributed random variable with mean b. Once unloaded, the stone is crushed at the rate of 25 tons per minute.

  a. Ignoring waiting times, determine the long-run arrival rate to the crusher of a randomly selected truck in tons per minute.

Table 2.1 Loading, dumping, and travel times (minutes)

Truck type	Loading time	Dumping time	Travel to or from shovel		
			1	2	3
20-ton	$3.5 + \mathcal{E}(1.5)$	$1.5 + \mathcal{E}(.5)$	$.25 + \mathcal{E}(.75)$	$2.5 + \mathcal{E}(1)$	$1.75 + \mathcal{E}(.5)$
50-ton	$8 + \mathcal{E}(2)$	$2.8 + \mathcal{E}(1.2)$	$.9 + \mathcal{E}(.30)$	$2.9 + \mathcal{E}(.8)$	$1.9 + \mathcal{E}(.55)$

b. Determine the long-run unloading rate at the crusher of a randomly selected truck in tons per minute.

c. If this were an open-loop system that generates stone to the crusher as fast as the shovels can load an unlimited number of trucks two-thirds of which are 20-ton and one-third of which are 50-ton, would it be stable? Explain your answer in detail.

d. Explain why travel to and from the crusher does or does not play a role in part c.

e. Model the problem using flowcharts based on the event-scheduling approach.

f. Model the problem using flowcharts based on the process-interaction approach.

✎ **2.2** A well-established demand exists for Hard Rock's crushed rock at A dollars per ton. Management knows that the crusher can process more stone than is currently supplied to it. To boost actual throughput, Hard Rock considers allowing an outside contractor to supply additional stone to the crusher in 50-ton loads. The interarrival times between successive loads would be i.i.d. from $\mathcal{E}(1/\lambda)$, where λ is to be determined by Hard Rock. Regardless of origin, 50-ton trucks receive priority over 20-ton trucks, among 50-ton trucks selection is based on first in first out, and among 20-ton trucks it is based on first in first out. External trucks have unloading times $1.5 + \mathcal{E}(1)$ in minutes.

Crushing costs $B = 0.50A$ dollars per ton. Hard Rock provides stone to the crusher at a cost of $C = 0.25A$ dollars per ton. The outside contractor offers to provide rock at $D = 0.26A$ dollars per ton. In order to maximize profit, Hard Rock management wants to know what value to assign to λ. Assume $A = 32$.

a. Determine the smallest rate λ_* such that the system would be unstable if $\lambda \geq \lambda_*$.

b. Model the problem with external as well as internal input using flowcharts based on the event-scheduling approach.

c. Model the problem with external and internal input using flowcharts based on the process-interaction approach.

LOADING TANKERS

✎ **2.3** Tankers take on crude oil in an African port for overseas shipment. The port has facilities for loading as many as three ships simultaneously. Interarrival times between

Table 2.2 Tanker Types

Type	Relative frequency	Loading time (hours)
1	0.25	$\mathcal{G}(1.5, 12)$
2	0.55	$\mathcal{G}(2.4, 10)$
3	0.20	$\mathcal{G}(.75, 48)$

tankers follow $\mathcal{E}(11)$, where $\mathcal{E}(\theta)$ denotes the exponential distribution with mean θ hours. Three types of tanker exist. Table 2.2 describes their characteristics, where $\mathcal{G}(\alpha, \beta)$ denotes the Gamma distribution with scale parameter α and shape parameter β (note that the mean is $\alpha\beta$).

One tug serves the port. All tankers require the tug to move them to and from the berths. When the tug is available, berthing (an empty tanker) time follows $\mathcal{E}(1)$, whereas deberthing (a full tanker) time follows $\mathcal{E}(1.25)$. When not towing a tanker, tug transit to pick up a tanker is called *deadheading* and takes time drawn from $\mathcal{E}(0.75)$.

Upon arrival at the berths the tug next services the first tanker in the deberthing queue. If this queue is empty, it services the first tanker in the berth queue in the harbor, provided that a berth is available. Upon arrival in the harbor, the tug services the first tanker in the berth queue. If the queue is empty, it services the first tanker in the deberthing queue. If both queues are empty, the tug remains idle.

To compete with other oil-loading ports, the port authority plans to institute a rebate program based on total *port time*. This is the summation of waiting, towing, and loading times. If a tanker's port time exceeds 48 but is no more than 72 hours, the authority will pay its owner a rebate of A dollars. If it exceeds 72 but is no more than 96 hours, the rebate is $3A$ dollars. In general, if it exceeds $24i$ but is no more than $24(i + 1)$ hours, the rebate is $(2i - 1)A$. Assume $A = 10,000$ dollars.

To estimate the cost of this program the authority commissions a simulation study to characterize port time. The first step is modeling.

a. Show that this delay system is stable.

b. Model the problem using flowcharts based on the event-scheduling approach.

c. Model the problem using flowcharts based on the process-interaction approach.

2.4 Additional Tankers. A shipper considers bidding on a contract to transport crude oil from the port described in Exercise 2.3 to the United Kingdom. He estimates that shipment would require five tankers of a particular type to meet contract specifications. Time to load one of these tankers follows $\mathcal{E}(21)$. After loading and deberthing, a tanker would travel to the United Kingdom, unload its crude oil cargo, and return to the port. Round-trip time follows $\mathcal{E}(240)$. Before the port authority agrees to accommodate the five new tankers, it wants to determine the effect of the additional port traffic on tanker waiting time, tug utilization, and berth utilization.

a. Describe how the additional tankers would affect the stability of the overall queueing system in the port.

b. Augment the model in Exercise 2.3b using flowcharts based on the event-scheduling approach.

c. Augment the model in Exercise 2.3c using flowcharts based on the process-interaction approach.

TERMINAL PACKET ASSEMBLY/DISASSEMBLY CONTROLLER MODEL (CHENG AND LAMB 1999)

✐ **2.5** The control of packet assembly and disassembly is a topic of major interest to all who provide communications network services to industry, government, and the public at large. The problem described here is considerably simplified to allow students to work on it without the impediments that very high transmission rates and network size impose. However, the issues that arise in the scaled-down version continue to be principal considerations in higher-speed problems as well.

A terminal packet assembly/disassembly (PAD) controller model contains r input buffers each of capacity m and one output buffer of infinite capacity. The PAD controller receives characters from $r = 10$ terminals, one for each input buffer, and assembles them into packets before transmitting them into a network. The characters from each terminal arrive at its input buffer and wait. A packet is formed when either an input buffer becomes full or a special control character arrives, whichever is first. If present, the control character is included in the packet. When a packet is formed, it is tagged with a fixed number of overhead characters and sent to the output buffer, where it waits in a FIFO queue. The output buffer serves all input buffers and, if there are any to transmit, puts out characters at a constant rate of one per time unit. The input characters into each buffer form a Poisson process with rate λ. Assume that r is constant, so that $\rho = r\lambda$ is the arrival rate of characters into the PAD controller.

The probability that an arriving character is a special control character is $q := .02$. Also, one overhead character is added to a packet when it moves from the input buffer to the output buffer.

The purpose for studying this model is to identify the rate λ and input buffer capacity m that minimize

$$\mu(m, \lambda) := \text{average character delay in the PAD controller model between} \atop \text{arriving at an input buffer and being transmitted to the network} \tag{2.6}$$

subject to the constraints $.30 \le \rho \le 1$ and $m \in \{8, 9, \ldots, 32\}$. Note that μ increases without bound as either $\rho \to 0$ or $\rho \to \rho_0 < 1$, where $\rho_0 :=$ system-saturation input rate.

a. Determine the mean packet size as a function of m and q.

  b. Determine ρ_0.

  c. Identify all elementary events and use them to form compound events.

  d. Using the event-scheduling approach, prepare flowcharts that describe the details of the events in this problem.

  e. Use the elementary events to form processes as in Figure 2.9.

  f. Using the process-interactive approach, prepare flowcharts that describe the details of the processes in part e.

MOBILE-TELEPHONE SERVICE (DALEY AND SERVI 1999)

✐ **2.6** Friendly Communications Incorporated provides mobile and standard telephone services by means of a collection of communication towers distributed throughout a region. A customer traveling through the region while using the mobile service automatically has her/his call handed off and managed (transmission and reception) by a sequence of towers en route. At any moment in time, the tower providing this service to the customer is the one geographically closest to him or her. Each tower can handle m calls at a time.

  Each tower has the capacity for handling a total of m standard and mobile calls simultaneously. For a traveling mobile caller, there is always the possibility that when her/his call is handed off from one tower to the next, that next tower is already serving m callers, and hence the mobile caller loses the connection. This loss is highly undesirable.

  Assume that the times between successive new standard calls for a tower are i.i.d. random variables from $\mathcal{E}(1/\lambda_s)$, whereas the times between successive handoffs of mobile calls to the tower are i.i.d. from $\mathcal{E}(1/\lambda_h)$. Also, assume that call-duration times are i.i.d. from $\mathcal{E}(1/\omega)$ and independent of type.

  To manage loss the company decides to dedicate r ($\leq m$) lines for handoffs. In particular, let $X_s(\tau)$ denote the number of standard call and $X_h(\tau)$ the number of mobile calls at time τ. Let τ_- denote the instant of time prior to a new demand. The operating policy is:

  - If a new request for standard service arrives at time τ, it is provided if $X_s(\tau) < m - r$. Otherwise, service is denied and the call is lost.

  - If a new mobile call handoff request occurs at time τ, it is assigned a dedicated line if $X_s(\tau_-) < r$. If $X_h(\tau_-) > r$ and $X_s(\tau_-) + X_h(\tau_-) < m$, it is assigned one of the $m - r$ nondedicated lines.

  - If $X_s(\tau_-) + X_h(\tau_-) = m$, the mobile call is lost.

  The purpose of this study is to evaluate how the choice of the number of dedicated lines r affects

$$\beta_h(m, r) := \text{probability of denying service to a handed-off mobile call,}$$

$\beta_s(m, r) :=$ probability of denying service to a new standard call,

and, in particular, the cost of losing calls. If the cost of losing a handed-off call is unity and the cost of denying service to a new standard call is $c(< 1)$, then it is of particular interest to determine the r that minimizes long-run average total cost

$$C(m, r, c) := \frac{\lambda_h}{\lambda_h + \lambda_s} \beta_h(r, m) + \frac{c\lambda_s}{\lambda_h + \lambda_s} \beta_s(m, r). \tag{2.7}$$

a. Determine the traffic intensity for this problem.

b. Identify all elementary events and form compound events from them.

c. Using the event-scheduling approach, prepare flowcharts that describe the details of the events.

d. Use the elementary events of part b to form processes as in Figure 2.9.

e. Using the process-interaction approach, prepare flowcharts that describe the details of the processes in part d.

2.9 REFERENCES

Cheng, R.C., and J.D. Lamb (1999). Making efficient simulation experiments interactively with a desktop simulation package, University of Kent at Canterbury, United Kingdom.

Daley, D. and L. Servi (1999). Personal communications.

Fishman, G.S. (1973). *Concepts and Methods in Discrete Event Digital Simulation*, Wiley, New York.

Gordon, G. (1969). *System Simulation*, Prentice-Hall, Englewood Cliffs, N.J.

Pritsker, A.A.B., J.J. O'Reilly, and D.K. LaVal (1997). *Simulation with Visual SLAM and AweSim*, Wiley, New York.

CHAPTER 3

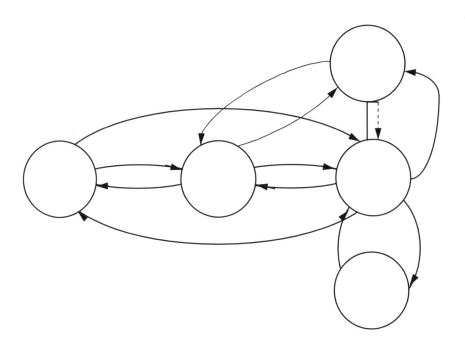

Data Collection and Averages

In simulating a discrete-event delay system, a principal objective is to evaluate its performance with regard to throughput, buffer occupancy, delay, resource utilization, and system loss. *Throughput* denotes the rate at which a system provides finished product. As examples, throughput in a manufacturing setting denotes the number of finished goods that a factory produces per unit of time. In a telecommunications network, the rate at which messages pass from origin to destination constitutes a measure of throughput. The number of vehicles that cross a bridge per unit time characterizes throughput in a particular type of transportation system. Buffer occupancy denotes the number of items in the system awaiting processing or service and is sometimes called *queue length*. Delay denotes the length of time that an item waits and alternatively is called *waiting time*. Occasionally, *system time* characterizes delay. It is the sum of waiting time and service time. *Resource utilization* denotes the proportion of system resources that are kept busy processing items.

System loss rate sometimes serves as an additional performance indicator. It measures the rate at which objects are lost to or leave the system without successfully completing their purposes when they encounter delays exceeding specified tolerances. For example, work-in-process within a manufacturing system may lose value unless delays between successive steps in the production process are kept within specified limits. This deterioration can arise in some forms of semiconductor wafer fabrication as a consequence of the properties of materials and the production process.

To evaluate performance, a discrete-event simulation uses data generated during execution to compute finite-sample time averages and arithmetic averages as approximations to the unknown true long-term averages that characterize system performance. Section 3.1 describes these averages and, in particular, the recursive forms that they customarily take. The remainder of the chapter provides the reader with a preview of the statistical issues that arise when assessing how well these finite-sample averages approximate their corresponding long-term averages. Among these issues, the most notable are:

- Choosing the initial state ($S(0)$) in which to begin simulation

- Choosing a warm-up interval ($k - 1$) that is to elapse before data collection begins in order to dissipate the influence of the initial state of the simulation on data used to compute the finite sample averages.

- Choosing a run length (t) that ensures that the finite-sample averages are representative of the unknown true long-term averages.

Sections 3.2 through 3.5 explain the significance of $S(0)$, $k - 1$, and t, leaving the issues of choice until Chapter 6, which gives a considerably more detailed account of these issues.

The account of computational forms in Section 3.1 aims to give the reader an understanding of the data-accumulation techniques that simulation programming languages often provide implicitly. For the cases where such techniques are not provided, the account provides sufficient information for a user to program the steps explicitly. Subsequent sections aim at sensitizing the reader to the important role that statistical issues play in assessing the quality of the numbers that a simulation generates as output. This topic has many facets, and any simulation user who contemplates presenting simulation output to an audience and convincing them that the output truly characterizes the sysem will benefit first from an understanding of what the statistical issues are (Sections 3.2 through 3.5) and second from an awareness of how to resolve them (Chapter 6).

3.1 PERFORMANCE STATISTICS

We again use the notion of Section 1.3. For $t > s \geq 0$, let

$$A(s, t) := \text{number of arrivals in time interval } (s, t], \qquad (3.1)$$

$$N(s, t) := \text{number of completions in time interval } (s, t], \qquad (3.2)$$

$$Q(t) := \text{queue length at time } t, \qquad (3.3)$$

$$B(t) := \text{number of busy servers at time } t, \qquad (3.4)$$

$$W_i := \text{waiting time of departure } i, \qquad (3.5)$$

from which follow the accounting identities

$$A(s, t) = A(0, t) - A(0, s),$$
$$N(s, t) = N(0, t) - N(0, s),$$

and

$$A(0, t) - N(0, t) = Q(t) + B(t). \tag{3.6}$$

Occasionally, we use the more concise notation $A(t) := A(0, t)$ and $N(t) := N(0, t)$.

A discrete-event simulation program is an engine for producing *sample paths* $\{A(\tau) - A(s), s \leq \tau \leq t\}$, $\{N(\tau) - N(s), s \leq \tau \leq t\}$, $\{Q(\tau), s \leq \tau \leq t\}$, $\{B(\tau), s \leq \tau \leq t\}$, and $\{W_i; i = N(s) + 1, \ldots, N(t)\}$. These sample paths, sometimes called *sample records* or *realizations*, provide histories of how the underlying processes in the simulation behave over the finite time interval $[s, t]$. It is these histories that provide the raw data for computing the summary sample statistics:

- average arrival rate

$$\bar{A}(s, t) := \frac{A(s, t)}{t - s}, \tag{3.7}$$

- average throughput

$$\bar{N}(s, t) := \frac{N(s, t)}{t - s}, \tag{3.8}$$

- average queue length

$$\bar{Q}(s, t) := \frac{1}{t - s} \int_s^t Q(u)\mathrm{d}u, \tag{3.9}$$

- average server utilization

$$\bar{B}(s, t) := \frac{1}{t - s} \int_s^t B(u)\mathrm{d}u, \tag{3.10}$$

- and average waiting time

$$\bar{W}(s, t) := \frac{1}{N(s, t)} \sum_{i=N(s)+1}^{N(t)} W_i. \tag{3.11}$$

Note that expressions (3.7) through (3.10) are time averages, whereas expression (3.11) is an arithmetic average. Sometimes $Q(\tau)$ and $B(\tau)$ are called *time-persistent variables*. Sections 3.1.1 and 3.1.2 explain why this name is applied.

The sample paths contain considerably more information than these averages alone can convey. A sequence of exceedance proportions often provides supplementary information of value. Let

$$
I_A(x) := \begin{cases} 1 & \text{if } x \in \mathcal{A}, \\ 0 & \text{otherwise.} \end{cases}
$$

Then the system also has the following sample performance indicators:

- proportion of time that queue length exceeds q

$$
\bar{P}_Q(q, s, t) := \frac{1}{t - s} \int_s^t I_{(q,\infty)}(Q(u))\mathrm{d}u, \qquad q = 0, 1, \ldots, \tag{3.12}
$$

- proportion of time that more than b servers are busy

$$
\bar{P}_B(b, s, t) := \frac{1}{t - s} \int_s^t I_{(b,\infty)}(B(u))\mathrm{d}u, \quad b = 0, 1, \ldots, \tag{3.13}
$$

- and proportion of customers who wait more than w time units

$$
\bar{P}_W(w, s, t) := \frac{1}{N(s, t)} \sum_{i=N(s)+1}^{N(t)} I_{(w,\infty)}(W_i), \qquad w > 0. \tag{3.14}
$$

While the averages (3.7) through (3.11) show *central tendencies*, the proportions (3.12) through (3.14) can provide a penetrating view of extreme system behavior. A table or graph of $\{\bar{P}_Q(q, s, t); \ q = 0, 1, \ldots\}$ reveals how likely queue length is to be large. It is particularly useful in sizing the buffer or waiting room where jobs wait. A table or graph of $\{\bar{P}_B(b, s, t); \ b = 0, 1, \cdots, m - 1\}$ reveals how well utilized all the servers are. A table or graph of $\{\bar{P}_W(w, s, t), w \in \{w_1 < w_2 < \cdots\}\}$ characterizes the delay that an arbitrary arrival to the system encounters. In practice, graphical display of these distributions enables one to draw meaningful conclusions more quickly than tabular displays do.

3.1.1 QUEUE LENGTH

While concise with regard to mathematical notation, expressions (3.7) and (3.14) give no clue as to how to compute these averages in practice. This section addresses this issue. Let

$$J_Q(t) := \text{number of changes in queue length in simulated time interval } (0, t],$$

$t_{Qi} :=$ time of the ith change in queue length for $i = 1, \ldots, J_Q(t)$,

$t_{Q0} := 0$.

Suppose a simulation is run for the simulated time interval $[0, s + \tau]$, where $s \geq 0$ and $\tau > 0$, and we wish to compute the average queue length over the interval $[s, s + \tau]$. Expression (3.9) implies

$$\bar{Q}(s, s + \tau) = \frac{1}{\tau} \int_s^{s+\tau} Q(u) du.$$

Moreover, the property

$$Q(u) = Q(t_{Q,i-1}), \qquad t_{Q,i-1} \leq u < t_{Qi}, \qquad i = 1, \ldots, J_Q(t),$$

reveals that $\{Q(u), s \leq u \leq s + \tau\}$ is piecewise constant, so that expression (3.9) has the equivalent form

$$
\begin{aligned}
\bar{Q}(s, s + \tau) &= \frac{1}{\tau} \left[\sum_{i=1}^{J_Q(s+\tau)} Q(t_{Q,i-1})(t_{Qi} - t_{Q,i-1}) + Q(t_{Q,J_Q(s+\tau)})(s + \tau - t_{Q,J_Q(s+\tau)}) \right. \\
&\quad \left. - \sum_{i=1}^{J_Q(s)} Q(t_{Q,i-1})(t_{Qi} - t_{Q,i-1}) - Q(t_{Q,J_Q(s)})(s - t_{Q,J_Q(s)}) \right] \\
&= \frac{1}{\tau} \left[\sum_{i=J_Q(s)+1}^{J_Q(s+\tau)} Q(t_{Q,i-1})(t_{Qi} - t_{Q,i-1}) + Q(t_{Q,J_Q(s+\tau)})(s + \tau - t_{Q,J_Q(s+\tau)}) \right. \\
&\quad \left. - Q(t_{Q,J_Q(s)})(s - t_{Q,J_Q(s)}) \right],
\end{aligned}
\tag{3.15}
$$

which leads to the recursion

$$
\begin{aligned}
\bar{Q}(s, t_{Q,J_Q(s)+1}) &= Q(s), \\
\bar{Q}(s, t_{Qi}) &= \left(1 - \frac{t_{Qi} - t_{Q,i-1}}{t_{Qi} - s}\right) \bar{Q}(s, t_{Q,i-1}) + \left(\frac{t_{Qi} - t_{Q,i-1}}{t_{Qi} - s}\right) Q(t_{Q,i-1}) \\
&\qquad\qquad\qquad\qquad \text{for } i = J_Q(s) + 2, \ldots, J_Q(s + \tau), \\
\bar{Q}(s, s + \tau) &= \frac{t_{Q,J_Q(s+\tau)} - s}{\tau} \bar{Q}(s, t_{Q,J_Q(s\,|\,\tau)}) + \frac{\tau + s - t_{Q,J_Q(s+\tau)}}{\tau} Q(t_{Q,J_Q(s+\tau)}).
\end{aligned}
\tag{3.16}
$$

In addition to average queue length, it is often enlightening to know the proportion of time that a queue exceeds a particular value during the time interval $[s, s + \tau]$. This is especially so when an objective of the simulation is to identify how large a buffer should be in order to ensure that overflow occurs with a frequency no greater than a specified quantity. In a manner analogous to that for average queue length, expression (3.12) admits the recursion

$$\bar{P}_Q(q, s, t_{Q, J_Q(s)+1}) = I_{(q,\infty)}(Q(s)),$$

$$\bar{P}_Q(q, s, t_{Qi}) = \left(1 - \frac{t_{Qi} - t_{Q,i-1}}{t_{Qi} - s}\right) \bar{P}_Q(q, s, t_{Q,i-1})$$

$$+ \left(\frac{t_{Qi} - t_{Q,i-1}}{t_{Qi} - s}\right) I_{(q,\infty)}(Q(t_{Q,i-1}))$$

$$\text{for } i = J_Q(s) + 2, \ldots, J_Q(s + \tau), \quad (3.17)$$

$$\bar{P}_Q(q, s, s + \tau) = \frac{t_{Q, J_Q(s+\tau)} - s}{\tau} \bar{P}_Q(q, s, t_{Q, J_Q(s+\tau)})$$

$$+ \frac{\tau + s - t_{Q, J_Q(s+\tau)}}{\tau} \times I_{(q,\infty)}(Q(t_{Q, J_Q(s+\tau)})).$$

Virtually all discrete-event-simulation programming languages use the recursions (3.16) and (3.17) to compute running time averages for queue length as simulated time evolves. For the time average (3.9), recursive computation requires retention of four values, $t_{Q,i-1}, t_{Qi}, s$, and $\bar{Q}(s, t_{Q,i-1})$, to calculate $\bar{Q}(s, t_{Qi})$. For exceedance proportions (3.12) corresponding to queue lengths $q_1 < \cdots < q_k$, it requires retention of $2 + k$ values, $t_{Q,i-1}, t_{Qi}, s, \bar{P}_Q(q_1, s, t_{Q,i-1}), \ldots, \bar{P}_Q(q_k, s, t_{Q,i-1})$, to calculate $\bar{P}_Q(q_1, s, t_{Qi}), \ldots, \bar{P}_Q(q_k, s, t_{Qi})$.

3.1.2 RESOURCE UTILIZATION

Computing averages for resource utilization is analogous to that for queue length. Let

$$J_B(t) := \text{number of changes in the number of busy RESERVATIONISTs}$$
$$\text{in simulated time interval } (0, t]$$

and

$$t_{Bi} := \text{time of the } i\text{th change in the number of busy RESERVATIONISTs for}$$
$$i = 1, \ldots, J_B(t).$$

Since

$$B(u) = B(t_{B,i-1}), \qquad t_{B,i-1} \le u < t_{Bi}, \qquad i = 1, \ldots, J_B(t),$$

we see that $\{B(u),\ s \le u \le s+\tau\}$ is also piecewise constant in time. Therefore, time average (3.10) has the alternative representation

$$
\bar{B}(s, s+\tau) = \frac{1}{\tau}\left[\sum_{i=J_B(s)+1}^{J_B(s+\tau)} B(t_{B,i-1})(t_{Bi} - t_{B,i-1}) + B(t_{B,J_B(s+\tau)}) \right.
$$

$$
\left. \times\, (s+\tau - t_{B,J_B(s+\tau)}) - B(t_{B,J_B(s)})(s - t_{B,J_B(s)}) \right], \qquad (3.18)
$$

so that

$$
\bar{B}(s, t_{B,J_B(s)+1}) = B(s),
$$

$$
\bar{B}(s, t_{Bi}) = \left(1 - \frac{t_{Bi} - t_{B,i-1}}{t_{Bi} - s}\right) \bar{B}(s, t_{B,i-1}) + \left(\frac{t_{Bi} - t_{B,i-1}}{t_{Bi} - s}\right) B(t_{B,i-1})
$$

$$
\text{for } i = J_B(s)+2, \dots, J_B(s+\tau),
$$

$$
\bar{B}(s, s+\tau) = \frac{t_{B,J_B(s+\tau)} - s}{\tau} \bar{B}(s, t_{B,J_B(s+\tau)}) + \frac{\tau + s - t_{B,J_B(s+\tau)}}{\tau} B(t_{B,J_B(s+\tau)}). \quad (3.19)
$$

For $b = 0, 1, \dots, m$, exceedance proportions over $[s, s+\tau]$ take the form

$$
\bar{P}_B(b, s, t_{B,J_B(s)+1}) = I_{(b,m]}(B(s)),
$$

$$
\bar{P}_B(b, s, t_{Bi}) = \left(1 - \frac{t_{Bi} - t_{B,i-1}}{t_{Bi} - s}\right) \bar{P}_B(b, s, t_{B,i-1}))
$$

$$
+ \left(\frac{t_{Bi} - t_{B,i-1}}{t_{Bi} - s}\right) I_{(b,m]}(B(t_{B,i-1}))
$$

$$
\text{for } i = J_B(s)+2, \dots, J_B(s+\tau), \quad (3.20)
$$

$$
\bar{P}_B(b, s, s+\tau) = \frac{t_{BJ_B(s+\tau)} - s}{\tau} \bar{P}_B(b, s, t_{B,J_B(s+\tau)})
$$

$$
+ \frac{\tau + s - t_{BJ_B(s+\tau)}}{\tau} I_{(b,m]}(B(t_{B,J_B(s+\tau)})).
$$

In closed-loop systems, resource utilization may also characterize throughput. For example, at time $u \ge 0$ let $B(u) = 1$ indicate that the crusher is busy crushing and $B(u) = 0$ otherwise in Excersie 2.1. Since the crusher is capable of crushing 25 tons per minute, $25\bar{B}(s, s+\tau)$ is the finite-sample average throughput over the interval $(s, s+\tau]$.

3.1.3 WAITING TIME

Computing the averages in expressions (3.11) and (3.14) for waiting time calls for a different approach than for queue length and resource utilization. Whereas the sample paths of queue length $\{Q(\tau),\ s \le \tau \le t\}$ and resource utilization $\{B(\tau),\ s \le \tau \le t\}$ are indexed on the continuous variable τ, the sample path of waiting times $\{W_i;\ i = N(s)+1, \ldots, N(t)\}$ is a discrete collection of observations. Moreover, whereas a simulation needs to keep track of the number in queue and number of busy servers to compute the averages (3.9), (3.10), (3.12), and (3.13), it must now keep track of the arrival time for each job in the system and subtract it from the simulated time at which the job enters service in order to compute its waiting time. Making arrival time an attribute of the temporary entity corresponding to job or CUSTOMER in the airline problem allows for this last calculation.

To conserve on space for data storage, one can again resort to recursion. Let

$$
V_i := \begin{cases} W_i & i = N(s)+1, \\ V_{i-1} + W_i & i = N(s)+2, \ldots, N(s+\tau), \end{cases}
\tag{3.21}
$$

and

$$
V_i(w) := \begin{cases} I_{(w,\infty)}(W_i) & i = N(s)+1, \\ V_{i-1}(w) + I_{(w,\infty)}(W_i) & i = N(s)+2, \ldots, N(s+\tau). \end{cases}
\tag{3.22}
$$

Then for each simulated time $t \in \{[s, s+\tau] : N(s+\tau) > N(s)\}$,

$$
\bar{W}(s,t) = \frac{V_{N(t)-N(s)}}{N(t) - N(s)}
\tag{3.23}
$$

and

$$
\bar{P}_W(w, s, t) = \frac{V_{N(t)-N(s)}(w)}{N(t) - N(s)}.
\tag{3.24}
$$

Section 3.5.1 describes a relationship known as *Little's law* that relates the arithmetic average (3.11) for waiting time to the time average (3.9) for queue length, thus reducing the amount of data required to compute both averages.

3.2 STATISTICAL ACCURACY

In principle, every simulationist wants:

- Estimates of long-term averages uninfluenced by the initial conditions prevailing at the beginning of simulation.

- Estimates whose statistical accuracy improves as the length of the sample record, on which the estimates are based, increases.

- An assessment of how well the sample averages approximate their corresponding unknown long-term averages.

Virtually every simulation programming language provides special capabilities for computing time and arithmetic averages and producing reports that display them. Indeed, some languages do this automatically for performance measures such as utilization, queue length, and waiting time, often producing histograms as well. While these conveniences reduce or eliminate the need for the simulation user to generate results, it in no way addresses the three just-mentioned statistical issues; nor for a simulation executed for two different sets of numerical values assigned to its input parameters does it provide a basis for evaluating the extent to which the difference between the two resulting time averages, say for queue length, is due to the differences in input values and the extent to which it is due to sampling variation. Without this ability to identify the source of output differences, a simulationist has no scientific basis for explaining how variation in input affects variation in output.

With rare exceptions, most simulation programming languages provide limited guidance for performing statistical analyses. The traditional reason for this limitation has been the difficulty in automating these analyses and making them accessible to simulation users who, by justifiable inclination, focus on final results rather than on the intricacies of method. This out-of-sight-out-of-mind approach can lull a simulationist into a false sense of confidence about how meaningful a simulation's output is. In others, it can arouse suspicion when they are asked to accept estimates of long-term averages without some indication of how well the estimates approximate the true values of these system parameters. However, techniques do exist that considerably facilitate one's ability to incorporate subprograms into a simulation to perform the desired assessment with minimal effort on the part of the simulation user.

Chapter 6 describes the statistical theory that underlies one of these approaches, the batch-means method, and shows how to incorporate it into a simulation program. Although we can safely defer detailed discussion of this topic until after we have an executable program, some familiarity with the statistical properties of stochastic time averages is essential at this point to understand the elements of the executable programs in Chapter 4. These properties relate to the *initial state*, the *warm-up interval*, and the *sample-path length*, and this chapter provides this familiarity.

3.2.1 SYSTEM STATE

Recall that the sample path for a simulation run results from the repeated operation of the timing routine on the simulation's future-event set (FES) as simulated time τ evolves. At time τ, the FES dictates which event occurs next and when to execute it. When, as frequently occurs, execution of an event includes the scheduling of a future event, it provides a new entry in the FES on which the timing routine eventually operates. While interaction between the timing routine and the FES and interaction between an executing event and the FES completely characterize the way in which simulated behavior evolves through time, it provides this characterization with an emphasis on computational issues. An alternative approach accomplishes this same objective by focusing on the *state* of the system. Introducing this approach here enables us to shift the focus of discussion from computational to statistical considerations while preserving all essential features of a simulation's dynamical behavior.

It is these statistical considerations that determine how well finite-sample averages, as in expressions (3.7) through (3.14), characterize long-term behavior.

The state of a simulated system at time τ_1 (> 0) conveys information that, together with one or more random innovations, determines the state of the system at each future time $\tau_2 > \tau_1$. More often than not, the probability laws governing these random innovations depend explicitly on the state of the system at time τ_1. As illustration, the airline reservation problem has the state vector

$$\mathcal{S}(\tau) := (\mathcal{I}_B(\tau), \mathcal{I}_C(\tau), \mathbf{S}(\tau)), \tag{3.25}$$

where

$$\mathcal{I}_B(\tau) := \{i_1, \ldots, i_{B(\tau)}\} = \text{set of busy RESERVATIONISTs}, \tag{3.26}$$

$$\mathcal{I}_C(\tau) := \{j_1, \ldots, j_{C(\tau)}\} = \text{set of CUSTOMERs listening to recorded} \tag{3.27}$$

message,

$$\mathbf{S}(\tau) = \begin{pmatrix} Q(\tau) & \text{queue length} \\ B(\tau) & \text{no. of busy RESERVATIONISTs} \\ C(\tau) & \text{no. of CUSTOMERs listening} \\ & \text{to recorded message} \\ S_A(\tau) & \text{remaining time to next CALL} \\ S_{Bi_1}(\tau) & \\ \vdots & \left.\begin{array}{c}\\\\\end{array}\right\} \text{remaining service times for} \\ & \text{the busy RESERVATIONISTs} \\ S_{Bi_{B(\tau)}}(\tau) & \\ S_{Cj_1}(\tau) & \\ & \left.\begin{array}{c}\\\\\\\end{array}\right\} \begin{array}{l}\text{remaining message times} \\ \text{for CUSTOMERs listening} \\ \text{to the recorded message,}\end{array} \\ \vdots & \\ S_{Cj_{C(\tau)}}(\tau) & \end{pmatrix}, \tag{3.28}$$

and

$$C(\tau) := A(\tau) - N(\tau) - Q(\tau) - B(\tau). \tag{3.29}$$

A little thought shows that a sample path $\{\mathcal{S}(s),\ 0 \le s \le \tau\}$ also provides sufficient information to construct the corresponding sequence of waiting times $\{W_1, \ldots, W_{N(\tau)}\}$.

If the last event occurred (executed) at time τ, then the next event or events execute at

$$\tau'(\tau) = \tau + \Delta(\tau), \tag{3.30}$$

where

$$\Delta(\tau) := \min[S_A(\tau), S_{Bi_1}(\tau), \ldots, S_{Bi_{B(\tau)}}(\tau), S_{Cj_1}(\tau), \ldots, S_{Cj_{C(\tau)}}(\tau)].$$

If

$$S_A(\tau) = \Delta(\tau),$$

then a CALL event occurs at time τ'. If

$$S_{Bi}(\tau) = \Delta(\tau),$$

then a COMPLETION event occurs at time τ' for RESERVATIONIST i for an $i \in \mathcal{I}_B(\tau)$. If

$$S_{Cj}(\tau) = \Delta(\tau),$$

then an END_MESSAGE event occurs for CUSTOMER j at time τ' for a $j \in \mathcal{I}_C(\tau)$.

We immediately see that the clock attributes of the event notices in the FES at time τ are in one-to-one correspondence with the elements of the set

$$\{\tau + S_A(\tau), \tau + S_{Bi_1}(\tau), \ldots, \tau + S_{Bi_{B(\tau)}}(\tau), \tau + S_{Cj_1}(\tau), \ldots \tau + S_{Cj_{C(\tau)}}(\tau)\}$$

and that the ordering of the FES for execution is identical with this set, ordered from small to large.

Hereafter, we frequently refer to $\mathcal{S}(\tau)$ as the state vector or the state at time τ. Note that the number of elemental entries in $\mathcal{S}(\tau)$ varies with τ. In particular, these changes in number can occur only at $\tau'(\tau)$ for each τ. Also, note that $\mathcal{S}(\tau)$ contains both continuous and discrete random variables as entries.

3.3   CONDITIONAL AND EQUILIBRIUM DISTRIBUTIONS

The centrality of $\{\mathcal{S}(s), 0 \le s \le \tau\}$ with regard to the future-event set and the determination of $\{B(s), 0 \le s \le \tau\}$, $\{Q(s), 0 \le s \le \tau\}$, and $\{W_i; i = 1, \ldots, N(\tau)\}$ in Section 3.2.1 enable us to unify our approach to understanding the statistical properties of the sample averages, as in expressions (3.7) through (3.14). Most critical in this assessment is the role of the probability distribution of the random state vector $\mathcal{S}(\tau)$ and the behavior of this distribution as a function of τ. As before, we use the airline reservation problem to illustrate concepts, but emphasize that the general case demands relatively minor modification.

3.3.1 INITIAL CONDITIONS

To execute a simulation run, we must first assign values to $S(0)$, using either a deterministic or a random rule. We do this indirectly by establishing *initial conditions* that are to prevail at simulated time zero. For example, suppose we start the airline reservation simulation with a CALL at $\tau = 0$ to an empty and idle system. Then the logic of this problem dictates the initial vector state

$$S(0) = \{\mathcal{I}_B(0), \mathcal{I}_C(0), \mathbf{S}(0)\},$$

where

$$\mathcal{I}_B(0) = \{i_1\},$$
$$\mathcal{I}_C(0) = \emptyset,$$

and $\mathbf{S}(0)$ has the entries

$$Q(0) = 0,$$
$$B(0) = 1,$$
$$C(0) = 0,$$
$$S_A(0) = \text{time at which second CALL occurs},$$
$$S_{Bi_1}(0) = \text{service time for first completion}.$$

Also,

$$\tau'(0) = \min[S_A(0), S_{Bi_1}(0)].$$

Alternatively, if we were to start the simulation with a CALL scheduled in a time units, three busy RESERVATIONISTs with remaining service times of b, c, and d time units, and no scheduled END_MESSAGEs, then $S(0)$ would have the entries

$$\mathcal{I}_B(0) = \{i_1, i_2, i_3\},$$
$$\mathcal{I}_C(0) = \emptyset,$$
$$A(0) = 3,$$
$$N(0) = 0,$$
$$Q(0) = 0,$$
$$B(0) = 3,$$
$$C(0) = 0,$$

$$S_A(0) = a,$$
$$S_{Bi_1}(0) = b,$$
$$S_{Bi_2}(0) = c,$$
$$S_{Bi_3}(0) = d.$$

Moreover,

$$\tau'(0) = \min(a, b, c, d).$$

Let $\mathcal{T}$ denote the set of all possible states that a simulation model can realize as simulated time elapses. Then the choice of initial conditions and the method by which a simulation user makes this choice induce an *initializing distribution*

$$\pi_0(\mathcal{A}) := \mathrm{pr}[\mathcal{S}(0) \in \mathcal{A}], \quad \mathcal{A} \subseteq \mathcal{T}, \tag{3.31}$$

denoted by π_0. In each of the two examples for the airline reservation problem, π_0 is degenerate, putting all its mass on a single state. These are special, but convenient, illustrations. However, if the simulationist in the second example elects to sample values for the remaining times to the CALL event and the three COMPLETION events, π_0 would be nondegenerate. Note that a simulationist does not literally sample from π_0; however, her/his method of assigning initial conditions is probabilistically equivalent to sampling from an implicit π_0. Recognizing this equivalence enables us to characterize sample-path behavior as time elapses.

For each $\tau \geq 0$ and $\mathcal{D} \in \mathcal{T}$, let $\pi_0(\cdot; \tau | \mathcal{D})$ denote the conditional distribution of the state set $\mathcal{S}(\tau)$, so that

$$\pi_0(\mathcal{A}; \tau \mid \mathcal{D}) := \mathrm{pr}[\mathcal{S}(\tau) \in \mathcal{A} \mid \mathcal{S}(0) = \mathcal{D})], \quad \mathcal{A} \subseteq \mathcal{T}. \tag{3.32}$$

To facilitate a meaningful statistical analysis on sample-path data, the simulation must endow this conditional distribution with at least two important properties:

- For every possible initial starting state $\mathcal{D} \in \mathcal{T}$,

$$\pi_0(\mathcal{A}; \tau | \mathcal{D}) \to \pi(\mathcal{A}) \quad \text{as } \tau \to \infty \quad \forall \mathcal{A} \subseteq \mathcal{T}, \tag{3.33}$$

where $\pi(\mathcal{A})$ is called the *equilibrium* or *steady-state distribution* for the subset of states $\mathcal{A}$ in $\mathcal{T}$. In particular, $\mathcal{A} \subseteq \mathcal{B} \subseteq \mathcal{T}$ implies $0 \leq \pi(\mathcal{A}) \leq \pi(\mathcal{B}) \leq \pi(\mathcal{T}) = 1$.

- This limit (3.33) must hold for every initial distribution π_0 on $\mathcal{T}$.

Expression (3.33) asserts that as simulated time τ elapses, the distribution of the state vector $S(\tau)$ converges to a distribution that is independent of the initial conditions $S(0) = \mathcal{D}$, regardless of which state $\mathcal{D} \in \mathcal{T}$ is chosen to start the simulation. Moreover, this occurs regardless of the distribution π_0. These are extremely useful properties. Were expression (3.33) not to hold, then the sample path generated during a simulation run would have statistical properties that depend on the initial conditions, a feature antithetic to our objective of approximating long-run averages for utilization, queue length, and waiting times, which are presumably unique for a given simulation model with specified numerical values for its input parameters. Unless otherwise explicitly stated, we hereafter assume that the design of the simulation model ensures that expression (3.33) holds.

Let $\pi := \{\pi(\mathcal{A}), \mathcal{A} \subseteq \mathcal{T}\}$. As with π_0, the interpretation of π must be in context. As τ increases, the dynamic logic of the simulation model implies that there exists a unique distribution π, independent of $S(0)$, that characterizes the behavior of $S(\tau)$. Moreover, this dynamic logic, not actual sampling from π, dictates the value that $S(\tau)$ assumes at time τ.

A simulation user frequently picks the initial state arbitrarily with no randomizing considerations at all so that π_0 is degenerate. Computational ease in assigning the initial state and limited knowledge usually dictate its choice. However, some states turn out to be better than others for the initial state inasmuch as they induce faster convergence to the limit (3.33) and thereby require a shorter warm-up interval before data collection begins. Section 6.1 returns to this topic.

3.3.2   WARM-UP INTERVAL AND CONVERGENCE

Since $\sup_{\mathcal{A} \subseteq \mathcal{T}} |\pi_0(\mathcal{A};\ s + \tau | \mathcal{D}) - \pi(\mathcal{A})|$ tends to decrease as s (> 0) increases, initial conditions tend to influence data collected over the interval $[s, s + \tau]$ less than data collected over the interval $[0, \tau]$. That is, both sample records have the same sample-path length τ, but the one collected after deleting the *warm-up interval* $[0, s)$ is affected less by the starting conditions of the simulation.

Recall that the output of a simulation run includes time averages, as in expressions (3.7) through (3.14), computed from sample-path data, that serve as approximations to the long-run averages, queue length, utilization and waiting time. These long-run averages characterize congestion. In particular, we say that congestion increases (decreases) as average queue length and waiting time increase (decrease). For fixed simulated times τ_1 and τ_2, the dependence between two state vectors $S(\tau_1)$ and $S(\tau_2)$ tends to increase (decrease) as congestion increases (decreases). For example, the dependence between queue lengths $Q(\tau_1)$ and $Q(\tau_2)$ increases (decreases) for fixed times τ_1 and τ_2 as the traffic intensity increases (decreases).

Since $S(0)$ denotes the conditions that prevail at the beginning of a simulation run, it is clear that the choice of an appropriate warm-up interval depends on the extent of congestion in the system. While we must defer a discussion of how to choose an appropriate warm-up interval until Chapter 6, we can at this point identify issues that should sensitize the reader to the role of congestion in analyzing data.

Recall that every discrete-event simulation model has one or more service time distributions to whose parameters a simulationist must assign numerical values before execution can begin. Moreover, an open-loop system has an interarrival time distribution whose parameters also require numerical assignments. We call these values *input* to the simulation. As the values assigned to the mean service time, the variance of service time, or in an open-loop system, to the arrival rate increases (decreases), long-run queue length and waiting time, and therefore congestion, increases (decreases). Hence, increases in these input values alert the simulationist to the need to increase the warm-up interval.

3.3.3   QUEUE LENGTH, RESOURCE UTILIZATION, AND CONVERGENCE

Since queue length $Q(\tau)$ and number of busy servers $B(\tau)$ at time τ both are entries in the state vector, the limit (3.33) has implications for these quantities. Let $\pi_Q := \{\pi_{Qq}; q = 0, 1, \ldots\}$ and $\pi_B := \{\pi_{Bb}; b = 0, 1, \ldots, m\}$ denote the marginal equilibrium probability mass functions (p.m.f.s) of π with respect to queue length and number of busy RESERVATIONISTs, respectively, where

$$\pi_{Qq} := \text{probability that } q \text{ CUSTOMERs awaiting service in the queue}$$

and

$$\pi_{Bb} := \text{probability that } b \text{ RESERVATIONISTs are busy.}$$

Then the limit (3.33) implies that for every fixed s,

$$\text{pr}[Q(s + \tau) = q | \mathcal{S}(s)] \to \pi_{Qq}$$

and   (3.34)

$$\text{pr}[B(s + \tau) = b | \mathcal{S}(s)] \to \pi_{Bb} \quad \text{as } \tau \to \infty.$$

While these limits are reassuring, they merely tell us that if one runs a simulation for a sufficiently long time τ, he/she can expect $Q(s + \tau)$ and $B(s + \tau)$ to exhibit statistical behavior that is independent of $\mathcal{S}(s)$. That is, the values that $Q(s + \tau)$ and $B(s + \tau)$ assume tell us something about congestion, uninfluenced by the conditions that prevailed at simulated time s. However, our principal interest is in the time averages $\{\bar{P}_Q(q, s, s + \tau); q = 0, 1, \ldots\}$, $\{\bar{P}_B(b, s, s + \tau); b = 0, 1, \ldots, m - 1\}$, $\bar{Q}(s, s + \tau)$, and $\bar{B}(s, s + \tau)$, which characterize the entire sample path. To ensure that these are meaningful approximations of their corresponding long-term averages, we need more than the limits (3.34).

Let

$$P_{Qq} := 1 - \pi_{Q0} - \pi_{Q1} - \cdots - \pi_{Qq}$$

and

$$P_{Bb} := 1 - \pi_{B0} - \pi_{B1} - \cdots - \pi_{Bb}.$$

Under relatively mild restrictions and for $q = 0, 1, \ldots$ and $b = 0, 1, \ldots, m - 1$,

$$\bar{P}_Q(q, s, s + \tau) \to P_{Qq} \quad \text{as } \tau \to \infty \quad \text{w.p.1.}$$

and   (3.35)

$$\bar{P}_B(b, s, s + \tau) \to P_{Bb} \quad \text{as } \tau \to \infty \quad \text{w.p.1.}$$

That is, the sample proportions $\{1 - \bar{P}_Q(q, s, s + \tau); q = 0, 1, \ldots\}$ converge to $\{1 - P_{Qq}; q = 0, 1, \ldots\}$ and the sample proportions $\{1 - \bar{P}_B(b, s, s + \tau); b = 0, 1, \ldots, m\}$ converge to $\{1 - P_{Bb}; b = 0, 1, \ldots m\}$.

The notation w.p.1 means "with probability one." It is equivalent to the notation a.e. for "almost everywhere," which is also common in the statistical literature. Both imply the strongest form of convergence that probability theory has to offer. When expression (3.35) holds we say that $\bar{P}_Q(q, s, s + \tau)$ is a *strongly consistent estimator* of P_{Qq} and $\bar{P}_B(b, s, s + \tau)$ is a *strongly consistent estimator* of P_{Bb}.

While the limits (3.34) are necessary for the sample paths $\{Q(\tau), \tau \geq s\}$ and $\{B(\tau), \tau \geq s\}$ to exhibit equilibrium behavior independent of S (0) as $\tau \to \infty$, the limits (3.35) assure us that the time averages $\bar{P}_Q(q, s, s + \tau)$ and $\bar{P}_B(b, s, s + \tau)$ tend to become increasingly better approximations for P_{Qq} and P_{Bb}, respectively, as the length of the sample path τ increases without bound. Were this not so, data collected on a single sample path would have little value.

To describe convergence for $\bar{Q}(s, s + \tau)$ and $\bar{B}(s, s + \tau)$, we rely on their relationships to $\{\bar{P}_Q(q, s, s + \tau), q = 0, 1, \ldots\}$ and $\{\bar{P}_B(b, s, s + \tau); b = 0, 1, \ldots, m - 1\}$, respectively. Let

$$\Delta\bar{P}_Q(q, s, s + \tau) := \bar{P}_Q(q - 1, s, s + \tau) - \bar{P}_Q(q, s, s + \tau), \qquad q = 1, 2, \ldots,$$
$$= \text{sample proportion of time } \tau \text{ that } q \text{ CUSTOMERs are in the queue}$$

and

$$\Delta\bar{P}_B(b, s, s + \tau) := \bar{P}_B(b - 1, s, s + \tau) - \bar{P}_B(b, s, s + \tau), \qquad b = 1, \ldots, m,$$
$$= \text{sample proportion of time } \tau \text{ that } b \text{ RESERVATIONISTs are busy.}$$

Then a minor amount of algebra shows that

$$\bar{Q}(s, s + \tau) = \sum_{q=1}^{\infty} q \Delta\bar{P}_Q(q, s, s + \tau) \tag{3.36}$$

and

$$\bar{B}(s, s + \tau) = \sum_{b=1}^{m} b \Delta\bar{P}_b(b, s, s + \tau). \tag{3.37}$$

To put limiting behavior for $\bar{Q}(s, s + \tau)$ and $\bar{B}(s, s + \tau)$ into perspective, we rely on yet another pair of equivalent representations,

$$\bar{Q}(s, s + \tau) = \sum_{q=0}^{\infty} \bar{P}_Q(q, s, s + \tau) \tag{3.38}$$

and

$$\bar{B}(s, s + \tau) = \sum_{b=0}^{m-1} \bar{P}_B(b, s, s + \tau), \tag{3.39}$$

which follow from expanding and summing like terms in expressions (3.36) and (3.37), respectively.

By definition,

$$\mu_B := \sum_{b=0}^{m} b\pi_{Bb} = \sum_{b=0}^{m} P_{Bb} \tag{3.40}$$

denotes the mean number of busy RESERVATIONISTs. If the limit (3.35) for $\bar{P}_B(b, s, s + \tau)$ holds for each $b = 0, 1, \ldots, m$, then for fixed s and finite m

$$\bar{B}(s, s + \tau) \rightarrow \mu_B \quad \text{as } \tau \rightarrow \infty \quad \text{w.p.1.}$$

That is, the sample time average $\bar{B}(s, s + \tau)$ converges to a constant that turns out to be the probability average called the mean number of busy RESERVATIONISTs.

For queue length, an additional property must hold. By definition,

$$\mu_Q := \sum_{q=0}^{\infty} q\pi_{Qq} = \sum_{q=0}^{\infty} P_{Qq} \tag{3.41}$$

denotes the mean queue length. If the limit (3.35) for $\bar{P}_Q(q, s, s + \tau)$ holds for each $q = 0, 1, \ldots$ and $P_{Qq} \rightarrow 0$ sufficiently fast as $q \rightarrow \infty$, then μ_Q is finite and

$$\bar{Q}(s, s + \tau) \rightarrow \mu_Q \quad \text{as } \tau \rightarrow \infty \quad \text{w.p.1.} \tag{3.42}$$

The condition that μ_Q be finite is an important one. For a system modeled with finite waiting room or buffers, it is clearly true that $P_{Qq} = 0$ for $q \geq c$, where c is the capacity of the buffer. In this case the condition always holds. However, if a system is modeled with infinite-capacity buffers, then the finiteness of μ_Q depends on the properties of the random input. If the interarrival times and service times all have finite moments, then μ_Q is finite, and as in expression (3.42), $\bar{Q}(s, s + \tau)$ approximates μ_Q with an error that tends to diminish as τ increases.

AN EXCEPTION

The preceding discussion assumes that an equilibrium distribution π exists and that μ_Q is finite, not unreasonable assumptions for many models of manufacturing, communications, and traffic systems. However, exceptions exist, and when they occur the concept of long-term average becomes conjectural and statistical inference based on finite-sample averages becomes suspect. An example illustrates the problem.

Consider a single-server queueing model with independent exponentially distributed interarrival times (finite mean and variance) and i.i.d. service times with finite mean but infinite variance. It is known that an equilibrium distribution (3.34) for queue length exists in this case but it has an infinite mean. Therefore, the limit (3.42) does not hold. The reason for this anomaly is not hard to find. While P_{Qq} decreases monotonically to zero with increasing q, it does so too slowly for the infinite sum (3.41) to converge.

3.3.4 FINITE SAMPLE-PATH LENGTH

Since τ is finite in practice, $\bar{B}(s, s + \tau)$ and $\bar{Q}(s, s + \tau)$ merely approximate μ_B and μ_Q, respectively, regardless of the value of s. To justify their use in subsequent analyses, one needs to assess how well they represent their respective long-run averages. For finite τ, two types of errors arise in these approximations, one *systematic* and the other *random*. As Section 3.3.1 indicates, initial conditions influence the sequence of states through which the simulation passes, especially near the beginning of a simulation. For example, for small τ a simulation that begins with an arrival to an empty and idle system generates sample paths $\{B(\tau), 0 \leq \tau \leq t\}$ and $\{Q(\tau), 0 \leq \tau \leq t\}$ that tend to lie below μ_B and μ_Q, respectively, so that $\bar{B}(0, \tau)$ and $\bar{Q}(0, \tau)$ often understate these quantities. Conversely, a simulation that starts with the arrival of a large number of jobs to a system whose servers are all busy generates sample paths $\{B(\tau), 0 \leq \tau \leq t\}$ and $\{Q(\tau), 0 \leq \tau \leq t\}$ that tend to exceed μ_B and μ_Q, respectively. Therefore, $\bar{B}(0, t)$ and $\bar{Q}(0, t)$ often overstate their respective long-run averages. Both examples illustrate *systematic error*.

To address this issue for a fixed τ and $t := s + \tau$, it is of interest to study the properties of time averages, as the length of the discarded warm-up interval, s, varies. Under relatively general conditions, including the limits (3.34) and the finiteness of μ_Q and μ_B,

$$\bar{Q}(s, s + \tau) - \mu_Q \to \Delta_Q(\tau) \quad \text{w.p.1} \quad \text{as } s \to \infty \tag{3.43}$$

and

$$\bar{B}(s, s + \tau) - \mu_B \to \Delta_B(\tau) \quad \text{w.p.1} \quad \text{as } s \to \infty, \tag{3.44}$$

where $\Delta_B(\tau)$ and $\Delta_Q(\tau)$ denote zero-mean random variables that are independent of initial conditions. Expressions (3.43) and (3.44) reveal that if one runs a simulation for $s + \tau$ simulated time units but uses the *truncated* sample paths $\{Q(x), s \leq x \leq s + \tau\}$ and $\{B(x), s \leq x \leq s + \tau\}$ to compute approximations $\bar{Q}(s, s + \tau)$ and $\bar{B}(s, s + \tau)$ of μ_Q and μ_B, respectively, then $\bar{Q}(s, s + \tau) - \mu_Q$ and $\bar{B}(s, s + \tau) - \mu_B$ converge with probability one to zero-mean random variables that are independent of the state in which the simulation run began execution. Most importantly, this property of *asymptotic unbiasedness* holds regardless of the initial state chosen from $\mathcal{T}$. Section 6.1 returns to this topic.

The relationships between $\bar{Q}(s, s+\tau)$ and $\Delta_Q(\tau)$ and between $\bar{B}(s, s+\tau)$ and $\Delta_B(\tau)$ yield additional properties of interest. In particular,

$$\tau \mathrm{E}\left[\bar{Q}(s, s+\tau) - \mu_Q\right]^2 \to \tau \mathrm{var}\Delta_Q(\tau) \to \sigma_\infty^2(Q) \quad \text{as } \tau \to \infty$$

and
$$\tag{3.45}$$

$$\tau \mathrm{E}\left[\bar{B}(s, s+\tau) - \mu_B\right]^2 \to \tau \mathrm{var}\Delta_B(\tau) \to \sigma_\infty^2(B) \quad \text{as } \tau \to \infty,$$

where $\sigma_\infty^2(Q)$ and $\sigma_\infty^2(B)$ denote positive quantities. Therefore, for large τ, $\bar{Q}(s, s+\tau)$ has approximate *mean-square error* $\sigma_\infty^2(Q)/\tau$, and $\bar{B}(s, s + \tau)$ has approximate mean-square error $\sigma_\infty^2(B)/\tau$, regardless of the length s of the warm-up interval. This observation raises a question as to why we should concern ourselves with truncation at all. The answer is that for two warm-up interval lengths $s_2 > s_1 \geq 0$, the mean-square errors converge more rapidly to $\sigma_\infty^2(Q)/\tau$ and $\sigma_\infty^2(B)/\tau$ for s_2 than for s_1, as τ increases.

The properties

$$\tau^{1/2}\left[\bar{Q}(s, s+\tau) - \mu_Q\right] \overset{\mathrm{d}}{\to} \mathcal{N}(0, \sigma_\infty^2(Q)) \quad \text{as } \tau \to \infty$$

and
$$\tag{3.46}$$

$$\tau^{1/2}\left[\bar{B}(s, s+\tau) - \mu_B\right] \overset{\mathrm{d}}{\to} \mathcal{N}(0, \sigma_\infty^2(B)) \quad \text{as } \tau \to \infty,$$

where $\overset{\mathrm{d}}{\to}$ denotes *convergence in distribution* and $\mathcal{N}(a_1, a_2)$ denotes the normal distribution with mean a_1 and variance a_2, provide the additionally essential ingredients for assessing how well $\bar{Q}(s, s + \tau)$ and $\bar{B}(s, s + \tau)$ approximate μ_Q and μ_B, respectively, for large τ. Expression (3.46) asserts that the distributions of the statistics $\tau^{1/2}\left[\bar{Q}(s, s+\tau) - \mu_Q\right]$ and $\tau^{1/2}\left[\bar{B}(s, s+\tau) - \mu_B\right]$ both converge to normal distributions as the sample-path length τ increases without bound. As illustration of their importance for error assessment for queue length, the first limit in expression (3.46) implies that the interval

$$\left[\bar{Q}(s, s+\tau) \pm \Phi^{-1}(1 - \delta/2)\sqrt{\sigma_\infty^2(Q)/\tau}\right] \tag{3.47}$$

includes the true unknown long-term average μ_Q with approximate probability $1 - \delta$ for large τ, where

$$\Phi^{-1}(\beta) := \left[z \in (-\infty, \infty): \ (2\pi)^{-1/2}\int_{-\infty}^{z} \mathrm{e}^{-x^2/2}\mathrm{d}x = \beta\right], \quad 0 < \beta < 1,$$

denotes the β-quantile of the standard normal distribution. A similar interval estimate exists for μ_B with $\bar{B}(s, s + \tau)$ and $\sigma_\infty^2(B)$ replacing $\bar{Q}(s, s + \tau)$ and $\sigma_\infty^2(Q)$, respectively.

In practice, $\sigma_\infty^2(Q)$ and $\sigma_\infty^2(B)$ are unknown. However, under specified conditions it is known that if $\hat{\sigma}_\infty^2(Q, \tau)$ and $\hat{\sigma}_\infty^2(B, \tau)$ are estimators of $\sigma_\infty^2(Q)$ and $\sigma_\infty^2(B)$, respectively, that satisfy

$$\hat{\sigma}_\infty^2(Q, \tau) \to \sigma_\infty^2(Q) \qquad \text{as } \tau \to \infty \quad \text{w.p.1}$$

and
$$(3.48)$$

$$\hat{\sigma}_\infty^2(B, \tau) \to \sigma_\infty^2(B) \qquad \text{as } \tau \to \infty \quad \text{w.p.1},$$

then for large τ, the interval $\left[\bar{Q}(s, s+\tau) \pm \Phi^{-1}(1 - \delta/2)\sqrt{\hat{\sigma}_\infty^2(Q, \tau)/\tau} \right]$ includes μ_Q with approximate probability $1 - \delta$ and $\left[\bar{B}(s, s+\tau) \pm \Phi^{-1}(1 - \delta/2)\sqrt{\hat{\sigma}_\infty^2(B, \tau)/\tau} \right]$ includes μ_B with approximate probability $1 - \delta$, where the errors in the probabilities diminish as τ increases. Here $\hat{\sigma}_\infty^2(Q, \tau)$ and $\hat{\sigma}_\infty^2(B, \tau)$ satisfying expression (3.48) are called strongly consistent estimators of $\sigma_\infty^2(Q)$ and $\sigma_\infty^2(B)$, respectively. The challenge is to employ strongly consistent estimators based on sample-path data and to keep this computation from being an onerous burden for the simulationist. Sections 6.6 and 6.7 provide the wherewithal to do this.

NOTE

> Collectively, the time-average properties of strong consistency, asymptotic unbiasedness with respect to warm-up interval, asymptotic normality, and the ability to compute consistent estimates of $\sigma_\infty^2(Q)$ and $\sigma_\infty^2(B)$ are crucial to any valid assessment of how well time averages approximate long-term averages.

Limiting properties analogous to those in expressions (3.43) through (3.48) also hold for $\bar{P}_Q(q, s, s+\tau)$ as an approximation to P_{Qq} for each $q = 0, 1, \ldots$ and for $\bar{P}_B(b, s, s+\tau)$ as an approximation to P_{Bb} for each $b = 0, 1, \ldots, m$.

3.4   TERMINATING EXECUTION

With regard to data collection, analysis, and run termination, simulationists frequently employ one of two options:

OPTION A

- Begin the simulation at time 0

- Begin data collection at specified time s (≥ 0)

- Complete data collection at a specified time $s + \tau$

- Terminate execution of the simulation at time $s + \tau$

- Calculate and display summary statistics based on sample-path data collected on the time interval $[s, s + \tau]$.

OPTION B

- Begin the simulation at time 0
- Begin data collection when the Mth job completes all service
- Complete data collection when the $(M + N)$th job completes all service
- Terminate execution of the simulation when the $(M + N)$th job completes all service
- Calculate and display summary statistics based on sample-path data collected over the time interval $(\tau_M, \tau_{M+N}]$, where

$$\tau_j := \text{time at which the } j\text{th job completes all service.}$$

Option A implies that the simulated time $[s, s+\tau]$ for data collection is predetermined but that the number of jobs, $N(s, s + \tau) = N(\tau) - N(s)$, completed in the time interval is random. Conversely, Option B implies that the length of the simulated time interval, $(\tau_M, \tau_{M+N}]$, is random but the number of jobs $N(\tau_M, \tau_{M+N}) = N$ is predetermined. Of the two, Option B has more appeal inasmuch as it allows the simulationist the freedom to specify how much completed work is to occur during the period of observation.

When Option B is employed, the sample average queue length is

$$\bar{Q}(\tau_M, \tau_{M+N}) = \frac{1}{\tau_{M+N} - \tau_M} \int_{\tau_M}^{\tau_{M+N}} Q(u)du, \tag{3.49}$$

where in addition to the sample path $\{Q(u), \tau_M \leq u \leq \tau_{M+N}\}$ being random, the length of the sample path is also random. For finite M and N, this new source of randomness contributes to the error of approximating μ_Q by $\bar{Q}(\tau_M, \tau_{M+N})$. Although the limiting properties that Sections 3.3.2 through 3.3.4 describe all assume Option A, it can be shown under mild additional qualifications that as $N \to \infty$ under Option B, limiting behavior analogous to expressions (3.42), (3.45), and (3.46) again holds, and as $M \to \infty$, limiting behavior analogous to expression (3.43) also holds.

Like properties hold for the sample average number of busy RESERVATIONISTs

$$\bar{B}(\tau_M, \tau_{M+N}) = \frac{1}{\tau_{M+N} - \tau_N} \int_{\tau_M}^{\tau_{M+N}} B(u)du \tag{3.50}$$

under Option B. Likewise, strong consistency, asymptotic unbiasedness, and asymptotic normality continue to hold for $\bar{P}_Q(q, \tau_M, \tau_{M+N})$ for each $q = 0, 1, \ldots$ and for each $\bar{P}_B(b, \tau_M, \tau_{M+N})$ for each $b = 0, 1, \ldots, m - 1$. Section 6.11 describes an indirect method for assessing the accuracy of $\bar{Q}(\tau_M, \tau_{M+N})$ and $B(\tau_M, \tau_{M+N})$ under Option B.

3.5   WAITING TIME

We now turn to the estimation of the long-term average waiting time. The analysis proceeds analogously to that for utilization and queue length, the differences being the discrete indexing of the sample path $W_1, \ldots, W_{N(\tau)}$, the continuous range $[0, \infty)$ for W_i, and the effect of alternative modes of terminating the simulation (Section 3.4). Recall that

$$\bar{W}(s, s + \tau) := \frac{1}{N(s + \tau) - N(s)} \sum_{i=N(s)+1}^{N(s+\tau)} W_i \qquad (3.51)$$

denotes the arithmetic average of waiting times and

$$\bar{P}_W(w, s, s + \tau) := \frac{1}{N(s + \tau) - N(s)} \sum_{i=N(s)+1}^{N(s+\tau)} I_{(w,\infty)}(W_i), \qquad 0 \le w < \infty,$$

the proportion of CUSTOMERs that wait longer than w time units for service. Most simulation languages either automatically compute or provide the means for computing the waiting-time *empirical distribution function* $\{1 - \bar{P}_W(w, s, s + \tau), w \in \{w_1 > w_2 > \cdots > w_r > 0\}\}$, for a user-specified integer r and $w_1, \ldots, w_r$.

Let $\{F_W(w), w \ge 0\}$ denote the equilibrium d.f. of waiting time. Under relatively general conditions for Option B and for each fixed M,

$$\bar{P}_W(w, \tau_M, \tau_{M+N}) \to 1 - F_W(w) \quad \text{as } N \to \infty \quad \text{w.p.1.} \qquad (3.52)$$

Moreover, provided that

$$\mu_W := \int_0^\infty w \, dF(w) = \int_0^\infty [1 - F(w)] dw < \infty,$$

it follows that

$$\bar{W}(\tau_M, \tau_{M+N}) \to \mu_W \quad \text{as } N \to \infty \quad \text{w.p.1.} \qquad (3.53)$$

Thus for each $w \ge 0$, $\bar{P}_W(w, \tau_M, \tau_{M+N})$ is a stongly consistent estimator of $1 - F(w)$ and $\bar{W}(\tau_M, \tau_{M+N})$ is a strongly consistent estimator of μ_W.

As in the case of queue length, the finiteness of μ_W holds for many simulation models. However, it does not hold when the interarrival-time or service-time distributions do not have finite first and second moments.

For finite N, an error arises in approximating $\{1 - F(w), w \ge 0\}$ by $\bar{P}_W(w, \tau_M, \tau_{M+N})$ and μ_W by $\bar{W}(\tau_M, \tau_{M+N})$. To facilitate the assessment of this error, we rely on additional limiting properties. We describe them for $\bar{W}(\tau_M, \tau_{M+N})$ and note that for each $w \geq 0$ analogous limiting properties hold for $\bar{P}_W(w, \tau_M, \tau_{M+N})$ as an approximation to $1 - F(w)$.

Under Option B, relatively general restrictions, and for finite N, we have

$$\bar{W}(\tau_M, \tau_{M+N},) - \mu_W \to \Delta_W(N) \quad \text{as } M \to \infty \quad \text{w.p.1}, \tag{3.54}$$

where $\Delta_W(N)$ is a zero-mean random variable that fluctuates independently of initial conditions. Moreover,

$$N \operatorname{var} \bar{W}(\tau_M, \tau_{M+N}) \to \sigma_\infty^2(W) \quad \text{as } N \to \infty, \tag{3.55}$$

where $\sigma_\infty^2(W)$ is a positive constant, and

$$\sqrt{N}[\bar{W}(\tau_M, \tau_{M+N}) - \mu_W] \xrightarrow{d} \mathcal{N}(0, \sigma_\infty^2(W)) \quad \text{as } N \to \infty. \tag{3.56}$$

These results facilitate the derivation and computation of confidence intervals for μ_W in Section 6.6 based on replacing $\sigma_\infty^2(W)$ with a strongly consistent estimator provided by the batch means method.

If Option A is employed for data collection, then expression (3.11) gives the sample average waiting time where in addition to the sample sequences $W_{N(s)+1}, \ldots, W_{N(s+\tau)}$ being random, the number of waiting times $N(s + \tau) - N(s)$ is random, contributing one more source of error to the approximation of μ_W for finite s and τ. Then under mild additional qualifications, limiting properties similar to expression (3.54) hold as $s \to \infty$ and properties similar to expessions (3.53) and (3.55) hold as $\tau \to \infty$. Also, a parallel development ensures strong consistency and asymptotic normality for $\bar{P}_W(w, s, s + \tau)$ for each $w \geq 0$, and the methods of Section 6.6 again apply.

3.5.1 LITTLE'S LAW

Let λ denote arrival rate. Then Little's law (e.g., Stidham 1974) states that for all queueing systems for which μ_Q, λ, and μ_W exist,

$$\mu_Q = \lambda \mu_W, \tag{3.57}$$

long-run average queue length = arrival rate × long-run average waiting time.

This idcntity suggests a way of eliminating the errors of approximation in $\bar{W}(s, s + \iota)$ due to a random number of elements, $N(s + \tau) - N(s)$, in the average under Option A. Since λ is known, $\bar{Q}(s, s + \tau)/\lambda$ provides an estimator of μ_W, free of this error. Conversely, under Option B, $\mu_N \bar{W}(\tau_M, \tau_{M+N})$ provides an estimator of μ_Q free of the error induced in $\bar{Q}(\tau_M, \tau_{M+N})$ by a random sample path length, $\tau_{M+N} - \tau_M$. Regrettably, no comparable result holds for the exceedance probabilities. Note that $E[N(s + \tau) - N(s)] = \tau/\lambda$.

3.6   SUMMARY

This chapter has introduced the reader to the mechanics of calculating time and arithmetic averages that serve as estimates of system performance measures. It has also alerted the reader to the presence of errors, to ways of limiting them, and to the need to assess the extent of remaining errors. In particular,

- Recursive expressions are available that minimize the need for data storage for computing time and arithmetic averages during simulation execution.

- Errors arise from several sources:

Sources of Errors of Approximation	To Reduce Errors
Influence of initial conditions on evolution of sample path	Truncate sample path near its beginning
Sampling fluctuations along sample path	Increase sample-path length, provided long-run average exists
Random number of elements, $N(\tau + s) - N(s)$, when estimating μ_W under Option A	Use Little's law
Random sample path length, $\tau_{M+N} - \tau_N$, when estimating μ_Q under Option B	Use Little's law.

- With regard to assessing the extent of error, strongly consistent estimators of the asymptotic quantities $\sigma_\infty^2(Q)$, $\sigma_\infty^2(B)$, and $\sigma_\infty^2(W)$ are needed. Chapter 6 describes these and software for calculating them using the batch-means methods.

3.7   PERIODIC DATA COLLECTION

Although computing the finite sample averages $\bar{Q}(s, s + \tau)$, $\bar{P}_Q(q, s, s + \tau)$ for each q and $\bar{B}(s, s + \tau)$, and $\bar{P}_B(b, s, s + \tau)$ for each b in Sections 3.1.1 and 3.1.2 provides us with estimates of their corresponding true long-run performance measures, these computations do not provide the necessary data to compute assessments of how well the sample averages approximate the true averages. If data collection begins at time s and ends at $s + \tau$, then the techniques of Chapter 6 rely on the availability of sample-path sequences for $j = 1, 2, \ldots,$

$$Q_j := \frac{1}{\Delta} \int_{s+(j-1)\Delta}^{s+j\Delta} Q(u)\mathrm{d}u,$$

$$P_{Qj}(q) := \frac{1}{\Delta} \int_{s+(j-1)\Delta}^{s+j\Delta} I_{(q,\infty)}(Q(u))\mathrm{d}u, \tag{3.58}$$

$$B_j := \frac{1}{\Delta} \int_{s+(j-1)\Delta}^{s+j\Delta} B(u)\mathrm{d}u,$$

$$P_{Bj}(b) := \frac{1}{\Delta} \int_{s+(j-1)\Delta}^{s+j\Delta} I_{(b,m]}(B(u))\mathrm{d}u \quad j = 1, \ldots, t := \lfloor \tau/\Delta \rfloor,$$

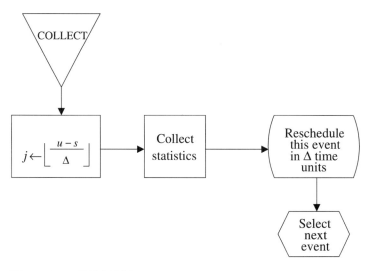

Figure 3.1   COLLECT event

where Δ is a positive time increment chosen, if possible, to make τ/Δ an integer. For example, $\{Q_1, \ldots, Q_t\}$ provides the data needed to compute an estimate $\hat{\sigma}_\infty^2(Q)$, of $\sigma_\infty^2(Q)$, which together with $\bar{Q}(s, s + \Delta t)$ is used to compute an approximating confidence interval for μ_Q.

To collect these data at simulated time u after a warm-up interval s, we rely on a periodically recurring event as in Figure 3.1, where we assume that the first COLLECT event is scheduled at time $u \geq s + \Delta$. For example, one approach gathers the data

$$
\begin{aligned}
Q'_j &:= Q(s + j\Delta), \\
P'_{Qj}(q) &:= I_{(q,\infty)}(Q(s + j\Delta)), \\
B'_j &:= B(s + j\Delta), \\
P'_{Bj} &:= I_{(b,m]}(B(s + j\Delta)),
\end{aligned}
$$

and at the end of the simulation run computes the averages

$$
\bar{Q}'_t := \frac{1}{t} \sum_{j=1}^{t} Q_j,
$$

$$
\bar{P}'_{Qt}(q) := \frac{1}{t} \sum_{j=1}^{t} P_{Qj}(q),
$$

$$
\bar{B}'_t := \frac{1}{t} \sum_{j=1}^{t} B_j,
$$

$$\bar{P}'_{Bt}(b) := \frac{1}{t} \sum_{j=1}^{t} P_{Bj}(b),$$

as approximations to their respective sample time-integrated averages $\bar{Q}(s, s + \tau)$, $\bar{P}_Q(q, s, s + \tau)$, $\bar{B}(s, s + \tau)$, and $\bar{P}_B(b, s, s + \tau)$. Since, for example,

$$\frac{\int_{s+(j-1)\Delta}^{s+j\Delta} Q(u)du}{\Delta \times Q(s + j\Delta)} \rightarrow 1 \text{ as } \Delta \rightarrow 0 \text{ w.p.1,}$$

the error of approximation diminishes as the interval between observations, Δ, decreases.

The appeal of this alternative approach is its simplicity, especially when the selected simulation language does not provide an automated procedure for collecting the time-integrated quantities (3.58).

3.8   LESSONS LEARNED

- A discrete-event simulation program is an engine for generating sequences of data points, called *sample paths*, on waiting or delay time, queue length, and resource utilization.

- The data are used to compute approximations, called *sample averages*, of system performance measures. These include *long-run average* waiting time, queue length, and resource utilization.

- The data can also be used to compute estimates of the probability distributions of waiting time, queue length, and resource utilization. In particular, estimates of *exceedance probabilities* allow assessment of extreme (rare) behavior.

- The *initial conditions* (state) that prevail at the beginning of a simulation run influence the appearance of the sample paths and therefore, the values of sample averages based on them. This influence tends to diminish as simulated time τ elapses.

- The rate of dissipation is inversely related to the level of *congestion* (delay) in the system.

- Most simulation languages provide a means for recursively updating the sample averages as simulated time elapses, thus reducing memory space required to store large data sets.

- For finite simulated time τ, the sample averages are approximations of true unknown long-run averages.

- Discarding data points on sample paths close to the beginning of a simulation run (*warm-up interval*) reduces the influence of the part of the error of approximation induced by the choice of initial conditions.

- As the length of a simulation run τ increases, the errors of approximation in sample averages, due both to initial conditions and *random sampling fluctuations*, decrease. That is, the values of the sample averages tend to converge to the values of the unknown corresponding long-run averages.

- As τ increases, a statistical theory based on the *normal distribution* becomes increasingly valid as a means of assessing how good the approximations are.

- Execution of a simulation terminates when either a prespecified simulation time has elapsed (Option A) or when a prespecified number of actions, such as job completions, has occurred (Option B).

- Under Option A, sample average waiting time and estimates of its exceedance probabilities are *ratio estimates*. Under Option B, sample averages of queue length and resource utilization and estimates of their exceedance probabilities are ratio estimates. Error analysis for ratio estimates calls for more detailed attention to sources of variation than for simple sample averages. See Section 6.11.

- *Little's law* (Section 3.5.1) establishes a proportional equivalence between the long-run averages of waiting time and queue length that can eliminate the issue of ratio estimation.

3.9 REFERENCES

Stidham, S., Jr. (1974). A last word on $L = \lambda W$, *Oper. Res.*, **22**, 417–421.

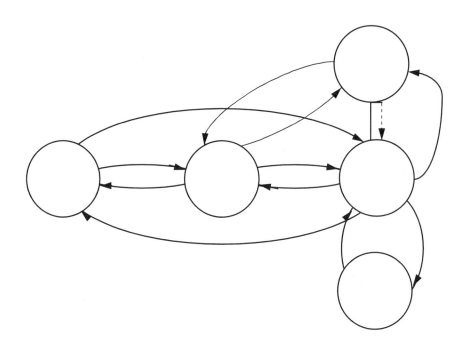

CHAPTER 4

Programming and Execution

Converting modeling logic as in Chapter 2 into executable code is the next step in simulation program development. To this logic must be added all the housekeeping functions of assigning values to input parameters and arranging for the desired output to emerge at the end of program execution. This chapter illustrates these steps, first using the SIMSCRIPT II.5 simulation programming language and then the Arena general-purpose simulation system based on the SIMAN simulation language. It also describes how the resulting programs are executed and displays their output.

A survey in Swain (1999) reveals at least 54 products for performing discrete-event simulation on a computer. These vary widely in the ways they allow program creation. Moreover, variations exist in how a program, regardless of program-preparation modality, is converted to executable code and how the mode of conversion affects *portability* and execution time. Portability denotes the extent to which a program is executable on different hardware platforms. Section 4.1 describes several modalities for program creation. When viewed as special cases of this account, subsequent sections based on SIMSCRIPT II.5 and Arena/SIMAN take on considerable more meaning for the reader.

4.1   MODALITIES FOR PROGRAMMING

Simulation software can be characterized in many ways. Does it operate under Unix-based operating systems (e.g., Sun and Dec workstations) and/or under Windows-based operating systems (e.g., IBM-compatible PCs)? Can a simulationist create a program using a textual editor for entry or using graphical entry? Does the software embrace traditional programming paradigms or is it object oriented (Section 1.9.2)? Does execution rely on a compiled program or is it performed in interpretive mode? Passing a program through a compiler leads to optimized code with respect to execution efficiency. Interpretive code avoids the compilation time but tends to execute more slowly.

Rather than answer these questions directly, we describe a general framework that encompasses most simulation software and briefly indicate how four frequently employed discrete-event simulation languages fit within this structure. The languages are GPSS/H (Schriber 1991), SIMAN (Pegden et al. 1995), SIMSCRIPT II.5 (CACI 1987, 1993a), and SLAM (Pritsker 1995). To varying degrees, each provides the essential features for discrete-event simulation programming as described in Section 1.9. In particular, all provide for a process-interaction orientation. SIMAN, SIMSCRIPT II.5, and SLAM also acconmodate the event-scheduling approach.

Three of these four languages are also embedded in general-purpose simulation systems: GPSS/H in SLX supported by Wolverine Software Inc (Henriksen 1997); SIMAN in Arena (Kelton et al. 1997), supported by Systems Modeling Inc.; and Visual SLAM in AweSim (Pritsker and O'Reilly 1999), supported by Symix Inc. These relatively comprehensive systems allow for program creation by simulationists of varying skill levels. They also facilitate animation and data transfer from and to external databases.

In today's PC-workstation environment, programming takes place either by textual entry via an editor or by some variation of *point, click, drag, drop*, and *connect*. Hereafter, we refer to this last option, concisely, as point and click.

Figure 4.1 displays five alternative modalities for creating an executable simulation program. In Modality (a), the simulationist enters statements via a textual editor. The resulting *source program* is compiled into binary machine code to form an *object program* to which required library routines are added to form executable code called a *load module*. This close relationship between simulation language and binary code tends to give this modality an advantage in execution time. However, it restricts portability by requiring a vendor to supply a separate compiler for each unique type of platform. SLX and the IBM and VAX versions of GPSS/H use this approach, the net result being highly efficient program execution.

Modality (b) also relies on textual entry but passes the resulting program through a parser written in C/C++ that translates it into binary code or converts the source code to C/C++ and then translates it into binary code. This approach increases portability across computing platforms. SIMSCRIPT II.5 embraces this option on Unix and VMS platforms. For IBM-compatible PCs and Sun workstations, GPSS/H uses a variant of Modality (b). It generates "P-code" for a virtual machine, and for selected hardware platforms provides an interpreter to carry out execution. This strategy again aims at efficiency in execution.

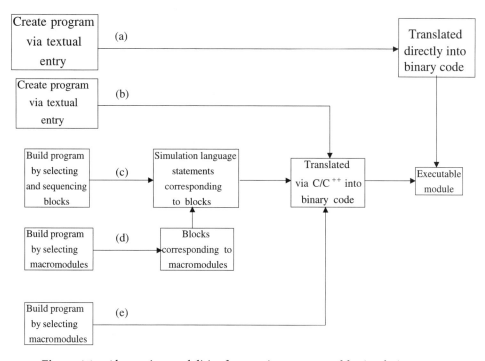

Figure 4.1   Alternative modalities for creating an executable simulation program

Modality (c) mimics Modalities (a) and (b) with one major difference. Templates of icons are displayed on the screen, one for each language statement or *block*. The simulationist merely points, clicks, drags, and drops the blocks into a "model window" and connects them to create a source program. Currently, this is the most commonly employed modality, and Arena/SIMAN, AweSim/Visual SLAM, and SIMLAB/SIMSCRIPT II.5 all permit this option on Windows-based PCs. Section 4.25 illustrates the concept with Arena/SIMAN.

In principle, Modality (d) performs as Modality (c) with one important difference. The icons in the tableaus correspond to *macromodules* each of which represents a sequence of simulation-language statements or blocks. For example, an "Arrive" macromodule includes all statements associated with bringing new items into a system. A "Server" macromodule includes all statements related to providing an item with service. Combining Modalities (c) and (d) enables a simulationist to flesh out a program that macromodules alone cannot create. Arena/SIMAN and AweSim/Visual SLAM offer this option, whose ostensible purpose is to reduce the required language familiarity that a simulationist must have to construct a program. Section 4.24 illustrates the concept for Arena/SIMAN.

Some macromodules in Modality (d) are relatively complex in content. For example, the availability of macromodules that represent the behavior of conveyors and automatic guided vehicles (AGV) to move items within factories and warehouses, automated storage and retrieval systems (ASRS), robots, and bridge cranes considerably reduces programming effort in building simulations of manufacturing and materials handling systems.

Modality (e) represents a strategy for packaging many of these special-purpose tasks that, for example, manufacturing requires into a moderately large collection of macromodules that correspond directly to off-the-shelf precompiled C^{++} subprograms or rely on a C^{++} translator for object-program creation. This eliminates any reliance on a simulation language per se. Many software offerings that cater to specialized modeling rely on this option. Its purpose is to attract potential users with limited or no knowledge of a simulation programming language by reducing the barriers to entry. While the benefit is immediately clear, a disadvantage emerges when a simulationist employs software that, while easy to use, does not have the flexibility to represent the exact flow logic in her/his model. Virtually all simulated software offers an option that allows a simulationist to merge user-created programs with user-written code in C, C^{++}, or other high-level language. Presumably this would be the option one would have to take to overcome the limitation of Modality (e). Since this approach requires a simulationist to be an able programmer in the selected language, it undermines the philosophy of "no programming required." Because of this barrier, some simulationists merely alter the original concepualization of their systems to make them amenable to representation in terms of the iconized blocks and modules that their selected simulation software offers. Although the extent to which this action undermines the quality of a study varies from problem to problem, the practice is lamentable, at a minimum, and jeopardizes scientific integrity, at worst. Section 4.27 describes potential pitfalls.

Rather than clarity of concept, today's highly interactive PC environment demands rapidity in moving from a less than completely thought-through concept to execution. While speed of creation is important, a large-scale discrete-event simulation with intricate flow-logic inevitably requires some awareness of the nuts and bolts of simulation programming. This is especially true if a program is to be used repeatedly in a computationally efficient way. The extent to which pedagogy that focuses on this rapid movement from concept to execution prepares the newcomer to simulation to address issues of model size and complexity is debatable. There is a more basic issue.

The point-and-click approach also encourages a simulationist to skip the preliminary steps in Sections 2.3 and 2.4 that explicitly identify events and flow logic in favor of creating the model directly on the PC screen and, at the click of the mouse, converting the representation to executable code. Although this protocol may work well for problems of small and moderate size, large problems with complex flows present a challenge that is most easily met with deliberate attention to the steps in Sections 2.3 and 2.4 prior to program creation. While especially true for the student just learning simulation methodology, experience indicates that more seasoned simulationists can also benefit from this focus on fundamental principles. Accordingly, our discussion of SIMSCRIPT II.5 relies heavily on the event-scheduling and process-interaction approaches of Chapter 2. Our discussion of Arena/SIMAN focuses on the process-interaction approach with passive resources.

Regardless of the extent of embeddedness, every discrete-event simulation programming language provides a user with a syntax that supports either the process-interaction or event-scheduling modeling approach. Some simulation programming languages, including the SIMAN, SIMSCRIPT II.5, and SLAM, offer both. Sections 4.2 through 4.21 use SIM-SCRIPT II.5 to illustrate both approaches, and Sections 4.24 and 4.25 use SIMAN to illustrate the process-interaction approach.

Of the two, the process-interaction approach has come to dominate, principally because it allows a syntax that relieves the simulationist of the considerable amount of code writing inherent in the event-scheduling approach and because iconization of this syntax on PC screens allows the convenience of the point-click-drag-drop mode of program creation.

To illustrate how syntax favors the process-interaction approach, consider a single-server system as in Figure 1.2a. Every arrival to the system demands service from server 1. If service is not available, the arrival waits. Otherwise, server 1 becomes busy and service begins. This sequence of actions characterizes a substantial portion of what occurs when service is demanded in a delay system. Because of the repeated occurrence of this sequence of actions in many systems in a process-interaction environment, all languages based on this paradigm contain a statement such as SEIZE 1 in GPSS/H that effects these actions. Moreover, the statement QUEUE 1, preceding the SEIZE statement, allows the simulation to collect statistics on queue length and waiting time. The structure of the event-scheduling approach does not allow equally comprehensive statements.

In general, the process-interaction approach leads to a more powerful syntax in the sense that far fewer statements per program are required than in one based on the event-scheduling approach. The conciseness of this syntax has induced simulationists to use block diagrams corresponding to its statements to model systems, thereby merging the modeling exercise into program development. The concept of simulation block diagramming based on syntax originated in 1962 with the IBM product GPSS developed by Geoffrey Gordon.

Several simulation languages, including GPSS/H, SIMAN, SIMSCRIPT II.5, and SLAM, contain provisions for improving the statistical quality of results at relatively little cost by manipulating their pseudorandom number generators. Section 4.22 describes an application of one of these options called *common pseudorandom numbers*. By including it here, we wish to encourage the reader to recognize the opportunities that these languages offer for statistical as well as programming refinements.

Section 4.27 returns to the issue of macromodules. As noted, these can save considerable time programming complex concepts such as conveyors, AGVs, ASRSs, robots, and bridge cranes. However, they should be used only when their implicit behavior is clearly understood. Section 4.27 describes the limitations of proceeding otherwise.

4.2   SIMSCRIPT II.5

Sections 4.3 through 4.21 describe how to convert the flowcharts for the airline reservation problem in Chapter 2 into computer programs written in SIMSCRIPT II.5. It also describes their input, execution, and output. Since our purpose is to acquaint the reader with the language, the subprograms illustrate concepts, but not necessarily the most efficient coding. The advantages of illustrative versus tightly written subprograms should not be underestimated.

SIMSCRIPT II.5 represents the latest in a sequence of discrete-event simulation languages that use the SIMSCRIPT title. Markowitz, Karr, and Hausner (1963) developed the original SIMSCRIPT at the Rand Corporation. SIMSCRIPT II, which represents a considerable extension of the original features and capabilities, was developed at Rand by Kiviat, Villanueva, and Markowitz (1969). SIMSCRIPT II.5 is a substantial enhancement of SIMSCRIPT II supported by CACI (1987, 1993a, 1993b, 1997). All examples described here are based on SIMSCRIPT II.5.

The language has a compiler and is divided into five levels:

LEVEL 1. A simple teaching language designed to introduce programming concepts to nonprogrammers.

LEVEL 2. A language roughly comparable in power to FORTRAN but departing greatly from it in specific features.

LEVEL 3. A language roughly comparable in power to C or PASCAL but again with many specific differences.

LEVEL 4. That part of SIMSCRIPT II.5 that contains the entity-attribute-set features of SIMSCRIPT. These features have been updated and augmented to provide a more powerful list-processing capability. This level also contains a number of new data types and programming features.

LEVEL 5. The simulation-oriented part of SIMSCRIPT II.5, containing statements for time advance, event processing, generation of statistical variates, and accumulation and analysis of simulation-generated data.

By providing the features in level 5, SIMSCRIPT II.5 relieves the user of the need to provide these special-purpose routines. Indeed, every simulation programming language provides these features, thereby giving them a considerable edge over general-purpose programming languages.

The principal building blocks of SIMSCRIPT II.5 are *entities, attributes,* and *sets.* Entities denote objects of a system; attributes, characteristics of the entities; and sets, entity groupings or lists. Figure 4.2 contains a SIMSCRIPT II.5 program that incorporates the logic specified in Figures 2.5 through 2.8 based on the event-scheduling approach. For convenience of exposition, we adopt the convention that all lowercase words denote SIMSCRIPT II.5 code, whereas all uppercase words denote user-defined concepts. In general, SIMSCRIPT II.5 makes no distinction between lower- and uppercase letters. Occasionally, we use quotation marks in the text to identify SIMSCRIPT II.5 keywords. Line numbers are for the benefit of the reader and should not appear in an executable version of this program. SIMSCRIPT II.5 statements can begin in any column.

Every SIMSCRIPT II.5 program written to execute a simulation begins with a nonexecutable subprogram called a *preamble* that defines the system's entities, attributes, sets, and the relationships among them. It also defines all global variables and functions accessible to all executable components (events and subroutines) within the program. In Figure 4.2, line 2 of the preamble indicates that unless otherwise specified all variables are assumed to have integer mode. Line 3 declares the existence of a set called QUEUE, which is the only nonSIMSCRIPT II.5 word there. We may call this set QUEUE, LIST, or any other name we like, provided that we avoid using the names of specified SIMSCRIPT II.5 system variables.

Upon encountering this set declaration for QUEUE in the preamble, SIMSCRIPT II.5 implicitly defines a global variable, n.QUEUE, that serves as the number of elements in the set. During execution, SIMSCRIPT II.5 updates n.QUEUE each time that either an addition or deletion is made. Initially, n.QUEUE is set to zero. More generally, for each set declared in a preamble, SIMSCRIPT II.5 defines a global variable with the same name prefaced by n. *Global variables* are accessible from anywhere in a SIMSCRIPT II.5 program, in contrast to *local variables*, which are accessible only from within the subprogram that defines them.

Figure 4.2   SIMSCRIPT II.5 AIR_EV.sim program

```
' '                       AIR_EV.sim (February 1998)
' '
' '-------------------------------------------------------------------
' '                            PREAMBLE
' '-------------------------------------------------------------------
 1 preamble
 2 normally mode is integer
 3 the system owns a QUEUE
   ' '
 4 permanent entities
 5    every RESERVATIONIST has a STATUS
 6    define NO.BUSY as a variable
 7    accumulate AVE.UTIL as the average, V.UTIL as the variance and MAX.UTIL
                      as the maximum of NO.BUSY
 8    accumulate UTIL.HISTO (0 to n.RESERVATIONIST by 1) as the histogram
                      of NO.BUSY
 9    accumulate AVE.Q as the average, V.Q as the variance and MAX.Q as the
                maximum of n.QUEUE
10    accumulate Q.HISTO (0 to 32 by 1) as the histogram of n.QUEUE
   ' '
11 temporary entities
12    every CUSTOMER has a TYPE,  a START and may belong to the QUEUE
13       define START and WAITING.TIME as double variables
14    tally AVE.WAIT as the mean, V.WAIT as the variance and
             MAX.WAIT as the maximum of WAITING.TIME
15    tally WAIT.HISTO (0 to 8 by .25) as the histogram of WAITING.TIME
16    define ZERO.WAIT as a variable
   ' '
17 event notices include CALL
```

Figure 4.2 (cont.)

```
18    every COMPLETION has a CLIENT and a WORKER
19    every END_MESSAGE has a CALLER
20    priority order is CALL,COMPLETION and END_MESSAGE
   ''
21 define LAMBDA, OMEGA and P as double variables
22 define M to mean n.RESERVATIONIST
   ''
   ''           LAMBDA := call rate
   ''            OMEGA := service rate for single-destination trips
   ''                P := multiple-destination probability
   ''                M := no. of RESERVATIONISTs
   ''
23 define NUM, K and T as variables
   ''
   ''          NUM := no. of CUSTOMERs that have or are receiving service
   ''          K-1 := no. of CUSTOMERS to skip before beginning data collection
   ''            T := no. of CUSTOMERs used for analysis
   ''
24 define RRHO to mean LAMBDA*(1+P)/(M*OMEGA)
   ''
25 define IDLE       to mean 0
26 define BUSY       to mean 1
27 define SUCCESS    to mean 0
28 define NO.SUCCESS to mean 1
29 define ONE        to mean 1
30 define MANY       to mean 2
   ''
31 define SEEDS  as a 1-dim array
32 define KIND   as a function
33 define S_TIME as a double function
34 end


   ''------------------------------------------------------------------------
   ''                                MAIN
   ''------------------------------------------------------------------------
1 main
2 open unit 10 for input
3 open unit 15 for output
4 use  unit 10 for input
5 use  unit 15 for output
6 read LAMBDA,OMEGA,M and P                            '' INITIALIZE
                                                       ''  PARAMETERS
7 if RRHO>=1.0      print 1 line thus
      ERROR-----ACTIVITY LEVEL>=1                      '' ERROR CHECK
8    stop                                              ''  PARAMETERS
9 otherwise
   ''
10 reserve SEEDS(*) as 4
   ''
   ''      intercall times use stream 1
   ''      single-destination service times use stream 2
```

Figure 4.2 (cont.)

```
   ''       second half of multiple-destination service times use stream 3
   ''       selection of TYPE uses stream 4
   ''
11 for I=1 to 4 do
12   read seed.v(I)                                    '' ASSIGN SEEDS
13   let SEEDS(I)=seed.v(I)                            ''   SAVE SEEDS
14 loop
   ''
15 read T and K
16 let NUM=1
   ''
17 schedule a CALL now                                '' ESTABLISH
18 create each RESERVATIONIST                          ''   INITIAL
19 start simulation                                    ''   CONDITIONS
20 end


   ''------------------------------------------------------------------
   ''                          event CALL
   ''------------------------------------------------------------------
 1 event CALL saving the event notice
 2 define ACTION as an integer variable
 3 define T.CALL to mean exponential.f(1./LAMBDA,1)
 4 create a CUSTOMER
 5 let TYPE=KIND
 6 let START(CUSTOMER)=time.v
 7 reschedule this CALL in T.CALL minutes
 8 call SERVICE yielding ACTION
 9 if ACTION=NO.SUCCESS
10   schedule an END_MESSAGE given CUSTOMER in .15 minutes
11 else
12   add 1 to ZERO.WAIT
13 always
14 return
15 end


   ''---------------------------------------------------------------
   ''                         routine KIND
   ''---------------------------------------------------------------
 1 routine KIND
 2 if random.f(4) < P
 3   return with MANY
 4 else
 5   return with ONE
 6 end


   ''---------------------------------------------------------------
   ''                       event END_MESSAGE
   ''---------------------------------------------------------------
 1 event END_MESSAGE given CALLER
 2 define ACTION and CALLER as integer variables
 3 let CUSTOMER=CALLER
 4 call SERVICE yielding ACTION
```

Figure 4.2 (cont.)

```
 5 if ACTION=NO.SUCCESS
 6    file CUSTOMER in QUEUE
 7 else
 8 always
 9 return
10 end


'' ------------------------------------------------------------
''                       event COMPLETION
'' ------------------------------------------------------------
 1 event COMPLETION given CLIENT and WORKER
 2 define ACTION,CLIENT  and WORKER as integer variables
 3 destroy CUSTOMER called CLIENT
 4 if QUEUE is empty
 5   let STATUS(WORKER)=IDLE
 6   let NO.BUSY=NO.BUSY-1
 7   return
 8 always
 9   remove the first CUSTOMER from QUEUE
10   let RESERVATIONIST=WORKER
11   call SERVICE yielding ACTION
12   return
13 end


'' ------------------------------------------------------------
''                       routine SERVICE
'' ------------------------------------------------------------
 1 routine SERVICE yielding ACTION
 2 define ACTION as an integer variable
 ''
 3 if event.v not equal to i.COMPLETION
 4    for each RESERVATIONIST with STATUS=IDLE find the first case
 ''
 5    if none
 6       let ACTION=NO.SUCCESS
 7       return
 ''
 8    else
 9       let STATUS=BUSY
10       add 1 to NO.BUSY
11 always
 ''
12 schedule a COMPLETION given CUSTOMER and RESERVATIONIST in S_TIME minutes
13 call ACCOUNTING
14 let ACTION=SUCCESS
15 return        end


'' ------------------------------------------------------------
''                       routine S_TIME
'' ------------------------------------------------------------
 1 routine S_TIME
 2 if TYPE= MANY
```

Figure 4.2 (cont.)

```
3   return with exponential.f(1./OMEGA,2)+exponential.f(1/OMEGA,3)
4 else
5   return with exponential.f(1./OMEGA,2)
6 end

  ,,-------------------------------------------------------------
  ,,                      routine ACCOUNTING
  ,,-------------------------------------------------------------
1 routine ACCOUNTING
2 define DEAD.TIME as a double saved variable
3 if NUM=K
4     reset totals of NO.BUSY,WAITING.TIME and n.QUEUE
5     let ZERO.WAIT = 0
6     let DEAD.TIME=time.v
7 always
8 if NUM >= K
9     let WAITING.TIME=1440.*(time.v-START(CUSTOMER))
10 always
11 if NUM=K+T-1
12     call SUMMARY(DEAD.TIME)
13     stop
14 always
15 add 1 to NUM
16 return        end

  ,,-------------------------------------------------------------
  ,,                      routine SUMMARY
  ,,-------------------------------------------------------------
1 routine SUMMARY(DEAD.TIME)
2 define DEAD.TIME as a double variable
3 define RUN.TIME to mean (time.v-DEAD.TIME)
4 define DUM1,DUM2 and DUM3 as double variables
5 start new page
6 skip 1 line
7 print 7 lines with M,LAMBDA,OMEGA,P,RRHO,K-1 thus
```

ANALYSIS OF SIMULATION RESULTS

NO. OF RESERVATIONISTS=**   CALL RATE=***.***   SERVICE RATE=***.***

     MULTIPLE
DESTINATION PROB.=*.***   ACTIVITY LEVEL=*.***   NO. TRUNCATED=******
```
 8 skip 1 lines
 9 print 1 line with NUM-K+1 and RUN.TIME thus
```

STATISTICS based on ********* waiting times and ****.**** elapsed days
```
10 skip 1 line
11 print 4 lines with AVE.WAIT,V.WAIT,MAX.WAIT,AVE.Q,V.Q,MAX.Q,AVE.UTIL,
                    V.UTIL,MAX.UTIL thus
```

Figure 4.2 (cont.)

```
                      AVERAGE          VARIANCE          MAXIMUM
     WAITING TIME     ............     .............     .............
     QUEUE LENGTH     ***.****        .............     *****
     UTILIZATION      ***.****        .............     *****
12 skip 1 line
13 let DUM1 = 1. - ZERO.WAIT/(NUM - K + 1)
14 let DUM2 = 1. - Q.HISTO(1)/RUN.TIME
15 print 4 lines with DUM1 and DUM2 thus
                           PROPORTION
        WAITING TIME       WHO WAIT        QUEUE      PROPORTION
         (minutes)         THIS TIME       LENGTH      OF TIME
          W >  0            *.****         Q >  0      *.****

16 let DUM1 = 1.0
17 for I=2 to 32 do
18    let DUM1 = DUM1 - WAIT.HISTO(I-1)/(NUM-K+1)
19    let DUM2 = max.f(0.,DUM2 - Q.HISTO(I)/RUN.TIME)
20    print 1 line with (I-1)/4,DUM1,I-1,DUM2 thus
                >= *.**         *.****         >***     *.****
21 loop
22 print 1 line with WAIT.HISTO(33)/(NUM-K+1),Q.HISTO(33)/RUN.TIME thus
                >= 8.00         *.****       Q >=32     *.****
23 skip 1 line
24 print 2 lines thus
                        NO. OF BUSY
                      RESERVATIONISTS        PROPORTION

25 let DUM3 = 1.0
26 for I=1 to M  do
27    let DUM3 = max.f(0.,DUM3 - UTIL.HISTO(I)/RUN.TIME)
28    print 1 line with I-1 and DUM3 thus
                          >***             *.****
29 loop
30 skip 1 line
31 print 6 lines with SEEDS(1),SEEDS(2),SEEDS(3),SEEDS(4),seed.v(1),seed.v(2),
               seed.v(3) and seed.v(4) thus



              RANDOM NO.     RANDOM NO.     RANDOM NO.     RANDOM NO.
               STREAM 1       STREAM 2       STREAM 3       STREAM 4
              ----------     ----------     ----------     ----------
   FIRST NO. ************   ************   ************   ************


    LAST NO. ************   ************   ************   ************
32 skip 3 lines
33 return   end
```

4.3 ENTITIES

As in line 4 of the preamble, the words "permanent entities" must preface declarations relating to such entities. Here the declared permanent entity RESERVATIONIST has an attribute STATUS. The significance of this attribute becomes apparent later when we search for an idle RESERVATIONIST. For each permanent entity, SIMSCRIPT II.5 implicitly defines a global variable n.*entity*. This quantity denotes the number of such entities in existence at a specific time during execution. In the present case, n.RESERVATIONIST serves this purpose. We assign it a positive integer value in the "main" program, prior to beginning a simulation run. The words "temporary entities" (line 11) preface declarations relating to such entities. Notice in line 12 that the temporary entity CUSTOMER has an attribute TYPE and may be a member of the set QUEUE, defined in line 3. TYPE allows for identification of a CUSTOMER as requesting a one-way or a round-trip reservation. Failure to assign any membership to a declared set results in a diagnostic error message during compilation.

4.4 EVENT NOTICES

The words "event notices" in line 17 of the preamble prefaces such declarations. However, as lines 17 through 19 indicate, two ways exist for declaring these notices. If the COMPLETION and END_MESSAGE event notices were to have no attributes, then we could replace lines 17 through 19 by

> event notices include CALL, COMPLETION and END_MESSAGE.

Similarly, if RESERVATIONIST and CUSTOMER have no attributes, the statements

> permanent entities include RESERVATION
> temporary entities include CUSTOMER

apply.

Note the use of the underscore in END_MESSAGE. This acceptable use in SIMSCRIPT II.5 provides a convenient way of making variable names more comprehensible to readers. Also note the use of the period in line 6 of the preamble, which improves comprehension. In this book we adopt the convention of using an underscore in the names of events and routines and using a period in entity, attribute, and variable names. Although entity names can have as many characters as one wishes, the first seven characters in the name of a routine, global variable, or attribute must be unique, as must be the first five characters in the name of an entity, event, or set.

Since two events may be scheduled for execution at the same time, the declaration in line 20 removes any ambiguity regarding the order of execution. This statement indicates that a CALL event has priority over COMPLETION and END_MESSAGE events and that a COMPLETION event has priority over END_MESSAGE events. If two CALL events are scheduled for the same time, SIMSCRIPT II.5 executes them in the order in which their corresponding events notices are created. Suppose that the priority declaration is

priority order is END_MESSAGE .

Then SIMSCRIPT II.5 *implicitly* assigns priority in the order END_MESSAGE, CALL, and COMPLETION. After ordering priorities as specified in the "priority order" statement, SIMSCRIPT II.5 goes back to the "event notices" statement and assigns successive priorities to events declared there but omitted from the "priority order" statement. These priorities are assigned in the order in which the events are declared. Note that in the present case omission of line 20 would not affect the priority order. Its value here is pedagogic.

4.5 SUMMARY STATISTICS

The "accumulate" statements in lines 7 through 10 and the "tally" statements in lines 14 and 15 of the preamble are declarations that cause summary statistics for utilization $\{B(s), s \geq 0\}$, queue length $\{Q(s), s \geq 0\}$, and waiting time $\{W_i, i \geq 0\}$ to be updated each time these processes change values during execution. See Section 1.3. To accomplish this in the present problem, the preamble declares two global variables, NO.BUSY (line 6) and WAITING.TIME (line 14). As a consequence of these declarations, SIMSCRIPT II.5 responds to a request for values of AVE.UTIL, AVE.Q, and AVE.WAIT at simulated time τ by computing (CACI, 1987, p. 269)

$$
\begin{aligned}
\text{AVE.UTIL} &= \bar{B}(0, \tau), \\
\text{AVE.Q} &= \bar{Q}(0, \tau), \\
\text{AVE.WAIT} &= \bar{W}(0, \tau),
\end{aligned}
\tag{4.1}
$$

the time averages that Section 3.1 describes. Section 4.14 shows how inclusion of a "reset" statement allows one to adjust these averages to account for the warm-up interval.

SIMSCRIPT II.5 responds to a request for values of MAX.UTIL, MAX.Q, and MAX.WAIT by computing

$$
\begin{aligned}
\text{MAX.UTIL} &= \max_{0 \leq s \leq \tau} B(s), \\
\text{MAX.Q} &= \max_{0 \leq s \leq \tau} Q(s), \\
\text{MAX.WAIT} &= \max_{1 \leq i \leq N(\tau)} W_i,
\end{aligned}
\tag{4.2}
$$

and to a request for the time variances V.UTIL and V.Q, and the variance V.WAIT, by computing

$$
\text{V.UTIL} = \frac{1}{\tau}\left[B^2(t_{J(\tau)})(\tau - t_{J(\tau)}) + \sum_{i=1}^{J(\tau)} B^2(t_{i-1})(t_i - t_{i-1}) \right] - (\text{AVE.UTIL})^2,
$$

$$
\text{V.Q} = \frac{1}{\tau}\left[Q^2(t_{J(\tau)})(\tau - t_{J(\tau)}) + \sum_{i=1}^{J(\tau)} Q^2(t_{i-1})(t_i - t_{i-1}) \right] - (\text{AVE.Q})^2, \qquad (4.3)
$$

$$
\text{V.WAIT} = \frac{1}{N(\tau) - 1}\left[\sum_{i=1}^{N(\tau)} W_i^2 - N(\tau)(\text{AVE.WAIT})^2 \right].
$$

By default, NO.BUSY has integer mode, whereas WAITING.TIME is a double precision floating-point variable. An alternative would be to define WAITING.TIME as a "real variable," which would make it a single-precision floating-point variable. To reduce numerical round-off error, we encourage double-precision declarations on 32-bit wordsize computers.

Line 8 of the preamble declares that SIMSCRIPT II.5 is to create a histogram of NO.BUSY, and line 10 does likewise for n.QUEUE. SIMSCRIPT II.5 responds to requests for values of these histograms by computing

$$
\text{UTIL.HISTO(I)} = \delta(B(\tau) - I + 1)(\tau - t_{J(\tau)}) + \sum_{i=1}^{J(\tau)} \delta(B(t_{i-1}) - I + 1)(t_i - t_{i-1}),
$$

$$
\text{I} = 1, \dots, \text{n.RESERVATIONIST} + 1, \qquad (4.4)
$$

and

$$
\text{Q.HISTO(I)} = \begin{cases} \delta(Q(t_{J(\tau)}) - I + 1))(\tau - t_{J(\tau)}) \\ \quad + \displaystyle\sum_{i=1}^{J(\tau)} \delta(Q(t_{i-1}) - I + 1)(t_i - t_{i-1}), & \text{I} = 1, \dots, 32, \\ \tau - \displaystyle\sum_{l=1}^{32} \text{Q.HISTO}(l), & \text{I} = 33, \end{cases} \qquad (4.5)
$$

where

$$
\delta(x) := \begin{cases} 1 & \text{if } x = 0, \\ 0 & \text{elsewhere.} \end{cases}
$$

Then UTIL.HISTO(I)/τ gives the proportion of time that I$-$1 RESERVATIONISTS are busy for $1 \le \text{I} \le 33$, and Q.HISTO(I) gives the proportion of time that I$-$1 CUSTOMERs await service for $1 \le \text{I} \le 32$. The remaining ordinate, Q.HISTO(33), contains the amount of time that 32 or more CUSTOMERs await service.

In terms of the exceedance proportions (3.12) and (3.13),

$$\bar{P}_B(b, 0, \tau) = 1 - \frac{1}{\tau} \sum_{I=0}^{b} \text{UTIL.HISTO(I)} \tag{4.6}$$

$$= \text{proportion of time that more than } b$$

$$\text{RESERVATIONISTs are busy}$$

and

$$\bar{P}_Q(q, 0, \tau) = 1 - \frac{1}{\tau} \sum_{I=0}^{q} \text{Q.HISTO(I)} \tag{4.7}$$

$$= \text{proportion of time that more than } q$$

$$\text{CUSTOMERs wait.}$$

These are the quantities that we print in the output.

In a similar manner, the declaration in line 15 of the preamble leads to

$$\text{WAIT.HISTO(I)} = \begin{cases} \displaystyle\sum_{j=1}^{N(\tau)} I_{[.25(I-1),.25I)}(W_j), & I = 1, \ldots, 32, \\[2ex] \displaystyle\sum_{j=1}^{N(\tau)} I_{[32,\infty)}(W_j), & I = 33, \end{cases} \tag{4.8}$$

where

$$I_{[a,b)}(x) := \begin{cases} 1 & \text{if } a \leq x < b, \\ 0 & \text{otherwise.} \end{cases}$$

If waiting times are recorded in minutes, then at time τ, WAIT.HISTO(I) contains the number of CUSTOMERs whose waiting times were at least $0.25(I-1)$ but less than $0.25I$ minutes for I=1, ..., 32, and WAIT.HISTO(33) contains the number of CUSTOMERs with waiting times of at least eight minutes. With regard to the exceedance proportion (3.14), recall that for all $w \in [0, \infty)$,

$$P_W(w, 0, \tau) = \text{proportion of waiting times that exceed } w, \tag{4.9}$$

whereas

$$1 - \frac{1}{N(\tau)} \sum_{I=1}^{\lfloor w/.25 \rfloor} \text{WAIT.HISTO(I)} = \text{proportion of waiting times} \geq w. \tag{4.10}$$

As line 12 of the CALL event shows, the integer variable ZERO.WAIT in line 16 of the preamble accumulates the total number of CUSTOMERs who do not wait, and therefore

$$1 - \frac{\text{ZERO.WAIT}}{N(\tau)} = \text{proportion of CUSTOMERs who wait.} \qquad (4.11)$$

We use expressions (4.10) and (4.11) in the output. Since waiting time has a continuous distribution function for all $w > 0$, the difference between the values that expressions (4.9) and (4.10) generate should be negligible.

4.6 TRAFFIC INTENSITY

The "define" statement in line 21 defines global variables whose modes are double, whereas line 23 defines global integer variables (by default). Table 4.1 relates SIMSCRIPT II.5 and problem equivalents. The "define to mean" statement in line 22 merely associates the global variable M with the SIMSCRIPT II.5 generated variable n.RESERVATIONIST. Although the "define to mean" statement in line 24 serves a similar purpose to that of lines 25 through 30, it illustrates good simulation procedure as well. For the moment, let us ignore the existence of the recorded message, and note that $1/\lambda$ is the mean intercall time and $1/\omega$ is the mean service time for one-way trips. In the simulation we assume that the mean service time for a round trip is $2/\omega$, so that the mean service time in general is $(1-p)/\omega + 2p/\omega = (1+p)/\omega$. Then the system *traffic intensity* is

$$\rho := \frac{\text{mean service time}}{\text{no. of reservationists} \times \text{mean intercall time}} = \frac{\lambda(1+p)}{m\omega}. \qquad (4.12)$$

Since we customarily consider alternative combinations of λ, ω, and ρ, the possibility arises that we may inadvertently select a setting such that $\rho \geq 1$. In a simulation with $\rho > 1$, the number of waiting calls increases without bound. For example, this would occur if one inadvertently specified the number of RESERVATIONISTs as $m < \lambda(1+p)\omega$, a potential scenario when one is trying to reduce resources. Since this *nonstationary* situation is contrary to intention, we regard a check on $\rho < 1$ as essential. This is done in the main routine, to

Table 4.1   Airline simulation

Problem variable	SIMSCRIPT II.5 equivalent
λ	LAMBDA
ω	OMEGA
p	P
m	M, n.RESERVATIONIST

be described shortly. The use of line 24 provides us with a convenient way of expressing RRHO, which corresponds to ρ in expression (4.12), concisely. When exact evaluation of ρ is not possible, computable lower and upper bounds ρ_L and ρ_U, respectively, can be useful. If $\rho_L > 1$, then $\rho > 1$; conversely, if $\rho_U < 1$ then $\rho < 1$.

We now consider the existence of the recorded message. If the simulation runs for a long period of time, then the number of calls that become available for service is independent of the time delay due to the recorded message. Then λ is the long-run arrival rate, and consequently the activity level defined in expression (4.12) continues to apply.

If one is simulating a closed-loop system, then the traffic intensity is always less than unity and no need exists for evaluating ρ. However, $\rho < 1$ in this case does not necessarily imply a unique equilibrium distribution.

Line 31 defines SEEDS as a global one-dimensional array having an integer mode because of the global declaration in line 2. Line 32 declares an integer function that determines whether a one-way or round-trip reservation is desired. Line 33 also defines a double-precision function S_TIME. We describe the contents of the KIND and S_TIME functions shortly. Functions are similar to, but not identical to, routines. In the particular case of lines 32 and 33 we note that the subprogram associated with S_TIME has a double mode as a result of the inclusion of "double" before "function." However, the function KIND has the integer mode because of the default to line 2. The "end" statement tells the SIMSCRIPT II.5 compiler that all global declarations have been made and what follows is the executable part of the program.

The "main" routine is the first subprogram that SIMSCRIPT II.5 executes. After defining input and output units in lines 2 through 5, it reads values (line 6) for LAMBDA, OMEGA, M, and P from the specified input unit using a free-form format. Lines 7 through 9 execute an error check using RRHO (line 24 of preamble) and demonstrate a simple "if" statement.

4.7   PSEUDORANDOM NUMBER GENERATION

Many algorithms exist for randomly generating samples from diverse probability distributions, and each simulation language provides programs in its collection of library routines for implementing many of these algorithms. All require a stream of *pseudorandom numbers* as input. Theorem 4.1 describes one property that accounts for this link between a sample from an arbitrary distribution and random numbers.

Theorem 4.1.

    Let X be a random variable with distribution function (d.f.) $\{F(x), a \leq x \leq b\}$, *denoted by F. Let*

$$G(u) := \inf [x : F(x) \geq u], \qquad 0 \leq u \leq 1,$$

which is called the inverse distribution function of X, and let U have the uniform distribution on $[0, 1]$*, denoted by* $\mathcal{U}(0, 1)$*. Then* $G(U)$ *has the d.f.* F*.*

Proof.   Let $Y := G(U)$. Then

$$\text{pr}(Y \leq y) = \text{pr}[G(U) \leq y] = \text{pr}[U \leq F(y)]$$
$$= F(y) = \text{pr}(X \leq y).$$

This result has immediate implications for sampling from F. For continuous F, one can sample U from $\mathcal{U}(0, 1)$ and solve $F(X) = U$ for X. For example, if $F(X) := 1 - e^{-\lambda x}$ for $\lambda > 0$ and $x \geq 0$, then $X = -\frac{1}{\lambda} \ln(1 - U)$ has the exponential distribution with rate λ. If F is discrete with points of increase at, say, $a, a + 1, \ldots, b - 1, b$ and $F(a - 1) := 0$, then $X = a + I$ is distributed as F, where

$$I := \min_{i=0,1,\ldots,b-a} [i : F(a + i - 1) < U \leq F(a + i)]. \qquad \square$$

Chapter 8 describes other methods for generating random samples, all of which presume the availability of a sequence of independent random samples from $\mathcal{U}(0, 1)$. To approximate this idealization, every simulation language provides a procedure for generating a sequence $\{U_i := Z_i/M, \ i = 1, 2, \ldots\}$ where $Z_0, Z_1, Z_2, \ldots$ are integers often derived from the one-step recursion

$$Z_i = AZ_{i-1} \pmod{M}, \qquad i = 1, 2, \ldots. \tag{4.13}$$

Here A and M are integers selected to make the points $\{U_i\}$ sufficiently dense in $(0, 1)$ and to have distributional properties that "closely" resemble those of the idealized i.i.d. uniform sequence. Expression (4.13) is called a *pseudorandom number generator* (png). GPSS/H, SIMAN, SLAM, and SIMSCRIPT II.5 all use pngs of this form.

To initialize the png at the beginning of a simulation experiment, one provides a *seed* Z_0 for expression (4.13). This seed dictates the subsequence of numbers that expression (4.13) generates during execution. SIMSCRIPT II.5 actually has 10 pseudorandom number streams, each of which requires a seed. The seed for stream I is stored in a system variable seed.v(I), which assumes a default value if none is assigned by the user. Although Chapter 9 deals with issues related to pseudorandom number generation in detail, it is important that the reader now recognize that well-designed computer simulation experiments call for explicit identification of the first and last pseudorandom numbers used in each stream that an experiment employs. One reason is reproducibility.

Recording the seeds, which are the first pseudorandom numbers used, enables one to run a related experiment with the exact same stream of numbers but different values for other input parameters, and then to compare experimental results with one major source of variation, the differences in pseudorandom numbers, removed. Section 4.22 returns to the topic of "common pseudorandom numbers" and shows how they can reduce the computing time to compare the output of two runs of a simulation, with different input values for one of its parameters, to within a given statistical accuracy.

Line 10 in the main program reserves space for the array SEEDS (line 31 of preamble). Lines 11 through 14 read four user-specified seeds into the SEEDS array and then assign SEEDS(I) to seed.v(I). Thus SEEDS(I) contains the first random number used in stream I. When the simulation terminates, seed.v(I) contains the seed to be used in the next generation from stream I. In particular, stream 1 generates pseudorandom numbers for interarrival times, stream 2 for service times for single-destination trips, stream 3 for the additional service time for multiple-destination trips, and stream 4 for determining values for TYPE. The "loop" statement in line 14 demarcates the end of the DO loop that begins in line 11.

For a given integer $I \geq 1$, the statements

release seed.v($\star$)
reserve seed.v($\star$) as I

cause SIMSCRIPT II.5 to resize seed.v. This would be required if $I > 10$ and should be inserted in the main program prior to assigning seeds to seed.v.

SIMSCRIPT II.5 allows coding of more than one statement on a single line. For example, we could have written lines 11 through 14 as

for I $=$ 1 to 4 do read SEEDS(I) let seed.v(I) $=$ SEEDS(I) loop ,

and SIMSCRIPT II.5 would interpret this in identical fashion.

By recording the last pseudorandom numbers one can run the simulation program again with these numbers as seeds but with the same values for other inputs, thereby creating an *independent replication* of the experiment.

4.8   WARM-UP AND PATH-LENGTH PARAMETERS

Line 15 reads T, the number of calls to be completed before the simulation terminates, and K, where $K-1$ specifies the number of completed calls to be discarded before data analysis begins for final reporting. This truncation of data reduces the influence of the *initial conditions* that prevail at the beginning of the simulation on the values of the data collected to estimate steady-state parameters. Initially we set $K = 1$ merely for illustration. We return to the important issue of selecting an appropriate K in Section 6.1 and show the error that potentially can arise by neglecting this issue.

The statement in line 18 creates n.RESERVATIONIST $=$ M records of RESERVATION-IST, where M receives its value in line 6. Remember that each record of a RESERVATIONIST includes a field for the attribute STATUS, which, by default, has the integer mode.

Line 17 creates an event notice CALL with a desired execution time equal to the current simulated time, time.v, a SIMSCRIPT II.5 system variable measured in double-precision decimal days. Since this action occurs at the beginning of execution and since SIMSCRIPT II.5 initializes all system and declared variables to zero, time.v = 0.0 at this point in simulated time.

4.9 EVENT MANAGEMENT

As we remarked earlier, event notices are filed in the future-event set in order of their desired execution times. SIMSCRIPT II.5 actually has a more elaborate structure for filing and retrieving event notices. For each class of events defined in the preamble, SIMSCRIPT II.5 establishes a distinct future-event set (e.g., Section 5.4.2). When an event notice of a particular class is created, it is filed in the set corresponding to that class in chronological order of desired execution time. When control returns to the SIMSCRIPT II.5 timing routine, it examines the first entry in each of the future event sets and determines which event to execute next. The priority ordering among events established in the preamble resolves ties. In the present problem there exist three classes of events, CALL, COMPLETION, and END_MESSAGE. This means that the timing routine examines the first entry in each of three sets each time it selects an event.

This multiple-event set feature caters to simulation with few events but many extant event notices per event. Conversely, the feature slows down the execution time of a simulation with many events most of which rarely occur. To circumvent this inefficiency one can define an omnibus event with an attribute whose value directs program flow within the event to one of several user-defined subprograms that contain the appropriate statements for effecting alternative types of state changes. This procedure simply reduces the number of event classes and increases the number of events in the future-event set corresponding to the omnibus event.

4.10 STARTING AND STOPPING EXECUTION

When SIMSCRIPT II.5 encounters line 19, it transfers control to SIMSCRIPT II.5's timing routine, "time.r," which selects the first event notice in its future event set, "ev.s." Once the simulation and concomitant analysis are completed, a programmer may choose to write code to terminate execution in one of several ways. One way returns control to the statement following "start simulation," which in this case leads to an end of execution in line 20. We illustrate an alternative approach shortly in line 12 of the ACCOUNTING subroutine.

4.11   CALL EVENT

Line 4 of the CALL event directs SIMSCRIPT II.5 to create a record corresponding to the record image declared in the preamble for a temporary entity CUSTOMER. This record includes space for the attributes TYPE and START. At the same time SIMSCRIPT II.5 treats the word CUSTOMER as a global variable and assigns to it the address within the computer of the newly created record. In effect, CUSTOMER becomes a *pointer* so that TYPE(CUSTOMER) is the attribute TYPE of this newly created CUSTOMER for the remainder of the execution of this CALL event.

Note that line 5 of the CALL event uses TYPE without a subscript. Doing so demonstrates the implicit subscripting feature of SIMSCRIPT II.5, which interprets TYPE to mean TYPE(CUSTOMER). If one wishes to access the TYPE attribute of another customer whose record address had been previously assigned during execution to a global variable CCUST, then TYPE(CCUST) would be that attribute. Line 5 assigns to TYPE(CUSTOMER) the value of KIND, an integer-valued function defined in line 31 of the preamble. The corresponding subprogram appears below the CALL event in Figure 4.2. In line 2, random.f(4) denotes a SIMSCRIPT II.5 library subprogram that produces a number from the uniform distribution on the interval (0, 1) using pseudorandom number stream 4. If this number is less than P, the probability of multiple destinations, then SIMSCRIPT II.5 assigns the value of MANY (see preamble) to KIND. Otherwise, it assigns the value of ONE. The demarcating word "else" is syntactically equivalent to "otherwise." The justification for this assignment procedure is easy to see. If U is uniformly distributed on $[0, 1]$, then

$$\mathrm{pr}(U \leq p) = \int_0^p \mathrm{d}u = p. \tag{4.14}$$

Line 6 of the CALL event assigns to START (CUSTOMER) the current simulated time time.v. This will be used later to compute the CALL's waiting time. Line 7 directs SIMSCRIPT II.5 to use the event notice that generated the current execution of this CALL event to *schedule* the execution of the next CALL event in T.CALL (line 3) minutes. Here "reschedule" is syntactically equivalent to "schedule." If "a" replaces "this," then SIMSCRIPT II.5 creates a *new* event notice. This difference is important with regard to program efficiency. Use of "a" requires SIMSCRIPT II.5 to find and allocate space to the new event notice. Use of "this" eliminates this often time-consuming search and allocation procedure. The reader should note that in order to use the CALL event notice again, the words "saving the event notice" must appear in line 1. Otherwise, SIMSCRIPT II.5 destroys the event notice when execution of the corresponding event is about to begin.

Note also in line 7 that the event is scheduled to occur in T.CALL minutes. Since time.v is in double-precision decimal days, the use of "minutes" causes SIMSCRIPT II.5 to schedule the next CALL in time.v+T.CALL/1440.0 days, since there are 1440 minutes in a day. For consistency CALL.RATE and SERVICE.RATE, read in the main subprogram, must be in calls per minute.

Line 8 indicates an attempt to arrange service by calling the SERVICE subprogram. It yields a value assigned to the variable ACTION, which is local to the CALL event. If no success occurs (line 9), then a recorded message is played for the CUSTOMER that ends in 9 seconds or, equivalently, 0.15 minutes (line 10). If success occurs, then the counter ZERO.WAIT is incremented by 1 in line 12. The statement "always" in line 13 denotes the end of the sequence of conditional statements beginning in line 9. Consequently, after line 13 is executed, a return to the timing routine occurs.

4.12   END_MESSAGE EVENT

When an END_MESSAGE event executes, its local variable CALLER contains the identification of the CUSTOMER assigned in the schedule statement in line 10 of the CALL event made at time time.v − .15/1440 days. Assignment of this identification to the global variable or entity CUSTOMER in line 3 is necessary because during the elapsed time a new CALL event may have occurred, and hence CUSTOMER may refer to a new entity. If an attempt at service (line 4) fails again (line 5), then the CUSTOMER is filed in the set called QUEUE.

If line 12 of the preamble had failed to mention that a CUSTOMER could belong to the set QUEUE, then the "file" statement here would induce an error message during execution indicating an attempt to file an entity in a set for which no corresponding membership declaration exists. In lieu of explicit declaration to the contrary in the preamble, CUSTOMERs are filed in the order in which they encounter this file statement.

4.13   FILING PRIORITIES

Suppose that we add the statement

define QUEUE as a set ranked by low TYPE

after line 12 in the preamble. Then an encounter of the file statement in line 6 of the END_MESSAGE event would cause an ordering first on TYPE, with one-way trips coming first, and then, if ties occur, an ordering on time of filing. In the problem under study the present preamble declares behavior closer to reality than the example given here does. However, other contexts may call for the use of alternatives similar to this example.

Several additional issues deserve attention. Suppose that we delete lines 11, 12, and 13 of the preamble and line 4 of the CALL event. Then the event notice CALL could serve as a record of a temporary entity instead of CUSTOMER, "a" would replace "this" in line 7 of the CALL event, and CALL would replace CUSTOMER in lines 3 and 6 of the END_MESSAGE event. The reader should recognize that an event notice is a specialized temporary entity and can function as such in many convenient ways. Although no great computational improvement accrues from this alternative approach here, situations may arise in practice where it is desirable to keep the number of temporary entities to a minimum. Note that set membership for CALL in QUEUE must also be declared.

The local variable ACTION in the END_MESSAGE event is not the same as the local variable ACTION in the CALL event. We chose the same labels to demonstrate that two or more subprograms can have local variables with the same names. In fact, a local variable can have the same name as a global variable, provided that the subprogram contains an appropriate "define" statement. For example, if we include

$$\text{define RRHO as a real variable}$$

in the END_MESSAGE event, the local variable RRHO would be distinct from the global variable RRHO defined in the preamble.

4.14   COMPLETION EVENT

Line 3 of the COMPLETION event instructs the simulation to release the space occupied by the record of the temporary entity CUSTOMER called CLIENT. Failure to release space in this manner may cause the simulation to run out of space when it needs to create records of new temporary entities. Line 4 indicates that if the set QUEUE is empty, then the RESER-VATIONIST called WORKER should have his attribute STATUS set to IDLE. Because of line 24 of the preamble, this action is equivalent to

$$\text{let STATUS(WORKER)} = 0 \,.$$

Lines 4 through 7 of the COMPLETION event imply that if no CUSTOMERs await service, then NO.BUSY should be reduced by 1, the corresponding RESERVATIONIST should be put in the idle mode, and control should return to the timing routine.

Recall that the prefix n. before a set name denotes the number of entries currently in that set. Therefore, n.QUEUE denotes the number of CUSTOMERs in the QUEUE, so that the statement

$$\text{if QUEUE is empty}$$

is equivalent to

$$\text{if n.QUEUE} = 0 \,.$$

The programmer has her or his choice of which to use.

If the set QUEUE is not empty in the COMPLETION event, then control passes to line 9, where the "first" CUSTOMER is removed, since he or she is about to receive service from the RESERVATIONIST with name WORKER who has just completed service. Failure to include line 9 results in an increasing list length even though CUSTOMERs receive service.

The assignment of the name WORKER to the entity name RESERVATIONIST is necessary if the SERVICE routine is to work correctly. The quantity event.v is a system variable that contains the *class number* i.event of the event that began execution last. Class numbers are assigned according to the priority order statement in the preamble. Class numbers for CALL, END_MESSAGE, and COMPLETION are i.CALL, i.END_MESSAGE, and i.COMPLETION, respectively, and assume values of 1, 3, and 2, respectively. Now all three classes of events call the SERVICE routine. In case of CALL and END_MESSAGE, a search for an idle RESERVATIONIST is necessary in lines 3 and 4 of the SERVICE routine. For a COMPLETION, control passes to line 9, since the RESERVATIONIST is already known. S_TIME is a function that we describe shortly.

If no RESERVATIONIST is idle (line 5), then the routine yields NO.SUCCESS, which is 1, according to the preamble. When success occurs, RESERVATIONIST denotes the idle server whose STATUS we want to change from IDLE to BUSY. Note that

let STATUS = BUSY

is equivalent to

let STATUS (RESERVATIONIST) = BUSY ,

again demonstrating the implicit subscripting ability of SIMSCRIPT II.5. However, note that substitution of the statement

let STATUS = IDLE

in line 5 of the COMPLETION event would probably cause an error, since there is little likelihood that RESERVATIONIST = WORKER because of the passage of time. Line 10 of the SERVICE routine merely increments the user-defined variable NO.BUSY by 1.

Line 12 indicates the scheduling of a COMPLETION for the current value of RESERVATIONIST. Line 13 calls a user-written subroutine ACCOUNTING that checks on the progress of the simulation with respect to truncation, computes WAITING.TIME for the CUSTOMER entering service if NUM $\geq$ K, and updates NUM. Line 14 of COMPLETION assigns SUCCESS to ACTION for the benefit of the calling event.

The S_TIME subprogram determines service time from the Erlang distribution in line 3 for a multiple-destination trip and from the exponential distribution in line 5 for a one-stop trip. Note that S_TIME is double precision. Failure to make this declaration in the preamble would result in S_TIME being treated as integer. SIMSCRIPT II.5's structured "if" statement has the form

if   a condition is true, perform the statements listed here

else if the condition is false, perform the statements listed here
always in either case continue execution after this key word.

If NUM $\geq$ K, the ACCOUNTING subroutine updates WAIT.TIME in line 9 with the waiting time, time.v–START(CUSTOMER), of the customer who is about to begin service. The factor 1440 converts time in double-precision decimal days to double-precision decimal minutes. Lines 3 through 7 concern truncation. Recall that we want to collect summary statistics only on customers K, K + 1, ..., K + T − 1. The global variable NUM contains the number of customers who have or are about to receive service. When NUM = K in line 3, SIMSCRIPT II.5 executes lines 4, 5, and 6. Line 4 resets the totals of all summary statistics, thereby removing any dependence on the data for the previous K − 1 customers. That is, for simulated time $\tau \geq \tau_K$ SIMSCRIPT II.5 computes utilization summary statistics as

$$\text{AVE.UTIL} = \bar{B}(\tau_{K-1}, \tau), \tag{4.15}$$

$$\text{V.UTIL} = \frac{1}{\tau - \tau_{K-1}} \left[B^2(\tau)(\tau - t_{J(\tau)}). \tag{4.16} \right.$$

$$\left. + \sum_{i=J(\tau_{K-1})+1}^{J(\tau)} B^2(t_{i-1})(t_i - t_{i-1}) \right] - (\text{AVE.UTIL})^2,$$

$$\text{MAX.UTIL} = \max_{\tau_{K-1} \leq s \leq \tau} B(s), \tag{4.17}$$

$$\text{UTIL.HISTO(I)} = \delta(B(\tau) - I)(\tau - t_{J(\tau)}) + \sum_{i=J(\tau_{K-1})+1}^{J(\tau)} \delta(B(t_{i-1}) - I + 1) \tag{4.18}$$

$$\times (t_i - t_{i-1}), \qquad I = 1, \ldots, \text{n.RESERVATIONIST},$$

summary queue length statistics as

$$\text{AVE.Q} = \bar{Q}(\tau_{K-1}, \tau) \tag{4.19}$$

$$\text{V.Q} = \frac{1}{\tau - \tau_{K-1}} \left[Q^2(\tau)(\tau - t_{J(\tau)}) \tag{4.20} \right.$$

$$\left. + \sum_{i=J(\tau_{K-1})+1}^{J(\tau)} Q^2(t_{i-1})(t_i - t_{i-1}) \right] - (\text{AVE.Q})^2,$$

$$\text{MAX.Q} = \max_{\tau_{K-1} \leq s \leq \tau} Q(s), \tag{4.21}$$

$$\text{Q.HISTO(I)} = \begin{cases} \delta(Q(\tau) - I + 1)(\tau - t_{J(\tau)}) \\ + \sum_{i=J(\tau_{K-1})+1}^{J(\tau)} \delta(Q(t_{i-1}))(t_i - t_{i-1}), & I = 1, \ldots, 32, \\ \tau - \sum_{l=1}^{32} \text{Q.HIST}(l), & I = 33, \end{cases} \tag{4.22}$$

and summary waiting time statistics

$$\text{AVE.WAIT} = \frac{1}{N(\tau) - N(\tau_K) + 1} \sum_{i=N(\tau_K)}^{N(\tau)} W_i, \tag{4.23}$$

$$\text{V.WAIT} = \frac{1}{N(\tau) - N(\tau_K)} \left\{ \sum_{i=N(\tau_K)}^{N(\tau)} W_i^2 - [N(\tau) - N(\tau_K) + 1](\text{AVE.WAIT})^2 \right\}, \tag{4.24}$$

$$\text{MAX.WAIT} = \max_{N(\tau_K) \leq j \leq N(\tau)} W_j, \tag{4.25}$$

$$\text{WAIT.HISTO(I)} = \begin{cases} \displaystyle\sum_{i=N(\tau_K)}^{N(\tau)} I_{[.25(I-1),.25I]}(W_i), & I = 1, \ldots, 32, \\ \displaystyle\sum_{i=N(\tau_K)}^{N(\tau)} I_{[32,\infty)}(W_i), & I = 33. \end{cases} \tag{4.26}$$

In the present case, note that $N(\tau_K) = K$ and $N(\tau) = K + T - 1$. Also note that line 5 reinitializes ZERO.WAIT.

Line 6 records the current time that the SUMMARY subroutine needs to normalize the time-integrated statistics properly. Note that line 2 defines DEAD.TIME as a double-precision saved variable. The qualifier "saved" ensures that DEAD.TIME retains its value between successive calls to ACCOUNTING. Omission of "saved" would cause DEAD.TIME to be initialized to zero at the beginning of each execution of ACCOUNTING.

If NUM $= K + T - 1$, then the simulation has been executed for the number of customers $K + T - 1$ specified in the main routine. It then calls the SUMMARY subroutine, which prints out all input parameter values in its line 7, all summary statistics in lines 9 through 21, and then prints the user-specified input seeds and the final numbers for each pseudorandom stream in line 22. The "stop" statement in line 13 of ACCOUNTING provides one way to terminate the simulation's execution.

4.15 GETTING AN EXECUTABLE PROGRAM

SIMSCRIPT II.5 is available for most mainframe computers, IBM-compatible personal computers, and Unix-based workstations. While each implementation has its own peculiarities, they all share common features an understanding of which is essential in trying to obtain an error-free executable program.

After preparing a source program in SIMSCRIPT II.5, a user submits it for *compilation*, the process of conversion from the *source program* in the syntax of SIMSCRIPT II.5 to an *object program* in the basic machine language of the computer. The SIMSCRIPT II.5 compiler scans the program. If the level of severity of the errors encountered is sufficiently high, the scan terminates, the compiler produces a list of error messages for the portion of the program it has scanned, and execution terminates. For example, an error in the preamble usually causes a scan termination before the executable portion of the program is reached.

Alternatively, the compiler scans the program, and, if the level of error is below that for scan termination, SIMSCRIPT II.5 produces an object program that is storable in a user-specified file. The compiler also produces a listing of the entire source program with appropriate error messages. A natural question to ask is: What happens to statements with errors? SIMSCRIPT II.5 ignores them during compilation, even though object programs are created for every subprogram. Consequently, the object programs corresponding to subprograms with errors are worthless and should be discarded. When the user examines her/his program listing, she/he identifies the errors, corrects them in the source program, and attempts to recompile the program successfully.

SIMSCRIPT II.5 also allows a user to compile and execute a program all in one submission. The appeal is apparent. If compiler-detected errors exist in the program, execution usually terminates with an execution error message. Such error messages also may occur when no compilation errors arise, since the user may have failed to construct an internally consistent executable program. During execution, stategically user-placed LIST statements in the source program can provide information about entities, attributes, sets, and other variables. In addition to the diagnostic at compilation, SIMSCRIPT II.5 also provides SIMDEBUG for tracing the sequence of routine calls during a simulation that lead to an execution error (CACI 1993b). Invoking the −d option at compilation enables this capability for subsequent program execution.

Unless a user has considerable confidence that the source program contains no errors detectable during compilation, he/she is well advised to use this combined compilation and execution procedure sparingly. This advice follows as a consequence of the mandatory *linkage editing* step that follows the compilation step and precedes the execution step. The linkage editing step links the object program together with a series of routines in the SIMSCRIPT II.5 library that are needed for execution but have not been previously incorporated. These routines differ from those automatically compiled into the object deck. For example, in our airline program there are more than 50 such routines, including those for Erlang, exponential, and uniform variate generation. The combined object program plus SIMSCRIPT II.5 routines form a *load module*. Linkage editing takes time, and a user who repeatedly uses the compilation and execution option with a program with probable compilation errors is wasting time.

We wish to point out that once a program has been successfully compiled and executed the linkage editing step can be avoided on future runs. To accomplish this a user submits his object program for linkage editing and then stores the resulting *load module* in a predesignated file. To execute the program one submits instructions to take the load module from the data set and execute it. This procedure takes negligible time.

This relatively brief description of SIMSCRIPT II.5 creates a familiarity, not an intimacy, with the language. The reader who intends to do the SIMSCRIPT II.5 exercises in this book will find regular access to the CACI references indispensable for success. Moreover, thorough understanding of the process of conversion of a *source program* written in

SIMSCRIPT II.5 into an *object program* coded in the *basic machine language* of the computer can lead to considerable cost saving and to reduced frustration in trying to get an executable object program. We describe several issues deference to which should reduce the needed effort.

4.16 A FIRST RUN

Before executing AIR_EV.sim, we need to specify numerical values for the input variables to be read in lines 6, 12, and 15 of the main program. Since that routine designates unit 10 as input in line 2, we create an input file called SIMU10 and load it with the input data. Suppose that past experience indicates that $\lambda = 1$, $\omega = .50$, and $p = .75$. Then expression (4.7) reveals that $m > 3.5$ is necessary for a stationary solution to exist for this problem. Since m must be integral, we assign $m = 4$. For the initial run SIMU10 contains

main program				
line 6:	1	.50	4	.75
line 12:	604901985	1110948479	1416275180	40579844
line 15:	1000	1		

Line 6 provides the input for the "read" statement in line 6 of the main program. Line 12 provides the seeds for the four pseudorandom number streams. These seeds come from entries in Table 4.2, which lists numbers that successively occur 1,000,000 numbers apart in the stream of numbers that the SIMSCRIPT II.5 pseudorandom number generator produces. The entries are from column 0 and rows 0, 1, 2, and 3, implying that successive numbers are 10,000,000 apart in the pseudorandom number sequence. For AIR_EV.sim, this choice of seeds implies that for runs made with $T + K \leq 10^7$ the four streams would not overlap. Nonoverlapping streams are essential for independent sampling with regard to each source of variation in this problem. "Line 15" of SIMU10 provides values for T and K.

Of necessity, virtually every discrete-event simulation begins with *guesses* for K and T and later revises them based on information that the preliminary run reveals. The present account follows this approach. By setting $K = 1$, AIR_EV.sim uses all the data generated during a run to compute statistics. Since this is a preliminary run, setting $T = 1000$ is merely a guess on how long to run the simulation to obtain meaningful data.

Figure 4.3 displays the output for these choices. It summarizes descriptive statistical behavior over the $0.6818 \times 24 = 19.632$ hours of simulated time that elapsed up to the moment at which the 1000th CUSTOMER entered service. Tableaus similar in format to this one are commonplace output for many discrete-event simulations. Unfortunately, they contain no information on how well these sample averages approximate the long-term

Table 4.2 Seeds for SIMSCRIPT II.5 Pseudorandom Number Generator (spaced 1,000,000 apart) $Z_{1,000,000}(10i+j)$

i \ j	0	1	2	3	4	5	6	7	8	9
0	604901985	1911216000	726466604	622401386	1645973084	1901633463	67784357	2026948561	1545929719	547070247
1	1110948479	1400311458	1471803249	1232207518	195239450	281826375	416426318	380841429	1055454678	711617330
2	1416275180	788018608	1357689651	2130853749	152149214	550317865	32645035	871378447	108663339	199162258
3	40579844	1357395432	829427729	1967547216	1955400831	2003797295	2094442397	2037011207	234502811	989756374
4	1912478538	740735795	1417788731	457273482	1315849725	270156449	440381980	1658090585	766876137	695181594
5	1140617927	1763098393	862536069	944596746	1832412366	1370121812	349487370	1490679974	1137328972	717010525
6	1768908897	214786952	1001221991	810390021	466350553	285688338	511678575	417093409	1857363469	833556803
7	1168765321	1578167688	904774509	1230498508	349030873	1280945183	411345101	1477279210	523541596	44009158
8	1034152236	171671287	1486370683	758021537	675302256	766501909	1570038230	127735274	531051296	481363190
9	36897498	377244078	632839225	1113362767	996065998	1127104879	1070367500	1992322523	548733793	1489852891
10	1719061561	3280042	1657019267	1741267743	691072269	1090974938	763220480	1468416723	111113372	129823184
11	705619618	224040976	749810472	2013253460	792658328	86620226	1386345831	1735570696	1896821842	278150891
12	331083792	1598034603	236635316	252926587	471825789	37034748	376315326	2146164618	1999725174	1243665337
13	1336631754	438764943	957241771	1713987536	1948054280	1998497269	575575728	677819252	501253489	1945486566
14	1245688777	2000276877	1503544457	464753035	200460239	1434443717	1353395916	133529216	1070327290	1928287458
15	1384322130	893932629	1181351847	606871877	523684234	1002300248	791546478	1328653688	246224997	48717991
16	2127710456	497489986	869470589	61678709	1145621181	1695821109	369707566	2110142430	763197278	11382857
17	1246856358	1714587858	873019592	1279964778	1396199241	1399930491	2078774631	985879465	486028376	1478854650
18	568269458	1657131937	1622687092	91313597	2080365494	1650386438	1871205625	563020663	1728323861	590715859
19	513501749	793321696	1443190012	1950895003	597186867	50824927	911425540	398409796	1672302183	1870294907
20	1827350617	727130390	774781557	1955524233	87547656	1061350608	1491284628	2029219172	1260286972	272221686
21	300327217	1787384331	2014193199	666978949	301608013	948450895	1453252697	419227541	1678821481	1765439952
22	177708831	1077885023	540589041	2007603355	1543903327	2017128524	1405189468	1003195553	337684924	1565288286
23	1745588809	485902835	1763594466	1082423631	194255479	1912029349	1219836142	1648556563	1508071873	549212365
24	1096265301	851153926	456184535	1408460673	1685622150	190036881	142510264	1465148875	459174572	597953519
25	606282883	1721436086	975289131	1376880454	1041829784	1596248327	2108842626	413468719	1204553786	9011429
26	626121779	203036030	1837216393	2086017536	697945492	1610367790	1512734966	925844033	1356793564	1549516128
27	649444368	510292152	1742847415	1285151791	1006048803	1538805236	1767548554	1446954807	1842805571	1969150158
28	2135785600	906045904	784888097	1048325198	416733559	993764136	1303965501	121604326	1780949928	1259010827
29	2102580471	741705624	1349641363	1376265619	653106395	271655202	1907239477	1015954941	1938162956	371946026
30	1842460388	543303725	1390660531	1138405358	31074996	1712456909	1247559532	1617264816	755962474	804894664
31	1186319334	494117518	1634124003	1027439985	319326476	1043817176	1323412056	668757796	406168599	1317790466
32	625592935	1520076665	1900759891	1684472723	209831417	2007811944	587647848	1424142232	6938939	427509950
33	1174409129	210461470	539786056	1629541058	1819341982	43434266	1391309379	244792174	750490187	423505883
34	904658495	245177844	841290082	1402203157	429993932	1836064376	1038894007	1923849909	1600794244	1950123859
35	960647611	974174009	1934369904	808615644	1879784225	795995555	1922561547	69238569	961625762	163574669
36	1095421583	554291019	432345723	855636598	971945436	881294334	1783677724	2088374054	1270902641	1182268061
37	2004539543	216813966	1304722830	629904199	834124981	898116231	864082217	389917225	1278587457	1869657489
38	1111329860	1594031833	571644852	1053085806	1019423302	482661750	1134475119	1104046462	1929367108	1097726063
39	295080740	1818474208	1129948309	75253501	1876330799	506183851	1692164827	873044608	1908879123	1897274087
40	1373006294	445158089	292594684	1305725195	847492301	1067706417	1718084174	133626496	971565385	1931931986
41	1677252989	312760713	891757911	776146802	462527160	1917080512	1349286825	909926550	1756656652	1570226967
42	105992480	1444798270	523658664	1024065462	299634195	884544980	1281114105	2106315124	869738253	1165668988
43	1036972300	462764735	314834790	222634789	1852359919	1085683702	115959636	695883743	1545554827	1372269350
44	287358587	2022531413	535827657	1657359673	904942635	317419570	576435644	1843235552	1761208654	912499785
45	1230888777	1724910189	1288149046	897051070	1758822711	1977931462	451692590	1554328303	1935881922	49510711
46	14291789	46830509	499758343	758082642	1018016755	1635908329	1577325787	423614699	1901481113	763520381
47	251686092	860899664	1253266716	487207054	1133665090	858964589	2104231773	554226602	126548109	568187190
48	71320167	1595587069	1886787081	883887920	1470032706	817226546	1572454264	766098912	1700253028	230254384
49	1488181869	1104477402	2087453595	1544780480	826720767	1926533288	23680406	97080449	2139633377	994728742

averages of interest. In particular, they contain not a clue as to how large K should be to make the influence of initial conditions negligible on the displayed summary statistics.

To make this issue of statistical accuracy more salient, AIR_EV.sim was rerun with only T changed from 1,000 to 1,000,000. This was motivated by the observation that since a run with T = 1000 took less than a second, a run with T = 1,000,000 would take no more than 17 minutes. Figure 4.4 shows the corresponding output, which differs substantially from that in Figure 4.3. For example, T = 1000 gives a sample average waiting time of 5.88179, whereas T = 1, 000, 000 gives 4.29191. Because the summary in Figure 4.4 is based on a larger sample path, one is inclined to believe that it is more representative of the long-run averages. However, it provides no other indication of how good the approximations are and, in particular, provides no guidance as to the choice of K to mitigate the effects of starting the simulation in the empty and idle state. Chapter 6 addresses both these issues in detail and describes techniques for estimating a suitable initial truncation and for assessing how well the time averages based on simulated data approximate the corresponding true long-run averages of interest.

```
                    ANALYSIS OF SIMULATION RESULTS
                    -----------------------------

NO. OF RESERVATIONISTS= 4     CALL RATE=  1.000       SERVICE RATE=   .500

   MULTIPLE-
DESTINATION PROB.= .750     ACTIVITY LEVEL= .875     NO. TRUNCATED=    0

STATISTICS based on     1000 waiting times and     .6818 elapsed days

                        AVERAGE        VARIANCE        MAXIMUM
        WAITING TIME   +5.88179E+000  +3.30388E+001  +2.86073E+001
        QUEUE LENGTH    5.8620        +3.55975E+001    30
        UTILIZATION     3.7135        +5.90100E-001     4

                        PROPORTION
        WAITING TIME    WHO WAIT       QUEUE     PROPORTION
        (minutes)       THIS TIME      LENGTH    OF TIME
        W >  0           .8550       Q >  0       .7825
           >=  .25       .8420          >  1      .7183
           >=  .50       .8270          >  2      .6475
           >=  .75       .8120          >  3      .5660
           >= 1.00       .7970          >  4      .4923
           >= 1.25       .7760          >  5      .4352
           >= 1.50       .7560          >  6      .3860
           >= 1.75       .7370          >  7      .3212
           >= 2.00       .7180          >  8      .2637
           >= 2.25       .6980          >  9      .2036
           >= 2.50       .6780          > 10      .1577
           >= 2.75       .6590          > 11      .1261
           >= 3.00       .6340          > 12      .1030
           >= 3.25       .6100          > 13      .0877
           >= 3.50       .5890          > 14      .0802
           >= 3.75       .5710          > 15      .0785
           >= 4.00       .5540          > 16      .0752
           >= 4.25       .5280          > 17      .0664
           >= 4.50       .5170          > 18      .0574
           >= 4.75       .4970          > 19      .0442
           >= 5.00       .4800          > 20      .0369
           >= 5.25       .4650          > 21      .0278
           >= 5.50       .4480          > 22      .0247
           >= 5.75       .4330          > 23      .0221
           >= 6.00       .4170          > 24      .0190
           >= 6.25       .3990          > 25      .0149
           >= 6.50       .3820          > 26      .0110
           >= 6.75       .3660          > 27      .0080
           >= 7.00       .3490          > 28      .0040
           >= 7.25       .3240          > 29      .0008
           >= 7.50       .3110          > 30      0.
           >= 7.75       .2950          > 31      0.
           >= 8.00       .2810       Q >=32      0.

                    NO. OF BUSY
                    RESERVATIONISTS      PROPORTION
                        >  0              .9925
                        >  1              .9616
                        >  2              .9064
                        >  3              .8530

                RANDOM NO.    RANDOM NO.    RANDOM NO.    RANDOM NO.
                 STREAM 1      STREAM 2      STREAM 3      STREAM 4
                ----------    ----------    ----------    ----------
FIRST NO.       604901985     1110948479    1416275180     40579844

LAST NO.        279098565     1319150004    1298427771    1408862871
```

Figure 4.3   AIR_EV.sim sample output (1000 callers)

```
                      ANALYSIS OF SIMULATION RESULTS
                      ------------------------------

NO. OF RESERVATIONISTS= 4      CALL RATE= 1.000        SERVICE RATE=  .500

   MULTIPLE-
DESTINATION PROB.= .750        ACTIVITY LEVEL= .875        NO. TRUNCATED=    0

STATISTICS based on   1000000 waiting times and   694.0694 elapsed days

                         AVERAGE         VARIANCE          MAXIMUM
      WAITING TIME    +4.29191E+000   +3.10938E+001   +5.57573E+001
      QUEUE LENGTH       4.1841       +3.51154E+001          64
      UTILIZATION        3.4963       +9.13621E-001           4

                       PROPORTION
      WAITING TIME     WHO WAIT       QUEUE      PROPORTION
       (minutes)      THIS TIME      LENGTH       OF TIME
       W >   0          .7336       Q >   0        .6161
          >=  .25       .7070         >   1        .5262
          >=  .50       .6794         >   2        .4484
          >=  .75       .6525         >   3        .3815
          >= 1.00       .6261         >   4        .3242
          >= 1.25       .6004         >   5        .2757
          >= 1.50       .5757         >   6        .2350
          >= 1.75       .5516         >   7        .2006
          >= 2.00       .5281         >   8        .1713
          >= 2.25       .5054         >   9        .1461
          >= 2.50       .4838         >  10        .1249
          >= 2.75       .4630         >  11        .1070
          >= 3.00       .4429         >  12        .0917
          >= 3.25       .4237         >  13        .0785
          >= 3.50       .4055         >  14        .0673
          >= 3.75       .3880         >  15        .0578
          >= 4.00       .3711         >  16        .0496
          >= 4.25       .3548         >  17        .0425
          >= 4.50       .3394         >  18        .0362
          >= 4.75       .3249         >  19        .0309
          >= 5.00       .3108         >  20        .0263
          >= 5.25       .2974         >  21        .0223
          >= 5.50       .2846         >  22        .0189
          >= 5.75       .2724         >  23        .0160
          >= 6.00       .2608         >  24        .0137
          >= 6.25       .2496         >  25        .0117
          >= 6.50       .2389         >  26        .0100
          >= 6.75       .2284         >  27        .0085
          >= 7.00       .2189         >  28        .0073
          >= 7.25       .2097         >  29        .0062
          >= 7.50       .2008         >  30        .0053
          >= 7.75       .1922         >  31        .0046
          >= 8.00       .1843       Q >=32        .0046

                    NO. OF BUSY
                  RESERVATIONISTS      PROPORTION
                      >  0              .9854
                      >  1              .9341
                      >  2              .8436
                      >  3              .7331

            RANDOM NO.      RANDOM NO.      RANDOM NO.      RANDOM NO.
            STREAM 1        STREAM 2        STREAM 3        STREAM 4
            ----------      ----------      ----------      ----------
FIRST NO.   604901985       1110948479      1416275180       40579844

LAST NO.    425804541       1400311458      1881615938      1163233127
```

Figure 4.4 AIR_EV.sim sample output (1,000,000 callers)

4.17 COLLECTING DATA

Although AIR_EV.sim produces a tableau of summary statistics by executing its SUMMARY subroutine at the end of a simulation run, it does not preserve a record of the actual sample sequences on which the summaries are based. For example, it averages the waiting times in AVE.WAIT in the ACCOUNTING subroutine but does not save the actual waiting times. As Chapter 6 shows, these data are essential for assessing the influence of initial conditions on AVE.WAIT and for estimating its variance. Adding the statements

define W as a double variable

to the preamble betweens lines 20 and 21,

reserve W(*) as T

to the main routine between lines 9 and 10, and

let W(NUM−K+1) = 1440.*(time.v−START(CUSTOMER))

to the ACCOUNTING routine between lines 8 and 9 directs SIMSCRIPT II.5 to save the waiting times in decimal days in the double-precision array W, which is 8T bytes in length. Therefore, the size of the array grows linearly with T, which conceivably can create a space problem as T becomes excessively large. Section 3.7 describes procedures for collecting sample-path data on queue length and resource utilization.

4.18 PROCESS-INTERACTION APPROACH

As Chapter 2 indicates, process interaction relies on a conceptually different representation of behavior than does event scheduling. In the event-scheduling approach, an event contains all decision flows and updating that relate to the change in state the execution of the event requires. Since time elapses between events, one can regard a simulation as the execution of a sequence of events ordered chronologically on desired execution times. However, no time elapses within an event.

By contrast, the process-interaction approach provides a process for each entity in a system. Each temporary entity moves through the system and consequently through time. Occasionally, a temporary entity encounters an impediment to progress and must wait. For example, if all reservationists (permanent entities) are answering calls when a new call (temporary entity) occurs, this call must wait for service. The inclusion of time flow in a procedure and in its corresponding flowchart distinguishes this approach from event scheduling. Moreover, it calls for a less direct form of computer programming in moving from flowchart to executable program. A process can also characterize a permanent entity. For example, a process can describe the behavior of each reservationist in the airline problem.

From the event-scheduling perspective of Section 2.3, the airline reservation problem concerns interactions between two types of entities, CUSTOMERs (temporary) and RESERVATIONISTs (permanent). One formulation of the process-interaction approach, as in Figures 2.10 and 2.11, provides a process for each CUSTOMER and a process for each RESERVATIONIST. Our SIMSCRIPT II.5 rendering of this problem follows the logic in these figures. Here, each CUSTOMER has an attribute TYPE that identifies whether a one-way or two-way reservation is desired. TYPE plays the same role here as in the event-scheduling approach in Section 2.3.

4.19 PROCESSES AND RESOURCES IN SIMSCRIPT II.5

SIMSCRIPT II.5 offers a capability for performing a simulation using the process-interaction approach using the executable *process* statements

> activate a *process name* in *time expression*
> create a *process name*
> work *time expression*
> wait *time expression*
> interrupt *process name* called *name*
> resume *process time* called *name*
> suspend
> reactivate *process name* called *name*
> destroy *process name* called *name*

and using the executable *resource* statements

> request *quantity* units of *resource name* (*index*) with priority *expression*
> relinquish *quantity* units of *resource name* (*index*).

Italics in this list denote numbers and variables that the simulation programmer assigns. In practice, one can use the process-interaction option in SIMSCRIPT II.5 to simulate a system in two different but equivalent ways. One relies on the process statements exclusively. The other exploits the resource statements "request" and "relinquish" as well. Section 4.20 describes the approach based exclusively on processes, and Section 4.21 describes the approach using processes and resources.

4.20 USING PROCESSES EXCLUSIVELY

Figure 4.5 lists a SIMSCRIPT II.5 program, based on Figure 4.2, which produces an output analysis identical to the SIMSCRIPT II.5 program in Figure 4.3. Line 3 of the preamble defines two sets, QUEUE and IDLE.LIST. The key word "processes" (line 4) indicates that subsequent statements define the processes in the simulation. In particular, a CUSTOMER has the attributes TYPE and START and can be linked to QUEUE. A RESERVATIONIST has an attribute DEMAND and can be linked to IDLE.LIST. If CUSTOMER had no attributes and set membership, then one could write

processes include CUSTOMER.

Since a process has no system variable n.*process* analogous to n.*entity* for a permanent entity, we use M (line 19) to serve this purpose globally for RESERVATIONISTs.

The main routine agrees with that in Figure 4.2, with two notable exceptions. Firstly, the statement in line 18

for I = 1 to M activate a RESERVATIONIST now

causes SIMSCRIPT II.5 to create a *process notice* for each RESERVATIONIST. This process notice contains fields for explicitly declared attributes in the preamble as well as fields to store scheduling information. In the event-scheduling approach, SIMSCRIPT II.5 assigns a future events set ev.s(I) to each event notice where the priority order declared in line 16 of the preamble determines I. SIMSCRIPT II.5 makes a similar assignment for process notices. In particular, I = 1 for RESERVATIONIST process notice and I = 2 for a CUSTOMER process notice.

In addition to allocating space for each RESERVATIONIST process notice, the afore-mentioned "activate" statement indicates in each time.a field that the desired execution time is time.v = 0 and files the notices in ev.s(1) in order of creation. In short, "activate" performs the same task for a process that "schedule" does for an event. The term "reactivate" is synonymous with "activate." Secondly, the statement

activate a CUSTOMER now

in line 19 of "main" causes creation and filing in ev.s(2) of a process notice for CUSTOMER with time.a = 0.

One issue regarding "activate" deserves explicit mention. Whereas the article "a" in the aforementioned "activate" statement causes creation and scheduling, the statement

activate $\genfrac{}{}{0pt}{}{\text{the}}{\text{this}}$ CUSTOMER now

Figure 4.5 SIMSCRIPT II.5 AIR_PR1.sim program: processes

```
''                    AIR_PR1.sim (February 1998)
''
''------------------------------------------------------------------
''                          PREAMBLE
''------------------------------------------------------------------
 1 preamble
 2 normally mode is integer
 3 the system owns a QUEUE and an IDLE.LIST
   ''
 4 processes
 5 every CUSTOMER has a TYPE,  a START and may belong to the QUEUE
 6    define START and WAITING.TIME as double variables
 7    define SERVER as a variable
 8    tally AVE.WAIT as the mean, V.WAIT as the variance and
           MAX.WAIT as the maximum of WAITING.TIME
 9    tally WAIT.HISTO (0 to 8 by .25) as the histogram of WAITING.TIME
10    define ZERO.WAIT as a variable
11 every RESERVATIONIST  has a DEMAND and can belong to the IDLE.LIST
12    define NO.BUSY and NO.RESERVATIONIST as variables
13    accumulate AVE.UTIL as the average,  V.UTIL as the variance and MAX.UTIL
                as the maximum of NO.BUSY
14    accumulate UTIL.HISTO (0 to 10 by 1) as the histogram
                of NO.BUSY
15    accumulate AVE.Q as the average, V.Q as the variance and MAX.Q as the
                maximum of n.QUEUE
16    accumulate Q.HISTO (0 to 32 by 1) as the histogram of n.QUEUE
17 priority order is RESERVATIONIST and CUSTOMER
   ''
18 define LAMBDA, OMEGA and P as double variables
19 define K,M,NUM and T as variables
20 define RRHO to mean LAMBDA*(1.+P)/(M*OMEGA)
21 define ONE  to mean 1
22 define MANY to mean 2
23 define SEEDS as a 1-dim array
24 define KIND  as a function
25 define S_TIME as a double function
26 end
   ''------------------------------------------------------------------
   ''                            MAIN
   ''------------------------------------------------------------------
 1 main
 2 open unit 10 for input
 3 use  unit 10 for input
 4 open unit 15 for output
 5 use unit 15 for output
   ''
 6 read LAMBDA,OMEGA,M and P                          '' INITIALIZE
                                                      ''  PARAMETERS
 7 if RRHO>=1.0      print 1 line thus
        ERROR-----ACTIVITY LEVEL>=1                   '' ERROR CHECK
 8    stop                                            ''  PARAMETERS
10 otherwise
   ''
11 reserve SEEDS(*) as 4
12 for I=1 to 4 do
```

```
13    read seed.v(I)                                    '' ASSIGN SEEDS
14    let SEEDS(I)=seed.v(I)                            ''  SAVE SEEDS
15 loop
16 read T and K
17 let NUM=1
18 for I=1 to M activate a RESERVATIONIST now
19 activate a CUSTOMER now
20 start simulation
21 end
  ''-------------------------------------------------------------
  ''                       process CUSTOMER
  ''-------------------------------------------------------------
1 process CUSTOMER
2 define PATRON as a variable
3 define T.CALL to mean exponential.f(1./LAMBDA,1)
4 let TYPE = KIND
5 let START = time.v
6 let PATRON = CUSTOMER
7 activate a CUSTOMER in T.CALL minutes
8 let CUSTOMER = PATRON
9 if the IDLE.LIST is not empty
10    remove the first RESERVATIONIST from the IDLE.LIST
11    let DEMAND = CUSTOMER
12    activate the RESERVATIONIST now
13    add 1 to ZERO.WAIT
14 else
15    wait .15 minutes
16    if n.IDLE.LIST = 0
17       file CUSTOMER in the QUEUE
18    else
19       remove the first RESERVATIONIST from the IDLE.LIST
20       let DEMAND = CUSTOMER
21       activate the RESERVATIONIST now
22    always
23 always
24 suspend
25 end
  ''-------------------------------------------------------------
  ''                       process RESERVATIONIST
  ''-------------------------------------------------------------
1 process RESERVATIONIST
2 'AAA' if QUEUE is empty
3         file RESERVATIONIST in the IDLE.LIST
4         suspend
5      else
6         remove  the first CUSTOMER from the QUEUE
7         let DEMAND = CUSTOMER
8      always
9         add 1 to NO.BUSY
10        call ACCOUNTING
11        work S_TIME minutes
12        subtract 1 from NO.BUSY
13        activate the CUSTOMER called DEMAND now
14        go to AAA
15    return    end
```

See Fig. 4.1 for KIND, S_TIME, ACCOUNTING, and SUMMARY routines.

causes the scheduling of the reactivation of an already existing process named CUSTOMER. Both the CUSTOMER and RESERVATIONIST statements offer illustrations of the "activate" statement in this context.

Once control passes to the timing routine via the "start simulation" statement in line 20 of "main," the routine scans the first entry in each list to select the next process, that is to continue execution of scheduled events. Since line 16 of the preamble indicates that RESERVATIONIST has priority over CUSTOMER, SIMSCRIPT II.5 first activates a RESER-VATIONIST. Control then passes to the RESERVATIONIST process, where a check of the QUEUE occurs. Since this set is empty at the beginning of the simulation, the RESER-VATIONIST is filed in the IDLE.LIST and then is "suspend"ed. The "suspend" statement deactivates this RESERVATIONIST. In particular, it terminates execution, sets the implic-itly assigned attribute sta.a (RESERVATIONIST) from "active," denoted by 1, to "passive," denoted by 0, and records the address of the next statement as the *reactivation point*. Note that failure to assign the RESERVATIONIST to a set, before invoking "suspend," can result in a loss of that process to the simulation. After the first RESERVATIONIST is suspended, control passes to the timing routine, where execution of the remaining M-1 RESERVATION-ISTs occurs in like manner. After all are suspended, the timing routine finds one remaining process, namely, the CUSTOMER, whose activation was scheduled in the main program.

The first executable statement in the CUSTOMER process (line 4) determines TYPE. Line 5 assigns the call time to START, and line 6 stores the name of the current CUSTOMER in a local variable PATRON. The need for this step is a consequence of the next activate statement, which schedules the occurrence of the next call. Once the activate statement has been executed, the global variable CUSTOMER refers to the *next* CUSTOMER, not the currently active one. However, the statement

let CUSTOMER = PATRON

restores the designation to the currently active CUSTOMER and facilitates implicit subscripting thereafter.

If the condition in line 9 is true, at least one RESERVATIONIST is idle. Then line 10 identifies the first of these, removes it from the IDLE.LIST, and line 12 signals the RESERVA-TIONIST to provide service to the CUSTOMER identified in line 11. Recall that DEMAND is an attribute of RESERVATIONIST. SIMSCRIPT II.5 then executes the "suspend" state-ment in line 24, which deactivates this CUSTOMER process. It is important to recognize that the actual execution of the RESERVATIONIST reactivation occurs after this "suspend" statement is executed.

If the IDLE.LIST is empty in line 9, then the program transfers to and executes line 14. Here, SIMSCRIPT II.5 sets the time.a attribute of the CUSTOMER process notice to time.v $+ 0.15/1440$ days and files the notice in ev.s(2). It also indicates in the notice that the subsequent line 16 is the *reactivation* point. After the wait, the CUSTOMER again checks for the availability of a RESERVATIONIST in line 16. If at least one is idle, SIMSCRIPT II.5 executes lines 19, 20, 21, and 24. If none are idle, it executes lines 17 and 24.

In the CUSTOMER process, line 12 or 21 causes SIMSCRIPT II.5 to set sta.(f.IDLE.LIST) to "active" and time.a (f.IDLE.LIST) to time.v, where f.IDLE.LIST is the address of the first entry in IDLE.LIST. Moreover, line 11 or 20 assigns the address of the executing CUSTOMER to the DEMAND attribute of identified RESERVATIONIST.

When the RESERVATIONIST process begins to execute in line 9, after being suspended in line 4, its DEMAND attribute identifies the CUSTOMER to which service is being provided. Like "wait," the "work" statement in line 11 denotes an unconditional wait of S_TIME minutes. After this time elapses, line 12 adjusts the value of NO.BUSY and line 14 signals the appropriate CUSTOMER that service has been completed. The RESERVATIONIST then transfers to line 2 to determine its next action. It next deactivates at either line 4 or 11. Then the CUSTOMER process, signaled in line 14 of the RESERVAIONIST process, resumes execution at its line 24.

The KIND and S_TIME functions and the ACCOUNTING and SUMMARY subroutines are identical to those in Figure 4.2. Although not used in the airline problem, two additional statements deserve attention at this point. The statement

interrupt CUSTOMER called HARRY

enables a process to deactivate a currently active CUSTOMER process whose address is stored in HARRY. In particular, sta.a(HARRY) is set to "passive," the time remaining until execution of HARRY is to continue is stored in his time.a attribute, and the process notice is removed from ev.s(2), the list of scheduled events corresponding to the CUSTOMER process. A subsequent

resume CUSTOMER called HARRY

enables a process to activate a currently inactive process called HARRY. In particular, it sets time.a(HARRY) ← time.a(HARRY) + time.v, sta.a(HARRY) ← "active" and files the process notice in ev.s(2) for eventual execution. Note the difference between "suspend" and "interrupt." The first allows self-inactivation; the second allows inactivation only of another process.

4.21 USING PROCESSES AND RESOURCES

By exploiting the "resource" feature of SIMSCRIPT II.5, one can create an entirely equivalent simulation of the airline reservation problem. Whereas a "process" behaves as a temporary entity, a "resource" behaves as a permanent entity. Moreover, implicitly executable statements associated with a "resource" can lead to reduced coding in some modeling simulations. Figure 4.6 describes a SIMSCRIPT II.5 program that incorporates a "process" and a "resource," CUSTOMER and RESERVATIONIST, respectively. This program again employs the same KIND and S_TIME functions as in Figure 4.2 but, because of a difference in implicitly created system variables, requires modified ACCOUNTING and SUMMARY subroutines.

Figure 4.6   SIMSCRIPT II.5 AIR_PR2.sim program: processes and resources

```
''                    AIR_PR2.sim  (February 1998)
''
''----------------------------------------------------------------
''                           PREAMBLE
''----------------------------------------------------------------
 1 preamble
 2 normally mode is integer
 3 processes
 4    every CUSTOMER has a TYPE and a START
 5       define START and WAITING.TIME as double variables
 6       tally AVE.WAIT as the mean, V.WAIT as the variance and
 7            MAX.WAIT as the maximum of WAITING.TIME
 8       tally WAIT.HISTO (0 to 8 by .25) as the histogram of WAITING.TIME
 9       define ZERO.WAIT as a variable
 ''
10 resources include RESERVATIONIST
11    accumulate AVE.UTIL as the average,  V.UTIL as the variance and MAX.UTIL
                        as the maximum of n.x.RESERVATIONIST
12    accumulate UTIL.HISTO (0 to 10 by 1) as the histogram
                        of n.x.RESERVATIONIST
13    define  NO.RESERVATIONIST as  a variable
14    accumulate AVE.Q as the average, V.Q as the variance and MAX.Q as the
             maximum of n.q.RESERVATIONIST
15    accumulate Q.HISTO (0 to 32 by 1) as the histogram of n.q.RESERVATIONIST
 ''
16 define LAMBDA,OMEGA  and P as double variables
17 define K,M,NUM and T as variables
 ''
18 define RRHO to mean LAMBDA*(1.+P)/(M*OMEGA)
19 define ONE  to mean 1
20 define MANY to mean 2
 ''
21 define SEEDS as a 1-dim array
22 define KIND  as a function
23 define S_TIME as a double function
24 end


    ''----------------------------------------------------------------
    ''                           MAIN
    ''----------------------------------------------------------------
 1 main
 2 open unit 10 for input
 3 use  unit 10 for input
 4 open unit 15 for output
 5 use  unit 15 for output
 ''
 6 let n.RESERVATIONIST=1
 7 create each RESERVATIONIST
 8 read LAMBDA,OMEGA,M and P                            '' INITIALIZE
 9 let u.RESERVATIONIST(1)=M
```

```
10 if RRHO>=1.0        print 1 line thus
        ERROR-----ACTIVITY LEVEL>=1                    '' ERROR CHECK
11   stop                                              '' PARAMETERS
12 otherwise
13 reserve SEEDS(*) as 4
14 for I=1 to 4 do
15   read seed.v(I)                                    '' ASSIGN SEEDS
16   let SEEDS(I)=seed.v(I)                            '' SAVE SEEDS
17 loop
18 read T and K
19 let NUM=1
20 activate a CUSTOMER now
21 start simulation
22 end


''----------------------------------------------------------------
''                         process CUSTOMER
''----------------------------------------------------------------
1 process CUSTOMER
2 define PATRON as a variable
3 define T.CALL to mean exponential.f(1./LAMBDA,1)
4 let TYPE=KIND
5 let START=time.v
6 let PATRON=CUSTOMER
7 activate a CUSTOMER in T.CALL minutes
8 let CUSTOMER=PATRON
9 if u.RESERVATIONIST(1) > 0
10   request 1 unit of RESERVATIONIST(1)
11   add 1 to ZERO.WAIT
12 otherwise
13   wait .15 minutes
14   request 1 unit of RESERVATIONIST(1)
15 always
16 call ACCOUNTING
17 work S_TIME minutes
18 relinquish 1 unit of RESERVATIONIST(1)
19 return   end


''----------------------------------------------------------------
''                         routine ACCOUNTING
''----------------------------------------------------------------
```

Replace NO.BUSY by n.x.RESERVATIONIST(1) and n.QUEUE by
n.q.RESERVATIONIST(1) in ACCOUNTING routine in Fig. 4.1.

```
''----------------------------------------------------------------
''                         routine SUMMARY
''----------------------------------------------------------------
```

Replace Q.HISTO(1) by Q.HISTO(1,1), Q(HISTO(I) by Q.HISTO(1,I),
Q.HISTO(33) by Q.HISTO(33,I), and UTIL.HISTO(I) by
UTIL.HISTO(1,I) in SUMMARY routine in Fig. 4.1.

See Fig. 4.1 for KIND and S_TIME routines.

The preamble now has a new key word "resources" in line 10. It prefaces the declaration of the list of resources or servers to be used in the program when all have attributes. Since in the present case we elect to define RESERVATIONIST without explicit attributes, the statement

resources include RESERVATIONIST

suffices. The preamble also contains modifications of the "accumulate" and "tally" statements. The need for these modifications becomes apparent to the reader as we describe the effects that using "resources" induces in the main routine and CUSTOMER process.

Although similar to those in Figures 4.2 and 4.5, the main routine differs in several important ways. In particular, n.RESERVATIONIST in line 6 of "main" refers to the number of different *classes* of RESERVATIONISTs here, not to the number of reservationists (see Section 4.3). This is a consequence of defining RESERVATIONIST as a resource instead of as a process. The implicitly generated variable u.RESERVATIONIST(1) denotes the number of *available* RESERVATIONISTs in class 1 and requires assignment of a value. Note that data structures are created for each resource class (line 7), not for each resource. A comparison of the CUSTOMER processes in Figures 4.5 and 4.6 indicates notable differences. In particular, in Figure 4.6 the system variable u.RESERVATIONIST(1) in line 9 of the CUSTOMER process enables a CUSTOMER to determine whether a RESERVATIONIST is available. If one is, the CUSTOMER executes the "request" statement in line 10. If one is not, the CUSTOMER executes the "wait" statement in line 13 and upon reactivation executes the "request" statement in line 14. The "request" and "relinquish" statements are powerful preprogrammed bodies of code designed to mimic the steps commonly encountered in demanding resources in a queueing simulation.

If a RESERVATIONIST is idle, the CUSTOMER executes the "request" statement in line 10. If not, the CUSTOMER executes the "wait" in line 13 and upon reactivation executes the "request" in line 14. The "request" statement invokes a SIMSCRIPT II.5 system routine designed to mimic the two alternative types of action that arise in a queueing simulation when resources are demanded. To understand the options, a knowledge of SIMSCRIPT II.5's implicit behavior is necessary.

Whenever a preamble declares a "resource," SIMSCRIPT II.5 creates two associated lists, q.*resource* and x.*resource*. At execution of a "request" statement SIMSCRIPT II.5 creates a record of a special temporary entity, qc.e, whose attributes include who.a, assigns the address of the requesting process to who.a(qc.e), and files it in an implicitly created list or set of resources owned by the requesting process. If u.*resource* is greater than or equal to the number of requested units of the resource, the qc.e record is also filed in x.*resource*. Otherwise, it is filed in q.*resource* to await service. If the desired number of units of the resource are available, the process continues execution. For example, "work S_TIME minutes" would be executed in line 17 of the CUSTOMER process. If the desired number of units of resource are unavailable, execution ceases and control passes to the timing routine with the next executable statement as reactivation point. For the airline reservation problem, the "request" statement incorporates blocks 8, 9, 10, and 11 in Figure 2.10.

The statement

relinquish 1 unit of RESERVATIONIST(1)

is unusually powerful in content, essentially corresponding to the logic in Figure 2.11. When executed in line 18 of the CUSTOMER process, SIMSCRIPT II.5 removes the temporary entity record qc.e from x.RESERVATIONIST(1) and from the list of resources owned by CUSTOMER who.a(qc.e) and files the CUSTOMER process notice for who.a(qc.e) in the corresponding list of scheduled processes with its time.a(who.a(qc.e)) = time.v, which is "now." Next, it destroys the qc.e record and checks to see whether q.RESERVATIONIST(1) is empty. If it is, then u.RESERVATIONIST(1) is augmented by 1. Otherwise, it removes the first qc.e from q.RESERVATIONIST(1), files it in x.RESERVATIONIST(1), and files the process notice corresponding to who.a(qc.e) in the single future event set with time.a(who.a) = time.v. In the present example, ev.s(1), the list of scheduled CUSTOMER processes, is the future-event set.

The programming convenience that the "request" and "relinquish" statements offer is self evident. They reduce programming effort by replacing a sequence of more elemental statements, as in Figure 4.5, that collectively would effect the same modeling logic. This saving in effort occurs k times in a simulation with k stations in series; it can become substantial as k increases. Every simulation that employs the process-interaction approach contains macrostatements analogous to "request" and "relinquish."

In exchange for this reduction in effort, a simulation user gives up explicit control over some parts of program flow. Although not necessarily detrimental, this loss of flexibility needs to be understood explicitly. For example, the steps SIMSCRIPT II.5 takes are not apparent from the program in Figure 4.6. In particular, note that the RESERVATIONIST automatically becomes committed when it is released if another process is waiting. This action can cause a problem in other simulations. For example, consider the tanker problem in Exercise 4.7, where a TANKER (process) may arrive in the harbor while the TUG (resource) is towing out another TANKER and yet another TANKER has completed loading and has requested the TUG. Here, the TANKER in the harbor would have to

request 1 unit of TUG(1) with priority A ,

where A is an integer larger than the priority with which the TANKER at the berth requested the TUG. Although this would work here, one needs to avoid this statement if the TANKER arrives while the TUG is heading in to the berths, since it would lead to an order of servicing contrary to that intended.

At least one other feature deserves attention. It is entirely possible to employ events, processes, and resources in the same SIMSCRIPT II.5 simulation. SIMAN and SLAM II also provide this option, which can introduce considerable flexibility in modeling a system with-

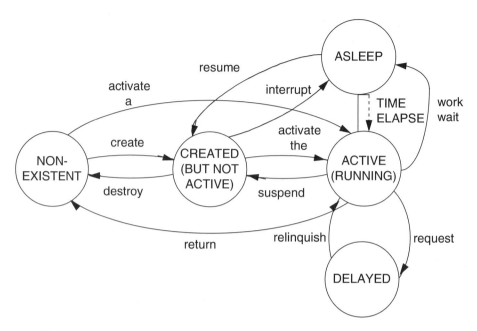

Figure 4.7 SIMSCRIPT II.5 process-interaction approach

out the restriction of adopting either the event-scheduling or process-interaction approach exclusively.

Figure 4.7 illustrates the effect of each SIMSCRIPT II.5 statement in using processes and resources. In particular, a process comes into existence in one of two ways. The "activate" statement creates a process and schedules the time of its first execution. Since the "create" statement only creates the process, at some later time an "activate" statement is needed to schedule its first execution. The "suspend" statement inactivates the process in which the statement occurs, but it remains extant. The statement "work" or "wait" simply reschedules continued execution for a later time. If a process is "interrupt"ed, its scheduled execution is canceled, and it becomes inactive. The "resume" statement restores the scheduled execution of the process. A "request" statement may lead to a delay. If so, the occurrence of a "relinquish" statement for the desired resources in another process causes SIMSCRIPT II.5 to schedule continued execution of the requesting process at the current time. The "return" statement obliterates an executing process from the simulation. The "destroy" statement does likewise for an inactive process.

4.22 COMMON PSEUDORANDOM NUMBERS

Recall the discussion of Section 4.7 regarding the seeding of pseudorandom number streams. Suppose we assign individual streams to each of r sources of variation and for each $i =$

$1, \ldots, r$ assign the same seed to stream i on each of two runs of a simulation program each with different values for an input parameter that does not affect sample generation. Then sample averages on each run are consequences of the same random input sequences. This implies that the difference between corresponding sample averages on the two runs arises principally, if not entirely, from the difference in the parameter's values. An example based on waiting time for the airline reservation problem illustrates the benefit that can arise from this property.

It is of interest to evaluate the reduction in waiting time that materializes when the number of reservationists increases from $m = 4$ to $m = 5$. A simulation for $m = 4$ generated a sample record of waiting times $X_k, X_{k+1}, \ldots, X_{k+t-1}$, and a simulation with $m = 5$ using the exact same seeds generated a sample record of waiting times $Y_k, \ldots, Y_{k+t-1}$, where the choice of k ensured that the influence of initial conditions was negligible (Section 6.1) and where $t = 100,000$.

The sample waiting times are

$$\bar{X}_{kt} = \frac{1}{t} \sum_{j=k}^{k+t-1} X_j = 4.072 \text{ minutes}$$

and

$$\bar{Y}_{kt} = \frac{1}{t} \sum_{j=k}^{k+t-1} Y_j. = 0.7347 \text{ minutes.}$$

Their difference is

$$\bar{Z}_{kt} = \frac{1}{t} \sum_{j=k}^{k+t-1} Z_j = \bar{X}_{kt} - \bar{Y}_{kt} = 3.337 \text{ minutes,}$$

where $Z_j := X_j - Y_j$. Under commonly encountered conditions (Section 3.5),

$$t \operatorname{var} \bar{X}_{kt} \to \sigma_\infty^2(X),$$
$$t \operatorname{var} \bar{Y}_{kt} \to \sigma_\infty^2(Y),$$
$$t \operatorname{var} \bar{Z}_{kt} \to \sigma_\infty^2(Z), \qquad \text{as } t \to \infty,$$

so that $\bar{X}_{kt}$, $\bar{Y}_{kt}$, and $\bar{Z}_{kt}$ have approximate large-sample variances $\sigma_\infty^2(X)/t$, $\sigma_\infty^2(Y)/t$, and $\sigma_\infty^2(Z)/t$, respectively. With LABATCH.2, as described in Sections 6.6 through 6.7.5, generated

$$\hat{\sigma}_{\infty}^{2}(X) = 3429,$$
$$\hat{\sigma}_{\infty}^{2}(Y) = 60.92,$$
$$\hat{\sigma}_{\infty}^{2}(Z) = 2847,$$

as estimates of $\sigma_{\infty}^{2}(X)$, $\sigma_{\infty}^{2}(Y)$, and $\sigma_{\infty}^{2}(Z)$, respectively.

If $\{X_i\}$ and $\{Y_j\}$ were independent, due to seeds that ensured nonoverlapping pseudorandom number streams on each run, then $\bar{Z}_{kt}$ would have asymptotic variance $[\sigma_{\infty}^{2}(X) + \sigma_{\infty}^{2}(Y)]/t$ estimated by $[\hat{\sigma}_{\infty}^{2}(X) + \hat{\sigma}_{\infty}^{2}(Y)]/t = 3490 \times 10^{-5}$. However, the estimated asymptotic variance, obtained by analyzing the sample record $Z_k, \ldots, Z_{k+t-1}$, is $\hat{\sigma}_{\infty}^{2}(Z)/t = 2847 \times 10^{-5}$. This difference implies that common random numbers require only

$$\frac{\hat{\sigma}_{\infty}^{2}(Z)}{\hat{\sigma}_{\infty}^{2}(X) + \hat{\sigma}_{\infty}^{2}(Y)} = 0.8158$$

as many observations to achieve the same variance for the sample average difference $\bar{Z}_{kt}$ as independent sample records. Moreover, this result occurs with no additional computing time per run.

Estimating the difference in the proportion of customers who wait reveals an even greater benefit of common random numbers. Let

$$X_j := \begin{cases} 1 & \text{if customer } j \text{ on run 1 waits,} \\ 0 & \text{otherwise,} \end{cases}$$

and

$$Y_j := \begin{cases} 1 & \text{if customer } j \text{ on run 2 waits,} \\ 0 & \text{otherwise.} \end{cases}$$

Then $\bar{X}_{kt}$ and $\bar{Y}_{kt}$ estimate the long-run proportion of customers that wait for $m = 4$ and $m = 5$ reservationists, respectively. The same sample records yielded

$$\bar{X}_{kt} = 0.7345, \qquad \bar{Y}_{kt} = 0.3748, \qquad \bar{Z}_{kt} = 0.3597$$

and

$$\hat{\sigma}_{\infty}^{2}(X) = 5.392, \qquad \hat{\sigma}_{\infty}^{2}(Y) = 2.624, \qquad \hat{\sigma}_{\infty}^{2}(Z) = 1.316.$$

Therefore, common random numbers require only

$$\frac{\hat{\sigma}_\infty^2(Z)}{\hat{\sigma}_\infty^2(X) + \hat{\sigma}_\infty^2(Y)} = 0.1642$$

more observations as independent sample records to achieve the same variance for the sample average difference $\bar{Z}_{kt}$.

Why do the savings in sample sizes vary so substantially for the difference of means and the difference of proportions? Recall that

$$\begin{aligned}
\text{var} \bar{Z}_{kt} &= \text{var}(\bar{X}_{kt} - \bar{Y}_{kt}) \\
&= \text{var} \bar{X}_{kt} - 2 \, \text{cov}(\bar{X}_{kt}, \bar{Y}_{kt}) + \text{var} \bar{Y}_{kt} \\
&\approx \sigma_\infty^2(Z)/t \\
&= [\sigma_\infty^2(X) - 2\sigma_\infty(X, Y) + \sigma_\infty^2(Y)]/t,
\end{aligned}$$

where

$$\sigma_\infty(X, Y) := \lim_{t \to \infty} t \, \text{cov}(\bar{X}_{kt}, \bar{Y}_{kt}).$$

Therefore,

$$\frac{\sigma_\infty^2(Z)}{\sigma_\infty^2(X) + \sigma_\infty^2(Y)} = 1 - \frac{2\rho(X, Y)}{\frac{\sigma_\infty(X)}{\sigma_\infty(Y)} + \frac{\sigma_\infty(Y)}{\sigma_\infty(X)}},$$

where

$$\rho(X, Y) := \text{corr}\,(\bar{X}_{kt}, \bar{Y}_{kt}) \approx \frac{\sigma_\infty(X, Y)}{\sigma_\infty(X)\sigma_\infty(Y)}.$$

Thus *variance reduction* is a function both of the correlation between sample averages and of the ratio $\sigma_\infty(X)/\sigma_\infty(Y)$. Low correlation diminishes the potential variance reduction, as does a large or small ratio $\sigma_\infty(X)/\sigma_\infty(Y)$.

For sample average waiting times,

$$\hat{\rho}(X, Y) = 0.7033$$

and

$$\frac{\hat{\sigma}_\infty(X)}{\hat{\sigma}_\infty(Y)} = 7.502.$$

For sample proportions,

$$\hat{\rho}(X, Y) = 0.8906$$

and

$$\frac{\hat{\sigma}_\infty(X)}{\hat{\sigma}_\infty(Y)} = 1.433.$$

The higher correlation and smaller relative difference between $\hat{\sigma}_\infty(X)$ and $\hat{\sigma}_\infty(Y)$ account for the greater benefit of common pseudorandom numbers for sample proportions than for sample average waiting times.

To realize these variance reductions calls for more than common pseudorandom numbers on each run. It requires that a separate pseudorandom number stream be dedicated to each source of random variation and that the seed for stream i be identical across runs. Most major simulation languages provide access to multiple pseudorandom number streams and allow user assignment of their seeds.

At least two additional issues deserve attention. The seeds for the streams must be chosen so that the sequences they generate do not overlap; otherwise, the sequence would be dependent, contrary to intention. The second issue concerns the number of pseudorandom numbers that stream i generates on each run and their use. The airline reservation simulation executed until exactly $k + t$ customers entered service. The structure of the problem ensured that each stream generated exactly the same number of pseudorandom numbers on the run for $m = 4$ and on the run for $m = 5$, and that each number was used for exactly the same purpose on both runs.

Common pseudorandom number streams may also induce a variance reduction when stochastic variation differs on the two runs. Suppose that the purpose for simulating the airline reservation model is to study the effect on waiting time of changing p from .75 to .90. If common pseudorandom number seeds are used, then both runs would have the same sequence of intercall times but different sequences of service times. Let

$$A_i := i\text{th exponential sample generated by stream 2,}$$
$$B_i := i\text{th exponential sample generated by stream 3,}$$
$$U_i := i\text{th uniform sample generated by stream 4.}$$

For specified p, the simulation then generates a sequence of service times

$$C_i(j) := A_i + I_{[0,p]}(U_i)B_{J_i(p)},$$

where

$$J_i(p) := \sum_{l=1}^{i} I_{[0,p]}(U_l), \qquad i \geq 1,$$

and $B_0 := 0$. Therefore, the difference between the ith service times on runs with $p = .75$ and $p = .90$ is

$$C_i(.75) - C_i(.90) = [B_{J_i(.75)} - B_{J_i(.90)}]I_{[0,.75]}(U_i) - B_{J_i(.90)}I_{(.75,.90]}(U_i),$$

which rarely is zero.

Although corresponding service times differ, the commonality of the intercall time sequence can induce a reduced variance for the difference of corresponding sample averages. As illustration for waiting time, two runs with $m = 4$ and $t = 100,000$ yielded $\bar{X}_{kt} = .7345$ as an estimate of $pr(W > 0 \mid p = .75)$ and $\bar{Y}_{kt} = .8851$ as an estimate of $pr(W > 0 \mid p = .90)$, thus giving $\bar{Z}_{kt} = \bar{X}_{kt} - \bar{Y}_{kt} = -.1506$ as an estimate of $pr(W > 0 \mid p = .75) - pr(W > 0 \mid p = .90)$. Moreover, LABATCH.2 (Section 6.7) gave

$$\hat{\sigma}_\infty^2(X) = 5.392, \quad \hat{\sigma}_\infty^2(Y) = 4.836, \quad \text{and} \quad \hat{\sigma}_\infty^2(Z) = 2.152.$$

Thus for each run, a sample size

$$\frac{\hat{\sigma}_\infty^2(X) + \hat{\sigma}_\infty^2(Y)}{\hat{\sigma}_\infty^2(Z)} = 4.752$$

times as large would be required to estimate this difference with the same estimated accuracy.

In the present case, relatively small changes in the S_TIME code in Figure 4.2 can make the two sequences of service time more similar. In particular:

Between lines 1 and 2, insert:

   define DUM1 and DUM2 as double variables

Between lines 4 and 5, insert:

   DUM1 = exponential.f(1/OMEGA,2)

   DUM2 = exponential.f(1/OMEGA,3)

   return with DUM1

Delete line 5.

These changes induce service times

$$C_i'(p) = A_i + B_i I_{[0,p]}(U_i), \qquad i \geq 0,$$

so that

$$
C_i'(.90) = \begin{cases} C_i'(.75) & \text{if } 0 \le U_i \le .75 \text{ or } .90 < U_i \le 1, \\ C_i'(.75) + B_i & \text{if } .75 < U_i \le .90. \end{cases}
$$

That is, only $100 \times (.90 - .75) = 15$ percent of the service times differ. Simulations run with these changes produced $\bar{X} = 0.7292$ for $p = 0.75$, $\bar{Y}_{kt} = 0.8806$, $\bar{X}_{kt} - \bar{Y}_{kt} = -0.1514$, $\hat{\sigma}_\infty^2(X) = 6.028$, $\hat{\sigma}_\infty^2(Y) = 5.789$, and $\hat{\sigma}_\infty^2(Z) = 1.590$, so that

$$
\frac{\hat{\sigma}_\infty^2(X) + \hat{\sigma}_\infty^2(Y)}{\hat{\sigma}_\infty^2(Z)} = 7.426,
$$

revealing a substantial variance reduction.

This improvement comes at a cost, although a relatively small one. A run using the original code generate pt times on average from stream 3. In the present case, the two runs generate $(.75 + .90)t$ exponential times on average. With the modified code, they generate $2t$ exponential times, $2/1.65 = 1.21$ times as many as previously required. Although this result implies higher simulation computing times with the modified code, this increase is relatively negligible when compared to the saving in sample size it induces to achieve a specified accuracy.

However, there is a more important consideration. Whereas LABATCH.2 (Section 6.7) computes $\hat{\sigma}_\infty^2(X)$ and $\hat{\sigma}_\infty^2(Y)$ without storing sample records generated on each run, computing $\hat{\sigma}_\infty^2(Z)$ requires storage of at least one of these records, as discussed in Section 6.11, to compute subsample averages based on these data. Therefore, assessing the statistical accuracy of $\bar{X}_{kt} - \bar{Y}_{kt}$ as an estimate of the true differences requires additional computer space and additional time to compute $\hat{\sigma}_\infty^2(Z)$. Section 6.11 returns to this topic.

4.22.1   OTHER VARIANCE-REDUCING METHODS

These examples illustrate a relatively easily implementable procedure to reduce the computing time required to obtain a difference in sample averages within a specified statistical accuracy. For its effectiveness, the method of common pseudorandom number seeds requires that a separate pseudorandom number stream feed each source of stochastic variation. To remain synchronized during the two runs, the method may also require that variates be generated but not used, as in the case for $m = 4$ and $p = .75$ versus $m = 4$ and $p = .90$. For a more extended discussion of the use of common pseudorandom number streams and its relationship to other variance-reducing techniques, see Sections 7.7 through 7.11.

Other methods exist for reducing variance, including *antithetic variates, control variates, importance sampling*, and *conditional sampling*. See Fishman (1996) and Law and Kelton (1991) for example. Although scenarios arise in which each of these can improve computational efficiency, all, with the exception of antithetic variates, demand substantially more effort than common pseudorandom number seeds to implement and to evaluate. This demand lowers their appeal in practice for simulators who prefer more easily implementable online methods.

4.23 POINT-CLICK-DRAG-DROP-CONNECT APPROACH

Some languages provide a screen-based block modeling approach to constructing a discrete-event simulation program. It relieves the simulationist of the onus of entering code via a textual editor, thereby reducing the frequency of typing errors and shortening the time it takes to move from concept to execution.

As a consequence of its emphasis on flow through a system, virtually every screen-based block-modeling approach embraces the process-interaction paradigm (Section 2.4). The concept originated with the GPSS simulation language (Gordon 1969). It allows a simulationist to string together macrostatements called *blocks* that correspond to the transformations of a temporary entity from its entry into a system through its completed processing and departure.

Although discrete-event simulation in the 1960s era was performed in a mainframe computing environment that allowed, at most, a primitive user/computer interaction during program development, the ease of program construction using blocks gave this approach considerable appeal. The advent of the PC with its emphasis on interactive programming and computing substantially enhanced block-oriented programming. Virtually all commercially available simulation software offers this mode of program construction.

Enhanced interactive capability has led to further software development. By packaging frequently occurring sequences of blocks into macromodules, vendors can offer a simulationist a menu of these modules that, in principle, he/she can string together in a logical sequence to produce a simulation program with little or no knowledge on the user's part of the underlying programming language's constructs. In practice, a blend of this approach and block-oriented programming becomes necessary as the logical complexity of the systems under study grows.

As illustration, this and the next several sections describe the point-click-drag-drop-connect approach to simulation programming using the Arena modeling system (Kelton et al. 1998; Systems Modeling Corporation 1995a, 1995b, 1995c) based on the SIMAN simulation programming language (Pegden et al. 1995). Arena allows simulationists with varying levels of familiarity with SIMAN to construct models and execute them. It does so by providing a collection of panels and allowing simulationists to construct models using entries in them. At the language level, the *Block* and *Elements panels* contain the SIMAN statements. The *Basic Process, Advanced Process*, and *Advanced Transfer panels* contain aggregates of these statements that correspond to frequently encountered discrete-event constructs. Hereafter, we refer to them collectively as rne BAAT panels. The BAAT three panels allow a modeler to create flowcharts using relatively standardized block representations on the screen. Like

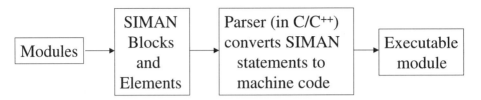

Figure 4.8   Arena discrete-event simulation program development

the previous three, the *Common, Support,* and *Transfer Process panels,* contain aggregates of SIMAN statements, again for frequently encountered constructs. Their representations on the screen attempt to identify them with their principal underlying activities. Hereafter, we refer to these three as the CST panels. Arena 4.0 contains these eight panels. Arena 3.51 contains the first two and the last three.

In principle, pointing, clicking, dragging, dropping, and connecting modules from, for example, the Common panel and dropping them into a model window allows a user with little to no familiarity with SIMAN to create a simulation program. Analogous operations applied to the Blocks and Elements panels allow a more knowledgeable user of SIMAN to construct a simulation program directly from its basic blocks or macromodules. Arena also offers the user the option to program directly in SIMAN via a textual editor. Figure 4.8 depicts program creation in the Arena environment.

As with all other simulation languages, SIMAN embraces the elementary concepts of entity, attribute, event, list, process, and future-event set described in the introduction to Section 2. It includes a grammar that caters to the process-interaction approach. Although it is not possible to create a SIMAN program that embraces the event-scheduling approach of Section 2.3, Section 4.28 describes SIMAN's EVENT block, which allows a simulationist to introduce elements of that approach within the framework of process interaction. The SIMAN literature often refers to event scheduling as event orientation and to process interaction as process orientation. Its event-scheduling capabilities can be traced to early work of Kiviat at U.S. Steel, who developed the GASP discrete-event simulation language (Pritsker and Kiviat 1969). Its process-interaction features can be traced back to GPSS (Gordon 1969).

Arena embraces SIMAN's process-interaction concepts as the basis for its modeling environment. SIMAN's time-advance protocol follows the next-event approach (Section 2.1). It maintains its future-event set (FES) in a single list as a heap (Section 5.4.3). Its pseudorandom number generator has the form (4.13) with $A = 16807$ and $M = 2^{31} - 1$, and it allows for multiple streams each with its own seed (Section 4.7). The seeds may be specified by the user. Otherwise, SIMAN employs its default seeds. Section 9.4 describes the properties of this generator. For reference, see Pegden et al. (1995).

As mentioned, the BAAT panels are available with Arena 4.0. Using them produces spreadsheets below the model display for entities, queues, resources, etc. The spreadsheets facilitate program construction. However, there are qualifications to using these and the CST panels. In a particular problem specification may exceed the capacity of these panels to model the problem completely. When this is so, the simulationist may augment the aggregates taken from these with statements from the Block and Elements panels. The "WHILE" statement illustrates one area in which this merging can be beneficial.

In addition to the modules provided by these panels, Arena's Professional Edition allows a user to create her/his own modules, out of SIMAN blocks, and collect them into a *template* (panel). Arena provides a capability for dynamic animation as simulated time elapses during execution and a means for postsimulation graphical display of sample paths, histograms and other selected concepts. Arena also allows for the incorporation of user-written code in Visual Basic, and C/C^{++}. However, the burden for this incorporation falls on the simulationist.

The description here is based on Arena Version 4.0. We illustrate Arena first using the high-level Common panel and then, at the other extreme, the Block and Elements panels. We then describe several features of the BAAT panels. In conformity with SIMAN reference material, we capitalize all SIMAN block, element, and module names. To distinguish between SIMAN code and user-specfied variables, etc., we capitalize only the first letters of the latter.

4.24 M/M/1 ARENA EXAMPLE

We first illustrate Arena program construction exclusively by means of the Common panel, some of whose modules are depicted in Figure 4.9. Consider a single-server queueing system with i.i.d. exponentially distributed interarrival times with mean 1.1111 and i.i.d. exponentially distributed service times with unit mean. Jobs receive service in first-come-first-served order. There are five types of state changes or elementary events (Section 2.3):

1. Arrival   4. Server becomes idle
2. Job begins service   5. Server becomes busy.
3. Job departs

These can easily be incorporated into a single process as in Section 2.4. In the Common panel, the ARRIVE module serves as arrival generator, the SERVER module as work processor, and the DEPART module as disposer of jobs after they complete service. Figure 4.10 shows a model constructed by pointing and clicking on these objects, one at a time, in the Common panel and dragging and dropping them in an opened model window labeled mm1.doe.

Double clicking with the left mouse key on the icon for the dropped ARRIVE module opens a dialog box that provides space for time of first arrival number of jobs per arrival and time between arrivals. The defaults for the first two are 0 and 1, respectively, and we allow those to prevail. Clicking on the time between arrivals brings up a list of distributional options for sampling interarrival times. The menu there lists 13 distributions, including two empirically based options, one for continuous random variables and the other for discrete random variables. Clicking on Expo(mean) and replacing "mean" by 1.1111,1 specifies the exponential distribution with mean 1.1111 using pseudorandom number stream 1.

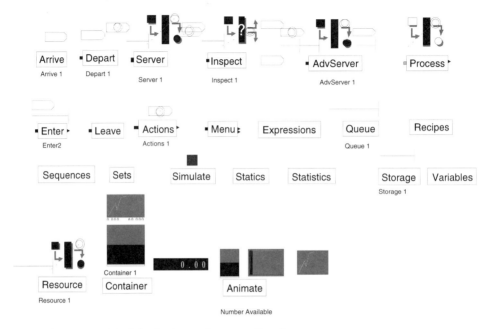

Figure 4.9   Modules in Arena Common Panel

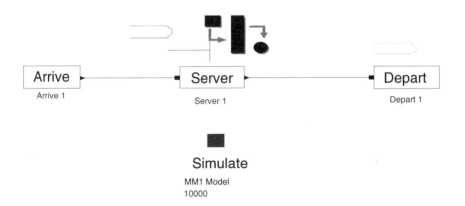

Figure 4.10   Arena/SIMAN mml.doe model

Clicking on the "connect" option in the dialog box sets the stage for linking the ARRIVAL and SERVER modules. Bringing down the "object" menu, clicking on "connect," and clicking once on the exit node of the ARRIVAL module and then once on the entry node to the SERVER module connects them. Similar steps for the SERVER module with mean replaced by 1,2 specifies exponential service times with unit mean using pseudorandom number stream 2. Connecting the SERVER and DEPART modules is accomplished by first

```
;
;
;   Model statements for module: Arrive 1
;                                                   PROJECT,    MM1 Model;

42$     CREATE,     1:expo(1.1111,1);               ATTRIBUTES:  QueueTime;

3$      STATION,    Arrive 1;                        QUEUES:     Server 1_R_Q,FIFO;
51$     TRACE,      -1,"-Arrived to system at station Arrive 1\n":;
6$      ASSIGN:     Picture=Default;                 PICTURES:    Default;
27$     DELAY:      0.;
55$     TRACE,      -1,"-Transferred to next module\n"::NEXT(0$);   RESOURCES:  Server 1_R,Capacity(1,),-,Stationary;

                                                    STATIONS:   Arrive 1:
;                                                               Depart 1:
;                                                               Server 1;
;   Model statements for module: Server 1
;                                                   TALLIES:    Server 1_R_Q Queue Time;

0$      STATION,    Server 1;                        DSTATS:     NQ(Server 1_R_Q),# in Server 1_R_Q:
158$    TRACE,      -1,"-Arrived to station Server 1\n":;                    MR(Server 1_R),Server 1_R Available:
121$    DELAY:      0.;
165$    TRACE,      -1,"-Waiting for resource Server 1_R\n":;
82$     QUEUE,      Server 1_R_Q:MARK(QueueTime);                        NR(Server 1_R),Server 1_R Busy;
83$     SEIZE,      1:
                    Server 1_R,1;                    REPLICATE,  1,0.0,10000,Yes,Yes,1000;
192$    BRANCH,     1:
                    If,RTYP(Server 1_R).eq.2,193$,Yes:
                    If,RTYP(Server 1_R).eq.1,95$,Yes;
193$    MOVE:       Server 1_R,Server 1;
95$     TALLY:      Server 1_R_Q Queue Time,INT(QueueTime),1;
202$    DELAY:      0.0;
        TRACE,      -1,"-Delay for processing time EXPO(1,2)\n":;
84$     DELAY:      EXPO(1,2);
166$    TRACE,      -1,"-Releasing resource\n":;
85$     RELEASE:    Server 1_R,1;
149$    DELAY:      0.,
172$    TRACE,      -1,"-Transferred to next module\n"::NEXT(2$);


;
;
;   Model statements for module: Depart 1
;

2$      STATION,    Depart 1;
233$    TRACE,      -1,"-Arrived to station Depart 1\n":;
203$    DELAY:      0.;
240$    TRACE,      -1,"-Disposing entity\n":;
232$    DISPOSE;
```

    (a) MOD file   (b) EXP file

Figure 4.11   Arena/SIMAN mm1.mod and mm1.exp files

clicking on the "connect" option in the dialog box and following steps analogous to those used previously.

A SIMULATE module dragged from the Common panel and dropped in the open mm1.doe window creates the basis for executing the model. Double clicking on it brings up its dialog box with entries for "title," "number of replications," "beginning time," and "length of replication." In the present setting, the last three of these are set to 1, 1000, and 10,000.

Clicking on "SIMAN" and then on "View" in the Run menu displays the corresponding SIMAN code. Figure 4.11 shows it partitioned into a *MOD file*, or *Model file* (left side), and an *EXP file*, or *Experiment Source file* (right side). The MOD file contains the executable SIMAN code whereas the EXP file contains all nonexecutable global declarations plus a list of final results to be tabulated. In the present case, Arena has automatically included these output quantities in TALLIES and DSTATS statements. TALLIES relates to output quantities computed from discrete-indexed sample-path data, for example, waiting time. SIMAN refers to these as *discrete-valued variables*. DSTATS relates to output computed from continuous-time indexed sample-path data; for example, queue length. SIMAN refers to these as *time-persistent variables*.

Of particular interest is the partition of the MOD file into three sets of SIMAN blocks that correspond to the ARRIVE, SERVER, and DEPART modules. The reader quickly appreciates the benefit that point and click from the Common panel offers.

Clicking on "Go" in the "Run" menu executes the program with animated activity in the mm1.doe window. Alternatively, clicking on "Fast Forward" suppresses the animation and induces considerably faster execution. When execution terminates, Arena asks the user whether he/she wishes to view its automatically generated output summary. We return to this topic in Section 4.25, where we examine the output for the airline reservation program.

Note that Arena has relieved the simulationist of the need to compile, link, and execute the program and, by default, has provided for the computation of summary statistics. It also uses default seeds for the two pseudorandom number streams.

What does the SIMAN code in the MOD and EXP files tell us? As a basis for answering this question, Section 4.24.1 describes a collection of SIMAN Block and Element statements and then uses the latter to explain the contents of the EXP file in Figure 4.11b. Then Section 4.25, which studies the Arena/SIMAN version of the airline reservation problem, use the Block elements to explain the contents of the MOD file for that model.

4.24.1 SIMAN STATEMENTS

SIMAN contains executable statements, or blocks, and nonexecutable global statements, or elements. Those in Figures 4.11 and 4.13 have the following generic forms:

BLOCKS

ASSIGN: $\underbrace{\textit{variable} \text{ or } \textit{attribute name} = \textit{expression} \text{ or } \textit{numerical value:}}_{\text{repeat;}}$

COUNT: *counter name, increment;*

CREATE: *batch size, offset time: interval, maximal number of batches to be created;*

DELAY: *time duration, storage name;*

DISPOSE;

ENDIF;

IF:

QUEUE, *queue name, capacity, transfer address when capacity exceeded*;

RELEASE: *resource name, quantity to release:*
repeat;

SEIZE, *priority: resource name, number of units:*
repeat;

TALLY: *tally name, name of quantity to be recorded, number of entries to be recorded*;

ELEMENTS

ATTRIBUTES: *number, attribute name (index), initial value:*
repeat;

COUNTERS: *number,*
counter name, limiting value, initial value, option, output file:
repeat;

DSTAT: *number, expression, name, output file:*
repeat;

EXPRESSIONS: *number, name (... ,...), expressions:*
repeat;

PROJECT: *title, analyst name, date, summary report*;

QUEUES: *number, name, ranking criterion:*
repeat;

REPLICATE: *number, beginning time, replication length, initialize system, initialize statistics, warm-up period*;

RESOURCES: *resource ID number, name, capacity* or *number of units:*
repeat;

SEEDS: *stream ID name* or *number, seed value, initialize option* (yes or no):
repeat;

TALLIES: *number, name, output file:*
repeat;

VARIABLES: *number, name(index), initial values,... :*
repeat;

The notation "repeat;" indicates that several records of multifield information may be entered by merely listing them sequentially, each followed by a colon, except for the last, which is followed by a semicolon. The "number" entry in each element with "repeat" designates a unique numerical identifier. Whenever the "name" entry in the same record is invoked in SIMAN, it refers to this unique identifier. For a complete listing of SIMAN Blocks and Elements, see Pegden et al. (1995), Systems Modeling Corporation (1995b), or the Help menu in the Arena screen display.

The content of the EXP file in Figure 4.11b provides a basis for familiarizing ourselves with several of the Elements statements. Recall that this file displays the nonexecutable global declarations that the user-specified model in Figure 4.10 automatically creates in the SIMAN code.

ATTRIBUTES identifies QueueTime as an attribute of the process. RESOURCES declares a single server, Server 1_R, Capacity (1,), and QUEUES associates a set Server 1_R_Q with this resource having the FIFO discipline. In the present context, STATIONS is self-explanatory but contains additional information in other models (e.g., Pegden et al. 1995, p. 214). We defer our discussion of the TALLIES and DSTATS statement until Section 4.25.

To follow an entity's progress through a system for diagnostic purposes, SIMAN provides the TRACE element. It allows a user to specify those entities and variables that are to be monitored during the simulation, presumably for the purpose of checking system flow for errors. Arena also provides for an automatic trace if one clicks on Run Control on the Run menu, then on Trace, and then on Enable Block Trace. Before using this option, the simulationist is encouraged to read the accompanying Help file.

4.25 ARENA/SIMAN AIRLINE RESERVATION EXAMPLE

For a simulationist with limited knowledge of SIMAN and for whom speed of modeling is a high priority, the appeal of working exclusively with the Common panel is apparent. If we try to use this panel to model the airline reservation problem, an obstacle arises when we try to represent the *branching at block* 5 in Figure 2.10 related to the busy message. Since the Common panel in Arena version 3.51 does not display a module for effecting branching, we could resort to making selections from it and from Arena's Support panel, which contains logic functions for effecting branching. However, we move away from the higher level of functionality that these panels offer and resort to modeling at the level of Arena's Blocks and Elements panels. Recall that the Blocks panel contains all SIMAN executable statements, whereas the Elements panel contains SIMAN nonexecutable global declaration and control statements.

Figure 4.12 shows the Arena/SIMAN airline reservation model derived from the Blocks and Elements Panel, and Figure 4.13 shows the corresponding Arena-generated MOD and EXP files.

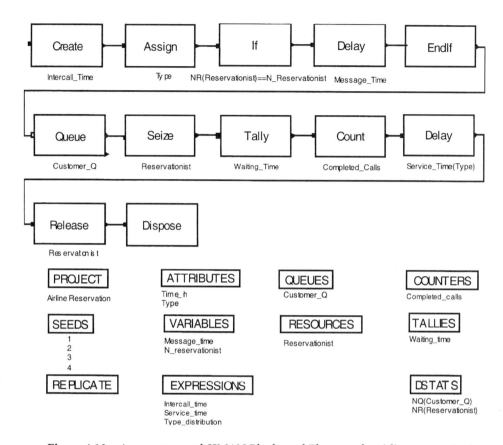

Figure 4.12   Arena generated SIMAN Blocks and Elements for airline reservationist problem

CREATE in Figure 4.13 iteratively brings new processes into existence, each of which we hereafter call Customer, for convenience of exposition. Furthermore, "batch size" = 1 indicates that these Customers are created one at a time. Creation starts at "offset time" = 0. Successive Customers are created at "interval" = EXPO(1,1) as defined in EXPRESSIONS in Figure 4.13. These are exponentially distributed samples with unit mean using pseudorandom number stream 1. MARK is a SIMAN-defined function that assigns the current time in the user-specified attribute Time-In of the Customer.

ASSIGN determines Type (of Customer), declared as its first attribute in ATTRIBUTES, by sampling from the user-specified Type_Distribution in EXPRESSIONS using stream 4. IF allows the subsequent DELAY to be executed for Message_Time_In VARIABLES if no Reservationist is available (NR(Reservationist) = = N_Reservationist;). ENDIF delimits the IF range.

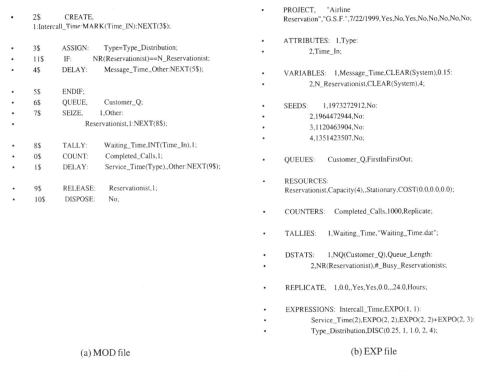

- 2$ CREATE,
  1:Intercall_Time:MARK(Time_IN):NEXT(3$);

- 3$ ASSIGN: Type=Type_Distribution;
- 11$ IF: NR(Reservationist)==N_Reservationist;
- 4$ DELAY: Message_Time,,Other:NEXT(5$);

- 5$ ENDIF;
- 6$ QUEUE,   Customer_Q;
- 7$ SEIZE,   1,Other:
             Reservationist,1:NEXT(8$);

- 8$ TALLY: Waiting_Time,INT(Time_In),1;
- 0$ COUNT: Completed_Calls,1;
- 1$ DELAY: Service_Time(Type),,Other:NEXT(9$);

- 9$ RELEASE: Reservationist,1;
- 10$ DISPOSE: No;

- PROJECT,   "Airline
  Reservation","G.S.F.",7/22/1999,Yes,No,Yes,No,No,No,No,No;

- ATTRIBUTES: 1,Type:
- 2,Time_In;

- VARIABLES: 1,Message_Time,CLEAR(System),0.15:
- 2,N_Reservationist,CLEAR(System),4;

- SEEDS: 1,1973272912,No:
- 2,1964472944,No:
- 3,1120463904,No:
- 4,1351423507,No;

- QUEUES: Customer_Q,FirstInFirstOut;

- RESOURCES:
  Reservationist,Capacity(4),,Stationary,COST(0.0,0.0,0.0);

- COUNTERS: Completed_Calls,1000,Replicate;

- TALLIES: 1,Waiting_Time,"Waiting_Time.dat";

- DSTATS: 1,NQ(Customer_Q),Queue_Length:
- 2,NR(Reservationist),#_Busy_Reservationists;

- REPLICATE,   1,0.0,,Yes,Yes,0.0,,24.0,Hours;

- EXPRESSIONS: Intercall_Time,EXPO(1, 1):
- Service_Time(2),EXPO(2, 2),EXPO(2, 2)+EXPO(2, 3):
- Type_Distribution,DISC(0.25, 1, 1.0, 2, 4);

(a) MOD file   (b) EXP file

Figure 4.13   Arena-generated SIMAN code for airline reservation problem

QUEUE files a Customer process in a user-specified list called Customer_Q, which has infinite "capacity." If the specified "capacity" were limited, and a Customer were denied entry because of space, the "balk label" would redirect the entity to an alternative block. SEIZE functions similarly to REQUEST in SIMSCRIPT II.5. Here "priority" = 1 denotes first-come-first-served for the Reservationist resource whose total number is N_Reservationist = 4 in VARIABLES. Once an entity flows from the SEIZE block, TALLY collects "value of tally" = INT(Time_In) and stores these data in "name of quantity" = Waiting_time for subsequent analysis and display. Time_In, the second attribute of Customer, declared in ATTRIBUTES, stores the simulated time at which the Customer was created in the CREATE block. INT(Time_In) is a SIMAN defined function that computes the elapsed time between current simulated time (at TALLY) and Time_In.

COUNT adds 1 to the user-declared Completed_Calls in COUNTERS. When the limit of 1000 completions, specified in the COUNTERS element, is reached at COUNT, SIMAN terminates the simulation run. Thus we have two ways of controlling run length, one based on elapsed time (Option A in Section 3.4) in the M/M/1 example and one based on number of Customers receiving service (Option B in Section 3.4) in the airline reservation problem.

The next DELAY provides service to the Customer using the service time distribution defined under EXPRESSIONS. If Type = 1 (single destination), it samples from EXPO(2,2), the exponential distribution with mean 2 using pseudorandom number stream 2. If Type = 2 (multiple destinations), it augments that sample by EXPO(2,3) using pseudorandom number stream 3. After time elapses at this second DELAY, SIMAN RELEASEs one unit of the resource Reservationist. Then DISPOSE eliminates the record of this Customer process from the simulation.

Recall that SIMAN uses a pseudorandom number generator of the form (4.27). The SEEDS element allows us to specify Z_0, $Z_{1000000}$, $Z_{2000000}$, and $Z_{3000000}$ for the four seeds using entries in Fishman (1978, Table A.2). Note that exclusive use of the Common panel does not allow this option. Also, note that the third field of each stream entry indicates that the streams are not to be reinitialized to the original seeds on successive replications. Since the REPLICATE element indicates 1 replication in its first field, this is not an issue here.

Recall that the DSTATS element allows us to specify the continuous-time sample paths on which data for time-persistent statistics are to be collected and analyzed. NQ and NR are SIMAN-defined system variables. In the present setting, NQ(Customer_Q) denotes queue length for the user-specified Customer_Q set and NR(Reservationist), the number of busy resources of the user-specified resource Reservationist. Note that the user-specified names, Queue Length and #_Busy_Reservationists, in field 3 refer to the respective Arena-assigned indices, 1 and 2, in field 1.

In the EXP file of Figure 4.13, DSTATS declares that data on two time-persistent phenomena are to be collected and summarized. One is NQ(Customer_Q), the number in the set designated by Customer_Q. It is to be named Queue_Length (field 3), which implicitly refers to index 1 in field 1. The other is NR(Reservationist) (field 2), the number of busy RESOURCES called by the user-designated name, Reservationist. It is to be named #_Busy_Reservationists, which implicitly refers to index 2 in field 1.

The executable TALLY statement in the MOD file in Figure 4.13 computes waiting times as INT(Time_In) in field 2 (see Section 4.22) and refers to them as Waiting_Time in field 1. The TALLIES statement must have this designation in field 2. In the present case, it also declares that these data are to be stored in the user-specified file, Waiting_time .DAT, in binary form. Double quotation marks must precede and follow the file name. The DSTATS element also allows for data storage. Section 6.9 describes how to use this file for a user-specified batch-mean analysis.

Figure 4.14 displays the user-specified statistical summary of output. The sample average waiting time 5.6114, queue length 5.8416, and number of busy reservationists 3.5064 are comparable to the values of their corresponding quantities, 5.8818, 5.862, and 3.7135, in Figure 4.3 for SIMSCRIPT II.5. Arena has a capability for computing an approximating confidence interval for the time long-run averages and displaying its half-length. The default significance level for the confidence level is 0.95. In the present case, the entry (Corr) indicates that the data are correlated so that a half-length cannot be computed by the Arena's built-in method. See Section 6.6 for a discussion of Arena's method of computing a confidence interval.

ARENA Simulation Results

Operations Research License: 20010064

Summary for Replication 1 of 1

Project:Airline Reservation   Run execution date :10/27/2000
Analyst:G.S.F.   Model revision date: 7/22/1999

Replication ended at time : 979.939

TALLY VARIABLES

Identifier	Average	Half Width	Minimum	Maximum	Observations
Waiting_Time	5.5749	(Corr)	.00000	32.076	1000

DISCRETE-CHANGE VARIABLES

Identifier	Average	Half Width	Minimum	Maximum	Final Value
Queue_Length	5.7490	(Corr)	.00000	37.000	21.000
#_Busy_Reservationists	3.5036	(Corr)	.00000	4.000	4.000

COUNTERS

Identifier	Count	Limit
Completed_Calls	1000	1000

OUTPUTS

Identifier	Value
System.NumberOut	.00000

Simulation run time: 0.03 minutes.
Simulation run complete.

Figure 4.14 Arena air.out output

After run termination, a simulationist can click on "Output" in the Tools menu and then on "Graph." This gives options for graphical and tabular displays. Figure 4.15 shows the table of the resulting waiting-time histogram. Clicking on "Data File" in the File menu brings up a dialog box, one of whose options is to export the saved data to a user-specified textual file for subsequent analysis.

Histories may be saved at the modular level using the STATS module in the Common panel. It provides options for DSTATS, TALLIES COUNTERS, and FREQUENCY variables.

Histogram Summary

Waiting time

Cell	Cell Limits		Abs. Freq.		Rel. Freq.	
	From	To	Cell	Cumul.	Cell	Cumul.
1	-Infinity	0	0	0	0	0
2	0	0.25	305	305	0.305	0.305
3	0.25	0.5	25	330	0.025	0.33
4	0.5	0.75	27	357	0.027	0.357
5	0.75	1	22	379	0.022	0.379
6	1	1.25	16	395	0.016	0.395
7	1.25	1.5	23	418	0.023	0.418
8	1.5	1.75	29	447	0.029	0.447
9	1.75	2	30	477	0.03	0.477
10	2	2.25	14	491	0.014	0.491
11	2.25	2.5	20	511	0.02	0.511
12	2.5	2.75	16	527	0.016	0.527
13	2.75	3	28	555	0.028	0.555
14	3	3.25	14	569	0.014	0.569
15	3.25	3.5	20	589	0.02	0.589
16	3.5	3.75	16	605	0.016	0.605
17	3.75	4	13	618	0.013	0.618
18	4	4.25	13	631	0.013	0.631
19	4.25	4.5	8	639	0.008	0.639
20	4.5	4.75	7	646	0.007	0.646
21	4.75	5	11	657	0.011	0.657
22	5	5.25	16	673	0.016	0.673
23	5.25	5.5	7	680	0.007	0.68
24	5.5	5.75	14	694	0.014	0.694
25	5.75	6	3	697	0.003	0.697
26	6	6.25	9	706	0.009	0.706
27	6.25	6.5	4	710	0.004	0.71
28	6.5	6.75	11	721	0.011	0.721
29	6.75	7	6	727	0.006	0.727
30	7	7.25	9	736	0.009	0.736
31	7.25	7.5	4	740	0.004	0.74
32	7.5	7.75	8	748	0.008	0.748
33	7.75	8	5	753	0.005	0.753
34	8	+Infinity	247	1000	0.247	1

Figure 4.15

User-specified Arena waiting-time histogram

Recall that SIMAN maintains its future-event set (FES) or *event calendar* as a single list. CREATE schedules arrivals by inserting the current Customer record into this FES with reactivation time being the next Inter_call time. The first DELAY does similar linkage with current time + 9 seconds as the reactivation time. The second DELAY performs similarly.

QUEUE effectively makes use of the lists defined in QUEUES to store records. SEIZE selects and removes records from the Customer_Q as resources become available at the RELEASE statement in another Customer process. Although SIMAN effects the same logic for its QUEUE-SEIZE-RELEASE sequence as SIMSCRIPT II.5 does for its q.resource-x.resource-request-relinquish sequence, its list-processing programming code differs substantially.

SIMAN's TRACE element allows a simulationist to generate a time history of events during a simulation run. This history can be helpful in debugging a model, especially for identifying unintended impediments to entity flow from node to node.

4.25.1   MODELING WITH THE BAAT PANELS

As already mentioned, the BAAT panels permit modeling at a higher level of code aggregation than exclusive reliance on the Blocks and Elements panels allow while retaining more modeling flexibility than the more aggregated models in the CAT panels allow. As with the other panels, each module in the BAAT panels has a dialogue box that enables a user to specify which of several actions are to be taken. However, the BAAT menus in these boxes offer more options and actions than those in the Block and Elements panels.

As illustration, the BAAT panels contain a multiple-purpose "process" module. We use quotation marks here merely to distinguish Arena's use of the word "Process" from our more generic use of the term to describe the principal concept in the process-interaction approach. See Chapter 2. The dialogue box of the "Process" module contains a menu with items

> Delay
> Size, Delay
> Size, Delay, Release
> Delay, Release

where Seize, Delay and Release are SIMAN statements with corresponding entries in the Blocks panel. A user clicks on the choice that fits the problem at hand. For example, a modeler of the airline reservation problem may use Seize, Delay, Release or Delay, Release, depending on how the time spent waiting is to be collected in the module. More generally, we emphasize the versatility of this and other BAAT modules that allow a modeler to make multiple Block selections in a single module.

Although space considerations limit what we can show here, we encourage the reader to model the airline reservationist problem using the BAAT panels with the same input specification depicted in Figure 4.13 based on Blocks and Elements. In principle, the outputs should be identical.

4.26   EVENTS IN SIMAN AND SLAM

As mentioned, both SIMAN and SLAM allow simulation modeling and programming using the event-scheduling approach. Unlike SIMSCRIPT II.5, however, for which the event-scheduling syntax is an integral component of the language, Arena/SIMAN requires skill in C (Pegden et al. 1995, Appendix F), and AweSim/Visual Slam requires skill in C or Visual Basic (Pritsker et al. 1995, Section 13.2). Regrettably, a consequence of these requirements is to limit accessibility to the event-scheduling approach in these languages.

From the point of view of pedagogy, this should in no way deter the reader from studying the concept of event scheduling in Chapter 2 and the programming constructs to which it leads in Sections 4.2 through 4.17. Familiarity with these ideas provides a considerably deeper understanding of the inner workings of discrete-event simulation than the higher-level functionality of the process-interaction approach allows. The familiarity can be of considerable help in trying to understand precedence relationships in complex models.

4.27 MACROMODULES

Macromodules can incorporate considerably more functionality than the elementary blocks of a simulation language. The M/M/1 example well illustrates this, where three macromodules in Figure 4.10 lead to the relatively detailed MOD and EXP files in Figure 4.11. Relying on macromodules clearly saves time and has great appeal for the experienced programmer. Moreover, the minimal programming to which it leads allows those with little if any familiarity with a language to model complex systems and execute simulations of them.

For both experienced and inexperienced programmers, there is an issue of correctness. In particular, the user of this approach needs to be assured that the generated code does, indeed, do exactly what he/she wishes. Studying the self-documenting code in Figure 4.11 provides one way to gain this assurance. In practice, many a simulationist pays scant attention to this code. Regardless of the simulation language, this can lead to a serious misrepresentation of modeling logic, as the next several examples show. This is especially so at the higher level, for example, in modeling conveyors, where many alternative logic protocols exist and only one of which may be embraced in the macromodules of a particular simulation language.

Three examples taken from Banks and Gibson (1997) illustrate the dangers inherent in this minimal approach to simulation programming based on macromodules:

WHAT IS THE SEQUENCING DISCIPLINE?

Suppose we use a popular programming simulation software package, which is oriented toward manufacturing; call it Software X. The goal is to simulate the loading of a trailer with 10 different rolls of carpet, to travel to the destination, then to unload the carpet in the reverse order of loading. Using the "LOAD" command accomplishes the placement of the carpet on the trailer. So far, so good. Using the "UNLOAD" command removes the carpet, but in the order that it was loaded. Obviously, the trailer should be unloaded in reverse order. To accomplish the desired result (provided that the modeler knew that the software was not reproducing reality) requires the use of "dummy" locations: fake entities used to allow work-arounds for such problems.

Dummy locations are used frequently in no-programming software to accomplish a desired result, and this can lead to much confusion. Not only is the frequent need to resort to this artifice confusing to the modeler, the use of dummy locations causes other aggravations. For example, in a complex process at a machine, dummy locations may have to be added to portray the various steps in the process. Then, when a report on the utilization of the machine is desired, even more dummy locations have to be added.

Using a different, "programming" simulation software, an order list construct can be used. This allows the modeler to specify to the model that the carpet rolls be unloaded in accordance with the value of an attribute that contains the time of their arrival. Unloading the rolls from lowest to highest time of arrival would accomplish the desired purpose. Although we would consider this a "programming" approach, it seems to us more straightforward and easier to build than the dummy location workaround. If the customer later decides that the truck should be unloaded in order of color instead, this too can be easily programmed using the same order list construct.

WHAT CAN THE BRIDGE CRANE DO?

The words sound right. You want to model a bridge crane. The no-programming software has a bridge crane! The model must include an automated guided vehicle (AGV). Okay, the no-programming software has an AGV. This looks easy, and the model is constructed rapidly. It takes only four hours to build with animation, and that's four-and-a-half days faster than with the package that doesn't advertise no-programming.

But you need to dig deeper into what the bridge crane module really does. You need to dig deeper into those AGV movement rules. What scheduling logic is being used? What parking rules are used? Is this the same as your bridge crane? Does your AGV system work this way?

There is much danger in treating these capabilities as a black box. You may find that you haven't accounted for the minimum spacing of the AGVs, and that is important. You may find that the control algorithm for the bridge crane is not the same as the one your system is using, and it can't be changed.

We might go as far as saying that by not requiring you to explicitly specify or "program" model logic, no-programming packages hide critical details that you may need to see in order to understand and verify the model's behavior and results. If you were told to design a machine, you wouldn't simply describe the machine's outside view, fill its surface with some pretty colors from your paintbrush type software, and then turn on the automatic inside machine design program. You would need to design the inside of that machine carefully, including all of the spindles, bearings, gears, races, and their interactions.

MODULE OPAQUENESS

Here's an example of the hidden dangers: A model was developed of a simple, high-speed packaging conveyer line using the approach promoted by Software Y. The modeler selected from among the high-level blocks provided, clicked and filled in the Windows "check boxes," and accepted some defaults. The modeler made one critical error, however, by not specifying a previous conveyer section resource to be released after gaining access to the next section; the modeler inadvertently left the default "release before" option selected. This model construct used an "internal" queue in the queue-size logic, which is hidden from the user. This queue actually allowed an unlimited number of package entities to build up when the downstream conveyer section filled, providing a totally erroneous result. The model's performance reports provided no clue about the problem, because the software doesn't report on these hidden queues. Had the modeler used a lower-level programming construct instead, this error would have been much easier to detect.

4.28 LESSONS LEARNED

- Major discrete-event simulation languages include GPSS/H, SLAM, SIMAN, and SIMSCRIPT II.5.

- All allow simulation programming using the process-interaction approach. Although several also allow programming via the event-scheduling approach, SIMSCRIPT II.5 is by far the easiest to use in this regard.

- Maintaining a separate pseudorandom number stream for each source of stochastic variation allows for control of stochastic variation within a run and across runs.

- Recording first and last seeds for each pseudorandom number stream allows for easy reproducibility of results. This is especially helpful in checking out a program. It also facilitates the use of common pseudorandom number streams.

- The method of *common pseudorandom number streams* is a technique that may lead to a variance reduction in comparing sample averages computed from data on two simulation runs, each based on a different assignment for parameter values or on different system operating rules.

- SIMSCRIPT II.5 is a full-service programming language that allows all the arithmetic and file management operations in languages like FORTRAN and C and, in addition, has a syntax catering to discrete-event simulation. Unix and Windows implementations are available, the latter allowing a *point-click-drag-drop* approach to program construction. Its principal concepts are events, processes, entities, attributes, sets, and event notices. It admits both the event-scheduling and process-interaction approaches to modeling. Its principal strength comes from the high level of generality it offers for modeling the competing demands for resources that arise in a wide variety of delay systems. However, this high generality puts more of a burden on the simulationist with regard to programming than do other languages, such as GPSS/H, SLAM, and SIMAN, that provide constructs considerably more focused on commonly encountered flow logics. Nevertheless, SIMSCRIPT II.5 offers many pedagogic benefits for teaching students how to model much of this logic, a particularly valuable asset when one encounters a nonstandard flow logic that some of these other languages either handle awkwardly or cannot handle.

- SIMAN is a discrete-event simulation language whose syntax principally consists of constructs that facilitate the programming of the competing demands for resources, so characteristic of demand-driven delay systems. By incorporating a considerable amount of commonly encountered system logic into these constructs, SIMAN reduces the effort required to create an executable simulation program for many delay systems encountered in practice. However, the tight coupling between its arithmetic and list-processing operations on the one hand and its demand-resource constructs on the other limits its ability to model some demand-driven delay systems. For these, the SIMAN coding may appear awkward or may not be possible.

- For the same problem, a program coded in SIMAN tends to execute faster than one coded in SIMSCRIPT II.5. SIMAN and SIMSCRIPT II.5 are C-based.

- Arena is a comprehensive simulation modeling system that relies on SIMAN as the means for programming a discrete-event simulation. Windows-based, it encourages a point-click-drag-drop approach to program construction. It contains *macromodules*, frequently required aggregates of SIMAN constructs, which further reduce programming effort. It also provides an input-data analyzer and automatic report generator. A user can easily tailor the content and form of these reports.

4.29 EXERCISES

GENERAL INSTRUCTIONS

These exercises provide opportunities for hands-on experience in writing simulation programs. As the specifications in each exercise reveal, considerable detail is necessary before a program can be executed. There is also the question of correctness. One way to address this last issue and to enhance the learning experience is to assign two sets of pseudorandom number seeds to each student. One set common to all students consists of the *class seeds*. If all students write logically equivalent simulation programs that preserve the same degree of numerical accuracy, then using the class seeds should result in the same numerical results for all. When this does not occur, students should compare their approaches to resolve differences. The instructor can assign seeds in a way that ensures nonoverlapping sequences for the streams assigned to each source of variation for each student as well as nonoverlapping sequences among students.

While this ability to compare answers is a classroom luxury not available in practice, its ability to sensitize students to how seemingly small coding differences can generate substantially different sample paths makes the approach worthwhile.

The second set of assigned seeds are unique to the student. Once differences in runs using the class seeds are resolved, each student uses the individual seeds to execute her/his program. For N students, this provides N independent sets of statistical results to the instructor for analysis.

In addition to a complete listing of the simulation program, each student's submission should contain a listing of the values of all input parameters. Each should contain two sets of output, one using class seeds and the other, the student's seeds. Each of these outputs should list initial and final seeds for each pseudorandom number stream. All results should be listed to four significant digits.

QUARRY-STONE CRUSHING (EXERCISES 2.1 AND 2.2)

✍ **4.1** Prepare a flowchart for an event that repeatedly executes at one-minute intervals and collects data on:

i. status of each shovel

ii. queue length at each shovel

iii. queue length at the crusher

iv. tons of stone unloaded at the crusher during the past minute (throughput).

For an illustration, see Figure 4.4.

✍ **4.2** This exercise establishes a basis for estimating profit without external deliveries. Prepare and execute a simulation program based on the event-scheduling approach using the flowcharts in Excercise 2.1e.

a. Begin the simulation with all trucks at their respective shovels and with the 50-ton truck ready to begin loading. Priority order is arrival at the crusher, arrival at a shovel, and data collection.

b. Table 4.3 shows how trucks are identified with shovels. Assign pseudorandom number stream i to truck i for $i = 1, \ldots, 9$ to generate times for loading, transit to crusher, dumping, and transit to shovels. At site 1 the ordering of trucks is 1, 2, and 3. At site 2 it is 4, 5, and 6. At site 3 it is 7, 8, and 9.

c. Run the simulation for six hours.

Table 4.3 Truck identification

Shovel Site	Truck	Tonnage
1	1	50
	2	20
	3	20
2	4	50
	5	20
	6	20
3	7	50
	8	20
	9	20

d. For debugging purposes, all students should use the same seeds for corresponding streams. For personal runs, each student should use distinct seeds taken from Table 4.2. Doing so allows us to regard personal runs as statistically independent replications.

  Note: Assigning distinct seeds within a row in Table 4.2 to the nine streams called for in part b ensures that the sequences generated by these streams do not overlap, provided that each truck makes no more than $1,000,000/4 = 250,000$ round trips from its shovel. Assigning a distinct row to each student and a distinct row for class seeds ensures that the sample paths generated by each student and by the class seeds are statistically independent.

e. Compute estimates of

  i. crusher throughput

  ii. shovel 1 utilization

  iii. shovel 2 utilization

  iv. shovel 3 utilization

  v. overall shovel utilization

  vi. mean queue at crusher

  vii. mean queue at shovel 1

  viii. mean queue at shovel 2

  ix. mean queue at shovel 3.

f. Prepare graphs of

  i. crusher throughput over five-minute intervals

    ii. crusher queue length at five-minute intervals.

  g. Describe the influence of initial conditions in the graphs of part f.

✎   **4.3** This exercise establishes a basis for comparing profits with and without external deliveries. Prepare and execute a simulation program based on the event-scheduling approach using the augmented flowcharts in Exercise 2.2b that incorporate stone delivered by the outside contractor.

  a. Use the initial conditions in Exercise 4.2a augmented by the arrival at the crusher of a truck from the external source.

  b. Assign stream 10 to interarrival times for external deliveries and stream 11 for corresponding unloading times.

     N.B. These assignments preserve the same stochastic variation associated with the original nine trucks, so that changes in performance are due to the addition of external trucks and the stochastic variation that their interarrival and service times induce.

  c. For your first run take $1/\lambda = 16$ minutes.

  d. Run the simulation using the same seeds for streams 1 through 9 as in Exercise 4.2b.

  e. Follow the same debugging practice as in Exercise 4.2d.

  f. Compute estimates of all performance measures listed in Exercise 4.2e.

  g. Prepare graphs of:

    i. crusher throughput per five minutes

    ii. crusher queue length at five-minute intervals.

  h. Describe the influence of initial conditions in the graphs of part g and compare them to what you observed in the corresponding graphs in Exercise 4.2f.

  i. Discuss the effects of external deliveries on profits and utilization of Hard Rock's equipment. See Exercise 2.2.

✎   **4.4 Independent Runs.** Rerun Exercise 3 using the last pseudorandom numbers produced there as the seeds for corresponding seeds in the new run. For $j = 1, \ldots, 72$, let

$$Z_{2j} := \text{crusher throughput during time } [5(j-1), 5j) \text{ for Exercise 4.2}$$
$$Z_{3j} := \text{crusher throughput during time } [5(j-1), 5j) \text{ for Exercise 4.3}$$
$$Z_{4j} := \text{crusher throughput during time } [5(j-1), 5j) \text{ for the current exercise.}$$

Plot $\{Z_{3j} - Z_{2j}; j = 1, \ldots, 72\}$ and $\{Z_{4j} - Z_{2j}; j = 1, \ldots, 72\}$ on the same graph and interpret the differences between them, ostensibly due to the use of common pseudorandom numbers.

✏ **4.5** Prepare and execute a simulation program based on the process-interaction approach using the flowcharts in Exercise 2.1f and Exercise 4.1. Follow the directions in Exercises 4.2a through 4.2e and show that this program produces the identical results as in Exercise 4.2e.

✏ **4.6** Prepare and execute a simulation program based on the process-interaction approach using the flowchart in Exercise 2.2c and Exercise 4.1. Follow the directions in Exercises 4.2a through 4.2e and show that this program produces the identical results as in Exercise 4.3f.

TANKER LOADING (EXERCISES 2.3 AND 2.4)

✏ **4.7** Using the flowcharts for Exercise 2.3b based on the event-scheduling approach, prepare and execute a simulation program for the tanker problem.

a. For events that occur at the same time, assume the *priority order:* tanker arrival, tanker berthing, loading completion, picking up a tanker at a berth, picking up a tanker in the harbor, and tanker departure.

b. Start the simulation with all berths idle, the tug in the harbor, and the arrival of a tanker in the harbor.

c. Run the simulation until the nth tanker has sailed from the harbor, where $n = 1000$ tankers.

d. Assign the following pseudorandom number streams:

Source of variation	Pseudorandom stream no.
Interarrival times	1
Towing in	2
Towing out	3
Deadheading in	4
Deadheading out	5
Type	6
Loading times	
type 1	7
2	8
3	9

Deadheading denotes the movement of the tug with no tanker in tow.

e. **Data Collection and Analysis.** Let τ_n denote the simulated time at which the nth tanker has sailed from the harbor. Use data from the simulated time interval $[0,\tau_n]$ to estimate system parameters. In addition to identifying input parameter values clearly, your output should contain averages, standard deviations, and maxima for:

i. Tanker waiting time

ii. Number of waiting tankers in harbor and for deberthing

iii. Number of loading berths

iv. Number of committed berths (reserved for tankers being towed in and holding loaded tankers waiting to be towed out). A "committed" berth is one that has a nonloading tanker occupying it or is awaiting a tanker that the tug is either towing or is deadheading to get.

Based on the port time definition in Exercise 2.3, it should also contain estimates of:

v. $\mathrm{pr}(\text{port time} > 24i \text{ hours})$ for $i = 1, \ldots, 10$

vi. Average rebate per tanker

vii. Average rebate per day.

4.8 Using the flowcharts for Exercise 2.3c based on the process-interation approach, prepare and execute a SIMSCRIPT II.5 simulation program for the tanker problem, treating tankers and tugs as processes. Include steps a through e of Exercise 4.7.

4.9 Based on the augmented flowcharts in Exercise 2.4b for the event-scheduling approach, prepare and execute a simulation program using the same input as in Exercise 4.7. For the additional tankers, use

Sources of variation for new tankers	Pseudorandom stream no.
Round-trip times	10
Deadheading in	11
Deadheading out	12
Towing in	13
Towing out	14
Loading times	15

Include steps a through e of Exercise 4.7.

✎ **4.10** Do the same as in Exercise 4.9 for the process-interaction approach as modeled in Exercise 2.4c.

✎ **4.11** Do the same as in Exercise 4.9 for the mixed approach modeled in Exercise 2.4d.

TERMINAL PAD CONTROLLER MODEL (EXERCISE 2.5)

The simulation programs developed in Exercises 4.12 and 4.13 establish a mechanism for generating sample-path data to be employed in Chapters 6 and 7 to estimate how mean delay in transmission $\mu(m, \lambda)$ in expression (2.6) varies as a function of character arrival rate λ and packet size m.

✎ **4.12** Using the flowcharts for Exercise 2.5b based on the event-scheduling approach, prepare and execute a simulation program for the terminal PAD controller problem.

a. Give precedence to arrivals at the output buffer over arrivals at input buffers and to transmission completions over arrivals at the output buffer.

b. Begin with a randomly scheduled arrival to a system of $r = 10$ empty input buffers, an empty output buffer, and an idle transmitter. Perusal of Section 5.1.2 suggests how to determine which input buffer receives an arriving character.

c. Assign pseudorandom number stream 1 to generate interarrival times and stream 2 to identify control characters across all input buffers. If the inverse transform method (Table 8.1) is used to generate interarrival times, this turns out to be a reasonably good way to control stochastic variation in character generation while varying arrival rate λ and maximal packet size m.

d. For a first run, take $\lambda = 1/20$ characters per time unit and $m = 32$ and run the simulation until the output buffer transmits the $t = 2500$th nonoverhead character.

e. Let

$T_i :=$ time delay for nonoverhead character i between arrival at an input buffer and transmission at the output buffer for $i = 1, \ldots, t$.

Collect and use $T_1, \ldots, T_t$, to estimate $\mu(m, \lambda)$ and the exceedance probabilities $\mathrm{pr}(T_i > w)$ for each $w \in \{350, 400, 450, 500, 550, 600\}$.

f. Graph $T_1, \ldots, T_t$. What does this reveal about the warm-up interval?

✐ **4.13** Using the flowcharts for Exercise 2.6 based on the process-interaction approach, prepare and execute a simulation program for the terminal PAD controller problem according to parts a through f in Exercise 4.12.

MOBILE-TELEPHONE SERVICE (EXERCISE 2.6)

The simulation programs developed in Exercises 4.14 and 4.15 provide a means to be employed in Chapters 6 and 7 for generating sample-path data for estimating the loss probabilities $\beta_h(m, r)$ and $\beta_s(m, r)$ in expression (2.7), as functions of number of transmission lines m and number of lines r dedicated to handoffs.

✐ **4.14** Using the flowcharts for Exercise 2.6 based on the event-scheduling approach, prepare and execute a simulation program for the mobile-telephone service problem.

   a. Give precedence to hand-off arrivals over standard arrivals and to call completions over all arrivals.

   b. Begin with one randomly scheduled hand-off and one randomly scheduled arrival to an otherwise empty and idle system.

   c. Assign pseudorandom number stream 1 to interarrival times of handoffs, stream 2 to standard interarrival times, stream 3 to duration times for handoffs, and stream 4 to duration times for standard calls.

   d. Run the simulation with $m = 11, r = 0, \lambda_h = 2, \lambda_s = 4, \omega = 1$ for a total of $t = 1000$ calls regardless of type. Hint: The quantity $\lambda_h/(\lambda_h + \lambda_s)$ denotes the probability that a new call is a mobile call.

   e. Devise and implement a data collection plan to estimate $\beta_h(m, r)$ and $\beta_s(m, r)$ in expression (2.7).

   f. Devise and implement a graphical analysis to assess the influence of initial conditions on the sample-path data of part e.

   g. Note any distinction between the statistical properties of the estimates of $\beta_h(m, r)$ and $\beta_s(m, r)$.

✐ **4.15** Using the flowcharts for Exercise 2.6e based on the process-interaction approach, prepare and execute a simulation program for the mobile telephone service problem according to parts a through g in Exercise 4.14.

4.30 REFERENCES

Banks, J., and R. Gibson (1997). Simulation modeling: some programming required, *IIE Solutions*, **29**, 26–31.

CACI (1987). *SIMSCRIPT II.5 Programming Language*, CACI Products Company, La Jolla, California.

CACI (1993a). *SIMSCRIPT II.5 Reference Handbook*, CACI Products Company, La Jolla, California.

CACI (1993b). *SIMDEBUG User's Manual: SIMSCRIPT II.5 Symbolic Debugger*, CACI Products Company, La Jolla, California.

CACI (1997). *SIMSCRIPT II.5 Reference Handbook*, CACI Products Company, La Jolla, California.

Fishman, G.S. (1996). *Monte Carlo: Concepts, Algorithms, and Applications*, Springer-Verlag, New York.

Gordon, G. (1969). *System Simulation*, Prentice-Hall, Englewood Cliffs, N.J.

Henriksen, J.O. (1997). *An introduction to SLX*, 1997 Winter Simulation Conference Proceedings, S. Andridóttir, K.J. Healy, D.H. Withers, and B.L. Nelson, editors, IEEE, Piscataway, N.J.

Kelton, W.D., R.P. Sadowski, and D.A. Sadowski (1998). *Simulation with Arena*, McGraw-Hill, New York.

Kiviat, P.J., R. Villanueva, and H.M. Markowitz (1969). *The SIMSCRIPT II.5 Programming Language*, Prentice-Hall, Englewood Cliffs, N.J.

Law, A.M., and W.D. Kelton (1991). *Simulation Modeling and Analysis*, McGraw-Hill, New York.

Markowitz, H.M., H.W. Karr, and B. Hausner (1962). *SIMSCRIPT: A Simulation Programming Language*, Prentice-Hall, Englewood Cliffs, N.J.

Pegden, C.D., R.E. Shannon, and R.P. Sadowski (1995). *Introduction to Simulation Using SIMAN*, second edition, McGraw-Hill, New York.

Pritsker, A.A.B. (1995). *Introduction to Simulation and SLAM II*, fourth edition, Wiley, New York.

Pritsker, A.A.B., and P.J. Kiviat (1969). *Simulation with GASP II*, Prentice-Hall, Englewood Cliffs, N.J.

Pritsker, A.A.B., and J.J. O'Reilly (1999). *Simulation with Visual SLAM and AweSim*, second edition, Wiley, New York.

Schriber, T.J. (1991). *An Introduction to Simulation Using GPSS/H*, Wiley, New York.

Swain, J.J. (1999). 1999 Simulation software survey, *ORMS Today*, **26**, 38-51.

Systems Modeling Corporation (1995a). *ARENA Template Reference Guide*, Sewickley, Pennsylvania.

Systems Modeling Corporation (1995b). *ARENA User's Guide*, Sewickley, Pennsylvania.

Systems Modeling Corporation (1995c). *SIMAN Reference Guide*, Sewickley, Pennsylvania.

CHAPTER 5

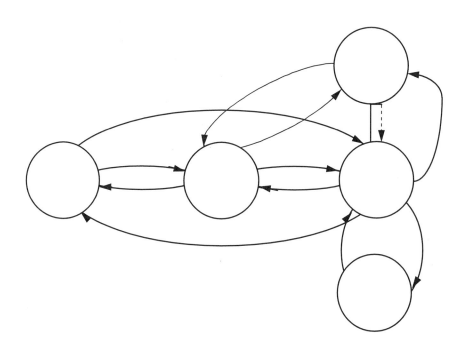

Search, Space, and Time

Every execution of a computer program uses memory space and consumes computing time. In particular, a discrete-event simulation expends a considerable proportion of its running time executing searches for new space and creating and maintaining order among the myriad of entity records and event notices it generates as simulated time evolves. In spite of their relative importance, current PC workstation environments, with their substantial memories and reduced, if nonexistent, emphasis on execution within a specified computing time constraint, make these topics appear less important to the simulationist than they were in the past. Moreover, every simulation programming language implicitly provides a means for managing space and performing searches during execution of virtually any program written in the language, further removing these issues from a simulationist's consciousness.

This computing environment is reassuring to a potential simulationist, for it relieves her/him of the need to be concerned about memory space and computing time for simulations of small and moderate sizes. Indeed, most advertising for simulation languages and packages focus on examples that illustrate this speed of execution. However, as the number of entity and event classes in a simulation increases, it is not unusual to see the corresponding computing time increase rapidly. What may have taken seconds for a small-size simulation now takes minutes for its moderate-size extension. What may have taken hours for a moderate-sized simulation now takes days for its large-size extension. When such an increase in computing time occurs, it often puzzles and frustrates the simulationist.

Simulation languages manage space and conduct searches in different ways. As a consequence, programs based on the same simulation logic but written in different languages vary in their execution times and space needs. It is not the purpose of this chapter to choose the one among these languages that performs best with respect to time and space conservation. Indeed, it is problematic to determine whether a single language dominates when one considers the widely varying characteristics of many simulations.

This chapter has a twofold purpose: to characterize the way in which memory space needs arise during a simulation and to describe prototypical modes of memory management and search that characterize the procedures that many simulation languages employ. An awareness of these concepts enables a simulationist to anticipate potential computing problems during the design phase of a study, rather than encountering them as a consequence of costly execution.

To establish a formalism for discussing space and time, Section 5.1 describes a series–parallel queueing model whose parameters we employ throughout the chapter to characterize how space and time needs vary. Section 5.1.1 addresses the space issue generically for the future-event set, and Section 5.1.2 illustrates how thoughtful analysis prior to modeling can significantly reduce space needs during execution.

This illustration has more general implications. Rather than regarding a problem's description as a recipe to be translated into programming code and then executed, the illustration encourages simulationists to apply their skills of mathematical analysis to transform the description into a probabilistic equivalent that increases computational efficiency.

To set the stage for a discussion of space management, Section 5.2 describes the essential features of list processing, with emphasis on predecessor–successor relationships. Section 5.3 introduces a succession of progressively more sophisticated procedures for this management. Section 5.4 focuses on the management of the future-event set. In particular Section 5.4.1 discusses single linear event lists, Section 5.4.2 multiple linear event lists, Section 5.4.3 heaps, and Section 5.4.4 Henriksen's Algorithm. An example, carried through the four subsections, provides a basis for understanding and comparing the time needs of each method of FES management.

Section 5.5 describes how the restricted formalism of Section 5.1, which Sections 5.1.1 through 5.4.5 employ to illustrate concepts, can be extended to a broader class of delay models.

5.1 MEMORY SPACE NEEDS AND COMPUTING TIME: A SIMPLIFIED PERSPECTIVE

While every discrete-event simulation model induces its own peculiar demands with regard to memory space and computing time, all share common properties. This section describes

arrivals

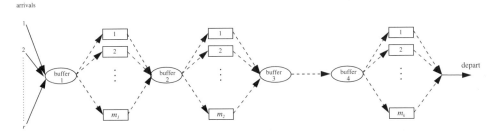

Figure 5.1   Series-parallel queueing model

several of these properties as they arise in the context of a commonly encountered class of delay problems. Once acquainted with these properties, the reader will have a perspective for understanding what motivates the discussion of alternative data structures for space search in Section 5.2 and for managing the FES in Sections 5.4.1 through 5.4.4. To broaden the range of applicability of these concepts, Section 5.5 relaxes several of the modeling assumptions made in the present section.

Consider a simulation model based on the k-station queueing system in Figure 5.1, where r statistically independent arrival processes feed jobs to station 1. Every job receives service at each station j from one of its m_j servers for $j = 1, \ldots, k$. Upon completion of service at station k, a job departs the system. For $i = 1, \ldots, r$, arrivals of type i occur at times $T_{i1} \leq T_{i2} \leq \cdots$ with identically distributed interarrival times $\{T_{i,l+1} - T_{il}; \quad l = 1, 2, \ldots\}$ with mean $1/\omega_{ai}$.

Each server at station j has identically distributed service times with mean $1/\omega_{sj}$ for $j = 1, \ldots, k$. Then one may think of $\omega_a := \omega_{a1} + \cdots + \omega_{ar}$ as the rate at which jobs arrive to the system at station 1 and of $m_j \omega_{sj}$ as the rate at which station j processes waiting jobs for $j = 1, \ldots, k$.

At each simulated time τ, interest focuses on

$$B(\tau) := B_1(\tau) + \cdots + B_k(\tau)$$

(number of busy servers)

and

$$Q(\tau) := Q_1(\tau) + \cdots + Q_k(\tau)$$

(number of jobs awaiting service),

where

$$B_j(\tau) := \text{number of busy servers at station } j$$

and

$$Q_j(\tau) := \text{number of jobs awaiting service at station } j.$$

For $Q(\tau) + B(\tau)$ to have an equilibrium probability distribution, we require that

$$\rho_{\max} := \max_{1 \leq j \leq k} \rho_j < 1,$$

where

$$\rho_j := \frac{\omega_a}{m_j \omega_{sj}}, \qquad j = 1, \dots, k.$$

At simulated time τ, the executing simulation's future-event set contains $r + B(\tau)$ event notices, one scheduling the next job arrival for each type i for $i = 1, \dots, r$ and $B(\tau)$ scheduling the completions of service for each of the busy servers.

At that same time τ, the executing simulation has also assigned memory space to the records of the $B(\tau) + Q(\tau)$ temporary entities called jobs. In the present setting each record has an attribute identifying the type of job. More generally, every model specification dictates the number of attributes that each temporary entity requires and, hence, its concomitant space needs. Associated with each station j is a set containing the records of the $Q_j(\tau)$ jobs awaiting service there. For present purposes, we refer to these k sets collectively as QUEUE. It occupies space proportional to $Q(\tau)$. We defer discussion of the computational complexity that alternative job selection rules from QUEUE imply until Section 5.2.1

In some cases, a record of a temporary entity can also serve as an event notice, thus reducing space requirements. We assume that these circumstances do not prevail here.

We first examine how space needs vary with numbers of servers and buffer capacities. As simulated time τ evolves, $B(\tau)$ and $Q(\tau)$ vary, and therefore space needs vary. In the present setting,

$$0 \leq B(\tau) \leq k\bar{m} \quad \text{w.p.1}, \tag{5.1}$$

where

$$\bar{m} := \frac{1}{k}(m_1 + \cdots + m_k). \tag{5.2}$$

Were station j to have a finite buffer capacity, c_j, for each $j = 1, \ldots, k$ and were jobs lost to the system when they arrive at station j and find its buffer full (i.e., $Q_j(\tau) = c_j$), then

$$0 \leq Q(\tau) \leq k\bar{c} \quad \text{w.p.1,} \tag{5.3}$$

where

$$\bar{c} := \frac{1}{k}(c_1 + \cdots + c_k).$$

Let

$$\rho_{\min} := \min_{1 \leq j \leq k} \rho_j,$$

and observe that for each fixed $\tau \geq 0$,

$$B(\tau) \to k\bar{m} \quad \text{as } \rho_{\min} \to 1 \quad \text{w.p.1} \tag{5.4}$$

and

$$Q(\tau) \to k\bar{c} \quad \text{as } \rho_{\min} \to 1 \quad \text{w.p.1.}$$

In a highly congested system, these properties imply that at an arbitrarily selected time τ:

- The number of event notices in the FES, $r + B(\tau)$, is most likely to be $r + k\bar{m}$.
- The number of extant temporary entities, $B(\tau) + Q(\tau)$, is most likely to be $k(\bar{c} + \bar{m})$.

Hereafter, we assume a highly congested system.

Suppose that traffic intensity ρ_j and service rate ω_{sj} at station j remain fixed as its number of servers m_j increase. These conditions imply that the arrival rate ω_a must also increase. Then for each $j = 1, \ldots, k$,

$$B_j(\tau) \to \infty \quad \text{as } m_j \to \infty \quad \text{w.p.1}$$

and

$$Q_j(\tau) \to 0 \quad \text{as } m_j \to \infty \quad \text{w.p.1,}$$

implying that:

- The number of event notices in the FES tends to grow without bound as $\bar{m}$ increases.

- The number of records in QUEUE becomes relatively small as $\bar{m}$ increases.

We next examine how space needs vary as a function of traffic intensity and the amount of stochastic variation in interarrival and service times. In a highly congested but uncapacitated system ($c_j = \infty$; $j = 1, \ldots, k$), memory space needed for event notices remains proportional to $r + k\bar{m}$. However, space needs to accommodate the records of temporary entities behave differently. As an illustration, suppose that $T_1 \leq T_2 \leq \cdots$ denotes the ordered interarrival times $\cup_{j=1}^{r}\{T_{jl}; l = 1, 2 \ldots\}$ for the r types of arrivals and assume that $\{T_{l+1} - T_l; l = 1, 2, \ldots\}$ is an i.i.d. sequence with mean $1/\omega_a$ and variance σ_a^2. Also, assume that the service times at station j have variance, σ_{sj}^2. Let

$$\gamma_a := \frac{\sigma_a}{1/\omega_a} = \text{coefficient of variation of interarrival times}$$

and

$$\gamma_{sj} := \frac{\sigma_{sj}}{1/\omega_{sj}} = \text{coefficient of variation of service times at station } j.$$

Small γ_a implies that interarrival times vary little from their mean, whereas large γ_a implies large deviations from this mean. An analogous characterization applies to γ_{sj} for service times at station j.

It is known (Köllerström 1974) that if interarrival times to station j are i.i.d. and traffic intensity ρ_j is less than but close to unity, then

$$\nu := \text{E}[B_j(\tau) + Q_j(\tau)] \approx \frac{\rho_j^2 \gamma_{sj}^2 + \gamma_a^2}{2(1 - \rho_j)}. \tag{5.5}$$

This expression reveals several important properties:

- For fixed γ_{sj} and γ_a, ν grows superlinearly with ρ_j.

- For fixed ρ_j and γ_{sj}, ν grows quadratically with γ_a.

- For fixed ρ_j and γ_a, ν grows quadratically with γ_{sj}.

As an example, a model with $\rho_j = .99$ and $\gamma_a = \gamma_{sj} = 1/\sqrt{2}$, as would be the case if interarrival and service times came from Erlang distributions with two phases (Section 8.13), gives $\nu \approx 49.50$. However, if interarrival and service times were exponentially distributed, then $\gamma_a = \gamma_{sj} = 1$, so that $\nu \approx 99.00$. The sensitivity to stochastic variation is apparent.

These results, based on mean-value behavior, convey only part of the story regarding space needs in an uncapacitated queueing system. Suppose one runs the simulation for exactly $\tau = t$ simulated time units. Then

$$G(t) := \sup_{0 \le z \le t} [B(z) + Q(z)],$$

where

$$G(t) \le r + k\bar{m} + \sup_{0 \le z \le t} Q(z) \quad \text{w.p.1}$$

gives the maximal number of temporary entities in the system during the simulation. For every fixed positive integer q, $\text{pr}[\sup_{0 \le z \le t} Q(z) > q]$ is an increasing function of t, implying that for given input $\omega_{a1}, \ldots, \omega_{ar}, \omega_{s1}, \ldots, \omega_{sk}, m_1, \ldots, m_k$, a long simulation run is likely to demand more space to store its records than a short run.

5.1.1 FES MANAGEMENT TIME PER JOB

Each job passing through the system induces $2(1 + k)$ interactions with the FES, one to schedule the next arrival of this type, k to schedule its completions at each of the k stations it must visit, and $1 + k$ to select the next event to execute when the job arrives at station 1 and completes service at stations $j = 1, \ldots, k$.

Let $g_a(q)$ denote the average computing time required to schedule an additional event when the FES contains q event notices and let $g_b(q)$ denote the average computing time required to select the next event from an FES containing q event notices. Then at simulated time τ at which an interaction with the FES occurs, each job passing through the system consumes

$$(1 + k)[Eg_a(r + B(\tau)) + Eg_b(r + B(\tau))] \le (1 + k)[g_a(r + k\bar{m}) + g_b(r + k\bar{m})] \quad (5.6)$$

computing time on average interacting with the future-event set. Each simulation language's protocol for managing the FES determines the particular form that $g_a(.)$ and $g_b(.)$ take. For example, Section 5.4.3 shows that simulation languages that structure their future-event sets as binary heaps have $g_a(B(\tau)) = O(\log_2 B(\tau))$ and $g_b(B(\tau)) = O(\log_2 B(\tau))$. Therefore, for a simulation of a highly congested system that uses these languages, each job's interaction with the FES takes $O((1 + k) \log_2(r + k\bar{m}))$ time on average. Regardless of the protocol, $\{g_a(q)\}$ and $\{g_b(q)\}$ both are increasing functions of q.

5.1.2 REDUCING THE SIZE OF THE FES

The last-mentioned property suggests that any reduction in the size of the FES reduces its average management time. Here we describe one opportunity for reduction. Suppose that for each arrival of type $i = 1, \ldots, r$, the interarrival times $\{T_{i,l+1} - T_{il}; \quad l = 1, 2, \ldots\}$

form a sequence of i.i.d. exponentially distributed random variables with exponential p.d.f.

$$f_i(z) = \omega_{ai} e^{-\omega_{ai} z}, \quad z \geq 0,$$

denoted by $cal\,E\,(1/\omega_{ai})$. Then

$$A_i(\tau) := \text{number of arrivals of type } i \text{ in the simulated time interval } (0, \tau]$$

has the Poisson distribution

$$\mathrm{pr}[A_i(\tau) = l] = \frac{(\omega_{ai}\tau)^l e^{-\omega_{ai}\tau}}{l!}, \quad l = 0, 1, \ldots,$$

denoted by $cal\,P(\omega_{ai}\tau)$. As we have described the model, one would proceed as follows:

PROTOCOL A

Each time τ at which an arrival of type i occurs, the simulation program:

- Creates a record for the arriving job and assigns i to its type attribute.
- Samples the next type i interarrival time ΔT from $\mathcal{E}\,(1/\omega_{ai})$.
- Creates and files an event notice in the FES for this next arrival with execution time, $\tau + \Delta T$.

Therefore, at every simulated time τ, the FES contains at least r event notices, one for each arrival type.

We can reduce this number by exploiting a property of the Poisson distribution. Since $A_1(\tau), \ldots, A_r(\tau)$ are statistically independent:

- $A(\tau) := A_1(\tau) + \cdots + A_r(\tau)$ has the Poisson distribution $\mathcal{P}(\omega_a\tau)$.
- The corresponding sequence of interarrival times for the aggregated arrival process are exponentially distributed from $(1/\omega_a)$.
- $\mathrm{pr}(\text{next arrival is of type } i) = \frac{\omega_{ai}}{\omega_a}, \quad i = 1, \ldots, r.$

As an immediate consequence, there exists an alternative procedure for generating arrivals:

PROTOCOL B

When an arrival occurs at time τ, the simulation program:

- Samples type I from the p.m.f. $\{\omega_{ai}/\omega_a; \; i = 1, \ldots, r\}$.

- Creates a record for the arriving job and assigns I to its type attribute.

- Samples the next interarrival time ΔT from $\mathcal{E}(1/\omega_a)$.

- Creates and files an event notice in the FES for this next arrival with execution time $\tau + \Delta T$.

Protocol B generates an arrival process with statistical properties identical to those induced by Protocol A. However, with regard to benefits and costs, Protocol A:

- Requires r memory spaces to store $\omega_{a1}, \ldots, \omega_{ar}$.

- Requires $O(r + k\bar{m})$ memory space to store the event notices needed to maintain the FES.

- Consumes $O((1 + k)[g_a(r + k\bar{m}) + g_b(r + k\bar{m})])$ average time per job interacting with the FES.

On the other hand, Protocol B:

- Requires one additional sample I of arrival type drawn from $\{\frac{\omega_{a1}}{\omega_a}, \ldots, \frac{\omega_{ar}}{\omega_a}\}$. This can be done in $O(1)$ time, independent of r, using memory space proportional to $2r$, either by the cutpoint method of Section 8.2 or the alias method of Section 8.4.

- Requires one memory space to store ω_a.

- Requires $O(1 + k\bar{m})$ memory space to store the event notices needed to maintain the FES.

- Consumes $O((1 + k)[g_a(1 + k\bar{m}) + g_b(1 + k\bar{m})])$ average time per job to interact with the FES.

The statistical equivalence of samples generated by Protocols A and B exists if and only if $\{T_{i,l+1} - T_{il};\ l = 1, 2, \ldots\}$ are i.i.d. exponentially distributed random variables for each type $i = 1, \ldots, k$. If these times have any other distributions, applying Protocol B would be inappropriate. However, for some simulation formulations, the assumptions of the underlying problem are that interarrival times for the aggregate arrival process are i.i.d. with known distribution and with known selection p.m.f. $\{q_i := \text{pr(that an arrival is of type } i);\ i = 1, \ldots, k\}$. Then we can effect the same procedure as in Protocol B as follows:

PROTOCOL C

When an arrival occurs at time τ, the simulation program:

- Samples type I from $\{q_i;\ i = 1, \ldots, k\}$.

- Creates a record for the arriving job and assigns I to its type attribute.

- Samples the next interarrival time ΔT from its interarrival time distribution.

- Creates and files an event notice in the FES for this next arrival with execution time $\tau + \Delta T$.

When it applies, Protocol C allows one to reap the same space and time benefits as in Protocol B.

If a modeler has empirical arrival-time data $\{T_{il};\ l = 1, \ldots, N_i\}$ for $i = 1, \ldots, r$, these revelations regarding the advantages of Protocol C should encourage her/him to follow these steps when modeling the arrival process (see Chapter 10):

- Form the sample sequence

$$\{T_l;\ l = 1, \ldots, N := N_1 + \cdots + N_r\} = \text{ordered} \cup_{i=1}^{r} \{T_{il};\quad l = 1, \ldots, N_i\}.$$

- Form the interarrival time sequence

$$\{Z_l := T_{l+1} - T_l;\quad l = 1, \ldots, N - 1\}.$$

- Test hypothesis H: $Z_1, \ldots, Z_{N-1}$ are i.i.d.

- If H is accepted:

    - Fit a p.d.f. to $Z_1, \ldots, Z_{N-1}$.

    - Estimate q_i by $\hat{q}_i = N_i/N$ for $i = 1, \ldots, r$.

5.2   LIST PROCESSING

Simulation languages usually employ one of two memory management techniques. One, as in SIMSCRIPT II.5, searches and allocates space each time the simulation calls for the creation of a record for a new temporary entity or for a new event notice. The other, as in GPSS/H, SIMAN, and SLAM, uses a block-array concept to accomplish the same purpose. At the start of execution it allocates a block of contiguous memory space, which the executing simulation uses to store new temporary entity records and event notices. Once a record or event notice is no longer needed, both approaches make previously committed space available for future use.

Regardless of the space allocation approach, all simulation languages rely heavily on the concept of *list processing* to keep order among records and event notices as simulated time elapses. List processing denotes a myriad of operations executable on a data structure in a digital computer. The data structure is called a *list*. The list contains records each of which has *fields* to store information. The operations include (Knuth 1973):

- Gaining access to the jth record of a list to examine or change its content.

- Inserting a new record after the jth but before the $(j + 1)$st record.

- Deleting the jth record from the list.

- Combining lists.

- Dividing one list into several.

- Determining the number of records of a particular kind in a list.

- Ordering records in a list in ascending order based on the values in specified fields.

- Searching the list for records with given values in certain fields.

In the nomenclature of discrete-event simulation, *list* is synonymous with *set*.

In discrete-event simulation, list processing arises so frequently that a highly experienced programmer, although familiar with these concepts, would find explicit programming of each occurrence an onerous task. Fortunately, computer programming languages exist that, in addition to having the list processing capability, contain libraries of subprograms that perform the specialized list processing that discrete-event simulation demands.

A concept fundamental to an understanding of list processing is the *predecessor–successor relationship*. Consider a simulation of ships arriving at a port to unload cargo. Since the port has limited facilities, a ship may have to wait before it can unload its cargo. A record characterizing each arriving ship contains fields in which relevant data are stored. Fields 1, 2, and 3 of the records in Figure 5.2a contain the ship's identification, its arrival time at the port, and its cargo tonnage, respectively. The address in the leftmost field of each record denotes the location of the first word of the record in the computer's memory bank. These addresses reveal that the four records are haphazardly located in memory. Usually this is due to the fact that the records were created at different simulated times and that the executing program searched for and allocated space at each of these times (Section 5.4.1). Section 5.4.2 describes an alternative memory management that relies on reserving a fixed block of contiguous memory at the beginning of simulation execution and then maintaining the list within this block.

Suppose only the four ships corresponding to the four records in Figure 5.2a are waiting to be unloaded. To keep track of them we put their records in a *list*. To form a list we need a criterion for ordering the records. Suppose this criterion is taken to be the ascending order of arrival times. A look at the records indicates that they should have order 14, 15, 16, and 17 in terms of identification. To make this ordering operational for the computer we introduce the concept of a predecessor–successor relationship.

Note that each record in Figure 5.2a has a predecessor address in field 4 and a successor address in field 5. A record's predecessor address identifies the record that precedes it in the list. Since we are ordering on arrival times, record 14 has no predecessor and is called the *header*. However, record 15 has a predecessor, whose record begins at address 01081. But this

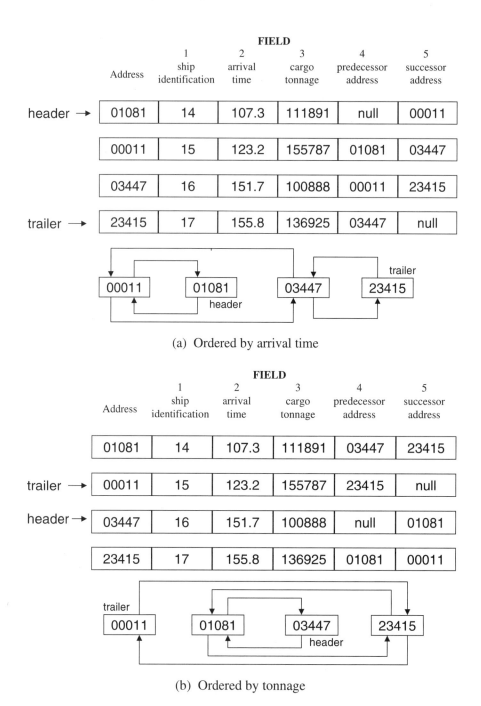

FIELD

Address	1 ship identification	2 arrival time	3 cargo tonnage	4 predecessor address	5 successor address
01081	14	107.3	111891	null	00011
00011	15	123.2	155787	01081	03447
03447	16	151.7	100888	00011	23415
23415	17	155.8	136925	03447	null

(a) Ordered by arrival time

FIELD

Address	1 ship identification	2 arrival time	3 cargo tonnage	4 predecessor address	5 successor address
01081	14	107.3	111891	03447	23415
00011	15	123.2	155787	23415	null
03447	16	151.7	100888	null	01081
23415	17	155.8	136925	01081	00011

(b) Ordered by tonnage

Figure 5.2 Predecessor–successor relationship

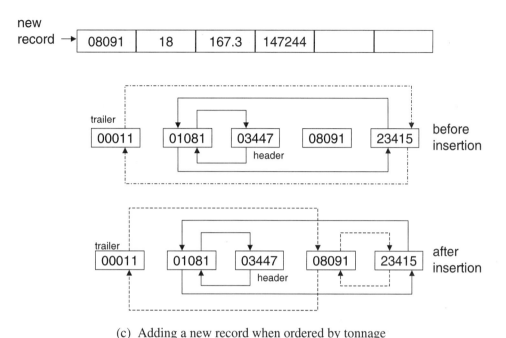

(c)   Adding a new record when ordered by tonnage

Figure 5.2   (continued)

is simply record 14. Similar predecessor relationships apply for records 16 and 17. A reverse relationship exists for the successor relationship. Record 17, the last record in the list, has no successor and is called the *trailer*.

Ordering the records in this way facilitates the selection of the next ship for unloading when a berth becomes idle. In particular, one selects the header record (1) and makes record 2 the new header. This selection presumes that ships receive service in the order of their arrival. Suppose that service is to occur in increasing order of cargo tonnage (field 3). Then Figure 5.2b shows the predecessor–successor relationships that prevail for the four waiting ships. Also suppose that a new ship, 18, arrives at time 167.3 (field 1) with a cargo of 147244 tons (field 3) and "enters" the queue. Figure 5.2c shows the new record and the old and new predecessor–successor relationships that prevail when cargo tonnage determines order. Note that in order to *chain* record 18 to the list one need change only the predecessor address of record 15, the successor address of record 17, and assign predecessor and successor addresses to record 18. These operations correspond to the concept of *inserting* a record into the list. *Deleting* a record from the list entails converse operations.

5.2.1   INSERTION AND DELETION

A discrete-event simulation maintains lists of records of entities waiting for resources to become available. The QUEUE list in Chapter 4 is an example. If first-in-first-out is the prevailing discipline for ordering the records upon insertion in the list, as in Figure 5.2a, then a single linear list has constant insertion and deletion times, regardless of its length.

If the ordering discipline depends on an attribute of the record other than simulated time at insertion as in Figure 5.2b, then relying on a single linear list becomes problematic. For example, suppose that ordering within the QUEUE in the airline reservation example is based first on TYPE (either ONE or MANY) and then on simulated time at insertion. If the list contains N records, insertion can take $O(N)$ computing time, as would occur when all N records have TYPE = MANY, the record to be inserted has TYPE = ONE, and the list is entered from its front end.

Maintaining two lists, one for each TYPE, would again lead to constant computing times for insertion and deletion, independent of list lengths. However, two lists create a need to check a record's TYPE to choose the appropriate list for insertion and a need to check both lists for deletion when the higher-priority list is empty.

More generally, the computing times to insert and delete records according to the values of an attribute other than simulated time of insertion can, in principle, be made independent of the lengths of the lists by using multiple lists as above. However, the computing time required to determine the list in which to insert a record or from which to delete a record tends to increase with the number of lists. If k lists hold N records in total and $N \gg k$, then multiple lists tend to offer a computational advantage. If $k > N$, as occurs when there are relatively few waiting customers virtually all of whom differ with regard to their values for the critical attribute, then a single list is likely to offer an advantage.

Every simulation language has built-in procedures for maintaining lists. Some use multiple linear lists as above, others use special data structures, such as *heaps* that do not maintain completely ordered lists but have $O(\log_2 N)$ insertion and deletion computing times. Section 5.4.3 describes binary heaps in the context of managing the future-event set, which is a list ordered on desired execution, rather than scheduling, time for an event.

While its supporting documentation rarely spells out the exact modality that a language has adopted for list processing, presumably for proprietary reasons, list processing times all have common tendencies. They grow with the number of sublists employed to accommodate the primary criterion for insertion and with the number of entries per sublist if the secondary criterion for insertion is other than simulated time of insertion. The present discussion is intended merely to sensitize the first-time user of a simulation language to a potentially major consumer of computing time.

5.3 DYNAMIC SPACE MANAGEMENT

During execution, a simulation must maintain records of temporary entities and of event notices in its FES to ensure continued evolution. As time elapses the number of these entity records and the number of these event notices vary, often substantially. Accommodating these records and event notices and managing their creation and eventual elimination account for a major portion of computing time.

Every computer program begins execution with an upper bound on the memory allocated to it. Sometimes this is literally at the individual program level. More often, today's computing environments impose this upper bound at the level of the user. That is, he/she is given some partition of memory within which to do computing in a stand-alone computing environment. This partition can be virtually the entire uncommitted memory capacity of the computer. In a network environment, the size of this partition usually reflects the competing needs of all network users.

At the beginning of a simulation run, some of this space becomes reserved for the program and data records that it needs to begin execution. Several languages also set aside large block arrays within which the simulation stores records of temporary entities and event notices as they arise during the simulation's execution. Others provide this space on demand. The present account describes both approaches.

Algorithms SPACE_FINDER and SPACE_RELEASER describe elementary prototypes of dynamic space management:

ALGORITHM SPACE_FINDER

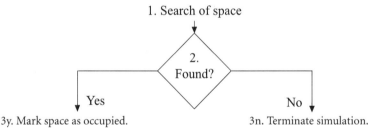

1. Search of space

2. Found?

Yes

3y. Mark space as occupied.
4y. Assign space.
5y. Continue simulation.

No

3n. Terminate simulation.

ALGORITHM SPACE_RELEASER

1. Mark space as unoccupied.

2. Merge with contiguous space.

3. Continue simulation.

As execution progresses, a discrete-event simulation makes successive requests for varying amounts of space. For example, as a simulation of the airline reservation problem evolves, it creates records of temporary entities in block 1 and event notices, to schedule future-event executions, in blocks 4 and 7 of Figure 2.5. Since the allocation of space for every creation reduces the space available to satisfy future space requests, a simulation program that merely consumes space could conceivably exhaust its availability and terminate prematurely.

The likelihood of this occurrence increases with the length of time for which the simulation program is to be run. To reduce the frequency of such an occurrence, an effective space management policy releases space as soon as it is no longer needed, so that it can be merged with contiguous uncommitted space and subsequently assigned to future demands as they arise. For example, the airline reservation simulation destroys temporary entity records for CUSTOMERs in block 4 of Figure 2.8 and destroys event notices for executed events in block 4 of Figure 2.5. Therefore, good programming practice encourages one to:

- Release space as soon as it is no longer needed.

Let *dynamic space manager* (DSM) denote the procedure that a simulation language employs to search for and assign space. For simplicity, one can regard the DSM as a sub-program to which a simulation program transfers control each time it requests or indicates that it no longer needs a space segment. Let an *uncommitted space segment* of length n denote a sequence of n successive words in computer memory that are not already assigned. Assume that the DSM maintains a *space roster*, a linked list of records each containing the address of the first word of a *noncontiguous* uncommitted space segment and its length. A noncontiguous uncommitted space segment of length n denotes a sequence of n successive uncommitted words in memory preceded and succeeded by previously allocated words.

Each attempt to assign space in Algorithm SPACE_FINDER requires a search of the space roster to identify the locations of uncommitted space segments and their lengths. Conversely, every attempt to merge space in Algorithm SPACE_RELEASER requires a search. Collectively, allocating and merging space consume a substantial portion of computing time in a simulation.

For a given simulation, let α_1 denote the mean computing time expended per space allocation and let α_2 denote the mean computing time expended per space merge. Suppose the simulation is executed until N jobs, or CALLs in the airline reservation problem, receive service. If f denotes the frequency with which new space is allocated per job or CALL during this simulation, and if we assume that a space merge eventually follows each space allocation, then the DSM consumes mean computing time $f \cdot (\alpha_1 + \alpha_2)$ per job or CALL. Clearly, choosing efficient space allocation and merge procedures makes α_1 and α_2 smaller than they would otherwise be, and that is clearly desirable. Moreover, adopting a simulation programming strategy that keeps f as small as possible is also desirable. To control f, we repeat our earlier admonition:

- Aggregate elementary events that occur at the same simulated time into compound events.

  Since allocating space for an n_2-word record takes more time in general than allocating space for an n_1-word record, where $n_2 > n_1$, the additional admonition:

- Keep the number of attributes of events and entities to the required minimum.

			predecessor address	successor address		
A(1,1)	A(1,2)	$\cdots$	A(1,$r-1$)	A(1,r)		D(1)
A(2,1)	A(2,2)	$\cdots$	A(2,$r-1$)	A(2,r)		D(2)
$\vdots$	$\vdots$	$\ddots$	$\vdots$	$\vdots$		$\vdots$
A(i,1)	A(i,2)	$\cdots$	A($i,r-1$)	A(i,r)		D(i)
$\vdots$	$\vdots$	$\ddots$	$\vdots$	$\vdots$		$\vdots$
A(n,1)	A(n,2)	$\cdots$	A($n,r-1$)	A(n,r)		D(n)

(a)

i	A(i,1)	A(i, 2)		A($i, r-1$)	A(i, r)		D(i)
1			$\cdots$	—	—		i_1
2			$\cdots$	j	n		i_2
3			$\cdots$	—	—		
$\vdots$	$\vdots$	$\vdots$	$\ddots$	$\vdots$	$\vdots$		$\vdots$
trailer→ $j-1$			$\cdots$	n			i_{n-4}
header→ j			$\cdots$	—	2		—
$\vdots$	$\vdots$	$\vdots$	$\ddots$	$\vdots$	$\vdots$		$\vdots$
$n-1$			$\cdots$	—	—		—
n			$\cdots$	2	$j-1$		

$i_1, i_2, \ldots, i_{n-4}$ is a permutation of $\{1, \ldots, n\} \setminus \{2, j-1, j, n\}$

(b)

Figure 5.3 Block arrays and ALIST

To effect space allocation and merge, a dynamic space manager can take alternative forms. For example, some simulation programming languages rely exclusively on linear linked lists to keep track of uncommitted space (e.g., Figure 5.2). With each successive space allocation, the computing time required to find an uncommitted segment of space of any given length tends to increase, usually because more search is required. Other languages rely on a block array approach, combined with a linked list, to reduce this search time. This alternative can significantly reduce search time.

For the moment, assume that all space requests are for segments of r contiguous words. Figure 5.3 depicts an $n \times r$ block array A and an $n \times 1$ block array D. Each row of A denotes a space segment of length r, including fields for predecessor and successor addresses that allow chaining linked committed rows of A into a doubly linked linear list called ALIST. We emphasize that this is merely one way to maintain a list. Section 5.4.3 describes the concept of a *heap*, which provides the basis for alternative list-maintenance procedures that become increasingly more efficient as the number of records in the list grows. The D array contains the addresses of uncommitted rows in A.

Since all rows in A are initially uncommitted, we begin with

$$D(i) \leftarrow i, \qquad 1 \leq i \leq n,$$

and

$$\#VACANT \leftarrow n.$$

Then Algorithm SPACE_FINDER* describes one way to satisfy a demand for space of length r in less computing time on average than Algorithm SPACE_FINDER. If A has no uncommitted rows (#VACANT $= 0$), the algorithm executes its right branch in computing time that varies with space availability in the remaining memory. However, if A has at least one uncommitted row (#VACANT > 0), the algorithm executes its left branch in *constant computer time*, independent of the availability of space in the remaining memory. This occurs because no need exists to search for space. As n increases relative to the number of space segments occupied at a moment in simulated time, the constant-time left branch comes to dominate. Figure 5.3b illustrates chaining in the ALIST where rows j, 2, n, and $j - 1$ comprise the ALIST, indicating that these are committed, whereas the remaining $n - 4$ rows listed in the D array are uncommitted. Note that the ALIST also may contain committed space segments not in the A array.

ALGORITHM SPACE_FINDER*

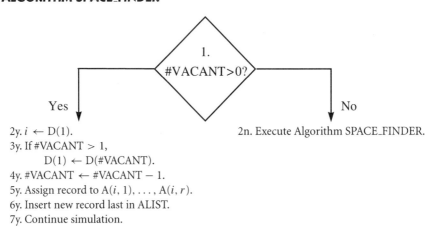

2y. $i \leftarrow D(1)$.

3y. If #VACANT > 1,

$\qquad D(1) \leftarrow D(\#VACANT)$.

4y. #VACANT $\leftarrow$ #VACANT $- 1$.

5y. Assign record to A$(i, 1)$, ..., A(i, r).

6y. Insert new record last in ALIST.

7y. Continue simulation.

2n. Execute Algorithm SPACE_FINDER.

Algorithm SPACE_RELEASER* shows one way of managing released space. In particular, it maintains the D array of uncommitted rows in A with computing time independent of n and r.

ALGORITHM SPACE_RELEASER*

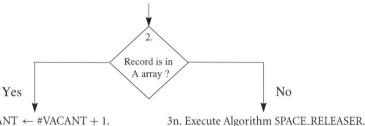

1. Remove record from ALIST.

2. Record is in A array ?

Yes

No

3y. #VACANT ← #VACANT + 1.
4y. D(#VACANT) ← row no. of
   released record.
5y. Continue simulation.

3n. Execute Algorithm SPACE_RELEASER.

For the left branch of Algorithm SPACE_FINDER*, let β_1 denote the mean computing time expended per space allocation and let β_2 denote the mean computing time expended per space release. Note that $\beta_1 \leq \alpha_1$ and $\beta_2 \leq \alpha_2$. Let p denote the probability that space is available in the A array to satisfy a space demand (left branch). Then mean computing time for Algorithm SPACE_FINDER* is

$$f \cdot p(\beta_1 + \beta_2) + f \cdot (1 - p)(\alpha_1 + \alpha_2) \leq f \cdot (\alpha_1 + \alpha_2),$$

encouraging a large initial space allocation n to make p close to unity.

5.3.1   AGGREGATING RECORD CLASSES

Collectively, Algorithms SPACE_FINDER* and SPACE_RELEASER* demonstrate the benefits of employing block arrays to manage space demands of fixed length r. Moreover, modifications of this logic allow one to accommodate k different types of space requests where type i requires r_i fields for $1 \leq i \leq k$. One approach applies Algorithms SPACE_FINDER* and SPACE_RELEASER* to k pairs of arrays $(A_1, D_1), \ldots, (A_k, D_k)$, where A_i has size $n_i \times r_i$ and D_i has size $n_i \times 1$ for $1 \leq i \leq k$. For $r_1 \geq r_2 \geq \cdots \geq r_k$, a second alternative maintains a single A array of size $n \times r_1$ and a single D array of size $n \times 1$. In this case, a separate ALIST of committed rows would not have to be maintained for each demand type. To compare these alternatives, let

$$p_i := \text{probability that the next request is for an } r_i\text{-field segment}$$

and

$$L_i := \text{number of committed rows in } A_i \text{ prior to next request.}$$

Since

$$p_1 \times \mathrm{pr}(L_1 \geq n_1) + p_2 \times \mathrm{pr}(L_2 \geq n_2) + \cdots + p_k \times \mathrm{pr}(L_k \geq n_k)$$
$$\geq \mathrm{pr}(L_1 + \cdots + L_k \geq n_1 + \cdots + n_k), \tag{5.7}$$

the second alternative allows one to choose an $n \leq n_1 + \cdots + n_k$ without necessarily increasing the frequency of search for space in remaining memory. Moreover, if $r_1 = \cdots = r_k$, then $nr_1 \leq (n_1 + \cdots + n_k)r_1$, implying a reduction in space as well.

5.3.2 RESIZING

Since the frequency of search for new space in Algorithm SPACE_FINDER* diminishes as n increases, computing time is reduced, but at the expense of reserving more, potentially unused, space for the A and D block arrays. To balance computing time and space considerations, some DSMs start with a small or moderate n but allow for an upward resizing of the arrays if the test #VACANT = 0 is true in Algorithm SPACE_FINDER*. Algorithm SPACE_FINDER** displays a prototypical procedure for performing these steps, where $\theta(n)$ $(> n)$ denotes an integer-valued function of n. For example, $\theta(n) := n$ doubles the numbers of rows in A and D each time it exhausts the already available space.

Using Algorithm SPACE_FINDER** instead of Algorithm SPACE_FINDER* replaces the need to search for space for individual records by a considerably less frequent need to search for block arrays to augment A and D. While each resizing of the block arrays may consume considerably more computing time, the lower rate at which resizing is necessary generally favors the approach. Again, we emphasize that applying Algorithm SPACE_FINDER** to each of k pairs of arrays $(A_1, D_1), \ldots, (A_k, D_k)$ allows considerable flexibility in how memory is allocated to different uses during the course of a simulation's execution.

Our exposition here is intended merely to make the reader aware of the fact that dynamic space management that incorporates block-array data structures generally executes more efficiently with respect to computing time and less efficiently with respect to space usage. Since computing time considerations usually dominate, one may well prefer a simulation programming language or package that employs these concepts. Since vendors rarely mention this topic in their promotional literature, it falls to the potential user of a language or package to elicit this information.

The greatest perceived benefit of Algorithm SPACE_FINDER** arises from the observation that it reserves blocks of space and requires little time when releasing space, and that the frequency with which it encounters the relatively high cost of resizing is inversely related to the block size n.

There remains the issue of unused space blocks. Suppose that k blocks of size $n \times r$ have been allocated. If the demand for space falls below k blocks during a simulation, virtually no language attempts to unallocate the unused space, since this would incur the relatively costly effort of relocating records spread throughout the k blocks into the retained $k - 1$ blocks.

Accordingly, there is a tendency for allocated space to increase as simulated time evolves. However, the elapsed simulated time between successive space increases tends to decrease, thus slowing the rate of increase.

ALGORITHM SPACE_FINDER**

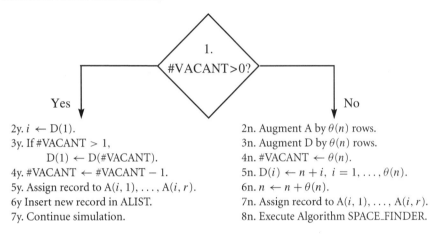

Yes	No
2y. $i \leftarrow D(1)$.	2n. Augment A by $\theta(n)$ rows.
3y. If #VACANT > 1,	3n. Augment D by $\theta(n)$ rows.
$\qquad D(1) \leftarrow D(\text{#VACANT})$.	4n. #VACANT $\leftarrow \theta(n)$.
4y. #VACANT $\leftarrow$ #VACANT $- 1$.	5n. $D(i) \leftarrow n + i$, $i = 1, \ldots, \theta(n)$.
5y. Assign record to $A(i, 1), \ldots, A(i, r)$.	6n. $n \leftarrow n + \theta(n)$.
6y Insert new record in ALIST.	7n. Assign record to $A(i, 1), \ldots, A(i, r)$.
7y. Continue simulation.	8n. Execute Algorithm SPACE_FINDER.

5.4 FUTURE-EVENT SET

Every discrete-event simulation that incorporates random behavior consumes computing time to:

- Search for space

- Manage the future-event set

- Perform other list processing

- Generate pseudorandom numbers

- Sample from probability distributions

- Perform arithmetic operations

- Perform statistical analysis

- Generate reports.

Of these, managing the future-event set accounts for the largest proportion of computing time. Although every simulation programming language provides special-purpose data structures to accommodate the FES and operations on it, individual differences do exist among languages, and as a consequence, the same simulation problem programmed in different languages can induce substantially different computing times. Nevertheless, all methods for managing the future-event sets share at least four capabilities in common:

a.   Insert a new event notice in the FES.

b.   Delete the next-event notice in the FES.

c.   After a deletion, find the next-event notice among those remaining on the FES.

d.   Find an event notice whose attributes assume particular values and delete it from the FES.

The next-event notice denotes the one whose clock attribute value is smallest among all event notices in the FES. Action a corresponds to scheduling an event, b to choosing the next event to execute, c to identifying which event notice is to be executed next, and d to canceling the scheduled execution of an event. Since the cancellation rate customarily is much smaller than the insertion, deletion, and search rates of actions a, b, and c, respectively, cancellations contribute little to the overall computing time of a simulation run. Accordingly, we say no more about them and, unless otherwise stated, hereafter subsume action c into action b.

Ideally, we prefer to employ the simulation programming language that manages the future-event set most efficiently with regard to space and time. Regrettably, empirical studies have repeatedly shown that no method demonstrates uniformly superior performance with regard to average computing time when tested on different types of simulation models (McCormack and Sargent 1981, Chung et al. 1993). Two factors principally determine the computing time required to perform actions a and b: the number of event notices in the FES and the distribution of simulated time between scheduling and executing an event. In particular, they influence the sample sequence of insertions and deletions observed during the simulation, the computing time devoted to action a, and, possibly, the computing time devoted to action b.

Let a and b denote executions of actions a and b, respectively. For a particular simulation with n events in the FES, recall from Section 5.1.1 that $g_a(n)$ and $g_b(n)$ denote the respective average times that these actions consume. At a particular moment in simulated time, suppose that the FES contains k event notices and experiences a sample sequence of 20 insertions and 20 deletions in the order

$$abababababababababababababababababababab. \tag{5.8}$$

Then the average computing time consumed is $20[g_a(k) + g_b(k+1)]$. Alternatively, suppose the 20 insertions and 20 deletions occur in the order

$$aaaaabbbbbaaaaabbbbbaaaaabbbbbaaaaabbbbb. \tag{5.9}$$

Then average computing time is

$$4[g_a(k) + g_a(k+1) + g_a(k+2) + g_a(k+3) + g_a(k+4) \tag{5.10}$$
$$+ g_b(k+5) + g_b(k+4) + g_b(k+3) + g_b(k+2) + g_b(k+1)],$$

revealing a potential difference in average computing time, depending on the form that the insertion/deletion sequence takes. In addition, two different types of simulation faced with the same insertion/deletion sequence have different average times $g_a(k)$ and $g_b(k)$ as a consequence of different distributions of elapsed time between scheduling and executing an event.

The absence of a universally dominant procedure with regard to average performance forces us to focus principally on a worst-case computing time measure of peformance that depends on the number of event notices in the FES but is usually independent of distributional properties. This measure reveals how computing time grows in the worst case as the size of the simulation grows. Its limitation is that it rarely informs us as to how a procedure performs relative to others for small and moderate size simulations. Nevertheless, the worst-case measure can be invaluable in explaining why a particular simulation software package peforms well for a relatively small demonstration problem but consumes dramatically more computing time when employed to simulate a moderate or large system.

5.4.1 SINGLE LINEAR EVENT LIST

As shown in expression (2.5), the record of an event notice has multiple fields for storing identification, time of execution (clock) other attributes, predecessor address, and successor address. Since the discussion of the next several sections focuses on the ordering of execution times, all depicted future-event sets show only the values of the clock attributes for event notices, and arrows point to their predecessors and successors.

Suppose that the FES is maintained as a doubly linked list as in Figure 5.4 with deletion from the low end, and insertion from the high end, where the number in each circle denotes the scheduled execution time for the corresponding event and where the simulation successively schedules four events with desired execution times 51, 42, 34, and 28 before it deletes the event notices with the most iminent execution time. Note that insertions made in this order take $4 + 7 + 12 + 15 = 38$ comparisons. More generally, inserting a new event notice in a doubly linked linear FES that already contains N event notices takes $O(N)$ worst-case computing time, but deletion times are constant, independent of N.

Recall from Section 5.3 that $g_a(N)$ denotes mean computing time expended per insertion, whereas $g_b(N)$ denotes mean computing time per deletion. Clearly, their values depend on the data structure adopted for the FES and how the scheduling procedure operates on it. In the present setting $g_a(N)$ is a constant, independent of the number of entries N in the FES. However, empirical studies have repeatedly shown that $g_b(N)$ varies with the form of the interevent-time distribution and cannot be easily characterized as a function of FES size, (e.g., McCormack and Sargent 1981 and Chung et al. 1993). Because of these limitations on

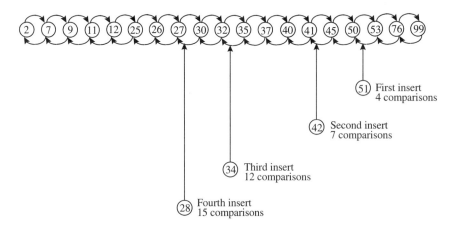

Figure 5.4　Inserting and deleting event notices in a doubly linked linear future-event set (entries are clock attribute values)

average behavior our assessment of alternative methods of managing the FES concentrates on worst-case analyses. In particular, we focus on how computing time is affected as the *size* of a simulation grows.

Consider the open-loop queueing system in Figure 5.1 with $r \geq 1$ types of arrivals, where interarrival times are random and where each arriving job has k tasks of which task i must be performed by one of the m_i servers of type i for $i = 1, \ldots, k$. Assume that the relationship between arrival and service rates is such that the simulation finds all servers busy almost all of the time. This high traffic intensity environment characterizes many manufacturing and telecommunications systems. Moreover, it implies that service completion at one station (deletion from the FMS) immediately induces the scheduling of service for this server and a task waiting in its buffer (insertion in the FMS).

During its tenure in the system, each job induces the scheduling of $1 + k$ events, one for the next arrival and k for completions at each of the k banks of servers. Suppose that the FES is a single linear doubly linked list of event notices ordered by desired execution times. Then, as previously stated, $g_b(N)$ is relatively constant, regardless of the number of event notices N in the FES. Since virtually all servers are busy at any moment in simulated time, the FES contains approximately $1 + m_1 + \cdots + m_k$ event notices. It also contains r event notices, one for each distinct arrival type. Therefore, the scheduling of an additional event notice takes at most $m_1 + \cdots + m_k + r$ comparisons. Since this occurs k times per job during the simulation, worst-case computing time per completion is proportional to $k^2 \bar{m} + kr$, where expression (5.2) defines $\bar{m}$.

For fixed k and r, this reveals that worst-case computing time grows linearly in $\bar{m}$, the mean number of servers per station; but, for fixed $\bar{m}$ and r, worst-case computing time grows as k^2, where k denotes the number of stations in series. For this type of list, series queueing problems clearly demand more computing time than parallel queueing problems do, as k grows relative to $\bar{m}^{1/2}$.

EFFICIENCY

As the remainder of this chapter shows, other methods of managing the FES can be expected to consume considerably less computing time than a single linear linked list as the number of simultaneously scheduled events in the FES grows. However, we cannot totally dismiss the value of this data structure. When there are few simultaneously scheduled events, it is known that this form of event management is computationally more efficient on average than all others that have been considered. For example, results in Jones (1986) indicate that a linear linked list is best for fewer than 10 simultaneously scheduled events.

5.4.2 MULTIPLE LINEAR EVENT LISTS

One approach to reducing list-processing time relies on multiple linear lists in the FES, one for each type or class of event. When scheduling an event of type i, one peforms:

SCHEDULING

1. Create an event notice.

2. Determine and assign desired execution time to its clock attribute.

3. Employ linear search to insert the event notice into list i.

4. If first in list, update the header address.

5. If last in list, update the trailer address.

When an event terminates execution, control passes to the timing routine, which performs:

SELECTING THE NEXT EVENT

1. Search the list of header addresses to identify which event type has the smallest value for its clock attribute.

2. Remove this event notice from its list.

3. Update the header address.

4. Advance simulated time to the value of this event's clock attribute.

5. Execute the event.

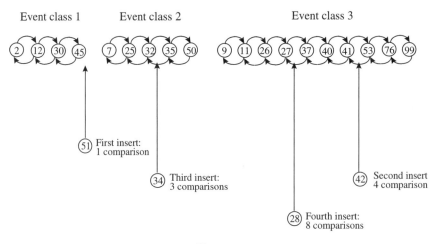

Figure 5.5   Inserting and deleting event notices in a doubly linked multiple linear FES (entries are clock attribute values)

These prototypes characterize SIMSCRIPT II.5's approach to event management.

Figure 5.5. displays the same entries as in the single list FES of Figure 5.4, but ordered in sublists by event class. The simulation again successively schedules events with desired execution times 51, 42, 34, and 28 before it executes the next event with execution time 2. Insertion takes $1 + 4 + 3 + 8 = 16$ comparisons in contrast to the 38 comparisons for the single list in Figure 5.4. However, deletion takes two rather than one comparisons because of the need to check which among the first entries in the three lists has the smallest execution time.

For the open-loop system in Figure 5.1 with $r \geq 1$, we can conceive of $1 + k$ event classes, list 1 for the r arrival event notices that must always exists and list $i + 1$ for each of the at most, m_i completions at station i for $i = 1, \ldots, k$. Therefore, scheduling for list 1 takes $O(r)$ comparisons, whereas scheduling for list $i + 1$ takes $O(m_i)$ comparisons. However, a deletion always takes $1 + k$ comparisons. Since each job passes through the k stations, the number of scheduling comparisons per job is $O(r + k\bar{m})$ but the number of deletion comparisons per job is $O(k + k^2)$.

For this multiple list formulation, worst-case number of comparisons per job is $O(k\bar{m} + r) + O(k + k^2)$. For fixed k and r, this bound grows linearly in $\bar{m}$, revealing an increasing benefit, due to a smaller coefficient, when compared to the single list approach. For fixed $\bar{m}$ and r, it grows as k^2 but its coefficient is now reduced from $\bar{m}$, in the single list FES of Section 5.4.1, to 1. Although substantial differences in computing times can occur as the number of stations in series k grows, the proportionality to k^2 in both the linear- and multiple-list approaches can require substantial computing times as k increases.

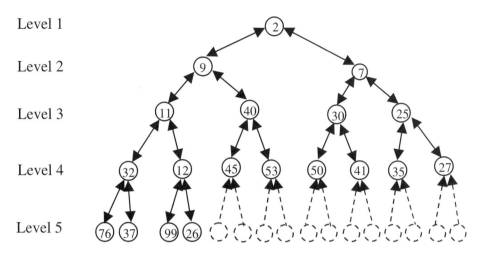

Figure 5.6   Binary heap displaying key(v) (doubly linked)

5.4.3 HEAPS

The concept of a binary heap provides another data structure for managing the FES (Williams 1964, Evans 1988). Consider a *directed tree* all of whose edges point toward its *root node l*. Figure 5.6 illustrates the concept, where for present purposes we again regard the numbers at each node as the execution times for events. The *children* of node v at level i are the nodes at level $i + 1$ connected to v by an edge, and v is the *parent* of each of its children. A node with no children is called a *leaf*. The depth of node v is the number of edges in the path connecting v to the root l. A *heap* is a rooted tree each of whose nodes v has a value key(v) that is the smallest among the keys of all nodes in its subtree. A heap in which each node has at most two children is called a *binary heap*.

The data structure associated with a binary heap is an array in which nodes are stored in order from top to bottom; its root occupies position 1, its children occupy positions 2 and 3, and, more generally, the nodes at depth i occupy positions $2^i, 2^i + 1, \ldots, 2^{i+1} - 1$. If the binary heap is maintained so that its N elements always occupy the first l left-justified positions as in Figure 5.6, then every element of the tree has depth no greater than $\lfloor \log_2 N \rfloor$.

The bidirectional edges in Figure 5.6 emphasize that the list maintains predecessor and successor addresses at each node and is thus doubly linked. In conformity with our definition of a binary heap and for visual clarity, Figure 5.7 displays unidirected edges.

In terms of the FES, a node represents an event notice and key(v) denotes the value of the clock attribute of event notice v. To insert or delete an event notice, one relies on procedures akin to Algorithms SIFTUP(x) and SIFTDOWN(x). The first takes an element x of the tree whose key(x) has been decreased and moves it up the tree to its appropriate place. The second takes an element x whose key has been increased and moves it down the tree to its appropriate place.

Algorithm SIFTUP(x)

Until SUCCESS:
    Repeat
        If x is the root, SUCCESS
        Otherwise:
            Let y be the parent of x
            If $\text{key}(y) \leq \text{key}(x)$, SUCCESS
            Otherwise:
                Swap entries in x and y.

Algorithm SIFTDOWN(x)

Until SUCCESS:
    If x is a leaf, SUCCESS
    Otherwise:
        Let y be the child of x with minimal key
        If $\text{key}(y) \geq \text{key}(x)$, SUCCESS
        Otherwise:
            Swap entries in x and y.

To insert event notice x into the FES, one:

- Inserts x as leaf with $\text{key}(x)$

- Executes Algorithm SIFTUP(x).

Before assigning elements x to position l, one can think of $\text{key}(l) = \infty$.

    An $n \times 1$ block array, as in Figure 5.3, can serve as a basis for maintaining an FES that employs the binary heap concept. If the FES holds $l - 1 < n$ event notices prior to an insertion, then D(1) points to the root or first event notice, and for $1 \leq i \leq (l-2)/2$, D($2i$) and D($2i + 1$) point to children whose parent's address is stored in D(i). Then insertion via Algorithm SIFTUP(x) makes x the last leaf in the heap by assigning the address of x to D(l).

    To delete the next event notice from the FES, one:

- Removes the root event notice from the heap.

- Moves the last leaf y to the root position.

- Performs SIFTDOWN(y).

Figure 5.7 shows a binary heap FES for the same event notices displayed in the single linear list of Figure 5.4 and in the multiple linear list in Figure 5.5. The successive insertion of event notices with times 51, 42, 34, and 28 takes $1 + 2 + 3 + 3 = 9$ comparisons and $0 + 1 + 2 + 2 = 5$ switches, in contrast to 38 comparisons and one link in the single linear list case and 16 and one link in the multiple linear list case. The single link comparisons operation in each of the linear list cases corresponds to the final step in the search, where the new event notice is chained to the list. Link and switch operations are comparable in computing time.

The subsequent deletion of the first event notice from the FES in Figure 5.7e takes one leaf-to-root transfer, three switches, and three comparisons. A leaf-to-root transfer is comparable in computing time to a switch. Although more work than for deletions in either Figure 5.4 or 5.5, the combined work for insertion and deletions for the binary heap favors this approach to list management over the single and multiple linear list approaches.

Since the particular clock times in Figures 5.7a through 5.7e dictate the numbers of comparisons per insertion and deletion, it is of interest to consider more global results. For four successive entries into a heap with 19, 20, 21, and 22 nodes, the worst-case numbers of comparisons are $\lfloor \log_2 19 \rfloor = 4$, $\lfloor \log_2 20 \rfloor = 4$, $\lfloor \log_2 21 \rfloor = 4$, and $\lfloor \log_2 22 \rfloor = 4$, respectively. For deletion following these four insertions, the worst-case number of comparisons is $\lfloor \log_2 23 \rfloor = 4$.

More generally, with D maintained as a binary heap with N entries, insertion and deletion each take $O(\log_2 N)$ computing time, an unequivocal advantage over single and multiple linear lists as $l \rightarrow \infty$. Also, canceling a particular event notice in a heap follows the same steps as a deletion and therefore has $O(\log_2 N)$ time. However, there is at least one limitation of this formulation that needs to be overcome. As just described, the binary heap makes no provision for priority order when two or more events have the same desired execution time. This can be overcome as follows. Let OLD denote the old event notice in the heap and let NEW denote the new event notice. If OLD and NEW have the same execution times but NEW has a priority order that favors it over OLD, swap the locations of OLD and NEW in the heap. Otherwise, do not swap them.

SPACE

Suppose that a simulation attempts to schedule an event when its FES already holds N event notices. Then the simulation programming language must provide for upwardly resizing the arrays and reconstituting the heap in the augmented array. While choosing N large reduces the frequency of resizing, it may also lead to considerably unused, but committed, space over long periods of simulated time. An alternative approach provides memory space incrementally, as needed.

A binary heap with N entries has $\lfloor \log_2 N \rfloor$ levels, where level i accommodates 2^{i-1} event notices for $i = 1, \ldots, N - 1$. Figure 5.6 illustrates the concept for $N = 19$, where $\lfloor \log_2 19 \rfloor = 5$ and levels 1 through 4 have 1, 2, 4, 8 entries, respectively, and level 5 has four entries and space to accommodate 12 additional entries. When the FES has exactly 2^{i-1}

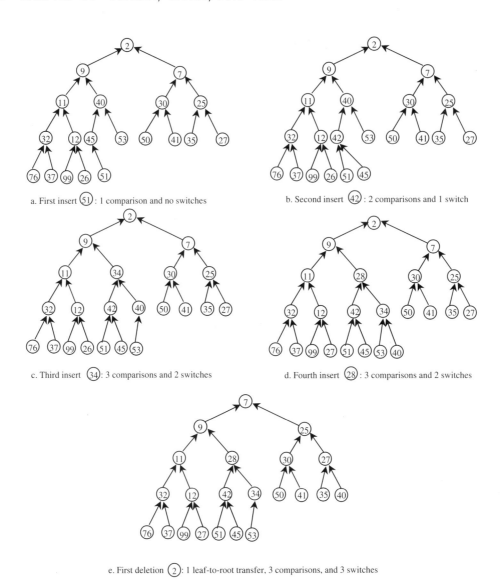

a. First insert (51): 1 comparison and no switches

b. Second insert (42): 2 comparisons and 1 switch

c. Third insert (34): 3 comparisons and 2 switches

d. Fourth insert (28): 3 comparisons and 2 switches

e. First deletion (2): 1 leaf-to-root transfer, 3 comparisons, and 3 switches

Figure 5.7 Inserting and deleting event notices in a binary heap future-event set (doubly linked; entries are clock attribute values)

entries in its i levels, and an event is to be scheduled, the dynamic space allocator finds a block array of length 2^{i+1} for level $i + 1$ and connects it pointerwise to level i, much as in the example in Figure 5.6. The address of the leftmost entry at level $i + 1$ now becomes the trailer for the list. Since the space at this level is a block array, the location of the next unoccupied record at level $i + 1$ is implicitly determined.

In summary, each time it runs out of space this FES doubles its size. Moreover, it takes $O(2^i)$ computing time to link the new space at level $i + 1$ to the old space.

For the prototypical simulation described in Figure 5.2 with $r \geq 1$ arrival types, each job passing through the system takes $O(2k \log_2(k\bar{m} + r))$ list processing time to maintain an FES structured as a binary heap, where the constant 2 merely emphasizes that deletion as well as insertion has a worst-case bound proportional to $k \log_2(k\bar{m} + r)$. As the number of visited stations increases, this computing time grows as $k \log_2 k$. Conversely, as the average number of servers increases, the bound grows as $\log_2 \bar{m}$. As k, $\bar{m}$, and r increase, these properties make clear the benefit that a binary heap offers for FES management when compared to the $O(k^2\bar{m} + kr)$ worst-case time for the single linear list of Section 5.4.1 and the $O(k\bar{m} + r) + O(k(k + 1))$ worst-case time of the multiple linear list approach in Section 5.4.2.

It is known that as N, the number of scheduled events in an FES maintained as a heap, increases, the average computing time required to maintain the heap is $O(\log_2 N)$ (e.g., Jones 1986). However, it is also known that for small N a heap is considerably less efficient than a linearly linked list for FES management. Presumably, the large setup time encountered each time a heap is modified accounts for its relatively poor performance for small N.

5.4.4   HENRIKSEN'S ALGORITHM

An alternative approach to event management in Henriksen (1977, 1983) also reduces the time complexity of FES managment as compared with either single or multiple linear lists. It formulates the FES as a single linear doubly linked list ordered from largest to smallest, but augments the list with two pseudo-event notices, one with clock attribute set to any negative value and the other with clock attribute set to ∞. In essence, these become the permanent trailer and header records, respectively, for the list.

The algorithm also maintains a binary search tree with a five-field record corresponding to each node. The fields contain a pointer to an event in the FES, the time at which this event is scheduled to be executed (clock), pointers to the event's predecessor and successor in the FES, and a pointer to the record in the tree with the next lowest scheduled execution time. The tree has $M := 2^K - 1$ nodes for some integer $K \geq 1$. For conciseness, let $D(1), \ldots, D(M)$ denote the pointers in the tree to M events in the FES with the property that $\text{clock}(D(i)) \leq \text{clock}(D(i + 1))$ for $1 \leq i \leq M - 1$ with $D(M) = \infty$. Here $\text{clock}(D(i))$ denotes clock attribute in the record corresponding to a node in the binary search tree and serves as the key (see Section 5.4.3) for maintaining the tree.

Today, most computers use *byte-addressable memory*. For example, a computer whose memory consists of 32-bit words allows a user to address each of the four bytes in a word in bits 1–8, 9–16, 17–24, and 25–32. However, the piece of memory needed to store the address of the byte must itself be a 32-bit word. When we use the term *word* hereafter, we usually mean a piece of memory that can contain a memory address, e.g., a 32-bit word.

Finding and deleting the event notice corresponding to the next event for execution takes constant computing time regardless of the number of entries in the FES. To insert a new event notice with desired execution time T in the FES (scheduling), one:

- Performs a binary search on D to find the pointer:

$$j = \min[i : \text{clock}(D(i)) \geq T, \ 1 \leq i \leq M].$$

- Scans the FES backwards starting from $D(j)$ and inserts the new event notice in its appropriate location.

During a scan, the procedure keeps a COUNT of the number of unsuccessful comparisons it has performed. If the COUNT reaches a predetermined number l and $j > 1$, the procedure alters $D(j-1)$ to point to the last scanned event notice, resets the COUNT to zero, and continues scanning the FES until success or until the COUNT again reaches l. Alternatively, if $j = 1$ and COUNT reaches l, the procedure increases the size of the D array to $M' = 2^{K+1} - 1 = 2M + 1$, reinitializes its entries, and continues scanning the FES until success.

The expanded tree is initialized by setting its leftmost leaf (i.e., $D(M')$) to point to the pseudo-event notice at the rightmost end of the FES and $D(1), \ldots, D(M'-1)$, the pointers at all other nodes of the tree, to point to the pseudo-event notice at the leftmost end of the FES. By assigning the "unused" pointers $D(1), \ldots, D(M'-1)$ to the header entry in the FES at the moment of resizing, the procedure can later reassign them, one at a time, whenever COUNT $= l$ and $j > 1$. Algorithm H-L describes the major steps for insertion.

ALGORITHM H-L

Purpose: To insert an event notice into the FES using Henriken's Algorithm.
Given: j, K, M, and $\{D(1), \ldots, D(M)\}$.
Method:

    Start scan at $D(j)$.
    COUNT $\leftarrow 0$.
    Repeat:
        COUNT $\leftarrow$ COUNT$+1$
        scan the next event notice
        if COUNT$=l$:
($*$)   if $j > 1$, reset $D(j-1)$ to point to the last scanned event notice
            otherwise:
            increase size of D from $M = 2^K - 1$ to $M' = 2^{K+1} - 1$
            set $D(M')$ to point to the trailer event in the FES and $D(1), \ldots,$
            $D(M'-1)$ to point to the header event in the FES
            $j \rightarrow M'$

COUNT ← 0
Until success.

Whereas a binary heap has the property that the key for each node is no greater than the key for any node *below* it in the tree, a binary search tree has the property that the key for each node is no greater than that for any element in the original node's left subtree but is no smaller than that for any element in the original nodes right subtree.

As l decreases the number of scans until success diminishes, as does the frequency of resizing D. However, decreasing l increases the likelihood of increasing M, the size of D, and thereby increasing the cost of reinitializing D. Henriksen (1977) suggests $l = 4$. Kingston (1986a,b) describes procedures that reduce the cost of initalization and, in finding the next event in the FES, for reducing the size of D.

Figure 5.8 displays the steps for insertion and deletion from an FES based on Henriksen's Algorithm using the four successive insertions with clock attribute times 51, 42, 34, and 28 followed by a deletion, as in Figures 5.4, 5.5, and 5.7. Each insertion induces at most $\lfloor \log_2 7 \rfloor = 2$ comparisons to find the pointer to the location at which to start the list search. In addition to the comparisons, insertion requires two pointer changes in Figures 5.8a and 5.8c. Note that no comparison is made with the trailer clock entry of ∞.

In an FES with N elements, Henriksen's Algorithm takes O(1) computing time to find the next event to execute and $O(\log(N/l)) + O(N^{1/2}) + O(l)$ worst-case time to insert a new event notice (Kingston 1986a, 1986b). However, substantial empirical evidence indicates that insertion takes mean computing time proportional to $\log_2 N$ for given l and that this result is insensitive to distributional properties for a wide variety of simulations (Henriksen 1994 and McCormack and Sargent 1981).

In the context of simulating a queueing system as in Table 5.1 with r arrival types, these results suggest $O((k+1)(k\bar{m}+r)^{1/2})$ worst-case computing time and an average computing time proportional to average $(k+1)\log_2(k\bar{m}+r)$ for fixed l. As the average number of servers per station, $\bar{m}$, increases for fixed k, this computing time grows as $\bar{m}^{1/2}$ at worst and as $\log_2\bar{m}$ on average. As the number of stations k increases for fixed $\bar{m}$, the computing time grows as $k^{3/2}$ at worst and as $k\log_2 k$ on average. When compared to the single or multiple linear list approaches with standard scanning, the benefits of Henriksen's Algorithm are apparent. Algorithm H-L is prototypical of the approaches to event management in SLX, GPSS/H, AweSim, and SLAM.

5.4.5   BEST FES MANAGER

Table 5.1 summarizes the bounds on computing time for managing the FES by the four data structures just described. However, at least one additional issue remains to be discussed, namely, an FES with few event notices. As Section 5.4.1 intimates, a linearly linked list is most efficient on average for these. Hence, a language that employs a binary heap or Henriksen's Algorithm may consume more time than necessary to manage an FES with few entries on average. However, we anticipate that any excess computing time will be small in absolute

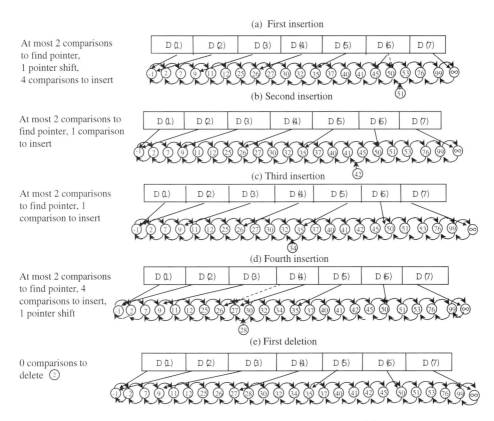

Figure 5.8   Inserting and deleting event notices in an indexed future-event set: Henriksen's Algorithm (entries in list are clock attribute values; entries in index array are ordered pointers; dashed arrows denote pointers before shift during search)

terms. Moreover, we can regard this excess as a relatively small penalty when compared to the efficiencies that heap-like structures and Henriksen's Algorithm offer for large future-event sets. The circumstances that are yet to be investigated fully are those in which an FES has few events notices most of the time and occasionally, but rarely, grows substantially in size. Do these rare but substantial excursions favor simple or sophisticated structures?

Since studies of more sophisticated heap structures for FES management fail to reveal unequivocal superiority for any one of them (e.g., Jones 1986 and Chung et al. 1993), it is not possible to choose a simulation programming language based on its universally accepted computing time superiority. Nevertheless, it is clear that for a given problem, GPSS/H and SLAM, based on Henriksen's Algorithm, and SIMAN, based on the binary heap concept, execute faster than SIMSCRIPT II.5, which relies on multiple doubly linked linear lists.

Table 5.1 Managing the future-event set for a saturated system ($k :=$ number of stations in series, $\bar{m} :=$ average number of servers per station, $r :=$ number of distinct arrival types)

Data structure	Insertion time/job	Deletion time/job
Linear linked list	$O(k^2\bar{m} + k(\bar{m} + 1) + r)$	$O(1)$
Multiple linear linked list	$O(k\bar{m} + r)$	$O(k^2 + k)$
Binary heap	$O((k + 1)\log_2(k\bar{m} + r))$	$O((k + 1)\log_2(k\bar{m} + r))$
Henriksen's Algorithm Indexed and linear lists	$O((k + 1)(k\bar{m} + r)^{1/2})$ $\propto (k + 1)\log_2(k\bar{m} + r)$ on average	$O(1)$

5.5 BROADENING THE CLASS OF DELAY MODELS

The prototypical delay model in Section 5.1 identifies the number of station, k, the average number of servers per station, $\bar{m}$, and in the capacitated case, the average buffer size, $\bar{c}$, as the major influences on interaction time per job with the FES. The model assumes that upon completing service at station j, a job moves instantaneously to station $j + 1$ for $j = 1, \ldots, k - 1$. We next investigate the FES interaction time for a more inclusive class of simulation models.

CAPACITATED TRAVEL

Suppose there are l_j pallets or conveyances at station j to move work to station $j + 1$ and that each job requires a conveyance to move from station j to station $j + 1$. After a pallet, cart, conveyance, etc., arrives at station $j + 1$, it must return to station j to pick up the next job. Let

$$\bar{l} := \frac{1}{k - 1} \sum_{j=1}^{k-1} l_j.$$

Since there are at most $(k - 1)\bar{l}$ event notices associated with the arrival of a job at a station or the arrival of the conveyance back at its home station, the FES contains at most $r + k\bar{m} + (k - 1)\bar{l}$ event notices.

If each pallet has unit capacity, then each job is responsible for $k - 1$ interactions with the FES for pallet arrivals and for an additional $k - 1$ for their return to their originating stations. Therefore, the total number of FES interactions is $k + 1 + 2(k - 1) = 3k - 1$, so that, for example, heap insertion and deletion times per job each are $O((3k - 1)\log_2(r + k\bar{m} + (k - 1)\bar{l}))$.

If pallet capacity at each station is $c > 1$ and a pallet moves from station i to station $i + 1$ only when full, then we conceive of each job as inducing $(k - 1)/c$ interactions for pallet arrivals and $(k - 1)/c$ for returns, so that, for example, heap insertion and deletion times per job each are $O(((1 + 2/c)k - 2/c) \log_2(r + k\bar{m} + (k - 1)\bar{l}))$.

UNCAPACITATED TRAVEL

If transit requires no pallet, cart, etc., the number of jobs in transit from station to station is unbounded, and therefore our analysis uses a different approach.

5.5.1 VARIABLE TRANSIT TIMES

Assume that upon completing service at station j, a job travels to station $j + 1$ with nonnegative transit time for $j = 1, \ldots, k - 1$. Two scenarios are possible, capacitated travel and uncapacitated travel. Let $L_j(\tau)$ denote the number of jobs in transit at time τ from station j to station $j + 1$. Then the arrival of a job at its destination constitutes an event and, hence, must be scheduled. Accordingly, at simulated time τ the FES contains $r + B(\tau) + L(\tau)$ event notices, where, as before, r denotes the number of scheduled arrival event notices and

$$L(\tau) := L_1(\tau) + \cdots + L_{k-1}(\tau).$$

Scheduling an event now takes $g_a(r + B(\tau) + L(\tau))$ average computing time, whereas selection takes $g_b(r + B(\tau) + L(\tau))$ average computing time, where, as before, each simulation language's procedure for scheduling and selection determines the forms of $g_a(\cdot)$ and $g_b(\cdot)$.

Unlike the model of Section 5.1, this broader model puts no upper bound on the number of event notices in the FES. Let $1/\omega_{tj}$ denote mean transit time for a job from station j to $j + 1$ for $j = 1, \ldots, k - 1$ and let

$$v_j := \frac{\omega_a}{\omega_{tj}}.$$

For an arbitrarily selected time τ, the mean number of jobs in transit in an uncapacitated buffer system ($c_j = \infty$, $j = 1, \ldots, r$) is

$$E(\tau) = (k - 1)\bar{v}, \tag{5.11}$$

where

$$\bar{v} := \frac{1}{k - 1} \sum_{j=1}^{k-1} v'_j.$$

Note that long transit times relative to interarrival times at state 1 imply a large mean number of jobs in transit.

Since jobs in transit from one station to the next are in an *infinite-server* system, we can think of this extended model as a sequence of $2k - 1$ stations where the odd-numbered stations each have finite numbers of servers and the even-numbered ones each have an infinite number of servers. At an arbitrarily chosen simulated time τ, the FES for the extended model has mean number of event notices

$$\mathrm{E}[r + B(\tau) + L(\tau)] \leq r - \bar{t} + k(\bar{m} + \bar{v}).$$

Moreover, each job passing through the system has $2k$ scheduling interactions and $2k$ selection interactions with the FES.

For the single linear list approach to FES management in Section 5.4.1,

$$\text{average scheduling time per job} = \mathrm{O}(2k(r - \bar{v} + k(\bar{m} + \bar{v})))$$
$$= \mathrm{O}(2k^2(\bar{m} + \bar{v}) + 2k(r - \bar{v})),$$

whereas
$$\text{average selection time per job} = \mathrm{O}(2k).$$

Since no competitive simulation language uses this data structure, these results are included merely for purposes of comparison with the more commonly employed data structures.

For the multiple linear list FES of Section 5.4.2, we treat all transit arrivals at station j as forming a separate class of events for $j = 2, \ldots, k$. Therefore, the extended model has $2k$ event classes, one for arrivals, k for service completions, and $k - 1$ for station-to-station transit completions. Then

$$\text{average scheduling time per job} = \mathrm{O}(k(\bar{m} + \bar{v}) + r - \bar{v}),$$

whereas

$$\text{average selection time per job} = \mathrm{O}(4k^2),$$

since the first entries in all $2k$ event lists must be compared for each job's $2k$ selections.

For an FES based on the binary heap concept of Section 5.4.3 and given the number of busy servers $B(\tau)$ and the number of jobs in transit $L(\tau)$, a single scheduling takes $\mathrm{O}(\log_2(r + B(\tau) + L(\tau)))$ average computing time, as does a single selection. The inequality

$$e^x \geq 1 + x, \qquad x \geq 0,$$

enables us to write

$$\log_2(r + B(\tau) + L(\tau)) \leq \log_2(r + k\bar{m}) + \frac{L(\tau)}{r + k\bar{m}} \cdot \frac{1}{\log_e 2}. \tag{5.12}$$

Therefore,

average scheduling time per job

$$= O\left(2k\log_2(r + k\bar{m}) + \frac{2k(k-1)\bar{v}}{r + k\bar{m}} \cdot \frac{1}{\log_e 2}\right)$$

$$= O\left(2k\log_2(r + k\bar{m}) + \frac{2k\bar{v}}{r/k + \bar{m}} \cdot \frac{1}{\log_e 2} - \frac{2\bar{v}}{r/k + \bar{m}} \cdot \frac{1}{\log_e 2}\right),$$

and

average selection time per job

$$= O\left(2k\log_2(r + k\bar{m}) + \frac{2k\bar{v}}{r/k + \bar{m}} \cdot \frac{1}{\log_e 2} - \frac{2\bar{v}}{r/k + \bar{m}} \cdot \frac{1}{\log_e 2}\right).$$

For fixed r, $\bar{m}$, and $\bar{v}$, these bounds grow as $k\log_2 k$. As $\bar{v}$ increases relative to $r/k + \bar{m}$ for fixed r, k, and $\bar{m}$, the bounds become increasingly larger than need be. Nevertheless, with regard to average scheduling and selection times per job, the binary heap approach maintains its edge over single and multiple linear list approaches.

Recall that empirical studies (Henriksen 1994) show that scheduling an event using Henriksen's Algorithm (Section 5.4.4) when N event notices are in the FES takes average computing time proportional to $\log_2 N$. Also, selecting an event takes O(1) time. Based on these properties, applying Henriksen's Algorithm for managing the FES for our $(2k - 1)$ station model has

average scheduling time per job

$$= O\left(2k\log_2(k\bar{m} + r) + \frac{2k\bar{v}}{r/k + \bar{m}} \cdot \frac{1}{\log_e 2} - \frac{\bar{v}}{r/k + \bar{m}} \cdot \frac{1}{\log_e 2}\right),$$

whereas

average selection time per job $= O(2k)$.

These results continue to imply an advantage for binary heaps and Henriksen's Algorithm for scheduling and selection times.

5.5.2 VARYING NUMBER OF TASKS PER JOB

An alternative extension of the model in Section 5.1 assumes that the number of stations that a job visits varies during its passage through the system. At least three modalities exist for this behavior:

1. Jobs make one pass through the system but skip some stations.

2. Jobs may visit some stations more than once.

3. For some or all stations j, failure of quality-assurance testing following completion at the station directs the job back to the station for rework.

Since Modality 1 induces k or fewer interactions per job with the FES, the results of Sections 5.4.1 through 5.4.4 continue to apply.

For Modality 2, let

$$w_{ij_l} := \text{probability that a job of type } i \text{ visits}$$
$$\text{station } j \text{ a total of } l \text{ times}, \qquad l = 0, 1, 2, \ldots.$$

If for a class of problems that incorporate Modality 2

$$\sum_{j=1}^{k} \sum_{l=0}^{\infty} l w_{ij_l} = \mathrm{O}(k)$$

as the number of stations grows, then the earlier results again continue to apply. In particular, the assumption holds if $w_{ijl} = 0$ for l greater than a specified integer.

Modality 3 requires more thought. Assume that quality-assurance testing follows each station's service. Thus the extended model has $2k$ stations where m'_j is the total number of testers after station j, each with mean testing time $1/\omega'_{tj}$, such that $\omega_a / m_j \omega'_{tj} < 1$. As a consequence, the FES has at most $k(\bar{m} + \bar{m}') + r$ event notices, where

$$\bar{m}' := \frac{1}{k}(m'_1 + \cdots + m'_k).$$

For convenience of exposition, assume that every job that completes service at station j has probability p of failing the subsequent quality-assurance test, independent of type and the number of times it has previously failed the test. Then the mean numbers of passes through station j and through its ensuing test are each $1/(1 - p)$. Therefore, the mean number of interactions per job with the FES is $1 + 2k/(1 - p)$.

For the multiple list approach to FES management in Section 5.4.2,

$$\text{average scheduling time per job} = \mathrm{O}(r + (\bar{m} + \bar{m}')/(1 - p)),$$

whereas

$$\text{average selection time per job} = \mathrm{O}((2k + 1)^2).$$

For an FES based on the binary heap structure or on Henriksen's Algorithm,

$$\text{average scheduling time per job} = \mathrm{O}((2k/(1 - p) + 1) \log_2(k(\bar{m} + \bar{m}') + r)).$$

Average selection time per job for binary heaps has the same time complexity, but Henriksen's Algorithm has $O(2k + 1)$ average selection per job.

5.6   LESSONS LEARNED

- Space management and list processing consume a relatively large proportion of the time expended in executing a discrete-event simulation program.

- Every simulation language incorporates a means for automatically managing space. Some rely on dynamic space allocation on demand. Others use reserved block-array structures that need to be resized and initialized as the demand for space increases.

- Every simulation language provides a *timing routine* for maintaining its *future-event set* (FES), the list of records of scheduled events to be executed in the future.

- In a highly congested system, space needs for the FES increase linearly with the number of stations in series and linearly with the average number of series per station.

- GPSS/H and SLAM use Henriksen's Algorithim (Section 5.4.4) to maintain their FESs. SIMAN uses an algorithim based on a heap concept (Section 5.4.3). SIMSCRIPT II.5 uses multiple linear lists (Section 5.4.2).

- When jobs of different types have i.i.d. interarrival times, an opportunity may exist for reducing the space the FES requires. This is especially true if the interarrival times have the exponential distribution.

- A job passing through all k stations in series in a simulation has interactions with the FES that take $O(k \log k)$ time for a heap-like structure and for Henriksen's Algorithm.

5.7   EXERCISES

TERMINAL PAD CONTROLLER MODEL (EXERCISE 2.5)

✐   **5.1** If one generates a character interarrival time separately for each of the r input buffers, the FES necessarily contains at least r event notices.

a. Using the exponential property, devise an alternative interarrival scheme that requires a single event notice in the FES to discount for these events. Justify your proposal.

b. For the two methods, what is the difference in the numbers of event notices?

✐   **5.2** If ω denotes the mean number of characters per packet, then direct sampling from $\mathcal{E}(1/\lambda)$ to generate interarrival times implies that an average of ω event notices are filed and removed from the FES to fill the packet and move it from its input buffer to the output buffer. The same average holds for the method of Excercise 5.1. Devise a sampling procedure that reduces the number of insertions and deletions from the FES to one each per packet while preserving the i.i.d. exponential property of the interarrival times and that allows sampling in time independent of ω, as ω becomes large. Hint: See Sections 8.13 and 8.19.

5.8 REFERENCES

Arguelles, M.C., and G.S. Fishman (1997). Reducing the frequency of future event set search in manufacturing simulation, Operations Research Department, University of North Carolina at Chapel Hill, in preparation.

Chung, K., J. Sang, and V. Rego (1993). A performance comparison of event calendar algorithms: an empirical approach, *Software-Practice and Experience*, **23**, 1107–1138.

Evans, J.B. (1988). *Structure of Discrete Event Simulation*, Ellis Horwood Limited, Chichester, England.

Henriksen, J.O. (1977). An improved events list algorithm, *Proceedings of the 1977 Winter Simulation Conference*, H.J. Highland, R.G. Sargent, and J.W. Schmidt, editors, 547–557.

Henriksen, J.O. (1983). Event list management, a tutorial, *Proc. Winter Simulation Conference*, IEEE and SCS, Piscataway, N.J., 543–551.

Henriksen, J.O. (1994). Personal communication.

Jones, D.W. (1986). An empirical comparison of priority-queue and event-set implementations, *Comm. ACM*, **29**, 300–310.

Kingston, J.H. (1986a). Analysis of Henriksen's algorithm for the simulation event set, *SIAM J. Comput.*, **15**, 887–902.

Kingston, J.H. (1986b). The amortized complexity of Henriksen's algorithm, *Bit*, **26**, 156–163.

Knuth, D. (1973). *The Art of Computer Programming: Sorting and Searching*, Addison-Wesley, Reading, MA.

Köllerström, J. (1974). Heavy traffic theory for queues with several servers. I, *J. Appl. Prob.*, **11**, 544–552.

McCormack, W.M. and R.G. Sargent (1981). Analysis of future event set algorithms for discrete event simulation, *Comm. ACM*, **24**, 801–812.

Williams, J.W.J. (1964). "Algorithm 232," *Comm. ACM*, **7**, 347–348.

CHAPTER 6

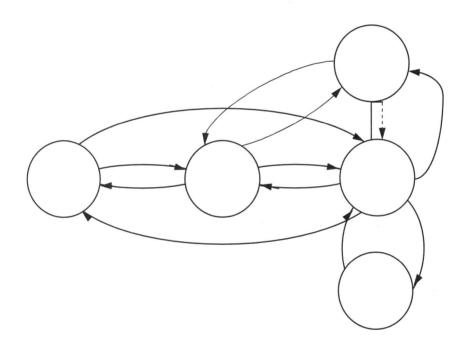

Output Analysis

Every discrete-event simulation experiment with random input generates random sample paths as output. Each path usually consists of a sequence of dependent observations that serve as the raw material for estimating the true values of one or more long-run performance measures, hereafter called *parameters*. Two phenomena, *sampling error* and *systematic error*, influence how well an estimate approximates the true value of a parameter. Random input induces sampling error, and the dependence among observations often enhances its severity. Systematic error arises from the dependence of the observations on the initially chosen state or conditions in the simulation. The presence of random and systematic errors implies that all estimates need to be interpreted with qualification.

This chapter addresses both types of error. It describes how to assess systematic error in point estimates and how to reduce its influence. It also shows how to gauge the extent of sampling error in the point estimates, principally, by constructing *asymptotically valid confidence intervals* for the corresponding true values of the parameters. A confidence interval is an *interval*, rather than a *point*, estimate of a performance measure. An asymptotically valid $100 \times (1 - \delta)$ percent confidence interval is one that includes the true parameter value with probability $1 - \delta$ for $0 < \delta < 1$ as the sample-path length $t \to \infty$.

While the width of a confidence interval indicates how well the corresponding point estimate approximates the true value, this interpretation has a justifiable statistical basis only if systematic error is negligible and the confidence interval is computed in a manner consistent with achieving the property of asymptotic validity. Accomplishing these objectives calls for considerably more effort than merely assessing error when the data consist of independent and identically distributed (i.i.d.) observations. As a consequence, simulation results are often reported with little if any supporting analysis, thus undermining conclusions based on the values of point estimates. This need not be so.

To make it easier for a simulationist to assess systematic and sampling error, this chapter focuses on easily implementable procedures, but grounded in statistical theory. This is especially so for confidence intervals, for which we describe readily accessible software, LABATCH.2, that performs the computation with minimal user effort. We also describe the automatic procedure used within Arena.

6.1   WARM-UP INTERVAL

Three factors influence the extent to which initial conditions affect how well an average based on sample-path data of finite length approximates its corresponding long-term system average. They are s, the state assigned to the system at the beginning of a simulation; k, the point on the sample path at which data collection for estimation begins; and t, the length of the sample path, used to compute the finite sample average.

In principle, selecting an s that induces convergence of the sample path to steady-state behavior for as small a k as possible is desirable. Since little is known a priori about the underlying stochastic structure of the sample path, this is rarely possible. Customarily, s is chosen for computational convenience. For example, beginning a simulation with an arrival to an empty and idle system relieves the simulation user of the need to compute initial queue levels, times to the completion of service, and scheduled execution times, as described in §4.

Suppose a simulation generates a sample path $X_1, \ldots, X_{t'}$ of length t'. Since the influence of initial conditions on X_j tends to decrease in importance as j increases, discarding data $X_1, \ldots, X_{k-1}$ for some $k \leq t'$ at the beginning of the path and using the remaining $t := t' - k + 1$ observations $X_k, \ldots, X_{t'}$ for estimation usually reduces systematic error. In principle, the more one discards of the sample path, beginning at its origin, the smaller is the systematic error in estimates based on the remaining data. This observation encourages substantial truncation; but doing so can be unnecessarily costly unless tempered by the need to reduce sampling error as well. In particular, since the width of a confidence interval tends to decrease as $t^{-1/2}$, one is encouraged to use a substantial portion of the sample path to reduce sampling error.

This last action runs counter to the tendency for excessive truncation. For a fixed total path length, this discussion makes clear the need to choose a *warm-up interval $k - 1$* in a way that makes systematic error negligible while providing a valid basis for computing a, hopefully, narrow confidence interval.

In a discrete-event simulation, the sequence $\{X_1, \ldots, X_{t'}\}$ may represent one of many phenomena. For example, for $i = 1, \ldots, t'$, X_i may be equal to:

- $W_i :=$ the waiting time of the ith customer

- $\bar{Q}((i-1)\Delta, i\Delta) = \Delta^{-1} \int_{(i-1)\Delta}^{i\Delta} Q(u)\mathrm{d}u$

- $\qquad\qquad\qquad\qquad =$ time-integrated queue-length average over $((i-1)\Delta, i\Delta]$

- $Q(i\Delta) :=$ queue length at time $i\Delta$, for some $\Delta > 0$

- $\bar{B}((i-1)\Delta, i\Delta) = \Delta^{-1} \int_{(i-1)\Delta}^{i\Delta} B(u)\mathrm{d}u$

- $\qquad\qquad\qquad\qquad =$ time-integrated average resource utilization over $((i-1)\Delta, i\Delta]$

- $B(i\Delta) :=$ number of busy servers at time $i\Delta$.

Alternatively, X_i may denote a function of one or more of these random quantities. For example, $\{X_i := I_{(w,\infty)}(W_i); i = 1, \ldots t'\}$ denotes a $0-1$ sequence, where $X_i = 0$ if $W_i \leq w$ and $X_i = 1$ if $W_i > w$; or

$$\left\{ X_i = \sum_{q=1}^{\infty} I_{(q-1,q]}(Q((i-1)\Delta)) I_{(q-1,q]}(Q(i\Delta))); \ i = 1, \ldots, t' \right\}$$

is a $0-1$ sequence, where $X_i = 0$ if queue lengths differ at times $i\Delta$ and $(i+1)\Delta$, whereas $X_i = 1$ if they do not.

To illustrate the problem of initial conditions, we focus on a sample path of waiting times generated in the SIMSCRIPT II.5 airline reservation example in Chapter 4, where X_i denotes the waiting time of the ith customer to begin service. Figure 6.1a displays the first 100 waiting times (in black) for a single sample path. Since the simulation began with all $m = 4$ reservationists idle and no callers waiting, the first four waiting times are necessarily zero. To include these data in the estimating procedure would induce a downward systematic error or *bias* in the sample waiting time time average as an approximation to the long-run waiting time time average. Figure 6.1a also displays (in bold black) the sequence of *truncated time averages*

$$\bar{X}_{j1,t'-j+1} := \frac{1}{t'-j+1} \sum_{i=j}^{t'} X_i, \qquad 1 \leq j \leq t', \tag{6.1}$$

where $\bar{X}_{j1,t'-j+1}$ omits the first $j-1$ waiting times. This graph shows a tendency for $\bar{X}_{j1,t'-j+1}$ initially to increase as j increases, reflecting the reduction in systematic error that truncation induces. However, as j continues to increase, $\bar{X}_{j1,t'-j+1}$ fails to stabilize in the

a. n=1, t'=100

b. n=10, t'=100

c. n=1000, t'=100

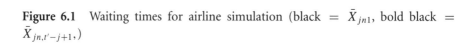

Figure 6.1 Waiting times for airline simulation (black $= \bar{X}_{jn1}$, bold black $= \bar{X}_{jn,t'-j+1}$)

neighborhood of any particular value. This behavior for large j is a consequence of the fact that random error in the time average $\bar{X}_{j1,t'-j+1}$ has variance proportional to $1/(t'-j+1)$ and, therefore, exhibits increasing variation as j approaches t'.

As Figure 6.1a shows, the presence of sampling error limits one's ability to detect systematic error from a single sample path. To reduce this sampling error, we can resort to *multiple independent replications*. Suppose that we generate a second run of the simulation with the same initial conditions but employ a pseudorandom number sequence disjoint from that used on the first sample path. This strategy allows us to regard the sample paths as statistically independent. Using the last pseudorandom numbers listed in Figure 4.3 as the seeds for the new sample path accomplishes this. Let us call each such run a replication and observe that one can generate n independent replications each of length t' by employing the last pseudorandom numbers on replication i as the seeds for replication $i+1$ for $i = 1, \ldots, n-1$.

To help us understand the value of multiple replications, let

$$\mathcal{S} := \text{set of all possible initial conditions for starting the simulation,}$$
$$X_j^{(i)} := j\text{th observation on replication } i,$$

and

$$S(\iota) := \text{state of the system at time } \tau.$$

For the moment, we take $X_j^{(i)}$ to be the jth waiting time on the ith of n replications each of length t', which are the rows in Table 6.1. Suppose that each replication begins in the same initial state $s \in \mathcal{S}$. Then

$$\bar{X}_{jn1} := \frac{1}{n} \sum_{i=1}^{n} X_j^{(i)}, \quad 1 \leq j \leq t', \tag{6.2}$$

define the column averages;

$$\bar{X}_{j,t'-j+1}^{(i)} = \frac{1}{t'-k+1} \sum_{l=j}^{t'} X_l^{(i)}, \quad 1 \leq i \leq n, \tag{6.3}$$

define the row averages based on observations $j, \ldots, t'$; and

$$\bar{X}_{j,n,t'-j+1} := \frac{1}{n} \sum_{i=1}^{n} \bar{X}_{j,t'-j+1}^{(i)} = \frac{1}{t'-j+1} \sum_{l=j}^{t'} \bar{X}_{ln1} \tag{6.4}$$

defines the grand truncated average, which omits observations $1, \ldots, j-1$ on each replication. As noted in Chapter 3, the distribution of $X_j^{(i)}$ converges to an equilibrium distribution

Table 6.1 n replications of length t'

Replication	Step j			
i	1	2 ...	t'	
1	$X_1^{(1)},$	$X_2^{(1)}, \ldots, X_{t'}^{(1)}$		$\bar{X}_{j,t'-j+1}^{(1)}$
2	$X_1^{(2)},$	$X_2^{(2)}, \ldots, X_{t'}^{(2)}$		$\bar{X}_{j,t'-j+1}^{(2)}$
$\vdots$	$\vdots$	$\vdots$	$\vdots$	
n	$X_1^{(n)},$	$X_2^{(n)}, \ldots, X_{t'}^{(n)}$		$\bar{X}_{j,t'-j+1}^{(n)}$
	$\bar{X}_{1n1},$	$\bar{X}_{2n1}, \ldots, \bar{X}_{t'n1}$		$\bar{X}_{jn,t'-j+1}$

as $j \to \infty$, say, with mean μ and variance σ^2 independent of s and j. However for finite $j \geq 1$:

$$\mu_{sj} := \mathrm{E}(X_j^{(1)} \mid S(0) = s) = \mathrm{E}(X_j^{(2)} \mid S(0) = s) = \cdots = \mathrm{E}(X_j^{(n)} \mid S(0) = s) \qquad (6.5)$$

and

$$\sigma_{sj}^2 := \mathrm{var}(X_j^{(1)} \mid S(0) = s) = \mathrm{var}(X_j^{(2)} \mid S(0) = s) = \cdots = \mathrm{var}(X_j^{(n)} \mid S(0) = s), \quad (6.6)$$

where μ_{sj} ($\neq \mu$) and σ_{sj}^2 ($\neq \sigma^2$) denote the conditional mean and variance, respectively, of waiting time j, given the initial state $S(0) = s$. Moreover, each column average in Table 6.1 has conditional mean

$$\mathrm{E}(\bar{X}_{jn1} \mid S(0) = s) = \mu_{sj},$$

because of identical initial states, and conditional variance

$$\mathrm{var}(\bar{X}_{jn1} \mid S(0) = s) = \frac{\sigma_{sj}^2}{n},$$

because of identical initial states and n independent replications.

Since $\mu_{sj} \to \mu$ as $j \to \infty$ for all $s \in \mathcal{S}$, the function $\{|\mu_{sj} - \mu|, \ j \geq 1\}$ characterizes systematic error due to initial conditions, and a graph of $\{\bar{X}_{jn1}, \ 1 \leq j \leq t'\}$ provides a way of assessing the extent of this error. Since $\mathrm{E}[\bar{X}_{jn1}|S(0) = s]$ is unaffected by n but the conditional standard error $\sqrt{\mathrm{var}(\bar{X}_{jn1}|S(0) = s)}$ decreases with increasing n, multiple replications enhance our ability to detect the presence of systematic error. Figure 6.1b based on $n = 10$ replications and $t' = 100$ illustrates the improvement over $n = 1$. It indicates the presence of systematic error for all $j \leq 20$. Figure 6.1c based on $n = 1000$ reveals even more. It shows that systematic error decreases monotonically as j increases from 1 to 100, suggesting that a not inconsequential systematic error may exist for initial path length $t' > 100$.

Figure 6.2 presents comparable information for the longer initial path length $t' = 1000$. For $n = 1$, Figure 6.2a offers little benefit over Figure 6.1a. However, for $n = 10$, Figure 6.2b suggests that systematic error becomes relatively unimportant in $\bar{X}_{jn1}$ for $j \geq 200$. Figure 6.2c confirms this for $n = 1000$. In practice, a warm-up analysis rarely employs more than 10 replications. Our use of $n = 1000$ is merely for illustration.

Figures 6.1 and 6.2 also reveal that the ability to identify a warm-up interval $k - 1$ such that systematic error is relatively incidental for $j \geq k$ depends crucially on taking a sufficiently long initial sample path length t'.

How does a simulation user choose suitable n and t' prior to simulation for a warm-up analysis? Occasionally, characteristics of the problem indicate the extent of variation across replications for a given j and the extent of dependence along a sample path. Since this is rarely so, an iterative solution often is the only alternative. One starts with an initial (n, t') 2-tuple and iteratively increases n and t' until the succession of graphs, as in Figures 6.1 and 6.2, suggests a warm-up interval $k - 1$ such that for $j \geq k$, $\bar{X}_{jn1}$ appears relatively uninfluenced by initial conditions.

We now summarize several important observations regarding $k, n, s,$ and t':

- As the warm-up interval k increases for a fixed t', systematic error decreases. However, random error increases, especially as k approaches t'.

- As t' increases for a fixed k, systematic and random errors decrease.

- Averages based on n independent replications $\{\bar{X}_{jn1}, 1 \leq j \leq t'\}$ provide a basis for discerning the extent of systematic error when all replications start in the same state.

- Random fluctuations in these averages decrease with increasing number of replications n.

- Systematic errors in these averages remain unaffected by increasing n.

- Starting the simulation in a state $s \in S$ for which the conditional mean $\{\mu_{sj}, \ j \geq 1\}$ is monotone in j enhances our ability to distinguish systematic error from sampling error in analyzing $\{\bar{X}_{jn1}, 1 \leq j \leq t'\}$. Although there is no guarantee that a state exists with this property, it is known that the empty and idle state induces the property for the conditional means of queue length and waiting time in many single-server queueing problems.

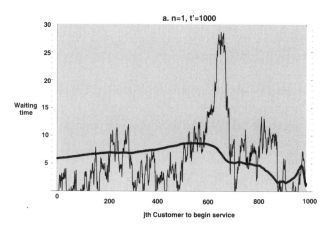

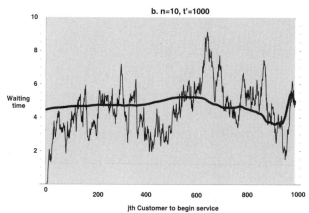

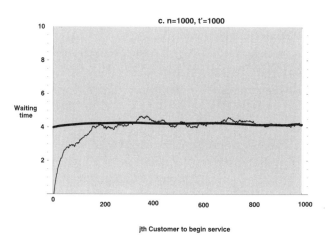

Figure 6.2 Waiting times for airline simulation (black $= \bar{X}_{jn1}$, bold black $= \bar{X}_{jn,t'-j+1}$)

6.1.1 MULTIPLE INITIAL STATES

Starting all n replications in the same initial state $s \in S$ enables us to draw inferences about an appropriate warm-up interval $k - 1$ from a sequence of estimates $\{\bar{X}_{jn1}\}$ of the conditional mean function $\{\mu_{sj}\}$. This approach can be especially revealing when $\{\mu_{sj}\}$ is monotone in j, particularly if a priori information enables a simulation user to select a state s that induces this monotonicity and that favors rapid convergence, relative to that from other possible initial states, to equilibrium.

When the monotone property is absent, the analyst faces the additional burden of distinguishing between fluctuations in $\{\bar{X}_{jn1}, j \geq 1\}$ attributable to local minima and maxima in $\{\mu_{sj}\}$ and fluctuations due to sampling error. In particular, some practitioners are reluctant to initialize each replication to a single state s for fear of perpetuating a large systematic error in each of them. Since $\bar{X}_{jn1}$ has mean-square error

$$\text{MSE}(\bar{X}_{jn1}; s, \ldots, s) = \frac{\sigma_{sj}^2}{n} + (\mu_{sj} - \mu)^2,$$

an uninformed choice of s can, for fixed j, result in a large contribution $(\mu_{sj} - \mu)^2$ to the mean-square error that remains regardless of how large n becomes. To limit this possibility, these practitioners resort to an alternative approach.

Suppose the ith independent replication begins in state $s_i \in S$ for $i = 1, \ldots, n$, where $s_1, \ldots, s_n$ are distinct and presumably *span* the state space S. For example, replication i of the airline reservation simulation might begin with an arrival to a system with q_i callers waiting for or in service, where $q_i < q_{i+1}$ for $i = 1, \ldots, n - 1$ and where $X_j^{(i)}$ denotes the waiting time of the $(q_i + j)$th caller on replication i for $j \geq 1$. Then $\bar{X}_{jn1}$ has expectation $\frac{1}{n} \sum_{i=1}^n \mu_{s_i j}$, variance $\frac{1}{n^2} \sum_{i=1}^n \sigma_{s_i j}^2$, and mean-square error

$$\text{MSE}(\bar{X}_{jn1}; s_1, \ldots, s_n) = \frac{1}{n^2} \sum_{i=1}^n \sigma_{s_i j}^2 + \frac{1}{n} \sum_{i=1}^n (\mu_{s_i j} - \mu)^2.$$

Although the contribution $\frac{1}{n} \sum_{i=1}^n (\mu_{s_i j} - \mu)^2$ is again independent of n for fixed j, it is the average of n squared deviations, presumably some small and some large, which hopefully make a smaller contribution than $(\mu_{sj} - \mu)^2$ when this latter quantity is large.

While this rationale has an intuitive appeal, this alternative has an advantage for detecting systematic error only if $\text{MSE}(\bar{X}_{jn1}; s_1, \ldots, s_n) < \text{MSE}(\bar{X}_{jn1}; s, \ldots, s)$ *and* the graphical depiction to which it leads gives a more definitive picture of conditional mean behavior as a function of j. In practice, the latter property is less likely to materialize. If s induces a monotone $\{\mu_{sj}\}$ but at least one of the s_i's does not, then the analyst faces the additional problem of distinguishing between changes in direction in $\{\bar{X}_{jn1}\}$ attributable to nonmonotonicity and changes due to sampling fluctuations.

There are systems for which a single initial state for all replications may limit the analyst's capacity to perform a comprehensive warm-up analysis: Suppose that the state space $\mathcal{S}$ is the union of two disjoint subsets $\mathcal{S}_1$ and $\mathcal{S}_2$ such that a simulation that begins in a state s in $\mathcal{S}_i$ tends to remain in $\mathcal{S}_i$ for an extraordinarily long simulated time before moving to a state in $\mathcal{S}_{i(\mathrm{mod}\ 2)} + 1$ for $i = 1, 2$. For example, a system with equipment whose times between failures and repair times are large relative to their processing times might have states with all equipment operating in $\mathcal{S}_1$ and states with at least one inoperative machine in $\mathcal{S}_2$. Unless t' is sufficiently long, starting all n replications with a state in $\mathcal{S}_1$ is likely to induce a warm-up analysis that fails to account for transition to $\mathcal{S}_2$.

6.1.2 HOW MANY WARM-UP ANALYSES?

Most discrete-event simulation runs generate sample paths on three generic types of stochastic processes, waiting time $\{W_i, i \geq 1\}$, queue length $\{Q(\tau), \tau \geq 0\}$, and resource utilization $\{B(\tau), \tau \geq 0\}$. If a warm-up analysis based on the waiting-time data concludes that deleting $W_1, \ldots, W_{k-1}$ from the data makes $W_k, W_{k+1}, \ldots$ relatively free of the influence of initial conditions, the analyst customarily deletes $\{Q(\tau), 0 \leq \tau \leq \tau_{k-1}\}$ and $\{B(\tau), 0 \leq \tau \leq \tau_{k-1}\}$ from their respective sample paths and treats all $Q(\tau)$ and $B(\tau)$ for $\tau > \tau_{k-1}$ as also independent of these initial conditions, where Section 3.4 defines τ_j.

Alternatively, if a warm-up analysis based on the queue-length data concludes that deleting $\{Q(\tau), 0 \leq \tau \leq \tau_*\}$ from the data makes $Q(\tau)$ for $\tau > \tau_*$ relatively free of the influence of initial conditions, then the analyst customarily deletes $\{B(\tau), 0 \leq \tau \leq \tau_*\}$ and $\{W_i;\ i = 1, \ldots, N(0, \tau_*)\}$ from their respective sample paths and treats $B(\tau)$ for $\tau > \tau_*$ and W_i for $i > N(0, \tau_*)$ as also independent of initial conditions.

Justification for these procedures comes from the implicit assumption that from an arbitrarily selected initial state the resource utilization and waiting-time (queue-length) processes exhibit convergence rates to equilibrium no smaller than that for the queue-length (waiting-time) process. If an equilibrium exceedance probability $\mathrm{pr}[Q(\cdot) > q]$ as well as a mean queue length μ_Q are to be estimated, it also implicitly assumes that starting from an arbitrarily selected state, $I_{(q,\infty)}(Q(\tau))$ converges at a rate at least as great as that for $Q(\tau)$ as $\tau \to \infty$. Similar assumptions hold when exceedance probabilities for resource utilization and waiting time are to be estimated. Theoretical results based on spectral structure for single-server queues support these assumptions.

6.1.3 STEPS FOR ESTIMATING THE LENGTH OF THE WARM-UP INTERVAL

We now summarize the steps to estimate the length of a suitable warm-up interval:

1. Choose the preliminary sample path length t'.

2. Choose the number of independent replications n.

3. Choose a set of initial pseudorandom number seeds.

4. Execute the simulation n times and collect $\{X_j^{(i)}, 1 \leq j \leq t'\}$ on replications i for $i = 1, \ldots, n$. On the first replication, use the initally chosen pseudorandom number

seeds. On each subsequent replication, use the final seeds of the previous replication as the starting seed.

5. Compute and plot $\{\bar{X}_{jn1}, 1 \leq j \leq t'\}$ and $\{\bar{X}_{jn,t'-j+1}, 1 \leq j \leq t'\}$.

6. If the graphs reveal a suitable warm-up interval $k-1$, use that choice in a new independent run to collect a larger sample path to assess sampling error. See Sections 6.6 through 6.8.

7. If the graph fails to reveal a suitable warm-up interval, increase t', either continue execution and data collection of the first n independent replications or simply execute n new independent replications with the new length, and go to step 5.

As illustration, an amended version of the SIMSCRIPT II.5 AIR_EV.sim program in Figure 4.2 can generate N independent sequences of waiting times where each replication begins in the empty and idle state. To save space, we list only the modifications. They are:

Step 1. preamble
  Between lines 33 and 34, insert:
    define N as a variable
    ,,   N := no. of independent replications

Step 2. main
  Between lines 15 and 16, insert:
    read N
    for I = 1 to N do
  Between lines 19 and 20, insert:
    loop

Step 3. ACCOUNTING
  In line 13, delete:
    stop
  Between lines 14 and 15, insert:
    if NUM >= K and NUM <= K+T-1 call WARM_UP_ANALYSIS
    always

Step 4. Merge WARM_UP_ANALYSIS in Figure 6.3 with AIR_EV.sim in Figure 4.2.

The output in SIMU15 will consist of n = N tableaus as in Figure 4.3 followed by the summary data $\{\bar{X}_{jn1}, 1 \leq j \leq t'\}$ and $\{\bar{X}_{jn,t'-j+1}, 1 \leq j \leq n\}$.

In addition to generating the data of interest, the WARM_UP_ANALYSIS routine in Figure 6.3 resets the state of the system back to the empty and idle state at the end of each replication. It does this by removing and destroying each event notice in the event lists and removing and destroying all CALLERs waiting in the QUEUE. Note that f.ev.s($\cdot$) denotes the first event notice in event list ev.s($\cdot$). To guarantee nonoverlapping pseudorandom number streams, the routine assigns the final pseudorandom numbers on the last replication as the seeds for the next replication. As currently written, the amended program produces a tableau as in Figure 4.3 for each of the N replications. We leave as an exercise the task of aggregating the N tableaus into a single comprehensive summary.

Figure 6.3 WARM_UP_ANALYSIS program for airline simulation

```
''(April 1998)

routine WARM_UP_ANALYSIS
''
''Computes {XBAR(J,N,1); J=K,K+1,...,K+T-1}
''        and
''        {XBAR(J,N,K+T-J); J=K,K+1,...K+T-1}.
''
define Z as a 1-dim double saved array
define COUNT as a saved variable
''     COUNT := currently executing replication
define Y as 1-dim double array
define FIRST as an integer variable
if COUNT=0
   let COUNT = 1
   reserve Z(*)  as  T
always
if NUM >= K and NUM <= K+T-1
   add WAITING.TIME to Z(NUM - K + 1)
always
if NUM < K + T -1    return
else
''
'' Empty the future event sets
''
   let FIRST = f.ev.s(i.CALL)
   cancel  the CALL called FIRST
   destroy the CALL called FIRST
   NJ =n.ev.s(i.END_MESSAGE)
   for J = 1 to NJ do
      let FIRST = f.ev.s(i.END_MESSAGE)
      cancel  the END_MESSAGE called FIRST
      destroy the END_MESSAGE called FIRST
   loop
''
   NJ = n.ev.s(i.COMPLETION)
   for J = 1 to NJ do
      let FIRST = f.ev.s(i.COMPLETION)
      cancel  the COMPLETION called FIRST
      destroy the COMPLETION called FIRST
   loop
''
'' Idle all RESERVATIONISTs
''
   for J = 1 to n.RESERVATIONIST
```

Figure 6.3 (cont.)

```
    let STATUS(J) =IDLE
,,
,, Empty the QUEUE
,,
   NJ = n.QUEUE
   for J = 1 to NJ  do
      remove the first PERSON  from QUEUE
      destroy the CUSTOMER called PERSON
   loop
,,
,, Assign the final pseudorandom numbers from this replication as the initial
,,        seeds for the next replication.
,,
   for J = 1 to 4 let SEEDS(J) = seed.v(J)
,,
,, Summarize sample path data for N independent replications
,,
   if COUNT = N
      reserve Y(*)  as T
,,
,,    Write {XBAR(J,N,1); J=K,K+1,...K+T-1} to unit 15.
,,
      for J = K to K+T-1 write J,Z(J-K+1)/N as i 7, e(16,6)
      let Y(T) = Z(T)/N
      for J = 2 to T do
         let Y(T - J + 1) = Y(T - J + 2) + Z(T - J + 1)/N
         let Y(T - J + 2) = Y(T - J + 2)/(J -1)
      loop
      let Y(1) = Y(1)/T
,,
,,    Write {XBAR(J,N,K+T-J); J=K,...K+T-1} to unit 15.
,,
      for J=K to K+T-1 write J,Y(J-K+1) as i 7, e(16,6)
   always
   add 1 to COUNT
   return   end
```

Arena/SIMAN provides a capability for automatically creating graphs for a warm-up analysis. Recall in Figure 4.13 that the TALLIES element declared that waiting times be collected. By specifying, say, 10 replications in the dialog box for the REPLICATIONS element, the simulationist instructs Arena to collect 10 independent waiting-time sample sequences. After simulation execution terminates, the "Output Analyzer" on the Tools menu allows the simulationist to plot all 10 sample paths on the same graph or to lump them together into a single plot.

The output in SIMU15 will consist of $n = N$ tableaus as in Figure 4.3 followed by the summary data $\{\bar{X}_{jn1}, 1 \le j \le t'\}$ and $\{\bar{X}_{jn,t'-j+1}, 1 \le j \le n\}$.

6.2   ASSESSING THE ACCURACY OF THE SAMPLE AVERAGE

After selecting a warm-up interval k, a simulator has at least three options:

A. Use $\bar{X}_{kn,t'-k+1}$ as a point estimate of μ and $\mathcal{A}_{kn,t'-k+1}(\delta)$ as an approximating $100 \times (1-\delta)$ percent confidence interval for μ, where

$$\mathcal{A}_{kn,t'-k+1}(\delta) := [\bar{X}_{kn,t'-k+1} \pm \tau_{n-1}(1 - \delta/2)\sqrt{s^2_{kn,t'-k+1}/n}\,], \qquad (6.7)$$

$$s^2_{kn,t'-k+1} := \frac{1}{n-1}\sum_{i=1}^{n}(\bar{X}^{(i)}_{k,t'-k+1} - \bar{X}_{kn,t'-k+1})^2, \qquad (6.8)$$

and $\tau_{n-1}(1-\delta/2)$ denotes the $1 - \delta/2$ critical value of Student's t distribution with $n-1$ degrees of freedom.

B. Select $t > t'$, restart the nth simulation in the state prevailing when $X^{(n)}_{t'}$ was collected, collect $k + t - t' - 1$ additional observations $X^{(n)}_{t'+1}, \ldots, X^{(n)}_{k+t-1}$, and compute the truncated sample average $\bar{X}^{(n)}_{kt}$ (see expression (6.3)) as a point estimate of μ, compute an estimate $V^{(n)}_t$ of $t\,\mathrm{var}\bar{X}^{(n)}_{kt}$ using the truncated sample-path data $X^{(n)}_k, X^{(n)}_{k+1}, \ldots, X^{(n)}_{k+t-1}$, and compute

$$\mathcal{C}^{(n)}_{ktl(t)}(\delta) := \left[\bar{X}^{(n)}_{kt} \pm \tau_{l(t)-1}(1 - \delta/2)\sqrt{V^{(n)}_t/t}\right] \qquad (6.9)$$

as an approximating $100 \times (1 - \delta)$ percent confidence interval for μ, where the degrees of freedom $l(t) - 1$ depend on the method employed for computing $V^{(n)}_t$.

C. Start a new independent simulation run, $n + 1$, in the same initial state as the preceding n replications, collect $X^{(n+1)}_k, \ldots, X^{(n+1)}_{k+t-1}$ after a warm-up interval $k - 1$, use $\bar{X}^{(n+1)}_{kt}$ as a point estimate of μ and $\mathcal{C}^{(n+1)}_{ktl(t)}(\delta)$ as an approximating $100 \times (1 - \delta)$ percent confidence interval for μ.

Computational and statistical properties determine the relative desirabilities of each option. Since Option B has no need to rerun the warm-up interval, it offers an advantage over Option C with regard to computational efficiency. However, if programming the simulation to restart in the last state presents too much of a challenge, then Option C becomes more appealing. Since the statistical properties of $\left(\bar{X}^{(n)}_{kt}, V^{(n)}_t, \mathcal{C}^{(n)}_{ktl(t)}(\delta)\right)$ and $\left(\bar{X}^{(n+1)}_{ktl(t)}, V^{(n+1)}_t, \mathcal{C}^{(n+1)}_{ktl(t)}(\delta)\right)$ are identical, we hereafter suppress the superscripts (n) and $(n + 1)$ unless required for clarity. Moreover, any discussion of Option B applies equally to Option C.

Because of i.i.d. replications, Option A easily computes an estimate, s^2_{kt}/n, of $\mathrm{var}\bar{X}_{kt}$. By contrast, computing V_t as an estimate of $t\,\mathrm{var}\bar{X}_{kt}$ with desirable statistical properties calls for considerably more demanding computational procedures. Section 6.6 describes these procedures, and Section 6.7 describes software, available on the Internet, for implementing the procedures with minimal effort for the simulationist.

Since the graphical method for selecting a warm-up interval k contains no formal statistical testing procedure for deciding whether or not systematic error remains after truncating $k - 1$ observations, it is clearly a subjective approach. Most importantly, picking k too small affects Options A and B in different ways. For Option B, the influence of any residual warm-up error diminishes as $t \to \infty$. By contrast, this error remains constant under Option A as $n \to \infty$, implying that Option B is more forgiving than Option A with regard to understating the warm-up interval.

Regardless of computational convenience, any assertions that a simulationist makes regarding the statistical errors in $\bar{X}_{knt}$ or $\bar{X}_{kt}$ are supportable only if the methodology for deriving $\mathcal{A}_{knt}(\delta)$ and $\mathcal{C}_{ktl(t)}(\delta)$ rests on well-established statistical theory. For fixed t, the interval estimate $\mathcal{A}_{knt}(\delta)$ is asymptotically valid if and only if

$$\mathrm{pr}[\mu \in \mathcal{A}_{knt}(\delta)] \to 1 - \delta \quad \text{as } n \to \infty. \tag{6.10}$$

For fixed k, $\mathcal{C}_{ktl(t)}(\delta)$ is asymptotically valid if and only if

$$\mathrm{pr}[\mu \in \mathcal{C}_{ktl(t)}(\delta)] \to 1 - \delta \quad \text{as } t \to \infty. \tag{6.11}$$

To address these and related issues, we need to introduce several formalisms.

6.3 FORMALISMS

Although the analysis in Section 6.1 relies exclusively on sample data for a particular simulation model, it offers an invaluable introduction to the assessment of systematic error. However, the perspective is a limited one. The next several sections describe a more formal setting for assessing error. The formalisms apply to data generated by many different types of discrete-event simulation, and this high level of generality overcomes the limitation of Section 6.1 by providing the reader with a considerably more comprehensive basis for error analysis.

Again consider a single sample path $X_1, \ldots, X_{k+t-1}$ with finite time average

$$\bar{X}_{kt} := \frac{1}{t} \sum_{i=k}^{k+t-1} X_i \tag{6.12}$$

as an approximation to the long-run average μ. Let $\mathcal{X}$ denote the sample space for the X_j and let $\{F_j(x|s), x \in \mathcal{X}\}$ denote the conditional d.f. of X_i given that the simulation starts in state $S(0) = s \in \mathcal{S}$ (Section 6.1), so that expressions (6.5) and (6.6) take the equivalent forms

$$\mu_{sj} = \int_{\mathcal{X}} x \, dF_j(x|s)$$

and

$$\sigma_{sj}^2 = \int_{\mathcal{X}} (x - \mu_{sj})^2 dF_j(x|s), \quad j \geq 1.$$

Suppose that X_i has the equilibrium d.f. $\{F(x), x \in \mathcal{X}\}$ and that the long-run average μ equates to the equilibrium mean of X_i; that is, $\mu = \int_{\mathcal{X}} x \, dF(x)$. For expression (6.12) to be a useful basis for inferring the true value of the unknown μ, the underlying stochastic process must have several properties. These include:

- For every initial state $s \in \mathcal{S}$, $\mu_{sj} \to \mu < \infty$ as $j \to \infty$.

- Let

$$\mu_{skt} := \mathrm{E}(\bar{X}_{kt}|S(0) = s) = \frac{1}{t} \sum_{j=k}^{k+t-1} \int_{\mathcal{X}} x \, dF_j(x|s). \tag{6.13}$$

For every initial state $s \in \mathcal{S}$ and each fixed integer $k \geq 1$,

$$\mu_{skt} \to \mu \qquad \text{as } t \to \infty. \tag{6.14}$$

- Let

$$\sigma_{skt}^2 := \mathrm{var}(\bar{X}_{kt}|S(0) = s).$$

For every $s \in \mathcal{S}$ and each fixed $t \geq 1$, there exists a positive constant σ_t^2 independent of s and k such that

$$\sigma_{skt}^2 \to \sigma_t^2 \quad \text{as } k \to \infty \tag{6.15}$$

and there exists a positive constant σ_∞^2 such that

$$t\sigma_t^2 \to \sigma_\infty^2 \quad \text{as } t \to \infty. \tag{6.16}$$

- For every $s \in \mathcal{S}$ and each fixed $k \geq 1$,

$$t\sigma_{skt}^2 \to \sigma_\infty^2 < \infty \qquad \text{as } t \to \infty. \tag{6.17}$$

Although convergence as in expressions (6.14) through (6.17) is essential, it is not sufficient for us to develop a distributional theory as the basis for computing an asymptotically valid confidence interval for μ. Speed of convergence is also an issue. For example, for

$$\text{pr}\left[\frac{|\bar{X}_{kt} - \mu_{skt}|}{\sqrt{\sigma^2_{skt}}} \leq z \,\Big|\, S(0) = s\right] = \text{pr}\left[\sqrt{\frac{\sigma^2_\infty}{t\sigma^2_{skt}}} \,\left|\frac{\bar{X}_{kt} - \mu}{\sqrt{\sigma^2_\infty/t}} + \frac{\mu - \mu_{skt}}{\sqrt{\sigma^2_\infty/t}}\right| \leq z | S(0) = s\right]$$

$$\rightarrow \text{pr}\left(\frac{|\bar{X}_{kt} - \mu|}{\sqrt{\sigma^2_\infty/t}} \leq z\right) \quad \text{as } t \rightarrow \infty \tag{6.18}$$

if and only if $\mu_{skt} \rightarrow \mu$ at a rate proportional to $t^{-1/2}$ or faster. For many stochastic processes encountered in discrete-event simulation,

$$\mu_{sj} = \mu + \text{O}(\beta^{\,j}) \quad \text{as } j \rightarrow \infty \tag{6.19}$$

for some $\beta \in [0, 1)$, so that

$$\mu_{skt} = \mu + \text{O}(\beta^{\,k}/t) \quad \text{as } k \rightarrow \infty \text{ and } t \rightarrow \infty, \tag{6.20}$$

which satisfies the requirement for convergence to μ.

6.4 INDEPENDENT REPLICATIONS

Our objective is to assess how well $\bar{X}_{knt}$ in Option A and defined in expression (6.4) approximates the unknown equilibrium mean μ. A confidence interval provides a basis for doing this. Four parameters affect the validity of this approach. They are the initial state s, the warm-up interval $k - 1$, the length of sample path t used for estimating μ, and the number of independent replications n. This section describes an approach based on independent replications for deriving confidence intervals that are asymptotically valid (as $n \rightarrow \infty$). Multiple independent replications allow us to employ standard, and relatively simple, statistical theory applicable to i.i.d. observations to compute confidence intervals. However, the approach does have limitations, as we show. Later, Section 6.5 describes an alternative approach that principally relies for its validity on $t \rightarrow \infty$ for fixed k and n.

Theorem 6.1.

Let $\mathcal{N}(\theta_1, \theta_2)$ denote the normal distribution with mean θ_1 and variance θ_2. Let $\{X_j^{(i)}, 1 \leq j \leq t' = k + t - 1\}$ for $1 \leq i \leq n$ denote n independent realizations or *sample paths of the process $\{X_j\}$ each of length $k + t - 1$. Suppose that expression (6.20) holds for the conditional mean and that the conditional variance has the representation*

$$\sigma^2_{skt} := var(\bar{X}^{(i)}_{kt} | S^{(i)}_0 = s) = \sigma^2_t + O(\beta^k/t^2), \tag{6.21}$$

where

$$\sigma^2_t = \sigma^2_\infty/t + O(1/t^2). \tag{6.22}$$

Then

i. *For fixed $k \geq 1$ and $t \geq 1$, $(\bar{X}_{knt} - \mu_{skt})/\sqrt{\sigma^2_{skt}/n} \xrightarrow{d} \mathcal{N}(0, 1)$ as $n \to \infty$.*

ii. *For fixed $t \geq 1$, $(\bar{X}_{knt} - \mu)/\sqrt{\sigma^2_t/n} \xrightarrow{d} \mathcal{N}(0, 1)$ if $k/\ln n \to \infty$ as $k \to \infty$ and $n \to \infty$.*

PROOF. Part i follows from the classical *Central Limit Theorem* for i.i.d. random variables. For part ii, let

$$A_{knt} := \frac{\bar{X}_{knt} - \mu}{\sqrt{\sigma^2_t/n}}$$

and

$$B_{knt} := \frac{\mu - \mu_{skt}}{\sqrt{\sigma^2_t/n}}, \tag{6.23}$$

so that

$$A_{knt} = \frac{\bar{X}_{knt} - \mu_{skt}}{\sqrt{\sigma^2_{skt}/n}} \sqrt{\frac{\sigma^2_{skt}}{\sigma^2_t}} - B_{knt}. \tag{6.24}$$

Substituting (6.20), (6.21), and (6.22) into (6.23) and (6.24) shows that the condition $k/\ln n \to \infty$ as $k \to \infty$ and $n \to \infty$ suffices for $B_{knt} \to 0$. Also, $\sigma^2_{skt} \to \sigma^2_t$ as $k \to \infty$. Therefore, A_{knt} converges to $(\bar{X}_{knt} - \mu_{skt})/\sqrt{\sigma^2_{skt}/n}$ w.p.1, which converges to a normal random variable from $\mathcal{N}(0, 1)$. $\square$

Although k is supposedly chosen to remove the ostensible influence of initial conditions on $X_k, X_{k+1}, \ldots$, it is nevertheless the case that $\mu_{skt} \neq \mu$ for fixed k and t; moreover, Theorem 6.1i implies that the interval

$$\left[\bar{X}_{knt} - \Phi^{-1}(1 - \delta/2)\sqrt{\sigma^2_{skt}/n}, \ \bar{X}_{knt} + \Phi^{-1}(1 - \delta/2)\sqrt{\sigma^2_{skt}/n} \right]$$

covers μ_{skt}, not μ, with approximate probability $1 - \delta$ for large n, where $\Phi^{-1}(1 - \delta/2)$ denotes the $1 - \delta/2$ point of the standard normal distribution $\mathcal{N}(0, 1)$. To achieve this same coverage rate for μ, Theorem 6.1 ii asserts that the warm-up interval k must grow faster than $\ln n$ as the number of replications n increases.

Expressions (6.20), (6.21), and (6.22) characterize the conditional means and variances of the sample means of many stochastic processes encountered in discrete-event simulation. This generality encourages us to adopt this characterization as we learn more about computing a valid confidence interval.

Since the variance σ_t^2 is unknown, Theorem 6.1 is of academic but not of operational interest. The availability of multiple independent replications proves especially helpful in making it operational.

Corollary 6.2.

For the specification in Theorem 6.1, fixed $t \geq 1$, and s_{knt}^2 as defined in expression (6.8),
$$(\bar{X}_{knt} - \mu)/\sqrt{s_{knt}^2/n} \xrightarrow{\text{d}} \mathcal{N}(0, 1) \text{ if } k/\ln n \to \infty \text{ as } k \to \infty \text{ and } n \to \infty.$$

PROOF. Let $Z_{knt} := (\bar{X}_{knt} - \mu_{skt})/\sqrt{s_{knt}^2/n}$ and note that

$$Z_{knt} - A_{knt} = A_{knt} \left(\sqrt{\frac{\sigma_t^2}{\sigma_{skt}^2}} \sqrt{\frac{\sigma_{skt}^2}{s_{knt}^2}} - 1 \right) + B_{knt} \sqrt{\frac{\sigma_t^2}{\sigma_{skt}^2}} \sqrt{\frac{\sigma_{skt}^2}{s_{knt}^2}}. \tag{6.25}$$

Since $s_{knt}^2/\sigma_{skt}^2 \to 1$ as $n \to \infty$ w.p.1 and $\sigma_t^2/\sigma_{skt}^2 \to 1$ as $k \to \infty$, then as in Theorem 6.1 $k/\ln n \to \infty$, as $k \to \infty$ and $n \to \infty$, ensure that Z_{knt} converges to A_{knt}, which has $\mathcal{N}(0, 1)$ as its limiting distribution. $\square$

Corollary 6.2, implies that

$$\left[\bar{X}_{knt} \pm \Phi^{-1}(1 - \delta/2)\sqrt{s_{knt}^2/n} \right] \tag{6.26}$$

provides an approximating $100 \times (1 - \delta)$ percent normal confidence interval for μ that is asymptotically valid for $k/\ln n \to \infty$ as $k \to \infty$ and $n \to \infty$ for fixed $t \geq 1$. Since $\tau_{n-1}(1 - \delta/2) > \tau_n(1 - \delta/2) > \Phi^{-1}(1 - \delta/2)$ for each fixed n, common practice uses the confidence interval (6.7), rather than (6.26), to compensate for the finiteness of n.

The most notable parametric dependence remains that between k and n. In particular, these results assert that as one takes more independent replications presumably to shorten the confidence interval, the warm-up interval k must increase faster than $\ln n$ in order for the desired limit to hold. This tie between k and n is inconvenient to implement in practice.

RECURSIVE COMPUTATION

The algebraic equivalences (e.g., Fishman 1996, p. 68)

$$
\bar{X}_{kit} =
\begin{cases}
\left(1 - \dfrac{1}{i}\right) \bar{X}_{k,i-1,t} + \dfrac{1}{i} \bar{X}_{kt}^{(i)}, & 1 \le i \le n, \\[2mm]
0, & i = 0,
\end{cases}
$$

and

$$
s_{kit}^2 =
\begin{cases}
\dfrac{i-2}{i-1} s_{k,i-1,t}^2 + \dfrac{1}{i}(\bar{X}_{kt}^{(i)} - \bar{X}_{k,i-1,t})^2, & 2 \le i \le n, \\[2mm]
0, & i = 1,
\end{cases}
$$

allow recursive computation of $\bar{X}_{kit}$ and s_{kit}^2, respectively, without the need to store $\bar{X}_{kt}^{(1)}, \ldots,$ $\bar{X}_{kt}^{(n)}$. This facility is especially useful when space is limited.

Since the recursion for s_{kit}^2 is the sum of two nonnegative quantities of roughly the same magnitude and both are independent of n, it induces a smaller numerical round-off error than the more commonly encountered space-saving formula (Neely 1966)

$$
s_{knt}^2 = \frac{1}{n-1}\left[\sum_{i=1}^{n}(\bar{X}_{kt}^{(i)})^2 - n\bar{X}_{knt}^2\right].
$$

This representation is the difference of two quantities each of which increases with n and thus can exhibit substantial round-off error when the two quantities are close in value.

6.4.1 ALLOWING t TO GROW

For Option A to generate an asymptotically valid confidence interval $\mathcal{A}_{knt}(\delta)$, as in expression (6.7) for fixed t, its computing time necessarily increases faster than $n \ln n$, principally due to the dependence of k on n. Under specified conditions, allowing t to increase for fixed n can reduce this truncation cost in relative terms. To do this effectively calls for an understanding of the distributional behavior of $(\bar{X}_{kt}^{(i)} - \mu)/\sqrt{\sigma_\infty^2/t}$ as $t \to \infty$.

BROWNIAN MOTION

Let $\{Z(\tau), \tau \ge 0\}$ denote a stochastic process in $\Re$ with continuous parameter τ and with the properties:

i. pr $[Z(0) = z_0] = 1$

ii. For $0 \leq t_0 \leq t_1 \leq \cdots \leq t_k$,

$$\mathrm{pr}[Z(t_j) - Z(t_{j-1}) \leq z_j, \ 1 \leq j \leq k] = \Pi_{j=1}^{k}\mathrm{pr}[Z(t_j) - Z(t_{j-1}) \leq z_j]$$

iii. For $0 \leq s < \tau$, the increment $Z(\tau) - Z(s)$ has the normal distribution with mean $\theta_1(\tau - s)$ and variance $\theta_2(\tau - s)$.

Then $\{Z(0) = z_0; \ Z(\tau), \ \tau > 0\}$ is called *Brownian motion* or a *Wiener process* starting at z_0. Most importantly, $\{Z_j := Z(j) - Z(j - 1); \ j = 1, 2, \ldots \}$ is a sequence of i.i.d. normal random variables with mean θ_1 and variance θ_2. If $\theta_1 = 0$ and $\theta_2 = 1$, then $\{Z(0) = z_0; \ Z(\tau), \ \tau > 0\}$ is called *standard Brownian motion* starting at z_0.

ASSUMPTION OF WEAK APPROXIMATION

If $\{X_j, \ j \geq 0\}$ is a stochastic sequence and there exist a standard Brownian motion $\{Z(\tau), \ \tau \geq 0\}$ and finite constants $a > 0$ and d such that $t^{1/2}(\bar{X}_t - d)/a$ converges in distribution to $Z(t)/t^{1/2}$, then $\{X_j, \ j \geq 0\}$ is said to satisfy the *assumption of weak approximation* (AWA).

This property provides an essential ingredient for deriving asymptotically valid confidence intervals as the sample-path length t increases.

Theorem 6.3.

For each $i = 1, \ldots, n$ and each fixed $k \geq 1$, suppose that the process $\{X_{k-1+j}^{(i)}, j \geq 1\}$ with $X_j^{(i)}$ as in Theorem 6.1 satisfies AWA with $a^2 = \sigma_\infty^2$ and $d = \mu$. Then for fixed $n \geq 1$ and $k \geq 1$ as $t \to \infty$,

i. $\sqrt{nt}(\bar{X}_{knt} - \mu)/\sqrt{\sigma_\infty^2} \overset{\mathrm{d}}{\to} \mathcal{N}(0, 1)$.

ii. $(n - 1)s_{knt}^2/\sigma_\infty^2 \overset{\mathrm{d}}{\to} \chi_{n-1}^2$, where χ_{n-1}^2 denotes the chi-squared distribution with $n - 1$ degrees of freedom.

iii. $(\bar{X}_{knt} - \mu)/\sqrt{s_{knt}^2/n} \overset{\mathrm{d}}{\to} \tau_{n-1}$, where τ_{n-1} denotes Student's t distribution with $n - 1$ degrees of freedom.

SKETCH OF PROOF. As a consequence of AWA, $\sqrt{t}(\bar{X}_{kt}^{(1)} - \mu)/\sqrt{\sigma_\infty^2}, \ldots, \sqrt{t}(\bar{X}_{kt}^{(n)} - \mu)/\sqrt{\sigma_\infty^2}$ converge w.p.1 to n independent random variables each from $\mathcal{N}(0, 1)$. The limits in parts i, ii, and iii follow from this property. $\square$

As a consequence of Theorem 6.3iii, $\mathrm{pr}[\mu \in \mathcal{A}_{knt}(\delta)] \to 1 - \delta$ as $t \to \infty$ for fixed n and k and $\mathcal{A}_{knt}(\delta)$ in definition (6.7). Simple in concept, and asymptotically valid as $t \to \infty$, this approach to confidence interval computation has a natural appeal, once one selects the number of replications n and the warm-up interval k.

For completeness, we describe the necessary and sufficient conditions that must exist between n and t for attaining desirable limiting behavior as these two quantities increase simultaneously. We also describe the necessary and sufficient conditions that must exist among k, n, and t as all three increase.

Theorem 6.4.

  Assume $\{X_j^{(1)}, 1 \leq j \leq k + t - 1\}, \ldots, \{X_j^{(n)}, 1 \leq j \leq k + t - 1\}$ *as in Theorem 6.1.*

i. *If for fixed* $k \geq 1, t/n \rightarrow \infty$ *as* $n \rightarrow \infty$ *and* $t \rightarrow \infty$, *then*

$$\sqrt{nt}(\bar{X}_{knt} - \mu)/\sqrt{\sigma_\infty^2} \xrightarrow{\mathrm{d}} \mathcal{N}(0, 1).$$

ii. *If* $k/\ln(n/t) \rightarrow \infty$ *as* $k \rightarrow \infty, n \rightarrow \infty$, *and* $t \rightarrow \infty$, *then*

$$\sqrt{nt}(\bar{X}_{knt} - \mu)/\sqrt{\sigma_\infty^2} \xrightarrow{\mathrm{d}} \mathcal{N}(0, 1).$$

PROOF.   For fixed k, t must increase faster then n is required to make

$$\frac{\mu_{skt} - \mu}{\sqrt{\sigma_{skt}^2/n}} = \frac{\mathrm{O}(\beta^k/t)}{\sqrt{\sigma_\infty^2/nt + \mathrm{O}(1/nt^2) + \mathrm{O}(\beta^k/nt^2)}}$$

vanish. For increasing k, this occurs if and only if $k \ln(n/t) \rightarrow \infty$.   □

Although Theorems 6.3 and 6.4 collectively offer a method of generating an asymptotically valid $100 \times (1 - \delta)$ percent confidence interval for μ, its implementation in practice is problematic. Given an upper bound on total computing time, any attempt to realize the asymptotic properties that accrue as t increases can only come at the expense of having n diminish. Conversely, any attempt to realize the asymptotic properties that accrue as n grows large can only come at the expense of having t diminish. If we were to adopt the latter course, we would have to ensure that the truncation interval k also grows faster than $\ln n$, so that Corollary 6.2 holds. If we adopted the former course (Option B), we would need an alternative way of estimating σ_∞^2 that ensure an asymptotically valid confidence interval for μ. Indeed, Sections 3.3.4 and 3.5 have already raised this issue for estimating long-run average utilization, queue length, and waiting time. In the current context, if one has an estimator V_t based on a single sample path of length t such that

$$V_t \rightarrow \sigma_\infty^2 \quad \text{as } t \rightarrow \infty \quad \text{w.p.1,} \tag{6.27}$$

then for $n = 1$ and fixed k, the distribution of $(\bar{X}_{kt} - \mu)/\sqrt{V_t/t}$ converges to $\mathcal{N}(0,1)$ as $t \rightarrow \infty$. Recall that a V_t satisfying expression (6.27) is called *strongly consistent*.

Many alternative methods exist for computing a strongly consistent estimator of V_t for Option B. All are less than direct methodologically and for large t can themselves pose a not inconsequential computing burden. Among these the *batch-means method* (Section 6.6) lends itself to relatively simple interpretation and has an algorithmic implementation, LABATCH.2 in Section 6.7, that induces $O(\log_2 t)$ storage space and $O(t)$ computing time, while producing a strongly consistent estimator of σ_∞^2 and thus an asymptotically valid confidence interval for μ. To the potential simulation user, the cost is to acquaint oneself with more advanced concepts of statistics than those associated with i.i.d. observations. The benefit is the need principally to concern oneself with *only* two design parameters, the warm-up interval k and the sample path length t, while fixing the number of replications, n, at unity.

6.5 STRICTLY STATIONARY STOCHASTIC PROCESSES

To understand the motivation that underlies the batch-means method for estimating σ_∞^2, it is necessary to become acquainted with the implications of dependence among $X_1, X_2, X_3, \ldots$. Let us momentarily assume that a warm-up interval k has been determined so that the corresponding error of approximation is negligible in regarding each $X_k, X_{k+1}, \ldots$ as having the desired equilibrium distribution. Interest then focuses on assessing the statistical accuracy of $\bar{X}_{kt}$ as an estimator of μ. This section begins this process by characterizing the structure of $\sigma_t^2 = \text{var}\bar{X}_{kt}$. It does this within the context of a *strictly stationary stochastic process* and, in particular, describes how dependence along the sample path affects σ_t^2.

Consider a stochastic process $\{X_j, j \geq 0\}$ with the property that for all positive integers i and for each positive integer l the i-tuples $(X_{j_1}, X_{j_2}, \ldots, X_{j_i})$ and $(X_{j_1+l}, X_{j_2+l}, \ldots, X_{j_i+l})$ have the same i-dimensional probability distribution. This property, called *strict stationarity*, immediately implies that for each $j \geq 0$, X_j has the same marginal distribution. Moreover, for $j \geq i \geq 1$ the *covariance* between X_i and X_j is

$$\text{cov}(X_i, X_j) = \text{E}_{X_i, X_j}[(X_i - \mu)(X_j - \mu)]$$

$$= \text{E}_{X_i}[(X_i - \mu)\text{E}_{X_j|X_i}(X_j - \mu)]$$

$$= \text{E}_{X_i}\{(X_i - \mu)[\text{E}_{X_j|X_i}(X_j) - \mu]\}$$

$$= \text{E}_{X_1}\{(X_1 - \mu)[\text{E}_{X_{j-i+1}|X_1}(X_{j-i+1}) - \mu]\}, \tag{6.28}$$

revealing the *conditional mean function* $\{\text{E}_{X_j|X_i}(X_j), j \geq 1\}$ and the *autocovariance function* $\{R_{j-i} := \text{cov}(X_i, X_j); i, j = 1, 2, \ldots\}$ to be functions of the index difference $j - i$, but not of the absolute locations i and j in the index set. Moreover, expression (6.28) shows that an intimate relationship exists between this conditional mean function and the covariance structure of $\{X_j, \ j \geq 1\}$. In particular, if for fixed $i < j$

$$E_{X_j|X_i}(X_j) \to \mu \quad \text{as } j \to \infty, \tag{6.29}$$

then $R_j \to 0$, as $j \to \infty$, revealing that the covariance between X_i and X_{i+j} vanishes as the index difference increases without bound.

Since $\text{cov}(X_1, X_{j+1}) = \text{cov}(X_i, X_{i+j})$ for all i, it suffices for us to examine the properties of the autocovariance function $\{R_j, \quad j = 0, \pm 1, \ldots\}$ for a strictly stationary process. Note that

$$\text{var}\, X_i = R_0, \tag{6.30}$$

$$R_j = R_{-j}, \tag{6.31}$$

and, from the inequality $E_{X_i, X_{i+j}}, (X_i - X_{i+j})^2 \geq 0$,

$$|R_j| \leq R_0, \quad j \ = \ 0, \pm 1, \pm 2, \ldots . \tag{6.32}$$

The correlation between X_i and X_j is

$$\text{corr}(X_i, X_j) = \rho_{i-j} := \frac{R_{i-j}}{R_0}, \quad i, j = 0, 1, \ldots, \tag{6.33}$$

where $\rho_0 = 1$, $\rho_j = \rho_{-j}$, and, as a consequence of expression (6.32), $|\rho_j| \leq 1$. Here $\{\rho_j; j = 0, \pm 1, \pm 2, \ldots\}$ denotes the *autocorrelation function* of $\{X_i\}$. Figure 6.4 shows the autocorrelation function, $\{\rho_j = \alpha^{|j|}\}$, for $\alpha = .90$ and $-.90$. As j increases one displays monotonically decreasing behavior and the other, alternating positive and negative values.

Suppose we observe the process in the steady state and collect a *sample record* or *realization* $X_1, \ldots, X_t$ for the purpose of estimating μ. Since data collection begins in the steady state, the sample mean, $\bar{X}_t = (X_1 + \cdots + X_t)/t$, is an unbiased estimator of μ. Also,

$$\sigma_t^2 := \text{var}\, \bar{X}_t = E(\bar{X}_t - \mu)^2$$

$$= \frac{1}{t^2} \sum_{i,j=1}^{t} \text{cov}(X_i, X_j)$$

$$= \frac{1}{t^2} \sum_{i,j=1}^{t} R_{i-j}$$

$$= \frac{1}{t^2} \sum_{j=1-t}^{t-1} (t - |j|)\, R_j. \tag{6.34}$$

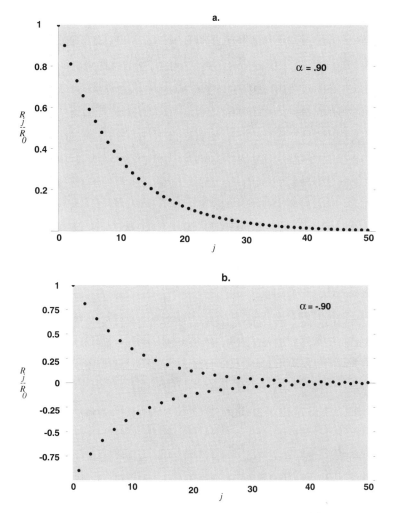

Figure 6.4 Autocorrelation function $\rho_j := R_j/R_0 = \alpha^{|j|}$

By the symmetry property (6.31),

$$\sigma_t^2 = \underbrace{\frac{R_0}{t}}_{\substack{\text{due to variance} \\ R_0 \text{ from each} \\ \text{observation}}} + \underbrace{\frac{2}{t} \sum_{j=1}^{t-1} (1 - j/t) R_j}_{\substack{\text{due to correlation} \\ \text{between observations}}}. \tag{6.35}$$

This partition reveals two distinct contributions to σ_t^2. If $X_1, X_2, \ldots$ were i.i.d., then the correlations between observations would be zero and $\sigma_t^2 = R_0/t$. However, congestion generally induces dependence in waiting time, queue length, and utilization sequences, and this dependence frequently manifests itself as positive correlation. As a consequence, $\sigma_t^2 > R_0/t$ for most queueing-related processes. Moreover, as a system becomes more congested, this positive correlation grows, usually increasing the relative contribution to σ_t^2 made by the underlying correlation structure.

The condition $\lim_{t \to \infty} t\sigma_t^2 = \sigma_\infty^2 < \infty$ in expression (6.16) implies that

$$R_0 \sum_{j=1-t}^{t-1} (1 - |j|/t)\rho_j \to \sigma_\infty^2 \quad \text{as } t \to \infty. \tag{6.36}$$

For this to hold, $\lim_{j \to \infty} \rho_j = 0$ is necessary but not sufficient. However, the condition

$$R_0 \sum_{j=1-t}^{t-1} \rho_j \to \sigma_\infty^2 < \infty \tag{6.37}$$

is both necessary and sufficient. This property implies that, in addition to the correlation between X_i and X_{i+j} converging to zero as $j \to \infty$, it must dissipate sufficiently fast so that the summation in expression (6.37) converges. An equivalent characterization is possible in terms of the conditional mean. See expression (6.28). Hereafter, we occasionally refer to σ_∞^2/R_0 as *variance inflation*.

To illustrate the implications of the limit (6.37), consider the autocorrelation function

$$\rho_j = \alpha^{|j|}, \quad -1 < \alpha < 1, \quad j = 0, \pm 1, \ldots, \tag{6.38}$$

for which

$$\sigma_t^2 = \frac{R_0}{t} \left[\frac{1+\alpha}{1-\alpha} - \frac{2\alpha(1-\alpha^t)}{t(1-\alpha)^2} \right] \tag{6.39}$$

and

$$\frac{\sigma_\infty^2}{R_0} = \frac{1+\alpha}{1-\alpha} = \frac{1+\rho_1}{1-\rho_1}. \tag{6.40}$$

Table 6.2 shows how $\text{var}\bar{X}_t$ becomes inflated as α increases. For example, a nearest-neighbor correlation of $\alpha = .10$ results in a 22 percent increase in the variance of $\bar{X}_t$ over that to be expected if the $X_1, X_2, \ldots$ were i.i.d. More generally, the entries in Table 6.2 reveal that treating $\bar{X}_t$ as the average of i.i.d. data when it is not can result in a substantial overstatement of the accuracy of $\bar{X}_t$.

Since $\sigma_\infty^2/R_0 < 1$ for $\alpha < 0$, $\text{var}\,\bar{X}_t$ is less than would occur if the data were i.i.d.

Table 6.2 Variance inflation

α	$\dfrac{\sigma_\infty^2}{R_0} = \dfrac{1+\alpha}{1-\alpha}$
.10	1.222
.20	1.500
.50	3.000
.75	7.000
.80	9.000
.85	12.333
.90	19.000
.95	39.000
.99	199.000

Although autocorrelation functions for utilization, queue length, and waiting time are rarely, if ever, literally of this geometric form, they often have geometric bounds. For example, a waiting time sequence, $X_1, X_2, \ldots$ for the stationary M/M/1 queueing model with arrival rate λ and service rate ω ($> \lambda$) has traffic intensity $v := \lambda/\omega$, mean $\mu = v/\omega(1-v)$, variance $R_0 = v(2-v)/\omega^2(1-v)^2$, and autocorrelation function (Blomqvist 1967)

$$\rho_j = \sum_{i=j+3}^{\infty} \left[\frac{v}{(1+v)^2}\right]^i \frac{(2i-3)!}{i!(i-2)!}(i-j-1)(i-j-2), \quad j = 0, 1, \ldots, \quad (6.41)$$

with tightest possible geometric upper bound

$$\rho_j \leq \left[\frac{4v}{(1+v)^2}\right]^j, \quad j = 0, 1, \ldots . \quad (6.42)$$

For example, $v = .90$ and $.95$ give $4v/(1+v)^2 = .9972$ and $.9993$, respectively, leading to relatively slow decreases in the bound (6.42) as the lag j between two waiting times increases. Moreover,

$$\lim_{t\to\infty} t \operatorname{var}\bar{X}_t = \sigma_\infty^2 = \frac{v}{\omega^2(1-v)^4}(v^3 - 4v^2 + 5v + 2), \quad (6.43)$$

revealing a variance inflation

$$\frac{\sigma_\infty^2}{R_0} = \frac{1}{(2-v)(1-v)^2}(v^3 - 4v^2 + 5v + 2) \quad (6.44)$$

that increases as $1/(1-v)^2$ as v increases.

6.6 BATCH-MEANS METHOD

To assess how well $\bar{X}_t$ approximates μ, we need an estimate of σ_∞^2. The batch-means method offers one option. Let $b(t)$ denote a positive integer $(< t)$, let $l(t) := \lfloor t/b(t) \rfloor$, and let $a(t) := l(t)b(t)$. Our version of the batch-means method fits within the context of Option B in Section 6.2. It partitions the sequence $X_1, \ldots, X_{a(t)}$ into $l(t)$ nonoverlapping batches each of size $b(t)$,

$$X_1, \ldots, X_{b(t)}, X_{b(t)+1}, \ldots, X_{2b(t)}, \ldots, X_{[l(t)-1]b(t)+1}, \ldots, X_{a(t)}, \tag{6.45}$$

computes the batch averages

$$Y_{jb(t)} := \frac{1}{b(t)} \sum_{i=1}^{b(t)} X_{(j-1)b(t)+i}, \quad j = 1, \ldots, l(t), \tag{6.46}$$

and an estimate of var $Y_{jb(t)}$

$$W_{l(t)b(t)} := \frac{1}{l(t) - 1} \sum_{j=1}^{l(t)} (Y_{jb(t)} - \bar{X}_{a(t)})^2, \tag{6.47}$$

and uses $V_{a(t)} := b(t)W_{l(t)b(t)}$ as an estimate of σ_∞^2. Then

$$\mathcal{C}_{l(t)}(\delta) := \left[\bar{X}_t \pm \tau_{l(t)-1}(1 - \delta/2)\sqrt{V_{a(t)}/t} \right], \tag{6.48}$$

as in expression (6.9), with subscripts k and t suppressed, provides an approximating $100 \times (1 - \delta)$ percent confidence interval for μ. The challenge is to choose the batching sequence $\{(l(t), b(t)), t \geq 1\}$ in a way that ensures asymptotic validity,

$$\text{pr}[\mu \in \mathcal{C}_{l(t)}(\delta)] \to 1 - \delta \quad \text{as } t \to \infty,$$

and, among all such choices, to select one that makes the width of the interval relatively narrow and that is computationally efficient. We first describe an application of the batch-means method. For a simulation of the M/M/1 queueing model begun in the steady state with interarrival rate $\nu = .90$, service rate unity, for $t = 10^7$ customers, and $\delta = .01$, Figure 6.5a shows $\{(\bar{X}_{t_j}, \mathcal{C}_{l(t_j)}(\delta)) : t_1 = 35, t_{j+1} := 2t_j; j = 1, \ldots, 18\} \cup \{(\bar{X}_t, \mathcal{C}_{l(t)}(\delta))\}$ for mean waiting time in queue ($\mu = 9$), and Figure 6.5b shows the corresponding statistics for the probability of waiting ($\mu := \text{pr}(W_i > 0) = .90$). These statistics were computed using LABATCH.2, an implementation of the batch-means method. It is available in C, FORTRAN, and SIMSCRIPT II.5 on the Internet at http://www.or.unc.edu/~gfish/labatch.2.html. LABATCH.2 can be applied either interactively to data generated during a simulation or after simulation termination to stored sample sequences generated by the simulation.

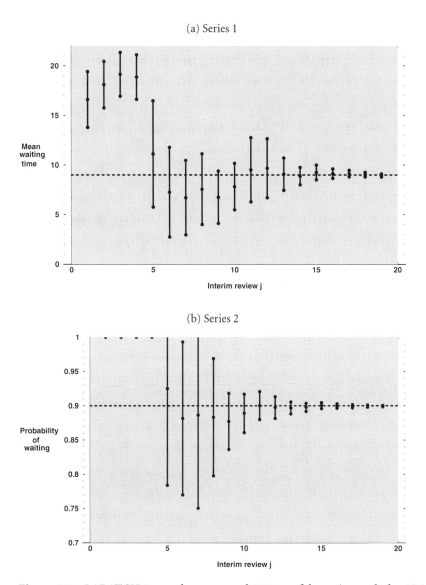

Figure 6.5 LABATCH.2 sample means and 99% confidence intervals for ABATCH rule; simulation starts in steady state

The graphs allow us to assess how well the point estimates $\{\bar{X}_{t_1}, \ldots, \bar{X}_{t_{18}}, \bar{X}_t\}$ approximate μ. For small j the $\bar{X}_{t_j}$ are unstable. However, the sequences stabilize with increasing j. At the same time, the confidence intervals first widen with j and then decrease. By $j = 14$ in each figure, stabilization of $\bar{X}_{t_j}$ and the width of the confidence intervals suggest that the error of approximation is relatively small. We return shortly to a discussion of this initial widening of confidence intervals.

6.6.1 LABATCH.2 TABLEAUS

Figure 6.6 shows LABATCH.2-generated tableaus that provided the data for the graphs in Figure 6.5. For each series, the "Final Tableau" presents results computed at the end of the simulation; in particular, the point estimate $\bar{X}$ of μ based on all $t = 10^7$ observations, the estimated standard error of $\bar{X}$, $\sqrt{B^*W(L, B)/t}$, where the numerator is based on the first $t'(10^7) = 9,175,040$ observations, a confidence interval [Lower, Upper] based on $C_{l(t)}(.01)$ in expression (6.48), and a rough measure of relative error $(\text{Upper}-\text{Lower})/|\bar{X}|$ for $\bar{X}$. Section 6.6.3 explains why $W(L,B)$ is based on $t'(10^7)$ rather than 10^7 observations.

Figure 6.6 reveals an estimated error of $\pm 100 \times .0302/2 = \pm 1.51$ percent relative to the sample mean waiting time and an error of $\pm 100 \times .002416/2 = \pm .1208$ percent relative to the sample exceedance probability. As a rule of thumb, a simulationist may regard an entry under $|\text{Upper} - \text{Lower}|/|\bar{X}| \leq .20$ as implying relatively high accuracy.

Readers concerned principally with results as in the Final Tableau may skip to Section 6.7 for a description of how to implement LABATCH.2 or to Section 6.6.5 for an account of how Arena/SIMAN uses the batch-means method. The remainder of this section together with Sections 6.6.2 through 6.6.9 describes details that influence the extent to which an application of the batch-means methods leads to asymptotically valid confidence intervals.

For each series, an "Interim Review Tableau" in Figure 6.6 lists results for each review j based on $t_j = L*B$ observations, where L denotes number of batches, B denotes batch size, and $t_{j+1} = 2t_j$. Section 6.6.5 describes the mechanism for increasing L and B. The entries under Sqrt[B*W(L,B)] provide a sequence of estimates $\sqrt{V_{t_j}}$ of σ_∞ that become increasingly stable as j increases. Figures 6.7a and 6.7b show the graphs for these entries for each series. Most notable is the initial growth followed by fluctuation about the theoretically known limits $\sigma_\infty = 189.5$ and $\sigma_\infty = 1.308$ in Figures 6.7a and 6.7b respectively (Blomqvist 1967). This initial growth pattern is responsible for the widening of confidence intervals for μ for successive small j. Section 6.8 provides an additional way of assessing convergence in Figure 6.12b.

6.6.2 SYSTEMATIC VARIANCE ERROR

Why do the estimates of σ_∞ behave as in Figures 6.7a and 6.7b? To the potential user of the batch-means method, this behavior and the initial widening of the confidence intervals can be puzzling. A description of the sampling properties of V_t explains these phenomena. When $X_1, \ldots, X_t$ are i.i.d., with $\text{var} X_i < \infty$, the assignment $b(t) = 1$ and $l(t) = t$ makes $C_{l(t)}(\delta)$ an approximating confidence interval, based on $t - 1$ degrees of freedom, whose error of approximation is generally negligible, say, for $t \geq 100$, regardless of the parent distribution of the X_i. When $\{X_i\}$ is a dependent sequence, as commonly occurs in discrete-event simulation, no comparable rule of thumb applies, since two properties now affect the error of approximation. As before, one is the marginal distribution of the X_i; the other is the correlation between X_i and X_j for all $i \neq j$ in the sample. An analysis of the properties of V_t reveals how these errors arise.

(a)

Final Tableau

Mean Estimation
****
(t = 10000000)

			99.0%			
	$\bar{X}$	Standard Error	Confidence Interval			
Series		Sqrt[B*W(L,B)/t]	Lower	Upper	(Upper-Lower)/|X|	
1	0.8949D+01	0.5681D-01	0.8802D+01	0.9097D+01	0.3290D-01	
2	0.8995D+00	0.4200D-03	0.8984D+00	0.9006D+00	0.2416D-02	

$\bar{X}$ is based on all t observations.
W(L,B) is based on first 91.75% of the t observations.

(b)

Interim Review Tableau   ABATCH Data Analysis for Series 1

					99.0%			
				$\bar{X}$	Confidence Interval			
Review	L*B	L	B		Lower	Upper	Sqrt[B*W(L,B)]	p-value
1	35	7	5	0.1660D+02	0.1379D+02	0.1940D+02	0.4475D+01	0.3804
2	70	10	7	0.1810D+02	0.1575D+02	0.2045D+02	0.6047D+01	0.1109
3	140	14	10	0.1914D+02	0.1693D+02	0.2135D+02	0.8680D+01	0.1524
4	280	20	14	0.1886D+02	0.1662D+02	0.2111D+02	0.1313D+02	0.0194
5	560	20	28	0.1111D+02	0.5748D+01	0.1646D+02	0.4432D+02	0.0000
6	1120	20	56	0.7254D+01	0.2741D+01	0.1177D+02	0.5278D+02	0.0001
7	2240	20	112	0.6706D+01	0.2960D+01	0.1045D+02	0.6196D+02	0.0005
8	4480	20	224	0.7556D+01	0.3996D+01	0.1112D+02	0.8328D+02	0.0358
9	8960	20	448	0.6747D+01	0.4111D+01	0.9383D+01	0.8721D+02	0.7293
10	17920	28	640	0.7817D+01	0.5478D+01	0.1016D+02	0.1130D+03	0.7095
11	35840	40	896	0.9513D+01	0.6280D+01	0.1275D+02	0.2260D+03	0.0655
12	71680	40	1792	0.9668D+01	0.6693D+01	0.1264D+02	0.2941D+03	0.8172
13	143360	56	2560	0.9073D+01	0.7449D+01	0.1070D+02	0.2304D+03	0.7200
14	286720	80	3584	0.8883D+01	0.8012D+01	0.9754D+01	0.1767D+03	0.2993
15	573440	112	5120	0.9248D+01	0.8511D+01	0.9985D+01	0.2128D+03	0.7163
16	1146880	160	7168	0.9126D+01	0.8653D+01	0.9600D+01	0.1945D+03	0.0993
17	2293760	160	14336	0.9138D+01	0.8809D+01	0.9468D+01	0.1913D+03	0.1852
18	4587520	224	20480	0.9032D+01	0.8805D+01	0.9259D+01	0.1874D+03	0.7882
19	9175040	320	28672	0.8949D+01	0.8802D+01	0.9097D+01	0.1796D+03	0.2362

If data are independent:

10000000	10000000	1	0.8949D+01	0.8941D+01	0.8957D+01	0.9821D+01	0.0000

0.10 significance level for independence testing.
Review 19 used the first 91.75% of the t observations for W(L,B).

Interim Review Tableau   ABATCH Data Analysis for Series 2

					99.0%			
				$\bar{X}$	Confidence Interval			
Review	L*B	L	B		Lower	Upper	Sqrt[B*W(L,B)]	p-value
1	35	7	5	0.1000D+01	0.1000D+01	0.1000D+01	0.0000D+00	0.0000
2	70	7	10	0.1000D+01	0.1000D+01	0.1000D+01	0.0000D+00	0.0000
3	140	7	20	0.1000D+01	0.1000D+01	0.1000D+01	0.0000D+00	0.0000
4	280	7	40	0.1000D+01	0.1000D+01	0.1000D+01	0.0000D+00	0.0000
5	560	7	80	0.9250D+00	0.7838D+00	0.1066D+01	0.9014D+00	0.0067
6	1120	7	160	0.8812D+00	0.7697D+00	0.9928D+00	0.1007D+01	0.0140
7	2240	7	320	0.8862D+00	0.7503D+00	0.1022D+01	0.1734D+01	0.1002
8	4480	10	448	0.8830D+00	0.7975D+00	0.9686D+00	0.1762D+01	0.8214
9	8960	14	640	0.8770D+00	0.8360D+00	0.9180D+00	0.1289D+01	0.9589
10	17920	20	896	0.8888D+00	0.8607D+00	0.9169D+00	0.1316D+01	0.9648
11	35840	28	1280	0.9001D+00	0.8797D+00	0.9206D+00	0.1398D+01	0.3029
12	71680	40	1792	0.8973D+00	0.8813D+00	0.9133D+00	0.1581D+01	0.9235
13	143360	56	2560	0.8967D+00	0.8879D+00	0.9055D+00	0.1250D+01	0.6376
14	286720	80	3584.	0.8976D+00	0.8916D+00	0.9036D+00	0.1220D+01	0.5008
15	573440	112	5120	0.9001D+00	0.8956D+00	0.9045D+00	0.1280D+01	0.1352
16	1146880	160	7168	0.8996D+00	0.8963D+00	0.9029D+00	0.1374D+01	0.5547
17	2293760	224	10240	0.8992D+00	0.8968D+00	0.9016D+00	0.1381D+01	0.6289
18	4587520	320	14336	0.8995D+00	0.8979D+00	0.9011D+00	0.1313D+01	0.7435
19	9175040	448	20480	0.8995D+00	0.8984D+00	0.9006D+00	0.1328D+01	0.6766

If data are independent:

10000000	10000000	1	0.8995D+00	0.8992D+00	0.8997D+00	0.3007D+00	0.0000

0.10 significance level for independence testing.
Review 19 used the first 91.75% of the t observations for W(L,B).

Figure 6.6   LABATCH.2 output for M/M/1 queueing simulation (Series 1: waiting time in queue, Series 2: 1:=wait, 0:=no wait)

(a) Series 1, $\sigma_\infty = 189.5$

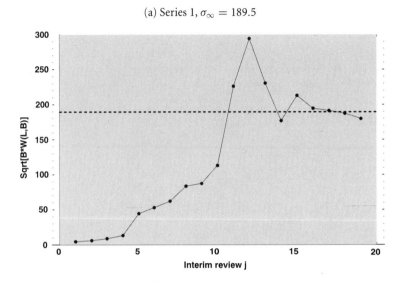

(b) Series 2, $\sigma_\infty = 1.308$

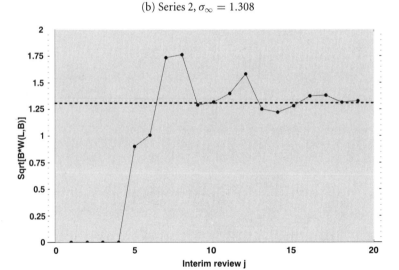

Figure 6.7 LABATCH.2 ABATCH estimates, $\sqrt{\mathrm{BW(L, B)}}$, of σ_∞

Observe that for $a(t) = t$, $b := b(t)$, and $l := l(t)$,

$$\mathrm{EV}_t = \frac{t}{l-1}(\sigma_b^2 - \sigma_t^2), \tag{6.49}$$

which for positively autocorrelated sequences is usually negative. If $\mathrm{E}(X_i - \mu)^{12} < \infty$ and $\{X_i\}$ is ϕ-*mixing* with $\phi_i = \mathrm{O}(i^{-9})$, then expression (6.49) takes the form (Goldsman and

Meketon 1986, Song and Schmeiser 1995)

$$EV_t = \sigma_\infty^2 + \gamma(l+1)/t + o(1/b), \qquad (6.50)$$

where

$$\gamma := -2\sum_{i=1}^{\infty} i \operatorname{cov}(X_1, X_{1+i}).$$

The sequence $\{X_i\}$ is said to be *ϕ-mixing* if there exists a sequence of real numbers $\{\phi_l\}$ with $\phi_l \to \infty$ as $l \to \infty$ such that for all subsets $\mathcal{A}$ and $\mathcal{B} \subseteq \mathcal{X}$ and all $i, j \geq 1$

$$|\operatorname{pr}(X_i \in \mathcal{A}, X_j \in \mathcal{B}) - \operatorname{pr}(X_i \in \mathcal{A})\operatorname{pr}(X_j \in \mathcal{B})| \leq \phi_{|i-j|}\operatorname{pr}(X_i \in \mathcal{A}).$$

The property implies that X_i and X_j become independent as $|i - j| \to \infty$.

In addition to the expectation (6.50), *ϕ-mixing* leads to (Chien et al. 1997)

$$\operatorname{var} V_t = \frac{2\sigma_\infty^4(l+1)}{(l-1)^2} + \mathrm{O}(1/lb^{1/4}) + \mathrm{O}(1/l^2), \qquad (6.51)$$

so that V_t has *mean-square error*

$$\operatorname{MSE}(V_t) := \operatorname{var} V_t + (EV_t - \sigma_\infty^2)^2$$

with the property that $\operatorname{MSE}(V_t) \to 0$ if and only if $l(t) \to \infty$ and $b(t) \to \infty$ as $t \to \infty$. This property is called *mean-square convergence*.

Expression (6.49) leads to the representation

$$V_t - \sigma_\infty^2 = \underbrace{t\sigma_t^2 - \sigma_\infty^2}_{\substack{\text{error due to} \\ \text{finite } t}} - \underbrace{t\sigma_t^2\left(\frac{1 - b\sigma_b^2/t\sigma_t^2}{1 - b/t}\right)}_{\substack{\text{error due to ignoring cor-} \\ \text{relation between batches}}} + \underbrace{\epsilon_t}_{\substack{\text{error due to} \\ \text{random} \\ \text{sampling}}}, \qquad (6.52)$$

where ϵ_t has mean zero and variance (6.51). Hereafter, we collectively refer to the errors due to finite t and to ignoring correlation as *systematic variance error*. Expression (6.50) shows that systematic variance error behaves as $\mathrm{O}((l+1)/t) = \mathrm{O}(1/b)$, whereas expression (6.51) shows that $\sqrt{\operatorname{var}\epsilon_t}$ behaves as $\mathrm{O}(1/l^{1/2})$, revealing the tradeoff between the two types of error that a choice of l and b induces. LABATCH.2 makes this choice automatically, thus relieving the user of a major burden. Sections 6.6.3 through 6.6.5 describe why this choice leads to a strongly consistent estimator of σ_∞^2 and an asymptotically valid confidence interval for μ.

Since $b\sigma_b^2 < t\sigma_t^2 < \sigma_\infty^2$ for a substantial number of stochastic processes, systematic error variance tends to induce a downward bias in V_t, as an estimator of σ_∞^2, that dissipates as t increases; hence the behavior in Figures 6.7a and 6.7b. Notice that after the systematic variance error becomes negligible, the remaining fluctuations in each graph tend to vary around some constant level and diminish in magnitude. This behavior signals that systematic variance error is negligible. Reduction of the magnitude of fluctuations reflects the var $\epsilon_t = O(1/l)$ property induced by the ABATCH rule (Section 6.6.5) that LABATCH.2 employed when choosing number of batches and batch size on each review.

Mean-square error convergence implies convergence in probability of V_t to σ_∞^2. In the current context, this means that for given $\epsilon > 0$

$$\mathrm{MSE}(V_t) \to 0 \quad \text{as } t \to \infty \Rightarrow \mathrm{pr}(|V_t - \sigma_\infty^2| > \epsilon) \to 0 \quad \text{as } t \to \infty.$$

We then say that V_t is a *weakly consistent estimator* of σ_∞^2. However, this property does not ensure that $C_{l(t)}(\delta)$ is an asymptotically valid confidence interval for μ. For this to be so, V_t must be a strongly consistent estimator of σ_∞^2. To achieve this property, we must impose additional restrictions on the evolution of $\{(l(t), b(t))\}$.

Assume that there exists a constant $\lambda \in (0, \frac{1}{2})$ such that

$$t^{1/2}(\bar{X}_t - \mu)/\sigma_\infty = Z(t)/t^{1/2} + \mathrm{O}(t^{-\lambda}) \text{ as } t \to \infty \text{ w.p.1,}$$

where $\{Z(s), s \geq 0\}$ denotes standard Brownian motion (Section 6.4). This is called the *Assumption of Strong Approximation* (ASA). A λ close to $\frac{1}{2}$ signifies a marginal distribution for the X_i close to the standard normal and low correlation between X_i and X_j for all $i \neq j$. Conversely, λ close to zero implies the absence of one or both of these properties. See Philipp and Stout (1975).

Theorem 6.5 (Damerdji 1994).

Assume that ASA holds. If $\{b(t)\}$ and $\{l(t)\}$ are deterministic nondecreasing functions in t, $b(t) \to \infty$ and $l(t) \to \infty$ as $t \to \infty$, $t^{1-2\lambda}(\log t)/b(t) \to 0$ as $t \to \infty$, and there exists a positive integer $q < \infty$ such that $\sum_{i=1}^{\infty} [l(i)]^{-q} < \infty$, then V_t is a strongly consistent estimator of σ_∞^2 and $(\bar{X}_t - \mu)/\sqrt{V_t/t} \xrightarrow{d} \mathcal{N}(0, 1)$ as $t \to \infty$.

Theorem 6.5 implies that $b(t)$ must grow at least as fast as t^θ, where $\theta > 1 - 2\lambda$. Conversely, $l(t)$ must grow sufficiently fast to ensure that the summation $\sum_{i=1}^{q} [l(i)]^{-q}$ converges. Since λ is usually unknown in practice, we are unable to confirm that our choice of $\{(l(t), b(t))\}$ satisfies the requirements. Nevertheless, the guidance is unequivocal. If X_i has a highly nonnormal distribution or $X_1, X_2, \ldots$ have high dependence (λ close to zero), batch size must grow faster than would be required if the distribution of X_i were close to the normal and $X_1, X_2, \ldots$ exhibited little dependence (λ close to $\frac{1}{2}$).

6.6.3 INTERIM REVIEW

If a simulationist computes only one estimate, $W_{l(t)b(t)}$, based on the total sample record of length t, then, regardless of the batch size assignment rule, the analysis provides no information about the extent to which the desired asymptotic properties hold for $V_t = b(t)W_{l(t)b(t)}$ as an approximation to σ_∞^2. As Figures 6.6 and 6.7 show, a sequence of *interim reviews* overcomes this limitation by computing and displaying results for overlapping segments of the sample record of successively increasing lengths $t_1 < t_2 < \cdots \leq t$. The approach has considerable value when the cost per observation is high and there is doubt, before experimentation, about how many observations to collect. Moreover, choosing $t_{j+1} = 2t_j$ for $j = 1, 2, \ldots$ with given t_1 allows LABATCH.2 to compute the sequence of estimates $W_{l(t_1)b(t_1)}, W_{l(t_2)b(t_2)}, \ldots, W_{l(t_j)b(t_j)}, W_{l(t)b(t)}$ for each series in $O(t)$ time. If LABATCH.2 is implemented dynamically, the space requirement is $O(\log_2 t)$. Maintaining an $O(t)$ computing time is essential for the single sample path approach to achieve greater statistical efficiency than the multiple independent replications approach. If generating the data takes $O(t)$ time, then no increase in computational time complexity arises. The $O(\log_2 t)$ space complexity is particularly appealing when t is large and interim reviews are desired for estimates of more than one mean.

To understand the statistical implications of our choice of $\{t_j\}$, we consider more general schemes $t_1 < t_2 < \cdots$. For a sequence of i.i.d. random variables $X_1, X_2, \ldots$, this choice implies $\operatorname{corr}(\bar{X}_{t_j}, \bar{X}_{t_{i+j}}) = \sqrt{t_j/t_{i+j}}$ for $i \geq 0$ and $j \geq 1$. More generally, if $\operatorname{corr}(X_1, X_{1+j}) = \alpha^{|j|}$, for some $-1 < \alpha < 1$, then $\lim_{j\to\infty} \operatorname{corr}(\bar{X}_{t_j}, \bar{X}_{t_{i+j}}) = \lim_{j\to\infty} \sqrt{t_j/t_{i+j}}$. These observations suggest that we choose $\theta := t_{j+1}/t_j$ sufficiently large so that estimates on successive reviews avoid substantial redundancy. We choose $\theta = 2$, which implies corr $(\bar{X}_{t_j}, \bar{X}_{t_{i+j}}) = 2^{-i/2}$ in the independent and asymptotic (as $j \to \infty$) cases. For example, $\operatorname{corr}(\bar{X}_{t_j}, \bar{X}_{t_{j+1}}) = \sqrt{2}/2 = .7071$ (in the i.i.d. case). Any growth factor less than 2 would induce higher correlation. Moreover, any factor other than 2 would substantially change the space complexity $\log_2 t$.

6.6.4 FNB RULE

Let $l_j := l(t_j)$ and $b_j := b(t_j)$ and recall that systematic variance error dissipates as $O(1/b_j)$. This implies that choosing $b_j \propto t_j$ would diminish this error most rapidly. We illustrate the benefit and the limitation of this approach using the FNB rule to define the batching sequence $\{(l_j, b_j), \ j = 1, 2, \ldots\}$. Given (l_1, b_1), this rule fixes $l_j = l_1$ for all j and doubles the batch size, $b_{j+1} = 2b_j$, on successive reviews; hence the title, FNB, for a fixed number of batches.

Appendix A describes the principal steps for implementing the rule with $t_{j+1} := 2t_j$ for $j = 1, \ldots, J(t)$, where

$$J(t) := 1 + \lfloor \log(t/l_1 b_1)/\log 2 \rfloor$$

$$= \text{total number of interim reviews.}$$

Under relatively weak conditions,

$$\text{pr}\big[\mu \in \mathcal{C}_{l_1}(\delta)\big] \to 1 - \delta \text{ as } t \to \infty, \tag{6.53}$$

implying that $\mathcal{C}_{l_1}(\delta)$ is an asymptotically valid confidence interval for μ. However,

$$\text{var} V_{t_{J(t)}} = \frac{2\sigma_\infty^4 (l_{(1+1)})}{(l_1 - 2)^2} + \text{O}(1/l_1^2) \qquad \text{as } t \to \infty, \tag{6.54}$$

implying that $V_{t_{J(t)}}$ does not converge in mean-square as $t \to \infty$ and, therefore, lacks the desired statistical property of consistency. Appendix A graphically illustrates how the absence of mean-square convergence results in a sequence of estimates of σ_∞^2 whose sampling fluctuations fail to dampen as $t \to \infty$.

6.6.5   ARENA'S AUTOMATIC USE OF THE BATCH-MEANS METHOD

Recall that Arena/SIMAN classifies sample-path data as time persistent or discrete change. Queue length and number of busy servers exemplify time-persistent phenomena, whereas waiting time exemplifies a discrete-change phenomenon. Upon termination of a simulation run, Arena automatically applies the batch-means method in an attempt to compute approximating 95 percent confidence intervals for the means of all time-persistent and discrete-change phenomena declared in DSTAT and TALLIES elements (Kelton et al. 1997, p. 232). The results appear in the half-width column in its output tableau (e.g., Figure 4.14).

For the discrete-change case based on number of observations $t < 320$, Arena enters (Insuf) in the column, signifying too small a sample size to implement its analysis. For $t \geq 320$, it dynamically maintains

$$l(\tau) = \lfloor \tau/16 \lfloor \tau/320 \rfloor \rfloor \tag{6.55}$$

batches, after collecting observation τ for $320 \leq \tau \leq t$ with

$$b(\tau) = \lfloor \tau/l(\tau) \rfloor = \lfloor \tau/\lfloor \tau/16 \lfloor \tau/320 \rfloor \rfloor \rfloor \tag{6.56}$$

observations per batch. As a consequence, $l(\tau)$ periodically cycles through the values $20, 21, \ldots, 39$ as τ increases, so that the number of batches always remains bounded. Therefore, the resulting $V_{a(t)}$ is not a consistent estimator of σ_∞^2 (Section 6.62 and Appendix A.1).

By way of comparison, columns "Lower" and "Upper" in the LABATCH.2 final-tableau display the lower (L) and upper (U) limits of the confidence interval (e.g., Figure 6.6), whereas column "Half Width" in the Arena final tableau displays $(U-L)/2$ (e.g., Figure 4.14). Although the latter quantity is usually called the half-length in statistical analysis, we retain the half-width notation merely for consistency with Arena labeling.

At the end of a run of t observations, Arena tests the batches for independence using the procedure in Section 6.6.8. If rejected, Arena enters (Corr) in the Half Width column, signifying that the batches are correlated. If accepted, Arena computes and enters the 95 percent half-width $\tau_{l(t)-1}(.975)\sqrt{V_{a(t)}/t}$, where $a(t) := l(t)b(t)$ denotes the number of observations used to compute $V_{a(t)}$. Note that the computation uses all the data if $\tau \geq 320$ and $\tau/16$ is integer. For $t \geq 320, a(t)/t > .955$; for $t > 1920, a(t)/t > .99$.

Let $M(\tau)$ denote the number of value changes in a time-persistent phenomenon such as queue length, in a simulated time interval $(0, \tau]$. For a sample path of t simulated time units, Arena enters (Insuf) in the Half Width column if $M(t) < 320$. Otherwise, it maintains

$$l(M(\tau)) = \lfloor M(\tau)/16\lfloor M(\tau)/320\rfloor\rfloor \qquad (6.57)$$

batches at simulated time $\tau \geq \inf[z \geq 0 : M(z) \geq 320]$ with

$$b(M(\tau)) = \lfloor M(\tau)/\lfloor M(\tau)/16\lfloor M(\tau)/320\rfloor\rfloor\rfloor \qquad (6.58)$$

value changes per batch. If H is rejected, it enters (Corr) in the Half Width column. Otherwise, it enters $\tau_{l(t)-1}(.975)\sqrt{V_{a(t)}/t}$, where $a(t) = \lfloor l(M(t))b(M(t))\rfloor$ is the number of observations used to compute $V_{a(t)}$. As before, $20 \leq l(M(\tau)) \leq 39$, and $t'(M(t))/M(t) \geq .955$ for $M(t) \geq 320$ and $\geq .99$ for $M(t) \geq 1920$.

For the bounded number of batches in the Arena method, expressions (6.50) and (6.51) imply that $V_{a(t)}$ has coefficient of variation

$$\gamma(V_{a(t)}) := \frac{\sqrt{\text{var}\,V_{a(t)}}}{\text{E}V_{a(t)}} \to \frac{2[l(t)+1]}{[l(t)-1]^2} \qquad \text{as } t \to \infty,$$

where

$$.2354 \leq \frac{2[l(t)+1]}{[l(t)-1]^2} \leq .3411.$$

This suggests persistent and relatively substantial fluctuations in $V_{a(t)}$ regardless of how large t is. An analogous result holds for $V_{t'(M(t))}$. Section 6.9 describes how an Arena user can specify batch size and confidence level and perform a post-simulation batch-means analysis on stored sample data. It also discusses the differences between Arena's and LABATCH.2's use of the batch-means method.

6.6.6   SQRT RULE

Within LABATCH.2, the SQRT rule offers an alternative that ensures convergence for V_t. It defines $\{(l_j, b_j), j \geq 1\}$ such that $l_{j+1}/l_j \doteq \sqrt{2}$ and $b_{j+1}/b_j \doteq \sqrt{2}$. More specifically, for given l_1 and b_1, it sets

$$l_2 = \tilde{l}_1 := \lfloor \sqrt{2}l_1 + .5 \rfloor, \tag{6.59}$$

$$b_2 = \tilde{b}_1 := \begin{cases} 3 & \text{if } b_1 = 1, \\ \lfloor \sqrt{2}b_1 + .5 \rfloor & \text{if } b_1 > 1, \end{cases} \tag{6.60}$$

$$l_{j+1} = 2l_{j-1},$$
$$b_{j+1} = 2b_{j-1} \quad j = 2, 3, \ldots.$$

Choosing (l_1, b_1) from $\mathcal{B}$ in Table 6.3 ensures that $2l_1 b_1 = \tilde{l}_1 \tilde{b}_1$, so that $t_j = l_j b_j = 2^{j-1} l_1 b_1$ and, therefore, $t_{j+1}/t_j = 2$, as desired.

Appendix A.2 describes the principal steps for the SQRT rule for $j = 1, \ldots, J(t)$. Under this rule, $V_{t_{J(t)}} \to 0$ as $t \to \infty$, ensuring the desired mean-square convergence. Also, the rule has an optimal property with regard to the standardized statistic

$$Z_{lb} := \frac{\bar{X}_t - \mu}{\sqrt{W_{lb}/l}} \tag{6.61}$$

with d.f. F_{lb}. Appendix A.2 shows that $l(t) \propto t^{1/2}$ and $b(t) \propto t^{1/2}$ ensures the fastest convergence of $F_{lb}(z)$ to $\Phi(z)$ for each $z \in (-\infty, \infty)$. Since for large t we use the normal d.f. Φ, in place of the unknown d.f. F_{lb}, to compute a confidence interval for μ, this last property implies that the actual coverage rate for confidence intervals converges to the desired theoretical coverage rate $1 - \delta$ as rapidly as possible.

Appendix A.2 also illustrates how the rule works for the M/M/1 example and compares its performance with that of the FNB rule. The comparison especially in Figure 6.8a reveals that the FNB rule offers an initial benefit but the SQRT rule ultimately offers the advantage of mean-square convergence.

If the specified sample path length is $t = 2^{\alpha-1} l_1 b_1$ for $\alpha > 1$ and (l_1, b_1) as in Table 6.3, then $a(t) = t$. Recall that $a(t)$ is the number of observations used by the batch-means method (Section 6.6). More generally, for any $t \geq 500$, LABATCH.2 automatically selects (l_1, b_1) such that $a(t)/t$ is maximal. This maximum always exceeds .89. See Appendix A.3.

6.6.7   LBATCH AND ABATCH RULES

The FNB and SQRT rules each have desirable and undesirable features. By employing a test for independence to choose between them on successive reviews, LABATCH.2 exploits the better properties of each and avoids their most serious limitations. It does this by means of two hybrid rules LBATCH and ABATCH.

Table 6.3 $\mathcal{B} := \{(l_1, b_1) : 1 \le b_1 \le l_1 \le 100, 2l_1 b_1 = \tilde{l}_1 \tilde{b}_1\}^a$

b_1	l_1	b_1	l_1
1	3	38	54
2	3,6	39	55
3	4	41	58,87
4	6	43	61
5	7,14,21,28	44	62, 93
7	10,15,20,25,30,35	46	65
8	11	48	51,68,85
9	13	49	69
10	14, 21,28	50	71
12	17,34,51,68,85	51	60,72,84,96
14	15,20,25,30,35	53	75
15	21,28	54	57
17	24,36,48,60,72,84,96	55	78
19	27,54	56	79
20	21, 28	58	82
21	25, 30, 35	60	68,85
22	31,62,93	61	86
24	34,51,68,85	62	66
25	28	63	89
26	37	65	92
27	38,57	66	93
28	30,35	67	95
29	41,82	68	72,84,96
31	44,66,88	70	99
32	45	72	85
33	47	82	87
34	36,48,60,72,84,96	84	85
36	51,68,85	85	96
37	52	88	93

[a] $\tilde{l}_1$ is defined in expression (6.59) and $\tilde{b}_1$ in expression (6.60).
For underlined quantities, see Section 6.7.4.

Let H_0 denote the hypothesis: On review j, the l_j batches $Y_{1b_j}, \ldots, Y_{l_j b_j}$ are i.i.d. The LBATCH and ABATCH rules both use the outcome of a test of H_0 to switch between the FNB and SQRT rules on successive reviews. The effect is to retain the desirable properties of each basic rule while reducing the influence of their limitations. The principal features of these hybrid rules are:

LBATCH RULE

- Start with the FNB rule on review 1.

- For $j \ge 1$, if H_0 is rejected on review j, use the FNB rule on review $j + 1$.

- Once H_0 is accepted on review j, use the SQRT rule on reviews, $j + 1, j + 2, \ldots$.

(a) Waiting time in queue, $\sigma_\infty = 189.5$

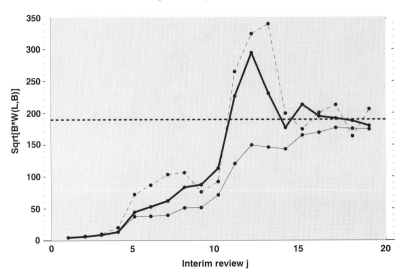

(b) Probability of waiting, $\sigma_\infty = 130.8$

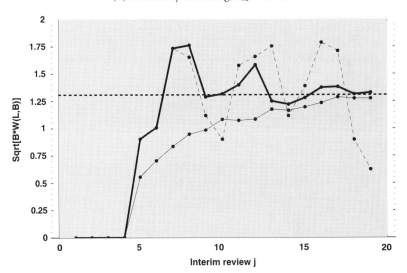

Figure 6.8 LABATCH.2 estimates of σ_∞ (ABATCH rule: heavy line, FNB rule: thin dashed line, SQRT rule: thin line, $\sigma_\infty = 189.5$: heavy dashed line)

and:

ABATCH RULE

- Start with the FNB rule on review 1.

- For $j \geq 1$, if H_0 is rejected on review j, use the FNB rule on review $j + 1$.

- For $j \geq 1$, if H_0 is accepted on review j, use the SQRT rule on review $j + 1$.

By initially fixing l, the LBATCH rule allows batch size, b, to increase at the maximal rate when H_0 is rejected, thus dissipating systematic error in V_{t_j} as fast as possible. Once H_0 is accepted, the rule switches to the SQRT rule to dissipate the error in coverage rate as rapidly as possible. By testing H_0 on every review, the ABATCH rule takes into consideration the possibility of Type II errors on successive replications. In the present context, Type II error denotes the random acceptance of H_0 when it is false. The LBATCH rule ignores this source of error, whereas the ABATCH rule allows the procedure to correct itself. Because of this self-correcting property, we favor the ABATCH rule.

As a consequence of this testing, batch size and number of batches on review j (> 1) are random. To acknowledge this property, we denote them by B_j and L_j respectively. Let K_j denote the number of test rejections of H_0 on reviews $1, \ldots, j$. Then the LBATCH and ABATCH rules induce

$$
L_j = \begin{cases} 2^{(j-K_j)/2} l_1 & \text{if } j - K_j \text{ is even,} \\ 2^{(j-K_j-1)/2} \tilde{l}_1 & \text{if } j - K_j \text{ is odd,} \end{cases} \tag{6.62}
$$

and

$$
B_j = \begin{cases} 2^{(j+K_j-2)/2} b_1 & \text{if } j - K_j \text{ is even,} \\ 2^{(j+K_j-1)/2} \tilde{b}_1 & \text{if } j - K_j \text{ is odd,} \end{cases} \tag{6.63}
$$

so that $t_j = L_j B_j = 2^{j-1} l_1 b_1$ for $j = 1, 2, \ldots$.

Since $\{(L_j, B_j), \; j \geq 1\}$ is a random rather than a deterministic batching sequence, Theorem 6.5 does not apply directly. We need additional qualifications to achieve the desired limiting results:

Theorem 6.6 (Yarberry 1993, Fishman and Yarberry 1997).
If $t\sigma_t^2 \to \sigma_\infty^2$ *as* $t \to \infty$ *and ASA holds,* $2l_1 b_1 = \tilde{l}_1 \tilde{b}_1$, *and there exist constants* $\Gamma < 1$ *and* $\theta > 1 - 4\lambda$ *such that* $\sum_{j=1}^{\infty} \mathrm{pr}(K_j/j > \Gamma) < \infty$ *and* $K_j/j \to \theta$ *as* $j \to \infty$ *w.p.1, then*

$$
B_j W_{L_j B_j} \to \sigma_\infty^2 \quad \text{as } j \to \infty \quad \text{w.p.1}
$$

and

$$\frac{\bar{X}_{t_j} - \mu}{\sqrt{W_{L_j B_j}/L_j}} \xrightarrow{\text{d}} \mathcal{N}(0,1) \quad as \; j \to \infty.$$

Recall that $a(t) = t_{J(t)}$, where $J(t)$ is defined in Section 6.6.4. Provided that its conditions are met, Theorem 6.6 gives a strongly consistent estimator of σ_∞^2 and the basis for the asymptotic validity of the confidence interval

$$\mathcal{D}_{L_{J(t)}}(\delta) = \left[\bar{X}_{a(t)} \pm \tau_{L_{J(t)}} - 1(1 - \delta/2)\sqrt{V_{a(t)}/a(t)} \right]. \tag{6.64}$$

We conjecture that there is a large class of problems that satisfy these conditions.

On review $J(t)$, LABATCH.2 actually reports the confidence interval $\mathcal{C}_{L_{J(t)}}(\delta)$ in expression (6.48), which differs from $\mathcal{D}_{L_{J(t)}}(\delta)$ principally by substituting $\bar{X}_t$, based on all the data, in place of $\bar{X}_{a(t)}$, based on the first $a(t)$ observations. If $a(t) = t$, then $\mathcal{C}_{L_{J(t)}}(\delta) = \mathcal{D}_{L_{J(t)}}(\delta)$ and Theorem 6.6 applies. If $t > a(t)$, then we need to reconcile the difference between $\mathcal{C}_{L_{J(t)}}(\delta)$ and $\mathcal{D}_{L_{J(t)}}(\delta)$ with regard to asymptotic validity. We do this in Appendix A.3.

For purposes of comparison, Figure 6.8 displays $\{\sqrt{B_j W_{L_j B_j}}; \; j = 1, \ldots, J(10^7)\}$ for the ABATCH, FNB, and the SQRT rules. The dissipation of systematic error under the ABATCH rule mirrors that under the FNB rule, but its sampling fluctuations are considerably more attenuated.

6.6.8   TESTING H₀

Because of stationarity (Section 6.5), $Y_{1b}, \ldots, Y_{lb}$ have the autocorrelation function

$$\omega_{|i-j|} := \mathrm{corr}(Y_{ib}, Y_{jb}), \quad i, j = 1, \ldots, l.$$

To test H_0 on the batch averages $Y_{1b}, \ldots, Y_{lb}$, we use the von Neumann ratio (von Neumann 1941, Young 1941, Fishman 1978, 1996)

$$\begin{aligned}
C_l :&= 1 - \frac{\sum_{i=1}^{l-1}(Y_{ib} - Y_{i+1,b})^2}{2\sum_{i=1}^{l}(Y_{ib} - \bar{X}_t)^2} \\
&= \hat{\omega}_1 + \frac{(Y_{1b} - \bar{X}_t)^2 + (Y_{lb} - \bar{X}_t)^2}{2\sum_{i=1}^{l}(Y_{ib} - \bar{X}_t)^2},
\end{aligned} \tag{6.65}$$

where, because of stationarity (Section 6.5),

$$\hat{\omega}_1 := \frac{\sum_{i=1}^{l-1}(Y_{ib} - \bar{X}_t)(Y_{i+1,b} - \bar{X}_t)}{\sum_{i=1}^{l}(Y_{ib} - \bar{X}_t)^2}$$

estimates the first lagged autocorrelation ω_1. Inspection reveals that C_l is itself an estimator of ω_1 adjusted for end effects that diminish in importance as the number of batches l increases. If $\{X_i, i \geq 0\}$ has a monotone nonincreasing autocorrelation function (e.g., Figure 6.4a), then in the present setting ω_1 is positive and decreases monotonically to zero as batch size b increases. If for some l and b, H$_0$ is true for $Y_{1b}, \ldots, Y_{lb}$, then $\omega_1 = 0$, and more generally, the monotone property implies corr$(Y_{ib}, Y_{jb}) = 0$ for all $i \neq j$.

If $\frac{Y_{1b}-\mu}{\sigma_b}, \ldots, \frac{Y_{lb}-\mu}{\sigma_b}$ are i.i.d. from $\mathcal{N}(0, 1)$, then under H$_0$, C_l has mean zero, variance $(l - 2)/(l^2 - 1)$, and a distribution that is remarkably close to normal for l as small as eight. For i.i.d. nonnormal $Y_{1b}, \ldots, Y_{lb}$ under H$_0$, C_l has mean zero. As l increases, $[(l^2 - 1)/(l - 2)] \times$ var $C_l \to 1$ and the *skewness*, E$(C_l - \text{E}C_l)^3/[\text{E}(C_l - \text{E}C_l)^2]^{3/2}$, and *excess kurtosis*, E$(C_l - \text{E}C_l)^4/[\text{E}(C_l - \text{E}C_l)^2]^2] - 3$, converge to zero.

Under the assumption of a monotone nonincreasing autocorrelation function for $\{X_i, i \geq 0\}$, a *one-sided test* of size β is in order. In particular, if $C_l \leq \Phi^{-1}(1 - \beta)\sqrt{(l - 2)/(l^2 - 1)}$, H$_0$ is accepted. Otherwise, it is rejected. The one-sided test protects against accepting H$_0$ when the sample autocorrelation, C_l, between Y_{ib} and Y_{jb} for all $|i - j| = 1$ is "too large" to be attributed to sampling variation.

For the ABATCH rule, Figure 6.9a shows $\{L_j; j = 1, \ldots, J(10^7)\}$ and Figure 6.9b displays

$$P_j := 1 - \Phi\left(\sqrt{\frac{L_j^2 - 1}{L_j - 2}}\, C_{L_j}\right), \qquad j = 1, \ldots, 19,$$

which we call the p-value sequence. In particular, P_j provides a credibility index on review j by estimating the probability of observing a C_{L_j}, under H$_0$, larger than the one computed from $Y_{1Bj}, \ldots, Y_{L_jBj}$. A small P_j implies low credibility for H$_0$. LABATCH.2 makes its decision on review j as follows. If $P_j < \beta$ on review j, for example for a user-specified $\beta = .10$, then LABATCH.2 rejects H$_0$ and uses the FNB rule on review $j + 1$. This implies $L_{j+1} = L_j$, as Figure 6.9a shows for $j = 4$ through 8, 11, and 16. If $P_j \geq \beta$, then LABATCH.2 uses the SQRT rule on review $j + 1$, so that $L_{j+1}/L_j \doteq \sqrt{2}$. Figure 6.9a illustrates this behavior for $j = 1, 2, 3, 9, 10, 12$ through 15, 17, and 18. The long run of successes beginning on review 12 suggests that batch sizes greater than $B_{12} = 10240$ make the residual correlation between batches negligible. When the LBATCH rule is used, LABATCH.2 makes its acceptance/rejection decisions according to that rule.

Suppose that $\{X_i\}$ has an autocorrelation function that exhibits damped harmonic behavior around the zero axis (e.g., Figure 6.4b), or that $\{X_i\}$ is an i.i.d. sequence. In the damped harmonic case, the one-sided test accepts H$_0$ when negative correlation is present.

(a) Number of batches

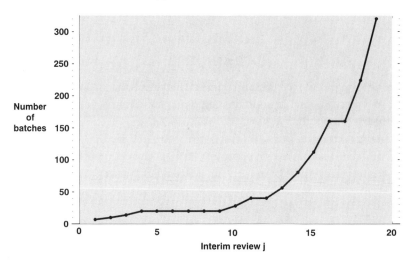

(b) p-value for testing H$_0$, $\beta = .10$

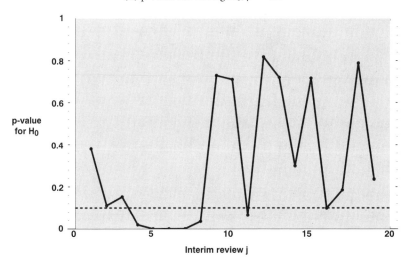

Figure 6.9 LABATCH.2 output for Series 1 using the ABATCH rule

In the i.i.d. case, review $J(t)$ produces final results with $B(t) > 1$ when a unit batch size suffices and gives a statistically better estimate of σ_∞^2. LABATCH.2 provides information that enables a simulationist to infer the presence of either of these cases and to take advantage of this knowledge.

Once LABATCH.2 completes its last review $J(t)$, it computes the p-value

$$P'_t := 1 - \Phi\left(\sqrt{\frac{t^2 - 1}{t - 1}} C'_t\right),$$

where

$$C'_t := 1 - \frac{\sum_{i=1}^{t}(X_i - X_{i+1})^2}{2(t - 1)W'_{t,1}},$$

$$W'_{t,1} := \frac{1}{t - 1}\sum_{i=1}^{t}\left(X_i - \bar{X}_t\right)^2,$$

for the hypothesis H_0 applied to the original observations $X_1, \ldots, X_t$. It also computes an approximating $100 \times (1 - \delta)$ percent confidence interval

$$\left[\bar{X}_t \pm \tau_{t-1}(1 - \delta/2)\sqrt{W'_{t,1}/t}\right] \tag{6.66}$$

and sample standard deviation $\sqrt{W'_{t,1}}$ and displays them in the Interim Review Tableau for a series below the title "If data are independent." For example, see Figure 6.6, where the p-value for each series is 0.0000, thereby rejecting H_0.

Whenever this final row displays $P'_t > \beta$, the specified significance level for testing H_0, three possibilities arise:

- $\{X_i\}$ is an i.i.d. sequence

- $\{X_i\}$ has a damped harmonic autocorrelation function

- This p-value merely exhibits a random aberration.

To investigate these possibilities we proceed as follows. If $P'_t > \beta$, and the sequence $\{\sqrt{B_j W_{L_j B_j}}; \ j = 1, \ldots, J(t)\}$ of estimates σ_∞, listed under Sqrt[B * W(L, B)], fluctuates around $\sqrt{W_{t,1}}$ displaying no discernible trend, then it is not unreasonable to assume that $\{X_i\}$ is an i.i.d. sequence, to H_0, and to use the confidence interval (6.66). This approach offers a statistically better estimate of σ_∞ and greater degrees of freedom.

If $P'_t > \beta$ but $\{\sqrt{B_j W_{L_j B_j}}; \ j = 1, \ldots, J(t)\}$ tends to decrease with increasing j and, in particular, $B_{J(t)} W_{L_{J(t)} B_{J(t)}} \gg W'_{t,1}$, then there is reason to believe that $\{X_i\}$ has an autocorrelation function that fluctuates around zero, implying the presence of negative as well as positive correlation. In this case, using the confidence interval for review $J(t)$ to reap the potential benefits of the negative correlation (e.g., see Section 6.5), usually inducing a shorter interval width.

If $P'_t > \beta$ but $\{\sqrt{B_j W_{L_j B_j}}; \ j = 1, \ldots, J(t)\}$ tends to increase with increasing j and, in particular, $B_{J(t)} W_{L_{J(t)} B_{J(t)}} \gg W_{t,1}$, then there is reason to believe that the p-value reflects a random aberration.

In summary, while LABATCH.2 automatically provides its Final Tableau based on the concept of asymptotically valid confidence intervals, a perusal of the entries in the row below "If the data are independent" for each series offers an opportunity for obtaining a tighter confidence interval.

6.6.9 ASSESSING THE ADEQUACY OF THE WARM-UP INTERVAL

LABATCH.2 also provides an assessment of the extent of warm-up bias in $\bar{X}_t$. We again use the M/M/1 simulation, but this time with an arrival to an empty and idle system and data collection beginning with $X_1 :=$ waiting time of the first arrival. For this scenario, $X_1 = 0$ w.p.1 and $X_1, X_2, \ldots, X_t$ is stationary, only asymptotically (as $t \to \infty$). This is a more biased environment than one would expect to encounter when data collection begins after a user-specified warm-up interval.

Figure 6.10 displays the 99 percent confidence intervals for mean waiting time in queue, taken from the LABATCH.2 interim review tableau for Series 1 for this run. It suggests little bias after review 11, which corresponds to $2^{11} \times L_1 \times B_1 = 2048 \times 7 \times 5 = 71,680$ observations.

Suppose the user had specified $t = 4480$ for the sample-path length. Then LABATCH.2 would have generated the same sample averages and confidence intervals for $L_1 = 7$ and $B_1 = 5$, but only for the first seven interim reviews. Moreover, a display of these results would have aroused considerably more concern about the dissipation of bias in $\bar{X}_t$. Interestingly, had we taken $t = 4480$ in our first simulation, which began in an equilibrium state, we might equally be suspicious of $\bar{X}_t$ based on the path it displays in Figure 6.5 for $j = 1, \ldots, 7$.

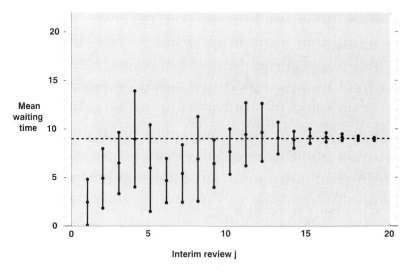

Figure 6.10 LABATCH.2 Sample means and 99 percent confidence intervals for Series 1 using the ABATCH rule; simulation starts in empty and idle state

However, this observation in no way mitigates the value of the assessment when we know for a fact that $\bar{X}_t$ may contain systematic bias as a result of starting in an arbitrarily chosen state and possibly choosing an inadequate warm-up interval.

6.7 LABATCH.2

LABATCH.2 is a collection of computer programs that perform statistical analyses on sample sequences collected on strictly stationary stochastic processes. It offers two modes of implementation: One integrates LABATCH.2 into an executing data-generating program (e.g., simulation) to analyze the evolving data on repeated subroutine calls; the other takes data from an existing file as input. In addition to minimizing user effort, the first option considerably reduces space requirements. It also allows user interaction with the executing program via screen displays of interim estimates. The second option permits statistical analysis of stored data, regardless of source and date of generation, thereby making LABATCH.2 applicable in a considerably wider range of data-generating environments.

LABATCH.2 is a revision of LABATCH (Fishman 1996, Fishman and Yarberry 1997) that considerably simplifies its implementation and use. A user merely inserts a single subroutine call statement in her/his main program and assigns values to several control arguments of the subroutine.

This section discusses several major features of LABATCH.2. C, FORTRAN, and SIM-SCRIPT II.5 implementations of LABATCH.2 are obtainable by anonymous file transfer protocol (ftp) at http://www.or.unc.edu/~gfish/labatch.2.html.

For any path length $t \geq 20$, LABATCH.2 automatically computes the number of batches, L_1, and the batch size, B_1, to be used in its first review. For example, it chose $L_1 = 7$ and $B_1 = 5$ for $t = 10^7$ for the M/M/1 example.

A call from a C FORTRAN SIMSCRIPT II.5 main program to

$$\text{BATCH\_MEANS(IN\_UNIT,OUT\_UNIT,T,S\_NUM,PSI\_VECTOR,} \tag{6.67}$$
$$\text{DELTA,RULE, BETA,L\_UPPER,SCREEN)}$$

implements LABATCH.2, where Table 6.4 defines its arguments. As an example, suppose they assume the values

$$\text{IN\_UNIT} = 0,$$
$$\text{OUT\_UNIT} = 15,$$
$$\text{T} = 1000,$$
$$\text{S\_NUM} = 2,$$
$$\text{PSI\_VECTOR} = \text{pointer to data vector,}$$
$$\text{DELTA} = .01, \tag{6.68}$$

Table 6.4 LABATCH.2 arguments

User-specified input	Definition	Notation in Chapter
IN_UNIT	:= 0 if data are repeatedly to be transferred from a main program	...
	:= 30 if data are to be read one vector at a time from input file c.30 fort.30 SIMU30 (30 is merely an example)	
OUT_UNIT	:= designated unit for writing output	...
T	:= total number of observations	t
S_NUM	:= number of sample series	...
PSI_VECTOR	:= pointer to vector of S_ NUM sample values	...
DELTA	:= desired significance level for confidence intervals (.01 or .05 suggested)	δ
RULE	:= 1 if ABATCH rule is used := 0 if LBATCH rule is used	
BETA	:= significance level for testing for independence (.10 suggested) If BETA = 1, each review uses the FNB rule = −1, each review uses the SQRT rule	β
L_UPPER	:= upper bound on initial number of batches (3 <= L_UPPER <= 100)	...
SCREEN	:= 1 if interim review estimates are to be displayed on the screen := 0 otherwise	

$$\text{RULE} = 1,$$
$$\text{BETA} = .10,$$
$$\text{L\_UPPER} = 30,$$
$$\text{SCREEN} = 0.$$

Then LABATCH.2 processes S_NUM=2 sequences in T = 1000 iterative calls from the user's main program, computes $100 \times (1 - \text{DELTA})$=99 percent confidence intervals for the sample averages, and writes the output to a file called   c.15 fort.15   (IN_UNIT=0, SIMU15

OUT_UNIT=15). RULE=1 causes LABATCH.2 to employ the ABATCH rule to determine batch size on each review, and BETA = .10 causes it to test for independence of batch averages at the .10 signifiance level. LABATCH.2 begins its first iteration with the number of batches no greater than L_UPPER=30. SCREEN=0 suppresses the interactive screen feature. In practice, all but the values of T and S_NUM can be set once and the subroutine used repeatedly in different settings.

Testing (Section 6.6.7) determines whether the batch size on review $j + 1$ increases by a factor of 2 (rejection) or approximately $\sqrt{2}$ (success). A doubling of batch size aims at reducing any residual systematic variance error detected on review j as fast as possible. A $\sqrt{2}$ increase signals that the number of batches is also increasing approximately by a $\sqrt{2}$ factor. This growth in both batch size and number of batches as the sample path length grows is a necessary condition for obtaining a strongly consistent estimator of σ_∞^2. Our subsequent description of LABATCH.2 uses the definitions in Table 6.4.

6.7.1 TWO MODALITIES

As already mentioned, LABATCH.2 provides two ways of accessing data for statistical analysis. One requires the user to insert a call statement into the data-generating program that executes the call each time it produces a new data vector. Calling and executing BATCH_MEANS each of T times that a data vector with S_NUM entries is generated results in O(S_NUM×T) computing time and O(S_NUM× $\log_2$ T) space being used to generate the LABATCH.2 output. Both complexities arise from choosing rules that cause either $B_{j+1} = 2B_j$ or $B_{j+1} \doteq \sqrt{2}B_j$ on successive reviews $j = 1, 2, \ldots$. The space bound is particularly appealing when T is large. Yarberry (1993) and Alexopoulos et al. (1997) describe the basis for these complexities.

The other option allows LABATCH.2 to read its data from a file, giving the software a considerably broader range of application than merely for in-line generated sample records. We illustrate how the in-line option works in the context of the M/M/1 example, but stress the applicability of the approach to stored sample data.

If IN_UNIT = 30 and OUT_UNIT = 15 (Table 6.4), then a main program needs to call BATCH_MEANS just once to cause LABATCH.2 to read its sample data from a file called $\begin{smallmatrix} c.30 \\ \text{fort.30} \\ \text{SIMU30} \end{smallmatrix}$ and to write its output to $\begin{smallmatrix} c.15 \\ \text{fort.15} \\ \text{SIMU15} \end{smallmatrix}$. While the O(S_NUM×T) time and O(S_NUM× $\log_2$ T) space bounds remain for LABATCH.2, they do not tell the whole story. In particular, the input file requires space proportional to S_NUM×T, which for S_NUM as small as 1 can be substantial for sufficiently large T.

Programs written in C, FORTRAN, or SIMSCRIPT II.5 can implement the second option without qualification. Any program that provides access to and execution of a C, FORTRAN, or SIMSCRIPT II.5 subroutine can take advantage of the first option. In a simulation environment, a user-written program in a language that provides standard linkages for incorporating a subroutine generally consumes less time calling BATCH_MEANS and analyzing data than a program generated at the icon level in a point-and-click environment.

6.7.2 INTERACTING WITH THE ANALYSIS

In queueing simulations, σ_∞^2 usually increases superlinearly with traffic intensity. In this setting, LABATCH.2 offers a way of viewing interim results that enables the user to assess the quality of the σ_∞^2 estimates during a run. If the interim results suggest that an acceptable level of accuracy has been achieved, then LABATCH.2 allows the user to terminate the sampling experiment at that point, thus saving computing time.

If SCREEN $= 0$ in expression (6.67), then LABATCH.2 performs as already described. However, if SCREEN $= 1$, then after executing interim review j, LABATCH.2 displays on the screen $\bar{X}_{t_j}$, the sample coefficient of variation of $\bar{X}_{t_j}$, sometimes called the relative error, and $\sqrt{B_j W(L_j, B_j)}$, for Series i in column $i + 1$ for $i = 1, \ldots, \min(6, \text{S\_NUM})$, where screen size dictates the upper limit on i. LABATCH.2 then asks the user whether he/she wishes to continue. Recall that $\sqrt{B_j W(l_j, B_j)}$ is an estimate of σ_∞.

Figure 6.11 displays the tableau for the steady-state M/M/1 simulation, where column 2 shows the sequence of estimates for Series 1 and column 3 does likewise for Series 2. If the user concludes that $\sqrt{B_j W(L_j, B_j)}$ has stabilized in each column, so that the systematic errors have become negligible, and that the coefficients of variation C.V.$(\bar{X}_{t_j})$ are sufficiently small, then he/she may terminate the simulation by typing "n". This action causes LABATCH.2 to compute confidence intervals for the S_NUM and to write both the final and interim review tableaus (Figure 6.7) to file OUT_UNIT. If this action occurs immediately after the on-screen display, for review j, then the final tableau, as well as the interim review tableaus, use $L_j B_j$ observations. If the screen display suggests that systematic error remains, then, provided that $L_j B_j < t$, typing "y" causes LABATCH.2 to collect additional data, to perform the next review, to display $\sqrt{B_{j+1} W(L_{j+1}, B_{j+1})}$ on screen, and to ask the user whether he/she wishes to continue. If $L_j B_j = a(t)$ and the user types "y", then the simulation goes to completion and LABATCH.2 uses all $t = T$ observations for the sample averages and the first $a(t)$ observations for the final $B_{J(t)} W(L_{J(t)}, B_{J(t)})$'s.

As illustration of how this procedure works in practice, consider review 11 in Figure 6.11, where C.V.(X_BAR) $= .1255$ for mean waiting time (Series 1) and C.V.(X_BAR) $= .008205$ for probability of waiting (Series 2) may encourage one to conclude that the sample averages (X_BAR) are statistically reliable. By contrast, the variability in the Sqrt[B*W(L,B)] estimates of σ_∞ for reviews 9 through 11 encourages continuation of the experiment; hence "y" in response to the query produces the output on review 12.

By way of qualification we encourage the reader to recognize that this approach differs substantially from the sequential mean estimation approach of Chow and Robbins (1965), as described in Fishman (1978) and Law and Kelton (1991). Whereas the objective there is to estimate μ to within a specified accuracy with specified probability, the objective of the interactive approach here is to allow a user to assess the quality of variance estimates while the sampling experiment is executing.

Kelton et al. (1998, p. 233) assert that a check is possible on half-width during Arena execution using the THALF(·) system function. However, the cited Section 11.5 and, in particular, Subsection 11.5.2 describe a different but related problem.

```
X_BAR, C.V.(X_BAR), and Sqrt[B*W(L,B)] for Series 1 Through  2
**************************************************************
          No. of
Review    Obs.                      1         2

  1         35 X_BAR           0.1660D+02 0.1000D+01
               C.V.(X_BAR)     0.4557D-01 0.0000D+00
               Sqrt[B*W(L,B)]  0.4475D+01 0.0000D+00
continue[y/n]? y
  2         70 X_BAR           0.1810D+02 0.1000D+01
               C.V.(X_BAR)     0.3993D-01 0.0000D+00
               Sqrt[B*W(L,B)]  0.6047D+01 0.0000D+00
continue[y/n]? y
  3        140 X_BAR           0.1914D+02 0.1000D+01
               C.V.(X_BAR)     0.3833D-01 0.0000D+00
               Sqrt[B*W(L,B)]  0.8680D+01 0.0000D+00
continue[y/n]? y
  4        280 X_BAR           0.1886D+02 0.1000D+01
               C.V.(X_BAR)     0.4158D-01 0.0000D+00
               Sqrt[B*W(L,B)]  0.1313D+02 0.0000D+00
continue[y/n]? y
  5        560 X_BAR           0.1111D+02 0.9250D+00
               C.V.(X_BAR)     0.1686D+00 0.4118D-01
               Sqrt[B*W(L,B)]  0.4432D+02 0.9014D+00
continue[y/n]? y
  6       1120 X_BAR           0.7254D+01 0.8812D+00
               C.V.(X_BAR)     0.2174D+00 0.3415D-01
               Sqrt[B*W(L,B)]  0.5278D+02 0.1007D+01
continue[y/n]? y
  7       2240 X_BAR           0.6706D+01 0.8862D+00
               C.V.(X_BAR)     0.1952D+00 0.4135D-01
               Sqrt[B*W(L,B)]  0.6196D+02 0.1734D+01
continue[y/n]? y
  8       4480 X_BAR           0.7556D+01 0.8830D+00
               C.V.(X_BAR)     0.1647D+00 0.2981D-01
               Sqrt[B*W(L,B)]  0.8328D+02 0.1762D+01
continue[y/n]? y
  9       8960 X_BAR           0.6747D+01 0.8770D+00
               C.V.(X_BAR)     0.1366D+00 0.1553D-01
               Sqrt[B*W(L,B)]  0.8721D+02 0.1289D+01
continue[y/n]? y
 10      17920 X_BAR           0.7817D+01 0.8888D+00
               C.V.(X_BAR)     0.1080D+00 0.1106D-01
               Sqrt[B*W(L,B)]  0.1130D+03 0.1316D+01
continue[y/n]? y
 11      35840 X_BAR           0.9513D+01 0.9001D+00
               C.V.(X_BAR)     0.1255D+00 0.8205D-02
               Sqrt[B*W(L,B)]  0.2260D+03 0.1398D+01
continue[y/n]? y
 12      71680 X_BAR           0.9668D+01 0.8973D+00
               C.V.(X_BAR)     0.1136D+00 0.6581D-02
               Sqrt[B*W(L,B)]  0.2941D+03 0.1581D+01
continue[y/n]?
```

Figure 6.11 LABATCH.2 screen display when SCREEN = 1

6.7.3 IMPLEMENTATION

As mentioned in the introduction, C, FORTRAN, and SIMSCRIPT II.5 versions of LABATCH.2 are available on the Internet. The statement

BATCH_MEANS(IN_UNIT,OUT_UNIT,S_NUM,T,DATA,DELTA,RULE,BETA,

L_UPPER,SCREEN);

in a C program calls BATCH_MEANS. It requires that a previously defined double S_NUM × 1 array, DATA, contain the latest data entries. Table 6.4 defines all remaining arguments. If numerical values replace IN_UNIT, OUT_UNIT, S_NUM, T, DELTA, RULE, BETA, L_UPPER, and SCREEN, then no need exists to define these variables in the calling program. This also applies to the FORTRAN and SIMSCRIPT II.5 calls.

The statement

call BATCH_MEANS(IN_UNIT,OUT_UNIT,S_NUM,T,DATA,

DELTA,RULE,BETA,L_UPPER,SCREEN)

in a FORTRAN program calls BATCH_MEANS. It requires that a previously defined double precision S_NUM × 1 array, DATA, contain the latest data entries. The character @ in column 6 indicates a continuation.

The statement

call BATCH_MEANS(IN_UNIT,OUT_UNIT,T,S_NUM,DATA(*),

DELTA,RULE,BETA,L_UPPER,SCREEN)

in a SIMSCRIPT II.5 program calls BATCH_MEANS. In addition to reserving space for the double S_NUM × 1 array DATA in the main program, a user must also include the statements

define DATA	as 1-dim double array
define PHI	as double function
define PHI_QUANTILE	as double function
define STUDENT_T_QUANTILE	as double function

in the preamble. See CACI (1983, p. 91).

6.7.4 USER-SPECIFIED l_1 AND b_1

LABATCH.2 also allows a user to choose any underlined (l_1, b_1) in Table 6.3 as the initial number of batches and the initial batch size by specifying a sample path length $T = 2^\alpha l_1 b_1$ for the selected l_1 and b_1 and where α is a positive integer. In this case, LABATCH.2 uses 100 percent of the data to compute the final BW(L,B).

6.7.5 CODING

As mentioned earlier, LABATCH.2 is a revision of the LABATCH statistical analysis package. The coding of LABATCH provides users with a considerable number of options. For example, it allows a user to specify whether headings are to be printed, and for each series it allows a user to specify different significance levels for testing H and different confidence levels for the means.

To reduce the burden of decision making that a potential user faces, these options have been severely curtailed in LABATCH.2. However, a perusal of the code makes clear that the code and data structures that support these wider choices remain in place, but are suppressed. This was done to reduce the possibility of error in modifying the code and to leave structures in place that may become useful once again in subsequent revisions of LABATCH.2.

6.8 AIRLINE RESERVATION PROBLEM

We now apply LABATCH.2 to the airline reservation problem. Our interest is in estimating the mean and 11 ordinates of the distribution function of waiting time. Hereafter, we refer to each sample sequence for estimating a parameter as a *series*. The 12 series of interest are:

Series	Parameter to be estimated	X_i Datum for ith observation
1	μ_W	W_i
2	$\mathrm{pr}(W > 0)$	$I_{(0,\infty)}(W_i)$
3	$\mathrm{pr}(W > 1)$	$I_{(1,\infty)}(W_i)$
4	$\mathrm{pr}(W > 2)$	$I_{(2,\infty)}(W_i)$
5	$\mathrm{pr}(W > 3)$	$I_{(3,\infty)}(W_i)$
6	$\mathrm{pr}(W > 4)$	$I_{(4,\infty)}(W_i)$
7	$\mathrm{pr}(W > 5)$	$I_{(5,\infty)}(W_i)$
8	$\mathrm{pr}(W > 6)$	$I_{(6,\infty)}(W_i)$
9	$\mathrm{pr}(W > 7)$	$I_{(7,\infty)}(W_i)$
10	$\mathrm{pr}(W > 8)$	$I_{(8,\infty)}(W_i)$
11	$\mathrm{pr}(W > 9)$	$I_{(9,\infty)}(W_i)$
12	$\mathrm{pr}(W > 10)$	$I_{(10,\infty)}(W_i)$.

To make the LABATCH.2 software operational within the AIR_EV.sim program in Figure 4.2, one modifies the program with:

Step 1. preamble
Between lines 33 and 34, insert:
"Vector for data transfer to batch_means
"

define DATA	as 1−dim double array

"Declarations for functions called by BATCH_MEANS
"

define PHI	as double function
define PHI_QUANTILE	as double function
define STUDENT_T_QUANTILE	as double function

Step 2. main

Between lines 1 and 2, insert:

reserve DATA(*)	as 12

Step 3. ACCOUNTING

Between lines 9 and 10, insert:

```
for I=1 to 12 DATA(I)=0.
DATA(I)= WAITING.TIME
    if DATA(1) > 0
      DATA(2) = 1
    endif
    if DATA(1) > 1
      DATA(3) = 1
    endif
    if DATA(1) > 2
      DATA(4) = 1
    endif
    if DATA(1) > 3
      DATA(5) =1
    endif
    if DATA(1) > 4
      DATA(6) = 1
    endif
    if DATA(1) > 5
      DATA(7) = 1
    endif
    if DATA(1) > 6
      DATA(8) = 1
    endif
    if DATA(1) > 7
      DATA(9) = 1
    endif
    if DATA(1) > 8
      DATA(10) = 1
    endif
    if DATA(1) > 9
```

DATA(11) = 1
endif
if DATA(1) > 10
    DATA(12) = 1
endif

call BATCH_MEANS(0,15,T,12,DATA(*),.01,1,.10,30,0)

The call to the BATCH_MEANS routine transfers the pointer to the DATA vector with 12 entries for analysis.

Figure 6.12a displays the final tableau for a run with warm-up interval K = 200 and sample path length T = 2×10^6. This choice of T induced LABATCH.2 to choose $l_1 = 13$ and $b_1 = 9$ and to produce results on $J(\text{T}) = 15$ interim reviews. Figure 6.13 shows graphs for $\sqrt{\text{BW(L,B)}}$ based on the interim review tableaus for Series 1 through 12. All graphs reveal the dissipation of systematic variance error well before review 15. Figure 6.14 shows the final point estimates and corresponding confidence intervals for $\{\text{pr}(W > w); w = 0, 1, \ldots, 10 \text{ minutes}\}$.

Figure 6.12b offers a particularly revealing illustration of how convergence occurs for $\bar{X}_{t_j} = \bar{X}$ and $\left[W(L_{(t_j)}, B_{(t_j)}) \right]^{1/2} = [\text{W(L,B)}]^{1/2}$ for $j = 6, \ldots, 15$ based on entries in the LABATCH.2 tableau for waiting time (series 1). It shows how both quantities converge to a relatively small neighborhood as sample-path length increases. Note that $t_6 = 9 \times 13 \times 2^5 = 3744$, whereas $t_{15} = 9 \times 13 \times 2^{14} = 1,916,928$. We began at review 6, rather than review 1, to focus on the ultimate convergence.

Our inclusion of Figures 6.12 through 6.14 here and Figures 6.5 through 6.8 in Section 6.6.1 is intended to inform the reader about the potential uses of LABATCH.2 output. Since Practical considerations may limit how much output for each data sequence can be presented and digested, we recommend, as a minimum, a display as in Figure 6.12a and, for each sequence, a display as in Figure 6.12b.

6.9 BATCH MEANS IN ARENA

Section 6.6.5 describes how Arena automatically employs the batch-means method with Student's t distribution to compute an approximating 95 percent half-width for each mean of interest. For $t = 2 \times 10^6$ waiting times, execution of the Arena airline reservation program in Figures 4.12 and 4.13 yielded a sample average waiting time of 4.229 and 95 percent half-width of .1309. For queue length it gave a sample average of 4.1189 and 95 percent half-width of .1295. For this t, expression (6.55) implies only $l(t) = 20$ batches but $b(t) = 100,000$ observations per batch.

Arena also provides an option for a post-simulation batch-means analysis for user-specified batch size and confidence level. As illustration, consider the Wait_Time.DAT file declared in Figure 4.13. At the end of a simulation run, a user clicks on "Output" on the Tools menu, on "Analyze," and on "Batch/Truncate Obs'vns . . . ," and then enters Wait_Time.DAT, as the file name, the desired number of observations to be truncated, the desired batch size, and the name of the file, suffixed by .flt, in which the batch averages are to be stored. The user again clicks on Analyze and then on Conf. Interval on Mean, and on Classical . . . , and then enters the name of the .stl file and the selected confidence level.

```
Final Tableau
                                    Mean Estimation
                                    **************
                                    (t =  2000000 )

                                               99.0%
                    _         Standard Error   Confidence Interval              _
     Series         X         sqrt[B*W(L,B)/t]   Lower        Upper      (Upper-Lower)/|X|

        1       +4.251E+000    +4.960E-002    +4.122E+000  +4.380E+000     +6.066E-002

        2       +7.339E-001    +1.559E-003    +7.298E-001  +7.379E-001     +1.099E-002

        3       +6.264E-001    +2.067E-003    +6.211E-001  +6.318E-001     +1.707E-002

        4       +5.282E-001    +2.454E-003    +5.219E-001  +5.346E-001     +2.405E-002

        5       +4.426E-001    +2.717E-003    +4.355E-001  +4.496E-001     +3.178E-002

        6       +3.708E-001    +2.849E-003    +3.635E-001  +3.782E-001     +3.976E-002

        7       +3.106E-001    +2.909E-003    +3.030E-001  +3.181E-001     +4.848E-002

        8       +2.603E-001    +2.895E-003    +2.528E-001  +2.677E-001     +5.756E-002

        9       +2.178E-001    +2.844E-003    +2.104E-001  +2.251E-001     +6.758E-002

       10       +1.829E-001    +2.762E-003    +1.757E-001  +1.900E-001     +7.816E-002

       11       +1.534E-001    +2.633E-003    +1.466E-001  +1.602E-001     +8.884E-002

       12       +1.288E-001    +2.498E-003    +1.223E-001  +1.352E-001     +1.004E-001


     _
     X(t) is based on all t observations.
     W(L,B) is based on first  95.85% of the t observations.
```

Figure 6.12 (a) LABATCH.2 final estimates for the airline reservation simulation using the ABATCH Rule

Since the approximating 99 percent confidence interval for waiting time (Series 1) in Figure 6.12a is based on an LABATCH.2 determined batch size of 4608 for the SIMSCRIPT II.5 airline reservation model, the batch-means analysis for the Arena data was performed for a batch size of $b(t) = 5000$, implying $l(t) = 400$ batches. It yielded an approximating 99 percent half-width of .1370 for waiting time, which compares to the LABATCH.2 half-width of .1290 in Figre 6.12a for Series 1. Although based on two independent sample paths generated by two different simulation programs, the estimates are reasonably close. However, LABATCH.2 automatically determined its batch size, whereas a user must provide a batch size for the Arena option.

Given the Wait_Time.DAT file, a user can overcome this limitation on knowing an appropriate batch size by using LABATCH.2. The Wait_Time.DAT file has binary mode. Arena provides a capability for converting these data to scientific notation and exporting them to a new file that can serve as input to LABATCH.2. To make this conversion, a user clicks on Tools, Output Analyzer, File, Data File, and Export, and then enters the name of the .DAT file and name of the receiving file.

For these data with maximal initial number of batches L_UPPER = 30 (Section 6.7), LABATCH.2 computed a final 99 percent confidence interval [4.039, 4.359], which gave a half-width of .130, based on $L(t) = 208$ batches and $B(t) = 9216$ observations per batch. The larger $L(t)$ compared to $l(t) = 20$ in the automatically generated Arena 95

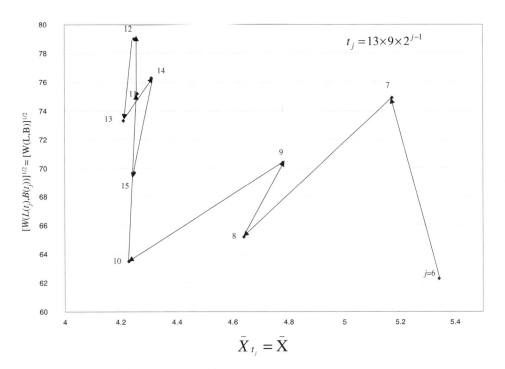

Figure 6.12   (b) Convergence for series 1

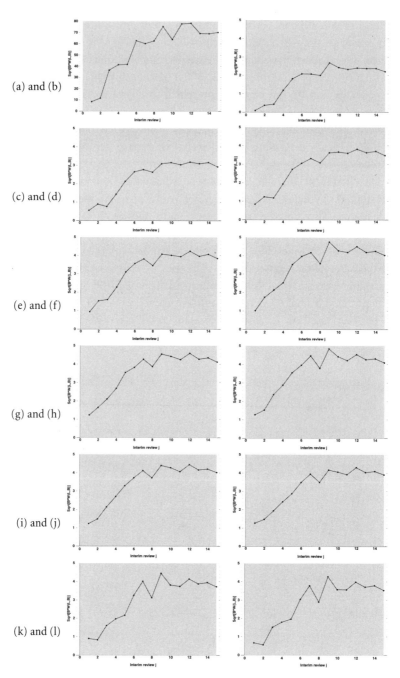

(a) and (b)

(c) and (d)

(e) and (f)

(g) and (h)

(i) and (j)

(k) and (l)

Figure 6.13 LABATCH.2 estimates of σ_∞ for Series 1 through 12 in AIR_EV.sim simulation using the ABATCH rule

percent analysis implies that in addition to accounting for the dependence among batches, the LABATCH.2 estimate of $t\sigma_\infty^2$ has smaller variance.

The .flt file requires about $16t$ bytes of space to store a record of observations containing t waiting times. This excessive space need arises because Arena also stores the times at which each observation was collected and retains eight-byte double-precision accuracy for each entry. In the present case $16t = 2 \times 8 \times 2 \times 10^6 = 32$ megabytes. In comparison, LABATCH.2 requires $O(8 \log_2(t))$ space to perform its analysis on t waiting times in double precision.

Had we also wanted to do an Arena analysis for queue length, at the .99 significance level, the additionally required space would have substantially exceeded that used for the waiting times because of the need to store data each time the time-persistent queue-length variable changes value. Therefore, we do this by using the Arena automatically generated queue-length statistics. Recall that these were a sample-average queue length 4.119 and a 95 percent half-width .1259. Multiplying the automatically displayed half-width by $\tau_{l(t)-1}(1 - \delta/2)/\tau_{l(t)-1}(.975)$ converts a 95 percent half-width to a $100 \times (1 - \delta)$ percent half-width, where equation (6.55) defines $l(t)$, the number of batches, and $\tau_{l(t)-1}(1 - \delta/2)$ denotes the $1 - \delta/2$ critical value of Student's t distribution with $l(t) - 1$ degrees of freedom. In the present case, $\delta = .01$ gives a 99 percent half-width .1771, but based on only $l(t) = 20$ batches.

Several additional differences distinguish the Arena and LABATCH.2 results with regard to interpretation. For each series, LABATCH.2 provides a Final Tableau, inclusive of all interim reviews, from which an analyst can assess the stability of the final estimate, BW(L,B), of σ_∞^2 and therefore the stability of the corresponding half-width. Arena's approach provides

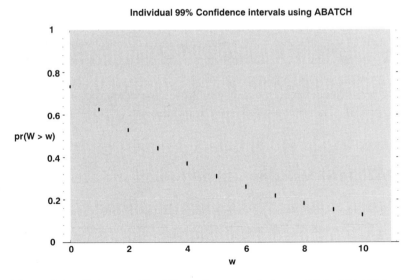

Figure 6.14   Estimate of $\{\mathrm{pr}(W > w); \quad w = 0, 1, \ldots, 10\}$ in AIR_EV.sim using the ABATCH rule

no comparable basis for assessment. For reasons of space, the Final Tableaus corresponding to the series in Figure 6.12 have been omitted here. However, the Final Tableaus in Figure 6.6 for the M/M/1 model, and their corresponding plots in Figure 6.7, attest to their value for assessing stability.

A second difference relates to ratio estimators. In a simulation, such as the airline reservation problem, executed for a fixed number of waiting times, the sample average queue length has a random denominator and thus is a ratio estimator. LABATCH.2 can be used to produce a confidence interval that accounts for this property (Section 6.11). Arena produces a half-width that does not account for this property.

6.10 PREBATCHING

The LABATCH.2 with IN_UNIT = 0 requires a call to the BATCH_MEANS subroutine after each vector of observations is collected. An alternative design can be formulated that would pass segments of each individual series at the moments of interim review, thereby reducing the number of calls to BATCH_MEANS substantially. However, this alternative requires a space allocation to store the segment for each series between reviews. A second alternative has more appeal. It reduces computing cost by *prebatching* the data. Let

$$\ddot{X}_j = c^{-1} \sum_{i=1}^{c} X_{c(j-1)+i}, \qquad c = \text{integer} \geq 1, \quad j = 1, 2, \dots. \tag{6.69}$$

Suppose that $\ddot{X}_1, \ddot{X}_2, \dots$ is the input sequence to LABATCH.2 in place of $X_1, X_2, \dots$. Then LABATCH.2 is executed no more than t/c times, thereby reducing computing time with no increase in required space. However, prebatching may induce a loss of statistical efficiency in estimating σ_∞^2 by inducing larger batches than the correlation, structure in $\{X_j\}$ warrants. Therefore, this approach has its greatest appeal in the presence of high correlation, which, regardless of prebatching, would inevitably induce large batches. Prebatching can reduce cost significantly when multiple sequences are being analyzed.

6.11 RATIO ESTIMATION

Recall Option B in Sections 3.4 and 3.5, where data collection begins when job M completes service and ends when job $M + N$ completes service. Then expression (3.51) with $s = \tau_M$ and $\tau = \tau_{M+N} - \tau_M$ gives the sample average waiting time

$$\bar{W}(\tau_M, \tau_{M+N}) = \frac{1}{N} \sum_{i=M+1}^{M+N} W_i, \tag{6.70}$$

where $N = N(\tau_{M+N}) - N(\tau_M)$, and expression (3.49) gives the sample average queue length as

$$\bar{Q}(\tau_M, \tau_{M+N}) = \frac{\int_{\tau_M}^{\tau_{M+N}} Q(u)du}{\tau_{M+N} - \tau_M}. \tag{6.71}$$

Since M and N are fixed, $\bar{W}(\tau_M, \tau_{M+N})$ is the ratio of a random quantity and a deterministic quantity, and LABATCH.2, operating on $W_{M+1}, \ldots, W_{M+N}$, provides an assessment of how well $\bar{W}(\tau_M, \tau_{M+N})$ approximates the long-run average waiting time μ_W. However, $\bar{Q}(\tau_M, \tau_{M+N})$ under Option B is the ratio of two random quantities, $\int_{\tau_M}^{\tau_{M+N}} Q(u)du$ and $\tau_{M+N} - \tau_M$. Therefore, direct application of the batch-means method does not provide an assessment of how well $\bar{Q}(\tau_M, \tau_{M+N})$ approximates the long-run average queue length μ_Q. However, an indirect application of LABATCH.2 can provide the ingredients for making this assessment.

Several indirect methods exist for assessing accuracy in the present case, and we describe them in the order of the increasing effort they require. First, there is Little's law (Section 3.5.1), which provides an alternative for estimation μQ using Series 1 in Figure 6.12a. Since $\lambda = 1$, that approach yields the point estimate 4.251 and the interval estimate [4.122, 4.380]. Relatively little effort is required to compute these quantities. However, Little's law provides no alternative if, for example, the objective is to estimate $\text{pr}[Q(\tau) > q]$ for some $q > 0$.

The remaining indirect methods make use of the observation that expression (6.70) has the equivalent form

$$\bar{Q}(\tau_M, \tau_{M+N}) = \frac{\dfrac{1}{N} \sum_{i=M+1}^{M+N} \int_{\tau_{i-1}}^{\tau_i} Q(u)du}{\dfrac{1}{N} \sum_{i=M+1}^{M+N} (\tau_i - \tau_{i-1})}, \tag{6.72}$$

which allows us to regard the numerator as the sample average of the N random variables $\int_{\tau_M}^{\tau_{M+1}} Q(u)du, \ldots, \int_{\tau_{M+N-1}}^{\tau_{M+N}} Q(u)du$ and the denominator as the sample average of the N random variables $\tau_{M+1} - \tau_M, \ldots, \tau_{M+N} - \tau_{M+N-1}$. To exploit this representation, we resort to a more general formulation.

Let $X_1, \ldots, X_t$ and $Y_1, \ldots, Y_t$ denote sample records collected on two phenomena of interest during a simulation. Let $\bar{X}_t$ and $\bar{Y}_t$ denote their respective sample means and $\bar{R}_t := \bar{Y}_t/\bar{X}_t$ their sample ratio. Assume $\mu_X \neq 0$. With reference to Option B, $Y_i = \int_{\tau_{M+i-1}}^{\tau_{M+i}} Q(u)du$ and $X_i = \tau_{M+i} - \tau_{M+i-1}$ for $i = 1, \ldots, t := N$. Our objective is to assess how well $\bar{R}_t$ approximates $R := \mu_Y/\mu_X$.

Under relatively mild conditions,

$$t \, \text{var}\bar{R}_t \to h_{XY} := R^2 \left[\frac{\sigma_\infty^2(X)}{\mu_Y^2} - \frac{2\sigma_\infty(X,Y)}{\mu_X\mu_Y} + \frac{\sigma_\infty^2(X)}{\mu_X^2} \right] \quad \text{as } t \to \infty,$$

where

$$\sigma_\infty^2(X) := \lim_{t\to\infty} t \operatorname{var}\bar{X}_t,$$

$$\sigma_\infty^2(Y) := \lim_{t\to\infty} t \operatorname{var}\bar{Y}_t,$$

$$\sigma_\infty(X, Y) := \lim_{t\to\infty} t \operatorname{cov}(\bar{X}_t, \bar{Y}_t).$$

Clearly, accounting for randomness in the denominator $\bar{X}_t$ as well as in the numerator $\bar{Y}_t$ adds a burden that one would prefer to avoid. Under certain conditions, this possibility exists. Let

$$\gamma_X^2 := \frac{\sigma_\infty^2(X)}{\mu_X^2},$$

$$\gamma_Y^2 := \frac{\sigma_\infty^2(Y)}{\mu_Y^2},$$

$$\rho_{XY} := \frac{\sigma_\infty(X, Y)}{\sigma_\infty(X)\sigma_\infty(Y)},$$

so that h_{XY} has the equivalent representation

$$h_{XY} = \frac{\sigma_\infty^2(Y)}{\mu_X^2} \left(1 - 2\rho_{XY}\frac{\gamma_X}{\gamma_Y} + \frac{\gamma_X^2}{\gamma_Y^2}\right).$$

Therefore, $\sigma_\infty^2(Y)/t\mu_X^2$ approximates $\operatorname{var}\bar{R}_t$ with relatively small error for large t if

$$\frac{1}{2}\left(\frac{\gamma_X}{\gamma_Y} - \frac{\gamma_Y}{\gamma_X}\right) \ll \rho_{XY} \ll \frac{1}{2}\left(\frac{\gamma_X}{\gamma_Y} + \frac{\gamma_Y}{\gamma_X}\right), \tag{6.73}$$

where $\ll$ denotes "much less than." For example, $\gamma_Y/\gamma_X = 10$ implies $-4.95 \ll \rho_{XY} \ll 5.05$, whereas $\gamma_Y/\gamma_X = 100$ implies $-49.995 \ll \rho_{XY} \ll 50.005$. Since $\rho_{XY} \in (-1, 1)$, the conditions are easily satisfied for these examples. In short, if the numerator $\bar{Y}_t$ has a considerably larger coefficient of variation than the denominator $\bar{X}_t$, then the error of approximation is relatively small in approximating $\operatorname{var}\bar{R}_t$ by $\sigma_\infty^2(Y)/t\mu_X^2$. If $\hat{\sigma}_\infty^2(Y)$ denote the batch-means estimate of $\sigma_\infty^2(Y)$, then $\hat{\sigma}_\infty^2(Y)/t\bar{X}_t^2$ is our approximation to $\operatorname{var}\bar{R}_t$. For the queue length example, $t\bar{X}_t = \tau_{M+N} - \tau_M$, so that the estimate is $t\,\hat{\sigma}_\infty^2(Y)/(\tau_{M+N} - \tau_M)^2$.

The photocopying problem in Section 1.4 illustrates ratio estimation in practice. For each scenario a simulation was executed for $t = 1000$ weeks, about 20 years of five-day weeks. Here we use data for Group 3 (Table 1.1) and for Scenario 1, the current operating policy. To estimate long-run average waiting two quantities were computed,

$$\bar{Y}_{1000} : \quad = \quad \text{sample average waiting time per week}$$
$$= \quad 676.0,$$
$$\bar{X}_{1000} : \quad = \quad \text{sample average number of users per week}$$
$$= \quad 124.9,$$

along with $\sqrt{\hat{\sigma}_\infty^2(Y)/t} = 39.96$ and $\sqrt{\hat{\sigma}_\infty^2(X)/t} = .3876$ via LABATCH.2, as estimates of $\sqrt{\text{var}\bar{Y}_t}$ and $\sqrt{\text{var}\bar{X}_t}$, respectively. Since

$$\hat{\gamma}_Y/\hat{\gamma}_X = \frac{39.96}{676.0} \times \frac{124.0}{.3876} = 19.05,$$

a negligible error arises by ignoring randomness in $\bar{X}_{1000}$.

We see that $\gamma_Y \gg \gamma_X$ implies that ignoring the randomness in the denominator introduces relatively little error. But what about moderate and small γ_Y/γ_X? If $\gamma_Y = \gamma_X$, then inequality (6.73) becomes $0 \ll \rho_{XY} \ll 1$, and the error vanishes for $\rho_{XY} = .50$. If $\gamma_Y/\gamma_X < \sqrt{2} - 1 \doteq .4142$, the inequality is

$$1 < \frac{1}{2}\left(\frac{\gamma_X}{\gamma_Y} - \frac{\gamma_Y}{\gamma_X}\right) \ll \rho_{XY} \ll \frac{1}{2}\left(\frac{\gamma_X}{\gamma_Y} + \frac{\gamma_Y}{\gamma_X}\right),$$

which cannot be satisfied. In summary, the approximation works best for $\gamma_Y \gg \gamma_X$. The batch-means inplementation in Arena implicitly uses this variance estimate in computing the half-length for sample averages that are ratio estimates.

A third indirect method of coping with randomness in denominator as well as numerator requires more effort but does not rely on condition (6.73). LABATCH.2-generated tableaus for the three input series $X_1, \ldots, X_t$; $Y_1, \ldots, Y_t$; and $Z_1, \ldots, Z_t$, where $Z_i := X_i - Y_i$, provide the basis for this assessment.

Assume that $\{X_i\}$ and $\{Y_i\}$ obey ASA (Section 6.6.2). Then as $t \to \infty$, $\bar{G}_t := \bar{Y}_t - R\bar{X}_t$ has the properties

$$\bar{G}_t \to 0 \quad \text{w.p.1},$$
$$t\,\text{var}\bar{G}_t \to \sigma_\infty^2(G) := R^2\sigma_\infty^2(X) - 2R\sigma_\infty(X, Y) + \sigma_\infty^2(Y)$$
$$= R^2\sigma_\infty^2(X) - R[\sigma_\infty^2(Z) - \sigma_\infty^2(X) - \sigma_\infty^2(Y)] + \sigma_\infty^2(Y)$$
$$< \infty,$$
$$\frac{\bar{G}_t}{\sqrt{\sigma_\infty^2(G)/t}} \xrightarrow{\text{d}} \mathcal{N}(0, 1),$$

where

$$\sigma_\infty^2(Z) := \lim_{t \to \infty} t \operatorname{var} \bar{Z}_t = \sigma_\infty^2(X) - 2\sigma_\infty(X, Y) + \sigma_\infty^2(Y).$$

For large t (Fieller 1954),

$$\operatorname{pr}\left\{ |\bar{Y}_t - R\bar{X}_t| \leq \Phi^{-1}(1 - \delta/2) \right.$$

$$\left. \times \sqrt{\{R^2\sigma_\infty^2(X) + R[\sigma_\infty^2(Z) - \sigma_\infty^2(X) - \sigma_\infty^2(Y)] + \sigma_\infty^2(Y)\}/t} \right\}$$

$$\approx 1 - \delta,$$

or, equivalently, with approximate probability $1 - \delta$,

$$R \in \left[\frac{2\bar{X}_t\bar{Y}_t - \theta^2[\sigma_\infty^2(X) + \sigma_\infty^2(Y) - \sigma_\infty^2(Z)]/t}{2[\bar{X}_t^2 - \theta^2\sigma_\infty^2(X)/t]} \pm \right.$$

$$\left. \frac{\sqrt{\{2\bar{X}_t\bar{Y}_t - \theta^2[\sigma_\infty^2(X) + \sigma_\infty^2(Y) - \sigma_\infty^2(Z)]/t\}^2 - 4[\bar{X}_t^2 - \theta^2\sigma_\infty^2(X)/t][\bar{Y}_t^2 - \theta^2\sigma_\infty^2(Y)/t]}}{2[\bar{X}_t^2 - \theta^2\sigma_\infty^2(X)/t]} \right],$$

$$(6.74)$$

where $\theta := \Phi^{-1}(1 - \delta/2)$ and where the argument of the square root is nonnegative as $t \to \infty$ w.p.1. Most importantly, this confidence interval remains asymptotically valid for R when strongly consistent estimates replace $\sigma_\infty^2(X)$, $\sigma_\infty^2(Y)$, and $\sigma_\infty^2(Z)$. The tableaus in LABATCH.2 for the three series $X_1, \ldots, X_t$; $Y_1, \ldots, Y_t$; and $Z_1, \ldots, Z_t$ provide these estimates.

Figure 6.15 shows the final tableau for Series 13 (numerator), Series 14 (denominator), and Series 15 (difference) for the airline reservation simulation, whose remaining output appears in Figure 6.12a. The average queue length μ_Q has point estimate $\bar{R}_t = 4.142 = 2.2876 \times 10^{-3}/(6.944 \times 10^{-4})$ and 99 percent approximating confidence interval $[4.103, 4.270]$ based on the interval (6.74), where $(3.362 \times 10^{-5})^2$, $(5.072 \times 10^{-7})^2$, and $(3.386 \times 10^{-5})^2$ are the estimates of $\sigma_\infty^2(Y)/t, \sigma_\infty^2(X)/t$, and $\sigma_\infty^2(Z)/t$, respectively. The relative length of the confidence interval is $(4.270 - 4.013)/4.142 = .0621$. Also, $\tau_{L'-1}(1 - \delta/2)$ replaces $\Phi^{-1}(1 - \delta/2)$, where L' denotes the minimum of the final number of batches in estimating $\sigma_\infty^2(X)$, $\sigma_\infty^2(Y)$, and $\sigma_\infty^2(Z)$.

For our first indirect method based on Little's law, recall the confidence interval $[4.122, 4.30]$. Since $\hat{\gamma}_Y = 16.53$ and $\hat{\gamma}_X = 1.033$ based on entries in Figure 6.15, our second indirect method seems worth applying. It leads to the confidence interval

$$\left[\frac{2.789}{.6944}, \frac{2.963}{.6944} \right] = [4.016, 4.267].$$

```
Final Tableau
                                  Mean Estimation
                                  ***************
                                  (t =  2000000 )

                                         99.0%
                  _     Standard Error   Confidence Interval              _
    Series        X     sqrt[B*W(L,B)/t] Lower       Upper     (Upper-Lower)/|X|
      .           .           .            .           .              .
      .           .           .            .           .              .
      .           .           .            .           .              .

     13     +2.876E-003   +3.362E-005  +2.789E-003 +2.963E-003    +6.051E-002

     14     +6.944E-004   +5.072E-007  +6.931E-004 +6.958E-004    +3.771E-003

     15     +2.181E-003   +3.386E-005  +2.094E-003 +2.269E-003    +8.035E-002


     _
     X(t) is based on all t observations.
     W(L,B) is based on first  95.85% of the t observations.
```

Figure 6.15   LABATCH.2 output for ratio estimation for airline reservation problem

In summary, all three methods give relatively similar confidence intervals in this case of mean queue-length estimation.

6.12 LESSONS LEARNED

- A simulation run of a delay system tends to generate sample paths made up of sequences of dependent events. The dependence increases with the level of congestion or delay in the system.

- As a consequence of this dependence, the initial conditions, or state $S(0)$, that prevail at the begining of a simulation run influence the course of the sample path that evolves as simulated time τ evolves. However, the influence of these initial conditions $S(0)$ on the state of the system $S(\tau)$ at time τ tends to diminish as τ increases.

- To generate sample-path data relatively free of the influence of initial conditions, a simulationist needs to identify a warm-up interval of length $k - 1$ that is allowed to elapse before data collection begins. The simulationist also needs to choose the length of the sample path t over which data are to be collected and the number of independent replications n to be run.

- Multiple replications are helpful in choosing an adequate k. Starting with the same initial conditions on each simplifies the analysis. Also, choosing an initial state that causes the influence of the initial state to diminish monotonically simplifies the analysis.

- If sample averages on n independent replications are used to estimate a long-run average, then the relatively elementary methods of statistical inference apply to assessing the error of approximation, provided that the warm-up interval of length $k - 1$ has elapsed on each replication before data collection begins.

- If statistical inference is based on a single $(n = 1)$ sample path of length t, the estimation of the variance of each sample average needs to account for the dependence among the t observations on the path. Failure to do so can lead to an overly optimistic assessment of accuracy. The single sample-path approach requires that one warm-up interval elapse before data collection begins. This contrasts with the need to let a warm-up interval elapse on each of n independent replications.

- The batch-means method provides one technique for incorporating this dependence.

- LABATCH.2 contains implementations of the batch-means method in C, FORTRAN, and SIMSCRIPT II.5 that provide a strongly consistent estimator of the variance of the sample mean and an asymptotically valid (as $t \to \infty$) confidence interval for the true long-run average of interest. In practice, it can be employed dynamically during a simulation to produce these results. Alternatively, it can process stored data after a simulation run terminates. Most importantly, LABATCH.2 provides the wherewithal for determining how well the estimating procedure has accounted for dependence among observations. It is available on the Internet.

- LABATCH.2 also can be used to infer the accuracy of ratio estimates (Section 3.4). See Section 6.11.

- Arena automatically provides a 95 percent confidence interval for each sample average. For a single sample path $(n = 1)$ it uses another variant of the batch-means method. However, its assessment of the adequacy of the estimate of the variance of the sample average is more limited than in LABATCH.2.

6.13   EXERCISES

You now have at least one working simulation program for one or more of the four problems described in the Exercise section of Chapter 2. The program relies on either the event-scheduling or process-interaction approach. The exercises described below are designed to estimate a suitable warm-up interval $k - 1$ and to assess how well finite-sample averages approximate corresponding unknown long-run averages. For the latter assessment, use LABATCH.2, if possible, either in-line or operating on data files after the simulation's execution has terminated. If an alternative procedure is used, it must provide an assessment of the stability of the variance estimate. See, for example, Exercise 6.2f.

The exercises specify sample-path lengths t' for the warm-up analysis and t for computing estimates. If either or both of these prove inadequate, increase them. Give supporting information in your reports as to why these changes were necessary.

QUARRY-STONE CRUSHING (EXERCISES 2.1, 2.2, 4.2, 4.3, 4.5, 4.6)

The objective is to estimate λ_{max}, the arrival rate for external trucks that maximizes profit as defined in Exercise 2.2. This calls for a simulation program developed in Exercise 4.2 and a sequence of simulation runs executed for different values of λ in $(0, \lambda_*)$, where $\lambda_* = \frac{1}{6}$ trucks per minute denotes the maximal allowable arrival rate at the crusher for external 50-ton trucks.

✎ **6.1 Initial Conditions.** Each distinct value of λ introduces a different level of congestion and, hence, a different interval for dissipating the influence of initial conditions. Since congestion increases with λ, the length of a suitable warm-up interval should also increase with λ and be largest for $\lambda_{**} :=$ maximal arrival rate used in experiments. Take $\lambda_{**} = .9\lambda_*$ to determine this maximal warm-up interval.

a. Execute $n = 10$ independent replications each for $t' = 100$ hours with $\lambda = \lambda_{**}$ and each starting with the same initial conditions and parametric values, specified in Exercises 2.1, 2.2, 4.2, 4.3, 4.5, and 4.6. For each stream j, use its last pseudorandom number on replication i as the seed for stream j on replication $i + 1$ for $i = 1, \ldots, n - 1$.

b. Using the methodology of Section 6.1, prepare graphs of sample average throughput per five minutes and crusher queue length and use the graphs to choose the k that makes the influence of initial conditions negligible. Write a short report justifying your choice of k. Use tables and graphs to support the choice. Label each table and graph with appropriate titles and identify all variables used therein.

✎ **6.2 Estimating Long-Run Average Profit.** Carry out a sequence of experiments to estimate the value of λ_{max}. Use the estimated warm-up interval $k - 1$ of Exercise 6.1 in all runs. For each distinct value of λ considered, generate a sample path of length $k + t - 1$ with $t = 1000$ hours.

a. On each run, compute the quantities listed in Exercise 4.2e and the sample average profit apart from the unknown factor A in Exercise 2.2.

b. On each run, compute 99 percent confidence intervals for long-run average profit and crusher queue length and relative assessments of the accuracy of their sample averages.

c. To identify the neighborhood of λ_{max}, use nine runs with

$$\lambda_i = \frac{i-1}{8} \lambda_{**} \quad \text{for } i = 1, \ldots, 9.$$

d. Let $[\lambda', \lambda'']$ denote the interval that your results to part c suggest contains λ_{max}. Continue to make runs for selected values in $[\lambda', \lambda'']$ to refine your estimate of λ_{max}. As guidance for when to stop, use your point and interval estimates of long-run average profit.

e. Prepare a report identifying your choice of λ_{max} using graphs and tables to support your choice.

f. In your report, describe the extent to which your estimates of σ_∞^2 are stable for sample average profit and crusher queue length for each λ.

LOADING TANKERS (EXERCISES 2.3, 4.7, 4.8)

To compete with other oil-loading ports, the port authority plans to institute a rebate program based on total *port time*. This is the summation of waiting, towing, and loading times. If a tanker's port time exceeds 48 but is no more than 72 hours, the authority will pay its owner a rebate of A dollars. If it exceeds 72 but is no more than 96 hours, the rebate is $3A$ dollars. In general, if it exceeds $24i$ but is no more than $24(i+1)$ hours, the rebate is $(2i-1)A$ dollars. From Exercise 2.3, $A = 10,000$ dollars.

✐ **6.3 Initial Conditions.** This exercise is designed to estimate a suitable truncation point $k-1$ for the port-time sequence based on multiple independent replications. In what follows, use your own pseudorandom number seeds and whichever of your two programs you wish.

a. Execute $n = 10$ independent replications each for 1000 tankers and each starting with initial conditions specified in Exercises 4.7 and 4.8. Use your orignal run as replication 1. For each stream j, use its last pseudorandom number on replication i as the seed for stream j on replication $i+1$ for $i = 1, \ldots, n-1$.

b. Using the methodology of Section 6.1, prepare graphs of sample averages of port times for tanker i for $i = 1, \ldots, 1000$ and use the graphs to choose the warm-up interval k that makes the influence of initial conditions "negligible." Write a short report justifying your choice of k. Use tables as well as graphs to support your choice. Label each graph and table with appropriate titles and identify all variables used therein.

✐ **6.4 Estimating Long-Run Average Rebate.** The purpose of this exercise is to estimate $\mu_r :=$ sample average rebate per tanker and $\mu_r' :=$ sample average rebate per tanker that received a rebate (Exercise 2.3).

a. Using the warm-up interval determined in Exercise 6.3, generate a sample path of length $k + t - 1 = 20,000$ tanker departures.

b. Compute point estimates of μ_r and μ'_r.

c. Compute a 99 percent confidence interval for μ_r.

d. Using the methodology of Section 6.11, compute a 99 percent confidence interval for μ'_r.

e. Prepare a report describing your findings and their statistical quality.

f. Your report should also describe the extent to which estimates of σ^2_∞ are stable for each sample sequence.

TERMINAL PAD CONTROLLER MODEL (EXERCISES 2.5, 4.12, 4.13)

The purpose of Exercises 6.5 and 6.6 is to generate sample-path data using the simulation program developed in either Exercise 4.12 or Exercise 4.13 to estimate the minimal mean delay $\mu(\lambda, m)$ in definition (2.6) and the λ at which it occurs for each $m \in \mathcal{M} := \{8, 16, 24, 32\}$. Recall that λ denotes character-arrival rate to an input buffer and m denotes maximal packet size. Recall that $.30 \le r\lambda \le 1$ and $m \in \{8, 9, \ldots, 32\}$, where $r = 10$ denotes the number of input buffers.

✐ **6.5 Initial Conditions.** For each fixed m, delay time varies with λ, but not necessarily in a monotone way. In particular, small λ implies that some characters wait a long time in their input buffers until being transferred to the relatively uncongested output buffer. Conversely, large λ implies that characters have a relatively short wait in their input buffers, but, because of congestion, have a long wait in the output buffer. While small λ induces dependence among no more than m characters in an input buffer, the low congestion in the output buffer induces little dependence among characters from different input buffers. However, congestion in the output buffer for large λ does create dependence among characters in different input buffers and needs to be taken into consideration when deciding on a warm-up interval.

Therefore, for each fixed m, the length of a suitable warm-up interval $k_m - 1$ can be expected to increase with λ and be largest for $\lambda_*(m)$, the maximal character-arrival rate used in experiments. Take $\lambda_*(m) = 0.95 \times \rho_0/r$, where ρ_0 is determined in Exercise 2.5b. Since monotone behavior for delay time is not apparent for varying m but fixed λ, the task of identifying a suitable warm-up interval becomes more challenging. To resolve this problem:

a. For each $m \in \mathcal{M}$, execute $n = 10$ independent replications each for $t' = 1,000$ transmitted nonoverhead characters with $(\lambda = \lambda_*(m))$ and each starting with the same initial conditions and parametric values as in Exercises 2.5 and 4.12. For each stream j, use its last pseudorandom number on replication $i + 1$ for $i = 1, \ldots, n - 1$.

b. Using the methodology of Section 6.1, prepare graphs of sample average delay time separately for each $m \in \mathcal{M}$ and use them to estimate suitable truncation points $\{k_m -$

1, $m \in \mathcal{M}$}. Write a report justifying your choices. Use tables and graphs to support your choice. Label each table and graph with appropriate titles and identify all variables used therein.

c. Comment about how k_m varies with m.

✐ **6.6 Estimating Long-Run Average Delay Time.** For each $m \in \mathcal{M}$, carry out a sequence of experiments to estimate the function {$\mu(\lambda, m), 0 < \lambda < \rho_0/r$} where ρ_0 is derived in Exercise 2.5b. Use the estimated warm-up interval $k_m - 1$ of Exercise 6.5 for all runs with size m input buffers. For each run, generate a sample path of length $k_m + t - 1$ where $t = 20,000$ transmitted nonoverhead characters.

a. Compute estimates of $\mu(\lambda, m)$ for each $m \in \mathcal{M}$ and $\lambda = \lambda_i$. where

$$\lambda_i := (0.05 + 0.1i)\rho_0/r \qquad i = 1, \ldots, 10.$$

b. Compute an approximating 99 percent confidence interval for each $\mu(\lambda, m)$.

c. If the objective is to compute a point estimate, $\hat{\mu}(\lambda, m)$ that is within ±10 percent of the unknown $\mu(\lambda, m)$, how well have you succeeded? Use quantitative evidence to justify your answer.

d. Use the results in part a to identify an interval $[a(m), b(m)] \subseteq (0, \rho_0/r)$ for λ in which you estimate $\min_\lambda \mu(\lambda, m)$ is located. Then for each $m \in \mathcal{M}$, compute estimates of $\mu(\lambda, m)$ for each

$$\lambda_i := \{a(m) + 0.1i[b(m) - a(m)]\}\rho_0/r \qquad i = 1, \ldots, 10.$$

e. Write a report identifying your point estimates of $\min_\lambda \mu(\lambda, m)$ and of

$$\lambda_{\min}(m) := \lambda \text{ at which } \mu(\lambda, m) \text{ has its minimum}$$

for each $m \in \mathcal{M}$. Use graphs and tables to support your choices, especially those {$\lambda_{\min}(m), m \in \mathcal{M}$}.

f. Provide evidence that your choices of σ_∞^2 for each (λ, m) combination is stable. If some are not, explain why.

g. For the data generated in part d, plot estimates of σ_∞ versus $\lambda_i = \{a(m) + 0.1i[b(m) - a(m)]\}\rho_0/r$ for $i = 1, \ldots, 10$ and each $m \in \mathcal{M}$. Explain the behavior in σ_∞ as functions of λ and m.

MOBILE-TELEPHONE SERVICE (EXERCISES 2.5, 4.14, 4.15)

The purpose of Exercises 6.7 and 6.8 is to generate sample-path data using the simulation program developed in either Exercises 4.14 or 4.15 and for each c in $C = \{.1i; i = 1, \ldots, 9\}$ to use these data to estimate the minimal long-run average total cost (2.7) as a function of m, the total number of available lines, and r, the number of lines dedicated to handed-off calls. Since $\lambda_h = 4$, $\lambda_s = 2$, and $\omega = 1$ calls per minute, m must exceed 6 and $m - r$ must exceed 2 in order for the mean number of lost calls per minute to remain bounded.

✐   **6.7 Initial Conditions.** For m transmission lines, r of which are dedicated to handed-off mobile calls, let

$$B(\tau) := \text{ number of busy lines at time } \tau$$

and

$$\mu(m, r) := \text{ long-run proportion of busy lines.}$$

Hereafter, we call $\mu(m, r)$ *utilization*. For fixed m, utilization decreases monotonically as r increases and, for fixed r, decreases monotonically as m increases. For given m, r, and initial conditions $S(0) = s$, we have

$$E[B(\tau) \mid S(0) = s] \to m\mu(m, r) \qquad \text{as } \tau \to \infty.$$

It is anticipated that the slowest convergence rate occurs for $m = 7$ and $r = 0$.

Let

$$Y_i = \begin{cases} 1 & \text{if call } i \text{ is a mobile call and it is handed off,} \\ 0 & \text{otherwise,} \end{cases}$$

and

$$Z_i = \begin{cases} 1 & \text{if call } i \text{ is a standard call and it is denied service,} \\ 0 & \text{otherwise.} \end{cases} \tag{6.75}$$

Then in equilibrium

$$EY_i = \frac{\lambda_h}{\lambda_h + \lambda_s} \beta_h(m, r)$$

and

$$EZ_i = \frac{\lambda_s}{\lambda_h + \lambda_s} \beta_s(m, r).$$

a. Execute $n = 10$ independent replications each for $t' = 1000$ calls with $m = 7, r = 0$, and other parametric values specified in Exercises 2.6 and 4.14. For each stream j, use its last pseudorandom number on replication i as the seed for stream j on replication $i + 1$, for $i = 1, \ldots, n - 1$.

b. Using the methodology of Section 6.1, prepare graphs of the sample average number of busy lines over five-minute intervals and use the graphs to determine a time at which to begin data collection for estimating $\beta_s(m, r)$ and $\beta_h(m, r)$. Write a report justifying your choice of warm-up interval $k - 1$ for $m = 7$ and $r = 0$.

c. Examine the extent to which this choice of k in part b suffices for all $r \in \{0, 1, \ldots, m - 2\}$ and $m \in \{7, 9, 11, 13, 15\}$. If you conclude that this choice is inadequate for these other (m, r) 2-tuples, perform additional warm-up analyses to identify more suitable intervals. Provide evidence to support your choices.

✎ **6.8 Estimating Loss Rates.** Carry out a sequence of independent sampling experiments to estimate the loss rates $\beta_s(m, r)$ and $\beta_h(m, r)$ for each $r \in \{0, 1, \ldots, m - 2\}$ and $m \in \{7, 9, 11, 13, 15\}$. Each experiment is to be run for $t = 10,000$ calls, regardless of type after the warm-up interval $k - 1$ chosen in Exercise 6.7b or 6.7c has elapsed. There are several available ways of estimating $C(m, r, c)$ in expression (2.7). One collects the data

$$
X_i = \begin{cases}
1 & \text{if a mobile call is handed off,} \\
c & \text{if a standard call is denied service,} \\
0 & \text{otherwise.}
\end{cases}
$$

Then

$$
\hat{C}(m, r, c) = \bar{X}_{kt} := \frac{1}{t} \sum_{i=k}^{k+t-1} X_i \tag{6.76}
$$

provides an estimate of $C(m, r, c)$. To estimate this quantity with these data requires 432 sample paths, one for each (m, r, c) triplet.

An alternative approach reduces the required number of sample paths. It relies on the Y_i and Z_i defined above. Since $X_i = Y_i + Z_i$, expression (6.76) is algebraically equivalent to

$$
\hat{C}(m, r, c) = \bar{Y}_{kt} + c\bar{Z}_{kt}, \tag{6.77}
$$

where

$$
\bar{Y}_{kt} = \frac{1}{t} \sum_{i=k}^{k+t-1} Y_i
$$

and

$$\bar{Z}_{kt} = \frac{1}{t} \sum_{i=k}^{k+t-1} Z_i.$$

However, for each fixed (m, r), $\left\{\hat{C}(m, r, c), c \in \mathcal{C}\right\}$ can be computed from a single replication, thus reducing the number of replications to 48.

If we merely want point estimates, the second data collection scheme would be unequivocally superior. However, we also want to assess the statistical error in each $\hat{C}(m, r, c)$. That can be done straightforwardly using LABATCH.2, for example, for each of the 432 runs. However, with a moderate amount of additional programming, it can also be done via LABATCH.2 for the second 48 sample-path data collection scheme. Since

$$\text{var}\bar{X}_{kt} = \text{var}\bar{Y}_{kt} + 2\,c\,\text{cov}(\bar{Y}_{kt}, \bar{Z}_{kt}) + c^2\text{var}\bar{Z}_{kt},$$

and since

$$\frac{\hat{C}(m, r, c) - C(m, r, c)}{\sqrt{\text{var}\bar{X}_{kt}}} \to \frac{\hat{C}(m, r, c) - C(m, r, c)}{\sqrt{\sigma_\infty^2(X)/t}} \xrightarrow{\text{d}} \mathcal{N}(0, 1) \qquad \text{as } t \to \infty,$$

where

$$\sigma_\infty^2(X) := \lim_{t\to\infty} t\,\text{var}\bar{X}_{kt}$$
$$= \sigma_\infty^2(Y) + 2\,c\sigma_\infty(Y, Z) + c^2\sigma_\infty^2(Z),$$
$$\sigma_\infty^2(Y) := \lim_{t\to\infty} t\,\text{var}\bar{Y}_{kt},$$
$$\sigma_\infty^2(Z) := \lim_{t\to\infty} t\,\text{var}\bar{Z}_{kt},$$
$$\sigma_\infty^2(Y, Z) := \lim_{t\to\infty} t\,\text{cov}(\bar{Y}_{kt}, \bar{Z}_{kt}),$$

we need estimates of $\sigma_\infty^2(Y)$, $\sigma_\infty^2(Z)$, and $\sigma_\infty(Y, Z)$. For example, using the data $\{(Y_i, Z_i, Y_i - Z_i), k \leq i \leq k + t - 1\}$ as input on each call to LABATCH.2 produces estimates of $\hat{\sigma}_\infty^2(Y)$, $\hat{\sigma}_\infty^2(Z)$, and $\hat{\sigma}_\infty^2(Y - Z)$, where the latter is an estimate of

$$\sigma_\infty^2(Y - Z) = \sigma_\infty^2(Y) - 2\sigma_\infty^2(Y, Z) + \sigma_\infty^2(Z).$$

Therefore,

$$\hat{\sigma}_\infty^2(Y, Z) = \frac{1}{2}\left[\hat{\sigma}_\infty^2(Y) + \hat{\sigma}_\infty^2(Z) - \hat{\sigma}_\infty^2(Y - Z)\right]$$

provides an estimate of $\sigma_\infty^2(Y, Z)$, and $\left[\hat{\sigma}_\infty^2(Y) + 2c\hat{\sigma}_\infty(Y, Z) + c^2\hat{\sigma}_\infty^2(Z)\right]/t$ provides the needed variance estimate. If $L(Y)$, $L(Z)$, and $L(Y - Z)$ are the numbers of batches used on the final interim reviews in LABATCH.2, then a conservative approach to interval estimation takes $\min[L(Y), L(Z), L(Y, Z)] - 1$ as the degrees of freedom.

a. On each run, compute estimates of $C(m, r, c)$ in expression (2.7) for each $c \in \mathcal{C}$.

b. On each run, compute an approximating 99 percent confidence interval for each $C(m, r, c)$.

c. For each fixed m and c, identify your choice of r that minimizes $C(m, r, c)$.

d. To what extent do the confidence intervals support your choice? Be specific in providing justification.

6.14   REFERENCES

Alexopoulos, C., G.S. Fishman, and A.T. Seila (1997). Computational experience with the batch means method, *Proceedings of the 1997 Winter Simulation Conference*, S. Andradóttir, K.J. Healy, D.H. Withers, and B.L. Nelson, eds., IEEE, Piscataway, NJ.

Blomqvist, N. (1967). The covariance function of the M/G/1 queueing system, *Skandinavisk Aktuarietidskrift*, **50**, 157–174.

CACI, Inc. (1983). *SIMSCRIPT II.5 Reference Handbook*, CACI Products Company, La Jolla, California.

Chien, C.H., D. Goldsman, and B. Melamed (1997). Large-sample results for batch means, *Management Science*, **43**, 1288–1295.

Chow, Y.S., and H. Robbins (1965). On the asymptotic theory of fixed-width sequential confidence intervals for the mean, *Ann. Math. Statist.*, **36**, 457–462.

Damerdji, H. (1994). Strong consistency of the variance estimator in steady-state simulation output analysis, *Math. Oper. Res.*, **19**, 494–512.

Fieller, E.C. (1954). Some problems in interval estimation, *J. Roy. Statist. Soc.*, Series **B**, **2**, 175–185.

Fishman, G.S. (1978). *Principles of Discrete Event Simulation*, Wiley, New York.

Fishman, G.S. (1996). *Monte Carlo: Concepts, Algorithms, and Applications*, Springer Verlag, New York.

Fishman, G.S., and L.S. Yarberry (1997). An implementation of the batch means method, *INFORMS J. on Comput.*, **9**, 296–310.

Law, A.M. and W.D. Kelton (1991). *Simulation Modeling and Analysis*, McGraw-Hill, New York.

Goldsman, D., and M.S. Meketon (1986). A comparison of several variance estimators for stationary increment stochastic process, Technical Report, School of Industrial and Systems Engineering, Georgia Institute of Technology.

Kelton, W.D., R.P. Sadowski, and D.A. Sadowski (1998). *Simulation with Arena*, McGraw-Hill, New York.

Philipp, W. and W. Stout (1975). Almost sure invariance principles for partial sums of weakly dependent random variables. *Mem. Am. Math. Soc.*, **2**.

Song, W.T. and B.W. Schmeiser (1995), Optimal mean-squared-error batch sizes, *Man. Sci.*, **41**, 110–123.

von Neumann, J. (1941). The mean square difference, *Ann. Math. Statist.*, **12**, 153–162.

Yarberry, L.S. (1993). Incorporating a dynamic batch size selection mechanism in a fixed-sample-size batch means procedure, unpublished Ph.D. dissertation, Dept. of Operations Research, University of North Carolina, Chapel Hill.

Young, L.C. (1941). On randomness of order sequences, *Ann. Math. Statist.*, **12**, 293–300.

CHAPTER 7

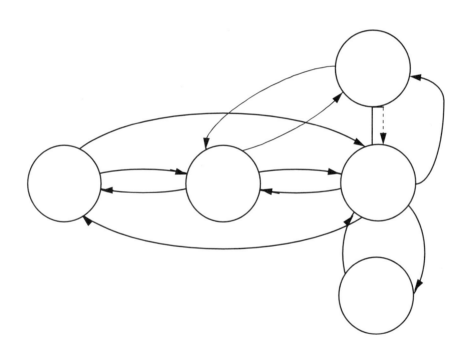

Making Sense of Output and Increasing Efficiency

A simulation run customarily generates sample-path data on user-specified dynamic processes that evolve as simulated time elapses. These often include delay time, queue length, and resource utilization. Sample averages summarize these data and usually become the focal point of reports and presentations to management. Since they merely approximate corresponding long-run averages that would be observed if the simulation were run for an infinite rather than a finite amount of time, some measure of error should accompany each sample average. Confidence intervals, standard errors, and relative errors provide three alternative means of assessing how well sample averages approximate true long-run averages. As illustration, the Final Tableau in Figure 6.12a displays 99 percent confidence intervals under the headings Lower and Upper, standard error under $\text{sqrt}[B^*W(L, B)/t]$, and relative error under $(\text{Upper} - \text{Lower})/|\bar{X}|$. A report that omits an error assessment can lead to serious misconceptions of how well sample averages characterize unknown long-run averages.

Providing an error assessment with each sample average facilitates informed decision making. Additional analyses, that focus on the particular problem to be resolved, can further enhance the quality of decision making. This chapter describes techniques for performing several of these analyses, in particular, ones that compare simulated system behavior for different operating scenarios. Recall that a scenario denotes a set of specific values assigned to input parameters and a collection of logical system-operating rules.

As illustration, consider the airline reservation problem and the waiting-time results in Section 4.22 for $m = 4$ and 5 reservationists with first-come-first-served queueing discipline. Suppose management is willing to increase the number of reservations from four to five if that action reduces mean waiting time by at least three minutes *and* if the proportion of customers who wait is cut by .30. To make this an informed decision, an analysis needs to assess how credible each difference is individually and how credible they are when taken simultaneously. Section 7.1 describes the details.

This example illustrates issues that arise in many studies that examine simulated behavior for different values of input parameters or different operating rules. The issues are the *comparison* of results generated on different experiments and the *simultaneity* of multiple confidence statements made about these results. This chapter describes statistical procedures for making these comparisions and simultaneous confidence statements. It presents them in the special context in which a simulationist may wish to use them when analyzing simulation output. Collectively, we refer to these procedures as *credibility analysis*, an elusion to the frequent need, in making a decision, to assess how plausible it is to assume that a difference of a particular magnitude actually exists.

Although space limitations restrict our discussion to the most basic of these issues, the exposition addresses problems that a simulationist frequently encounters when preparing a report for a management choosing among alternative resource configurations or operating modes. Within the considerable statistical literature on these topics, Bechhofer et al. (1995) provides a comprehensive account of the technical issues with special emphasis on selecting the *best among alternative systems*. Matejcik and Nelson (1995a, 1995b) give one type of analysis of multiple comparisons within the special context of discrete-event simulation.

When a simulation study focuses on the difference between two sample averages each generated under a different scenario, Section 4.22 shows how common pseudorandom numbers (cpn) can induce a higher level of statistical accuracy than would arise for independent runs. The discussion there concentrates on the mechanics of implementing cpn. Section 7.7 and, in particular, 7.7.1 return to this subject, focusing on how some differences in input scenarios tend to favor substantial variance reductions whereas others do not.

Sections 7.8 through 7.10 then introduce the reader to a second variance-reducing technique. It is based on optimal-sample-path length (opt) for independent runs. The account reveals when opt induces a smaller variance for the difference of sample averages than cpn and when the reverse occurs. In particular, a large disparity in individual sample-average variances tends to favor opt over cpn.

Many other methods exist for reducing variance, including control variates, stratified sampling, and antithetic variates; see, e.g., Fishman (1996) and Law and Kelton (1991). The present account discusses only cpn and opt for several reasons. First, they are the two that address differences in averages. Second, they each require a modest amount of effort to implement. Third, space considerations limit what can be covered under this topic.

In what follows, we repeatedly make use of d.f.s and inverse d.f.s. Recall that $\Phi(\cdot)$ denotes the d.f. of a random variable from $\mathcal{N}(0, 1)$, and $\Phi^{-1}(\cdot)$ denotes its inverse. Also, define $G_d(\cdot)$ as the d.f. of a random variable from Student's t distribution with d degrees of freedom and $\tau_d(\cdot)$ as its inverse.

7.1 SIMULTANEOUS CONFIDENCE STATEMENTS

We begin with the concept of a multiple confidence statement and how it relates to individual confidence statements. Let $A_1, \ldots, A_r$ denote r independent random events. If $A_1, \ldots, A_r$ are independent, then the probability that they occur simultaneously is

$$\mathrm{pr}(A_1, \ldots, A_r) = \Pi_{i=1}^{r} \, \mathrm{pr}(A_i). \tag{7.1}$$

As an example, suppose that r parameters $\mu_1, \ldots, \mu_r$ have been estimated by confidence intervals. If A_i denotes the event that the ith confidence interval includes the ith unknown μ_i with probability

$$\mathrm{pr}(A_i) = 1 - \alpha,$$

then

$$\mathrm{pr}(A_1, \ldots, A_r) = (1 - \alpha)^r.$$

To achieve a $1 - \delta$ confidence level for simultaneous coverage, α needs to be chosen as

$$\alpha = 1 - (1 - \delta)^{1/r}. \tag{7.2}$$

How does simultaneity affect the lengths of the confidence intervals? For large numbers of degrees of freedom (e.g., Section 6.6), interval length is proportional to $\Phi^{-1}(1 - \alpha/2)$. Then

$$v(\delta, r) := \frac{\Phi^{-1}\left((1 + (1 - \delta)^{1/r})/2\right)}{\Phi^{-1}(1 - \delta/2)} \tag{7.3}$$

gives the ratio of simultaneous to individual interval lengths. For $r = 2$, $v(.01, 2) = 1.089$ and $v(.05, 2) = 1.141$. For $r = 11$, $v(.01, 11) = 1.287$ and $v(.05, 11) = 1.444$. These numbers indicate that claiming simultaneous coverage without correcting interval lengths results in an overestimate of accuracy that increases with r for a given δ. For example, $v(.01, 11) = 1.287$ implies that the simultaneous interval lengths should be 28.7 percent longer.

When two or more confidence intervals are computed from data generated on the same simulation run, they are rarely independent, and expression (7.1) does not apply. Section 6.8 illustrates this case, where Figure 6.12a displays a 99 percent confidence interval for each of the 11 exceedance probabilities, $\mathrm{pr}(W > w)$, for waiting time for the airline reservation problem using Series $= w + 2$ for $w = 0, 1, \ldots, 10$ minutes. While characterizng the quality of each corresponding point estimate separately, these intervals do not make a single statement regarding all 11 point estimates taken together. Moreover, since they are based on the same data, the intervals are dependent. To address this issue, we resort to simultaneous confidence intervals based on *Bonferroni's inequality*.

Let $\bar{A}_i$ denote the event that A_i does not occur. In the present case, we may think of A_1 as the event that the confidence interval for Series 2 on review 15 in Figure 6.12a covers $\mathrm{pr}(W > 0)$, A_2 as the event that the confidence interval for Series 3 on review 15 covers $\mathrm{pr}(W > 1)$, etc. From the law of total probability,

$$\mathrm{pr}(A_1, A_2) + \mathrm{pr}(A_1, \bar{A}_2) + \mathrm{pr}(\bar{A}_1, A_2) + \mathrm{pr}(\bar{A}_1, \bar{A}_2) = 1,$$

and from the observation that $\mathrm{pr}(A_1, \bar{A}_2) + \mathrm{pr}(\bar{A}_1, \bar{A}_2) = \mathrm{pr}(\bar{A}_2)$ and $\mathrm{pr}(\bar{A}_1, A_2) \leq \mathrm{pr}(\bar{A}_1)$, we have

$$\mathrm{pr}(A_1, A_2) \geq 1 - \mathrm{pr}(\bar{A}_1) - \mathrm{pr}(\bar{A}_2).$$

This is a lower bound on the probability that both A_1 and A_2 occur simultaneously.

Let $B_2 := A_1 \cap A_2$. Then

$$\mathrm{pr}(A_3, B_2) + \mathrm{pr}(A_3, \bar{B}_2) + \mathrm{pr}(\bar{A}_3, B_2) + \mathrm{pr}(\bar{A}_3, \bar{B}_2) = 1,$$
$$\mathrm{pr}(A_3, \bar{B}_2) + \mathrm{pr}(\bar{A}_3, \bar{B}_2) = \mathrm{pr}(\bar{B}_2),$$

and

$$\mathrm{pr}(\bar{A}_3 B_2) \leq \mathrm{pr}(\bar{A}_3),$$

so that

$$\mathrm{pr}(A_3, B_2) \geq 1 - \mathrm{pr}(\bar{B}_2) - \mathrm{pr}(\bar{A}_3)$$
$$= 1 - 1 + \mathrm{pr}(B_2) - \mathrm{pr}(\bar{A}_3)$$
$$= 1 - \mathrm{pr}(\bar{A}_1) - \mathrm{pr}(\bar{A}_2) - \mathrm{pr}(\bar{A}_3).$$

More generally, for each j and $B_j := \cap_{i=1}^{j} A_i$,

$$\mathrm{pr}(B_j) \geq 1 - \sum_{i=1}^{j} \mathrm{pr}(\bar{A}_i), \qquad j = 1, \ldots, r, \tag{7.4}$$

which is called Bonferroni's inequality. Since $\text{pr}(B_r) = \text{pr}(A_1, \ldots, A_r)$, we have a lower bound for all r events occurring simultaneously.

Again, let A_i denotes the event that the ith confidence interval includes the ith unknown mean μ_i with $\text{pr}(A_i) = 1 - \alpha$. Then inequality (7.4) implies that simultaneous coverage for $j = r$ is at least $1 - r\alpha$. To achieve a $1 - \delta$ simultaneous confidence level, we choose α as

$$\alpha = \delta/r, l \tag{7.5}$$

leading to an asymptotic ratio of simultaneous and individual interval lengths

$$v'(\delta, r) := \Phi^{-1}(1 - \delta/2r)/\Phi^{-1}(1 - \delta/2).$$

For $r = 2$, $v'(.01, 2) = 1.090$ and $v'(.05, 2) = 1.144$. For $r = 11$, $v'(.01, 11) = 1.288$ and $v'(.05, 11) = 1.448$. The ratios reveal that the interval lengths increase at roughly the same relative rates as their corresponding ratios in the independent case. However, it is easily seen that $\Phi^{-1}(1 - \delta/2r) - \Phi\left((1 + (1 - \delta)^{1/r})/2\right)$ increases with r.

For the 11 confidence intervals for exceedance probabilities in Figure 6.12a, $\text{pr}(\bar{A}_i) = .01$, so that

$$\text{pr}(B_{11}) := \text{probability that all 11 intervals simultaneously}$$
$$\text{cover their respective true probabilities} \geq .89.$$

If a simulationist wants a $1 - \delta = .99$ simultaneous confidence for the $r = 11$ ordinates of the above example, one would set $\alpha = .01/11 = .0009091$ in the original runs. As a final point, note that for $0 \leq \alpha < 1/r$, $(1 - \alpha)^r/(1 - r\alpha)$ decreases to its lower bound of unity as α decreases.

Returning to the issue, raised in the introduction, of increasing m from four to five reservationists in the airline problem, we recall that management is amenable to this change if mean waiting time is reduced by three minutes and the proportion of customers who must wait decreases by .30. Section 4.22 estimates the reduction in mean waiting time as $\bar{Z} = 3.337$ minutes with standard error $\sqrt{\hat{\sigma}_\infty^2(Z)/t} = .1687$. It also estimates the reduction in the proportion who wait as $\bar{Z}' = .3597$ with standard error $\sqrt{\hat{\sigma}_\infty^2(Z')/t} = 3.628 \times 10^{-3}$.

The point estimates $\bar{Z}$ and $\bar{Z}'$ support the decision to increase the number of reservationists. However, these are estimates of the unobserved true changes, and hence a decision maker would benefit from knowing how credible these changes are. The standardized statistics

$$\bar{Z}(\theta) := \frac{\bar{Z} - \theta}{\sqrt{\hat{\sigma}_\infty^2(Z)/t}}, \tag{7.6}$$

$$\bar{Z}'(\phi) := \frac{\bar{Z}' - \phi}{\sqrt{\hat{\sigma}_\infty^2(Z')/t}}, \tag{7.7}$$

with $\theta := E\bar{Z}$ and $\phi := E\bar{Z}'$, each have degrees of freedom in excess of 100 (determined by LABATCH.2), so that treating each as from $\mathcal{N}(0, 1)$ introduces negligible error.

Under the hypothesis $H_1(\theta') : \theta = \theta'$, the probability of the event

$A_1(\theta') :=$ event of observing a $\mathcal{N}(0, 1)$ random variate no less than $\bar{Z}(\theta')$ in value

is

$$\mathrm{pr}\left[A_1(\theta')\right] = \Phi(\bar{Z}(\theta')).$$

Under the hypothesis $H_2(\phi') : \phi = \phi'$, the probability of the event

$A_2(\phi') :=$ event of observing a $\mathcal{N}(0, 1)$ random variate no less than $\bar{Z}'(\phi')$ in value

is

$$\mathrm{pr}\left[A_2(\phi')\right] = \Phi(\bar{Z}'(\phi')).$$

Then for the joint event Bonferroni's inequality gives the lower bound

$$\mathrm{pr}\left[A_1(\theta'), A_2(\phi')\right] \geq \Phi(\bar{Z}(\theta')) + \Phi(\bar{Z}'(\phi')) - 1.$$

For $\theta' = 3$ and $\phi' = .30$,

$$\mathrm{pr}\left[A_1(3), A_2(.30)\right] \geq .9772 + 1.000 - 1 = .9772.$$

This lower bound gives considerable credibility to $H_1(3) \cap H_2(.30)$. However, suppose that management is willing to increase the number of reservationists by one only if $\theta' = 3$ and $\phi' = .36$. Then

$$\mathrm{pr}\left[A_1(3), A_2(.30)\right] \geq .9772 + .4670 - 1 = .4442,$$

which gives considerably less support for the joint hypotheses $H_1(3) \cap H_2(.36)$. Most importantly, it can alert management to the greater risk of error in assuming $\theta' = 3$ and $\phi' = .36$ than assuming $\theta' = 3$ and $\phi' = .30$.

The next section extends this form of credibility analysis as a function of increasing mean difference.

7.2 COMPARING TWO MEANS

The airline example of the last section compares mean waiting times for four and five reservationists. The illustration takes advantage of the dependence induced by common pseudorandom numbers to reduce standard errors. This section addresses the more general case of independence of results across experiments and replications. Section 7.5 further extends these ideas to the case of single replications under each scenario. Section 7.7 returns to the approach based on common pseudorandom numbers.

Recall that Section 6.1 defines $\bar{X}^{(i)}_{k_1 t_1}$ as the sample average of t_1 observations collected sequentially on sample path i after deleting $k_1 - 1$ observations to reduce the influence of initial conditions. Moreover, for n_1 independent replications expression (6.4) defines $\bar{X} := \bar{X}_{k_1 n_1 t_1}$ as the arithmetic average of $\bar{X}^{(1)}_{k_1 t_1}, \ldots, \bar{X}^{(n_1)}_{k_1 t_1}$, and expression (6.8) defines $s^2(X) := s^2_{k_1 n_1 t_1}$ as the sample variance of an $\bar{X}^{(i)}_{k_1 t_1}$, so that $s^2(X)/n_1$ is an estimate of $\mathrm{var}\, \bar{X}_{k_1 n_1 t_1}$.

By analogy, let $\bar{Y}$ denote the grand sample average and $s^2(Y)$ the sample variance of the sample averages $\bar{Y}^{(1)}_{k_2 t_2}, \ldots, \bar{Y}^{(n_2)}_{k_2 t_2}$ collected on n_2 independent replications of the simulation generated under a different scenario than for the first n_1 replications. Note that the first set of replications each delete $k_1 - 1$ observations and use t_1 observations to compute each $\bar{X}^{(i)}_{k_1 t_1}$, whereas the second set deletes $k_2 - 1$ observations and thereafter uses t_2 observations to compute each $\bar{Y}^{(i)}_{k_2 t_2}$. Presumably, k_1 and k_2 reflect differences in the rates at which the influences of initial conditions dissipate under the two different scenarios, and t_1 and t_2 reflect differences associated with computing costs and possibly differences due to sampling variation along sample paths. Assume that t_1 and t_2 are sufficiently large so that we may regard each $\bar{X}^{(i)}_{k_1 t_1}$ and each $\bar{Y}^{(i)}_{k_2 t_2}$ as approximately normally distributed.

Comparing the sample averages $\bar{X}$ and $\bar{Y}$ is a commonly encountered objective in simulation studies. Recall that $\bar{X}$ and $\bar{Y}$ are estimates of long-run averages μ_X and μ_Y, respectively, and it is of interest to assess the extent to which μ_X and μ_Y differ as a consequence of different scenarios in each experiment. An approximating $100 \times (1-\delta)$ percent confidence interval for $\mu_X - \mu_Y$ is

$$\left[\bar{X} - \bar{Y} \pm \tau_d(1 - \delta/2)\sqrt{s^2(X)/n_1 + s^2(Y)/n_2} \right], \tag{7.8}$$

where

$$d = d(\hat{c}, n, n_2) := \left[\frac{\hat{c}^2}{n_1 - 1} + \frac{(1 - \hat{c})^2}{n_2 - 1} \right]^{-1} \tag{7.9}$$

with

$$\hat{c} := \frac{s^2(X)/n_1}{s^2(X)/n_1 + s^2(Y)/n_2}$$

gives the degrees of freedom for the critical value of Student's t distribution.

If $\mathrm{var}\, \bar{X}^{(\cdot)}_{k_1 t_1} = \mathrm{var}\, \bar{Y}^{(\cdot)}_{k_2 t_2}$ and $n_1 = n_2$, then $(\bar{X} - \bar{Y})/\left[s^2(X)/n_1 + s^2(Y)/n_2 \right]^{1/2}$ has Students's t distribution with $n_1 + n_2 - 2$ degrees of freedom, and no issue of approximation arises other than the assumption of normality. When either or both of these conditions are absent, this statistic does not have Student's t distribution with $n_1 + n_2 - 2$ degrees of freedom. Then, expressions (7.8) and (7.9) comprise Welch's suggested approach to approximating the distribution of $(\bar{X} - \bar{Y})/\left[s^2(X)/n_1 + s^2(Y)/n_2\right]^{1/2}$. See for example, Kendall and Stuart (1961, vol. 2) and Scheffé (1970). Section 7.3 describes the basis for this approximation.

❏ EXAMPLE 7.1

As illustration, consider two variants of a queueing system, the first of which produces a sample average waiting time $\bar{X} = 12.94$ hours with sample variance $s^2(X) = 30.43$ based on $n_1 = 15$ independent replications and the second of which produces a sample average waiting time $\bar{Y} = 8.734$ hours with sample variance $s^2(Y) = 15.07$ based on $n_2 = 10$ independent replications. Then $\bar{X} - \bar{Y} = 4.206$ hours, and expression (7.8) yields an approximating 99 percent confidence interval $[-1.094, 9.506]$ for $\mu_X - \mu_Y$ based on $d = 22.89$ degrees of freedom.

Cost considerations, external to the simulation study, lead management to conclude that the operating environment that generated $\bar{Y}_{k_2t_2}^{(1)}, \ldots, \bar{Y}_{k_2t_2}^{(n_2)}$ would be preferable to the one that generated $\bar{X}_{k_1t_2}^{(1)}, \ldots, \bar{X}_{k_1t_1}^{(n_1)}$ if $\mu_X - \mu_Y \geq 2$. Therefore, management wants to know the extent to which the simulation output provides evidence that this inequality indeed holds. We first study this issue in a more general setting.

Let

$$Z(\mu_X - \mu_Y) := \frac{\bar{X} - \bar{Y} - \mu_X + \mu_Y}{[s^2(X)/n_1 + s^2(Y)/n_2]^{1/2}}$$

and take the Student t distribution with $d = 22.89$ degrees of freedom as the approximating distribution for $Z(\mu_X - \mu_Y)$. Given $\bar{X}$, $\bar{Y}$, $s^2(X)/n_1$, and $s^2(Y)/n_2$, suppose it is of interest to determine how credible it is to assume that $\mu_X - \mu_Y \geq v$ for some specified v. Under the hypothesis $H(\theta) : \mu_X - \mu_Y = \theta$, let

$A(\theta) :=$ event of observing a Student t random variate, with d degrees of freedom

no greater than $Z(\theta)$.

Then

$$\text{pr}\,[A(\theta)] = G_d(Z(\theta)),$$

where $G_d(Z(\theta))$ gives the probability of observing a standardized statistic no greater than $Z(\theta)$. Moreover, $G_d(Z(\theta))$ decreases monotonically with increasing θ. Hence, $G_d(Z(\theta))$ provides a measure of credibility for $\mu_X - \mu_Y \geq \theta$. The graph of this function in Figure 7.1 is revealing. It indicates $G_d(\theta) = .87$ for $\theta = 2$ hours, thus providing support for management's action. However, if the decision point had been $\theta = 4$ hours, Figure 7.1 would offer considerably less support for the scenario that generated $\bar{Y}_{k_2t_2}^{(1)}, \ldots, \bar{Y}_{k_2t_2}^{(n_2)}$.

To evaluate the credibility of $\mu_X - \mu_Y \leq \theta$, we use $1 - G_d(Z(\theta))$, the probability of observing a standardized statistic no less than $Z(\theta)$ under $H(\theta)$. This probability monotonically decreases with decreasing θ.

◻

Let us now reflect on this form of credibility analysis. Graphically, it provides considerably more *food for thought* than a report that merely displays point estimates and, possibly, standard errors. In particular, it allows management to assess the plausibility of conjectured differences in means that are pivotal to its decision making. Most importantly, the graphics are easily created with application software such as Microsoft Excel.

7.3 RATIONALE FOR DEGREES OF FREEDOM

Expression (7.9) for degrees of freedom may well arouse the reader's curiosity. This section describes its rationale. A reader willing to accept these d.f.s may proceed directly to Section 7.4.

Since we assume that t_1 and t_2 are sufficiently large so that $\bar{X}_{k_1t_1}^{(1)}, \ldots, \bar{X}_{k_1t_1}^{(n_1)}$ and $\bar{Y}_{k_2t_2}^{(1)}, \ldots, \bar{Y}_{k_2t_2}^{(n_2)}$ are normally distributed, $\bar{X}$ and $s^2(X)$ are statistically independent, $(n_1 - 1)s^2(X)/\text{var}\bar{X}_{k_1t_1}^{(1)}$ has the chi-squared distribution with $n_1 - 1$ degrees of freedom, denoted by $\chi^2(n_1 - 1)$, and $(\bar{X} - \mu_X)/\sqrt{s^2(X)/n_1}$ has Student's t distribution with $n_1 - 1$ degrees of freedom, denoted by $\tau(n_1 - 1)$. Likewise, $\bar{Y}, s^2(Y)$ are independent, $(n_2 - 1)s^2(Y)/\text{var}\bar{Y}_{k_2t_2}^{(i)}$ is from $\chi^2(n_2 - 1)$, and $(\bar{Y} - \mu_Y)/\sqrt{s^2(Y)/n_2}$ from $\tau(n_2 - 1)$.

But what about the standardized difference

$$\left[\bar{X} - \bar{Y} - (\mu_X - \mu_Y)\right]/\sqrt{s^2(X)/n_1 + s^2(Y)/n_2}?$$

If $\text{var}\bar{X}_{k_1t_1}^{(i)} = \text{var}\bar{Y}_{k_2t_2}^{(l)}$ and $n_1 = n_2$, it is from $\tau(n_1 + n_2 - 2)$. However, this equality of variances rarely arises in simulation practice. Welch's solution to this problem relies on properties of the chi-squared distribution; in particular, a chi-squared variate Z with l degrees of freedom has $EZ = l$ and $\text{var}Z = 2l$.

Therefore,

$$l = \frac{2E^2Z}{\text{var}Z}. \tag{7.10}$$

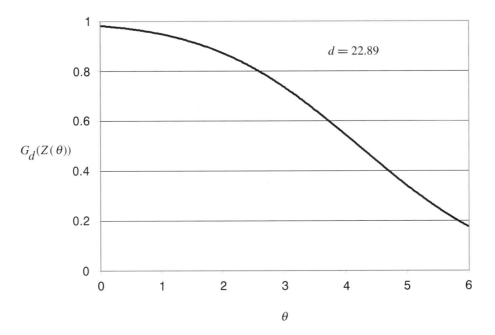

Figure 7.1 Credibility analysis for $\mu_1 - \mu_{11} \geq \theta$ in the photocopying problem (Section 1.4)

Welch employs this equivalence to derive approximate degrees of freedom. Let $a := \operatorname{var} \bar{X}^{(i)}_{k_1 t_1}$ and $b := \operatorname{var} Y^{(i)}_{k_2 t_2}$. Since

$$
\mathrm{E}\left(\frac{s^2(X)}{n_1} + \frac{s^2(Y)}{n_2}\right) = \frac{a}{n_1} + \frac{b}{n_2}
$$

and

$$
\operatorname{var}\left(\frac{s^2(X)}{n_1} + \frac{s^2(Y)}{n_2}\right) = \frac{2a^2}{(n_1 - 1)n_1^2} + \frac{2b^2}{(n_2 - 1)n_2^2},
$$

expression (7.10) gives

$$
d(c, n_1, n_2) = \left[\frac{c^2}{n_1 - 1} + \frac{(1 - c)^2}{n_2 - 1}\right]^{-1},
$$

$$
c := \frac{a/n_1}{a/n_1 + b/n_2},
$$

and $d(\hat{c}, n, n_2)$ in expression (7.11) is an estimate of this quantity with $s^2(X)$ and $s^2(Y)$ replacing a and b, respectively.

7.4   RATIO ANALYSIS

Assume that $\bar{X}$ and $\bar{Y}$ are independent and normally distributed with variances a/n_1 and b/n_2, respectively. Occasionally, it is of interest to estimate the relative difference $R :=$ μ_Y/μ_X, assuming $\mu_X \neq 0$. Since $\bar{Y} - R\bar{X}$ is normally distributed with mean zero and variance $b/n_2 + R^2 a/n_1$, we have by analogy with the development in Section 6.11

$$\mathrm{pr}\left(R \in \frac{\bar{X}\bar{Y} \pm \nu\sqrt{\bar{X}b/n_2 + \bar{Y}a/n_1 - \nu^2 ab/n_1 n_2}}{\bar{X}^2 - \nu^2 a/n_1}\right) \approx 1 - \delta, \qquad (7.11)$$

where $\nu := \Phi^{-1}(1 - \delta/2)$.

This result has meaning only if $\bar{X}b/n_2 + \bar{Y}a/n_1 - \nu^2 ab/n_1 n_2 \geq \nu$ and $\bar{X}^2 - \nu^2 a/n_1 > 0$. It can be shown that both conditions hold asymptotically w.p.1 as $\min(n_1, n_2) \to \infty$.

Since a and b are unknown, we replace them by $s^2(X)$ and $s^2(Y)$, respectively, and $\Phi^{-1}(1 - \delta/2)$ by $\tau_d(1 - \delta/2)$, where $d = d(\tilde{c}, n_1, n_2)$ and

$$\tilde{c} := \frac{s^2(X)/\bar{X}^2 n_1}{s^2(X)/\bar{X}^2 n_1 + s^2(Y)/\bar{Y}^2 n_2}, \qquad (7.12)$$

and compute an approximating $100 \times (1 - \delta)$ percent confidence interval for R. The rationale for this choice of d comes from the observation that

$$\mathrm{E}\left[\frac{s^2(Y)}{n_2} + \frac{R^2 s^2(X)}{n_1}\right] = \frac{b}{n_2} + \frac{R^2 a}{n_1}$$

and

$$\mathrm{var}\left(\frac{s^2(Y)}{n_2} + \frac{R^2 s^2(X)}{n_1}\right) = \frac{2b^2}{(n_2 - 1)n_2^2} + \frac{2R^4 a^2}{(n_1 - 1)n_1^2},$$

which, by analogy with the equivalence (7.10), leads to an approximating number of degrees of freedom

$$\frac{2\mathrm{E}^2\left[\dfrac{s^2(Y)}{n_2} + \dfrac{R^2 s^2(X)}{n_1}\right]}{\mathrm{var}\left(\dfrac{s^2(Y)}{n_2} + \dfrac{R^2 s^2(X)}{n_1}\right)} = \frac{\left(\dfrac{b}{\mu_Y^2 n_2} + \dfrac{a}{\mu_X^2 n_1}\right)^2}{\dfrac{b^2}{\mu_Y^4(n_2 - 1)n_2} + \dfrac{a^2}{\mu_X^4(n_1 - 1)n_1}},$$

which we estimate by $d = f(\tilde{c}, n_1 n_2)$.

7.5   COMPARISON OF MEANS ON TWO SAMPLE PATHS

For single replications ($n_1 = n_2 = 1$), the comparison of experiments proceeds like that for multiple replications but with several differences. We assume that t_1 and t_2 are sufficiently large so that $\mathrm{var}\bar{X}_{k_1 1 t_1} \approx \sigma^2_\infty(X)/t_1$ and $\mathrm{var}\bar{Y}_{k_2 1 t_2} \approx \sigma^2_\infty(Y)/t_2$ (Section 6.3). Then we employ LABATCH.2 (Section 6.7) to compute a strongly consistent estimate $\hat{\sigma}^2_\infty(X) = B_{J(t_1)} W_{L_{J(t_1)} B_{J(t_1)}}$ for $\sigma^2_\infty(X)$ from the sample record $X_{k_1}, \ldots, X_{k_1+t_1-1}$. For $\sigma^2_\infty(Y)$ we again use LABATCH.2 to compute a $\hat{\sigma}^2_\infty(Y) = B_{J(t_2)} W_{L_{J(t_2)} B_{J(t_2)}}$ based on the sample-path data $Y_{k_2}, \ldots, Y_{k_2+t_2-1}$.

Recall that $L_{J(t_1)}$ denotes the estimated number of independent batches used in the final review of $X_{k_1}, \ldots, X_{k_1+t_1-1}$ and $L_{J(t_2)}$, the estimated number of independent batches used in the final review of $Y_{k_2}, \ldots, Y_{k_2+t_2-1}$. By analogy with the case of independent replications, this suggests the approximating $100 \times (1 - \delta)$ percent confidence interval

$$\left[\bar{X} - \bar{Y} \pm \tau_d(1 - \delta/2)\sqrt{\hat{\sigma}^2_\infty(X)/t_1 + \hat{\sigma}^2_\infty(Y)/t_2} \, \right],$$

where

$$d = d(\ddot{c}, L_{J(t_1)} - 1, L_{J(t_2)} - 1) \tag{7.13}$$

and

$$\ddot{c} := \frac{\hat{\sigma}^2_\infty(X)/t_1}{\hat{\sigma}^2_\infty(X)/t_1 + \hat{\sigma}^2_\infty(Y)/t_2},$$

and expression (7.9) defines $d(\cdot, \cdot, \cdot)$.

As before, d varies with $\ddot{c}$, from a minimum of $\min(L_{J(t_1)}, L_{J(t_2)}) - 1$ to a maximum of $L_{J(t_1)} + L_{J(t_2)} - 2$. Setting $d = \min(L_{J(t_1)}, L_{J(t_2)}) - 1$ would again lead to a wider confidence interval and thus militate against any tendency to induce a coverage rate less than $1 - \delta$ because of the inherent approximations in the degrees of freedom.

7.6   MULTIPLE COMPARISONS

Occasionally, a simulation study addresses differences between corresponding performance measures that materialize in a system run under more than two different scenarios. The purpose of the study is to make inferences about the relative desirability of each scenario based on the estimates of long-run averages and their statistical errors. The simulationist then has the task of reporting these inferences to decision makers. In addition to being scientifically defensible, the collection of assertions in the report needs to be organized into a way that *weaves a story* that *makes sense* of the many results and can act as a guide for decision making. When two scenarios are being compared, the accounts in Sections 7.2 and 7.4 provide a basis for accomplishing this. When there are more than two, organizing the results becomes a greater challenge.

Formalisms exist in the statistical literature for making these multiple comparisons. Banks et al. (1996, Chapter 13) and Law and Kelton (1991, Chapter 10) address these issues within the special context of discrete-event simulation. The present account takes a more illustrative approach, focusing on a particular example to describe concepts. The presumption is that the particular issues raised in the example carry over to other problems studied by discrete-event simulation. Our account is based on the concepts of simultaneity (Section 7.1) and the analysis of pairwise differences (Sections 7.2 and 7.4).

7.6.1 LIBRARY PHOTOCOPYING

Recall that the purpose of the library photocopying simulation study in Section 1.4 was to assess the extent to which alternative operating policies affect long-run patron waiting time. At current demand rates, these included:

SCENARIO

1 Current operating system

3 Current operating system with no copier breakdowns

5 Scenario 1 plus one additional copier on the third floor

7 Scenario 5 plus one additional copier on the fourth floor

9 Move the copier on the fifth floor to the fourth floor

11 Replace two copiers on the third floor and two on the fourth floor with new copiers. The replaced copiers have the highest failure rates.

The study also examined these scenarios with a fifty percent increase in demand (these are Scenarios 2, 4, 6, 8, 10, 12). Here we address only the current-demand case. Moreover, we focus on Group 3 (Table 1.1), the time interval from 11 a.m. to 4 p.m., the busiest on a daily basis.

Let μ_i denote long-run patron waiting time for Scenario i, so that interest focuses on estimating μ_1, μ_3, μ_5, μ_7, μ_9, μ_{11} and comparing their point estimates. For each scenario, a single run of $t = 1000$ weeks was made. This corresponds roughly to 20 years of weekday operation. For each Scenario $2i - 1 \in \{1, 3, 5, 7, 9, 11\}$, a sample average waiting time $\hat{\mu}_{2i-1}$ was estimated. Because the elapsed run time was fixed, the number of sample waiting times in this average was random, so that $\hat{\mu}_{2i-1}$ was a ratio estimate (Section 3.4). However, its denominator had a coefficient of variation considerably smaller than that of its numerator, and hence ignoring the randomness in the denominator introduced negligible error. This was true for all scenarios. See Section 6.11 for details of ratio estimation and, in particular, on how to assess the importance of randomness in a denominator.

Table 7.1a displays summary results for $\hat{\mu}_{2i-1}$ for $i = 1, \ldots, 6$. The quantity $\hat{\sigma}_{\infty,2i-1}^2$ denotes an estimate of $t \, \text{var} \hat{\mu}_{2i-1}$ computed by LABATCH.2 and degrees of freedom are one less than the number of batches determined by LABATCH.2. See Section 6.6. Although the run length of 1000 weeks resulted in relative errors considerably larger than the .20 that Section 6.6.1 regards as desirable, the large number of runs dictated this limitation on length. For visual comparison, Figure 7.2 displays the approximating 99 percent confidence intervals.

NATURAL ORDERING

The structure of a particular problem often allows a simulationist to anticipate a natural ordering among some, if not all, long-run averages prior to output analysis. In the present study, the scenarios imply

$$\mu_1 > \mu_{11} > \mu_3,$$
$$\mu_1 > \mu_5 > \mu_7,$$

merely based on the knowledge that waiting time diminishes with decreasing failure rate and/or adding resources. Therefore, the analysis principally focused on how large the difference $\mu_1 - \mu_j$ is for $\mu_1 > \mu_j$.

Table 7.1b gives estimates of differences between long-run average waiting on the baseline Scenario 1 and the remaining five scenarios (Section 7.4). Of principal interest is $\mu_1 - \mu_3$, the greatest possible reduction in long-run average waiting time that can materialize without adding or reassigning resources. Of course, this is an ideal, since all copiers eventually break down. Nevertheless, it is instructive to note that $\hat{\mu}_1 - \hat{\mu}_3 = 3.635$ minutes with an approximating 99 percent confidence interval [2.656, 4.613].

Table 7.1b also reveals that reassigning old copiers and adding a single new copier to the fifth floor (Scenario 7) induced an estimated reduction of $\hat{\mu}_1 - \hat{\mu}_7 = 2.092$ minutes with approximating 99 percent confidence interval [.4242, 3.759]. Also, adding two new copiers on the third and also on the fourth floor (Scenario 11) induced a reduction of $\hat{\mu}_1 - \hat{\mu}_{11} = 2.828$ minutes with approximating 99 percent confidence interval [1.806, 3.849]. Recall that the structure of the problem allows no a priori inference regarding Scenarios 7 and 11. Although the point estimates seem to favor Scenario 11 over Scenario 7, the overlap in their confidence intervals raises a question.

The entries in Table 7.1c for $\hat{\mu}_7 - \hat{\mu}_{11}$ suggest that with current sample sizes a significant difference between μ_7 and μ_{11} is not supportable. We return to this issue presently. First, we note that Table 7.1d estimates the difference $\mu_3 - \mu_{11}$ by $\hat{\mu}_3 - \hat{\mu}_{11} = -.8089$ minutes with approximating 99 percent interval $[-1.221, -.3923]$, suggesting that it is not unreasonable to expect Scenario 11 to reduce long-run average waiting time by at least 2 of the estimated 3.635 minute reduction achievable under the perfect-copier Scenario 3.

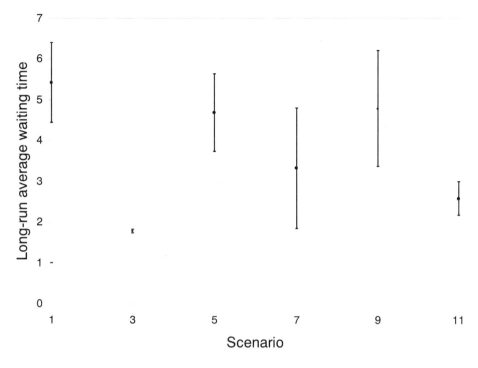

Figure 7.2 Waiting time for photocopying problem in Section 1.4

Implementing Scenario 7 would result in a smaller immediate capital cost than Scenario 11. However, there is an issue of how much better it is than Scenario 1 and how much benefit one gives up by not choosing Scenario 11. To shed light on this problem, consider two hypotheses:

$$H_1 : \mu_1 - \mu_7 \geq 2 \text{ minutes;}$$
$$H_2 : \mu_7 - \mu_{11} \leq 1 \text{ minute.}$$

If H_1 and H_2 were true, then Scenario 7 would induce a reduction of at least two minutes in long-run average waiting time and would be within one minute of the average that would obtain with Scenario 11. We now examine the credibility of these hypotheses.

For $(i, j) = (1, 7)$ and $(7, 11)$, let

$$Z_{ij}(\mu_i - \mu_j) := \frac{\hat{\mu}_i - \hat{\mu}_j - \mu_i + \mu_j}{\sqrt{(\hat{\sigma}^2_{\infty\,i} + \hat{\sigma}^2_{\infty\,j})/t}},$$

whose approximating distribution is Student's t distribution with d_{ij} degrees of freedom computed as in expression (7.13) using the output in Table 7.1.

Table 7.1 Long-run average waiting-time analysis for photocopying problem in Section 1.4

(time in minutes)

(a)

i	Sample mean $\hat{\mu}_i$	Standard error $\sqrt{\hat{\sigma}^2_{\infty i}/t}$	d.f.	99% Confidence interval Lower	Upper	Relative error $\dfrac{\text{Upper}-\text{Lower}}{|\hat{\mu}_i|}$
1	5.412	.3199	12	4.435	6.389	.3611
3	1.778	.01559	25	1.734	1.821	.04893
5	4.676	.3280	17	3.725	5.626	.4066
7	3.321	.5097	17	1.844	4.798	.8896
9	4.784	.4645	12	3.365	6.203	.8023
11	2.585	.1478	25	2.173	2.997	.3190

(b)

i	$\hat{\mu}_1 - \hat{\mu}_i$	$\sqrt{\hat{\sigma}^2_{\infty 1} + \hat{\sigma}^2_{\infty i})/t}$	$\ddot{c}$				$\dfrac{\text{Upper}-\text{Lower}}{|\hat{\mu}_1-\hat{\mu}_i|}$
3	3.635	.3203	.9976	12.06	2.656	4.613	.5384
5	.7363	.4582	.4876	28.37	−.5298	2.002	3.439
7	2.092	.6018	.2827	27.08	.4242	3.759	1.594
9	.6282	.5640	.3218	21.29	−.9688	2.225	5.663
11	2.828	.3524	.3524	17.29	18.06	3.849	.7224

(c)

$\hat{\mu}_7 - \hat{\mu}_{11}$	$\sqrt{(\hat{\sigma}^2_{\infty 7} + \hat{\sigma}^2_{\infty,11})/t}$					$\dfrac{\text{Upper}-\text{Lower}}{|\hat{\mu}_7-\hat{\mu}_{11}|}$
.3618	.5307	.8642	22.38	−1.134	1.858	8.270

(d)

$\hat{\mu}_3 - \hat{\mu}_{11}$	$\sqrt{\hat{\sigma}^2_{\infty 3} + \hat{\sigma}^2_{\infty,11})/t}$					
−.8069	.1487	.01099	25.56	−1.221	−.3925	1.027

For $\hat{\mu}_1$, $\hat{\mu}_7$, $\hat{\mu}_{11}$, $\hat{\sigma}^2_{\infty 1}$, $\hat{\sigma}^2_{\infty 7}$, and $\hat{\sigma}^2_{\infty 11}$, let

$A_1(v)$:= event of observing a standardized statistic no greater than $Z_{1,7}(\theta)$
under the assumption that $\mu_1 - \mu_7 = \theta$

and

$A_2(v')$:= event of observing a standardized statistic no less than $Z_{7,11}(\phi)$
under the assumption that $\mu_7 - \mu_{11} = \phi'$.

Then

$$\text{pr}\,[A_1(\theta)] = G_{27.08}(Z_{1,7}(\theta))$$

and

$$\text{pr}\,[A_2(\phi)] = 1 - G_{22.38}(Z_{7,11}(\phi)),$$

and using Bonferroni's inequality,

$$\text{pr}\,[A_1(\theta), A_2(\phi)] \geq G_{27.08}(Z_{17}(\theta)) - G_{22.38}(Z_{7,11}(\phi)).$$

Table 7.2 gives these probabilities for selected θ and ϕ. While there is some support for $\mu_1 - \mu_7 \geq 2$ and $\mu_7 - \mu_{11} \leq 1$, there is considerably more for $\mu_1 - \mu_7 \geq 1.5$ and $\mu_7 - \mu_{11} \leq 1$.

Table 7.2   Credibility analysis for photocopier problem for long-run average waiting time (minutes: seconds)

$\theta \backslash \phi$	$G_{27.08}(Z_{1,7}(\theta)) - G_{22.38}(Z_{7,11}(\phi))$			$G_{27.08}(Z_{1,7}(\theta))$
	.30	45	1:00	
1:30	.4345	.5969	.7121	.8329
1:45	.3141	.4765	.5917	.7125
2:00	.1614	.3238	.4390	.5599
$1 - G_{22.38}(Z_{7,11}(\phi))$	.6015	.7640	.8791	

7.7 COMMON PSEUDORANDOM NUMBERS REVISITED

Recall that Section 4.22 describes how the technique of common pseudorandom numbers (cpn) applied to two simulation runs, each based on a different scenario, can induce $\mathrm{var}(\bar{X} - \bar{Y}) < \mathrm{var}\bar{X} + \mathrm{var}\bar{Y}$. Since confidence-interval length tends to be proportional to standard error, this reduction in variance implies a shorter interval length for the difference of two means than would arise for independent runs. This property motivates us to identify circumstances in which a simulationist may wisely employ cpn. We begin with a review and extension of the formalisms in Section 4.22. Section 7.7.1 then describes which differences in scenarios can be expected to reduce $\mathrm{var}(\bar{X} - \bar{Y})$ and which are less likely, if at all, to do so. Section 7.8 then compares the benefits of cpn with those of another variance-reducing technique, based on optimal path lengths, which has particular appeal when one scenario induces considerably larger sampling fluctuations than the other.

Recall from Section 4.22 that using cpn leads to

$$v_{\mathrm{cpn}} := \left[\sigma_\infty^2(X) - 2\rho(X, Y)\sigma_\infty(X)\sigma_\infty(Y) + \sigma_\infty^2(Y)\right]/t,$$

where $\rho(X, Y) = \mathrm{corr}(\bar{X}, \bar{Y})$. Hereafter, we use the more concise notation

$$h := \frac{\sigma_\infty(X)}{\sigma_\infty(Y)} \quad \text{and} \quad \rho := \rho(X, Y). \tag{7.14}$$

The induced variance reduction is

$$\mathrm{VR}_{\mathrm{cpn}} := \frac{\text{variance of } X - Y \text{ with cpn}}{\text{variance of } X - Y \text{ for independent runs}} \tag{7.15}$$

$$= 1 - \frac{2\rho}{h + 1/h},$$

revealing that $\max(h, 1/h)$ and ρ determine the level of variance reduction. In particular, for given ρ, $\mathrm{VR}_{\mathrm{cpn}}$ increases monotonically as $\max(h, 1/h)$ increases, thus limiting the appeal of cpn. If $\rho < 0$, then variance inflation occurs, and cpn is to be avoided. Hereafter, we assume $\rho \geq 0$.

The quantity $100 \times \left(1 - \sqrt{\mathrm{VR}_{\mathrm{cpn}}}\right)$ gives the percentage reduction in the length of a confidence interval when cpn, rather than independent runs with $t_1 = t_2$, is used. Table 7.3 displays $1 - \sqrt{\mathrm{VR}_{\mathrm{cpn}}}$ as a function of selected h and ρ. For equal asymptotic variances ($h = 1$), it reveals a 68 percent reduction for $\rho = .90$. However, $\max(h, 1/h) = 10$ leads to only a 9.3 percent reduction for this same ρ. For $\rho = .50$ the reductions are 29.3 and 5.1 percent for $\max(h, 1/h) = 1$ and 10, respectively.

Recall that cpn in Section 4.22 estimates $tv_{\mathrm{cpn}} = \sigma_\infty^2(Z)$ by $\hat{\sigma}_\infty^2(Z)$, computed by LABATCH.2 using the combined sample-path data $Z_i := X_i - Y_i$ for $i = 1, \ldots, t$. Using the sequence of differences in this way captures the correlations between $X_1, \ldots, X_t$ and $Y_1, \ldots, Y_t$ in a single variance correlation. However, expression (7.13) no longer applies for degrees of freedom. In this case treating $(\bar{X} - \bar{Y})/\sqrt{\hat{\sigma}_{\infty}^2(Z)/t}$ as a Student t variate with degrees of freedom one less than the number of batches used to compute $\hat{\sigma}_{\infty}^2(Z)$ allows one to compute an asymptotically valid $100 \times (1 - \delta)$ percent confidence interval for $\mu_X - \mu_Y$.

7.7.1 HOW SCENARIOS AFFECT CORRELATION

Although it falls to the simulationist to match pseudorandom number streams and seeds on the two runs, it is the properties induced by the two scenarios that ultimately dictate the correlation ρ and the ratio h. This section addresses the way in which scenarios affect ρ. The next addresses the way in which h affects variance reduction in cpn and how one can take advantage of a rough estimate of h to reduce variance in yet another way.

Recall that a scenario consists of a set of numerical values assigned to input parameters and a collection of logical operating rules. For example, the parameters for the airline reservation problem include the intercall rate λ, the service rate ω for a single destination, the proportion of customers p who plan multidestination trips, and the number of reservationists m. The set also includes distinct seeds for the four sources of pseudorandom variation and the number of customers t to be simulated. The problem has a single operating rule: Calls are serviced in the order they arrive.

The first example in Section 4.22 shows how cpn works when $m = 4$ and 5 and all other parameters and rules remain the same. By assigning separate pseudorandom number streams to each source of variation in that problem, customer i was assigned the same intercall time and the same service time on both runs for each $i = 1, \dots, t$, thus inducing the sample correlation $\hat{\rho}(X, Y) = .7033$ for the sample average waiting time and $\hat{\rho}(X, Y) = .8906$ for the sample exceedance probability.

The second example keeps $m = 4$ but considers the effect of varying p from $p_1 = .75$ to $p_2 = .90$. This assignment results in customer i being assigned the same intercall time on both runs but not necessarily the same service times. However, by following the strategy described in Section 4.22, $100\,[\min(1 - p_1, 1 - p_2) + \min(p_1, p_2)] = 85$ percent of the customers can be expected to be assigned the same service times on both runs. With regard to exceedance probability, the $(m = 4, p_1 = .75)$ and $(m = 4, p_2 = .90)$ assignments induced a sample correlation

$$\hat{\rho}(X, Y) = \frac{\hat{\sigma}_{\infty}^2(X) + \hat{\sigma}_{\infty}^2(Y) - \hat{\sigma}_{\infty}^2(Z)}{2\sqrt{\hat{\sigma}_{\infty}^2(X)\hat{\sigma}_{\infty}^2(Y)}} = .8656,$$

a relatively small loss from .8906. Note that in this particular problem at least 50 percent of the service times match regardless of the value of p_1 and p_2.

Suppose the two runs were made with the same values for ω, p, and m but two different values of λ, namely λ_1 and λ_2. Then customer i has the same service time on each run, but its intercall times differ. The extent to which this difference militates against a variance reduction depends on the particular distribution being sampled and on the method used to generate intercall times. Recall that intercall times in the example come from $\mathcal{E}(1/\lambda)$.

Let V_1 and V_2 denote customer i's intercall times on the runs with λ_1 and λ_2, respectively. If the inverse transform method (Section 8.1) were used to generate these quantities with the same uniform deviate $U \in (0, 1)$, then

$$V_1 = -\frac{1}{\lambda_1} \ln(1 - U),$$

$$V_2 = -\frac{1}{\lambda_2} \ln(1 - U),$$

so that

$$V_2 = \frac{\lambda_1}{\lambda_2} V_1,$$

implying

$$\text{corr}(V_1, V_2) = 1.$$

We see that corresponding intercall times differ on the two runs but are nevertheless perfectly correlated. While not as favorable as inducing the same intercall times, perfect positive correlation contributes to a positive correlation between the output on the two runs, thus continuing to give a benefit.

But what if a method other than the inverse transform method were used to generate intercall times? With the exponential distribution, we are in luck. Recall from Section 8.9 that all algorithms for sampling from $\mathcal{E}(1/\lambda)$ actually sample from $\mathcal{E}(1)$ and then multiply the result by $1/\lambda$, thus ensuring perfect correlation. More generally, perfect correlation between V_1 and V_2 arises regardless of distribution if the number of uniform deviates required to generate V_1 is independent of the intercall time parameters and V_2 is computed as $V_2 = a V_1 + b$ for some $a > 0$ and b. This includes sampling from $\mathcal{E}(1/\lambda)$ with varying λ, $\mathcal{N}(\mu, \sigma^2)$ with varying μ and σ^2, $\mathcal{G}(\alpha, \beta)$ with varying β but the same α, and $\mathcal{W}(\alpha, \beta)$ with varying β but the same α. It also applies to the inverted Gamma and inverse Weibull distributions for varying β.

Suppose intercall times are drawn from $\mathcal{G}(\alpha_1, \beta)$ and $\mathcal{G}(\alpha_2, \beta)$ on the two runs using Algorithm GKM3 (Section 8.13) based on the ratio-of-uniforms methods (Section 8.6). Recall that this method uses a random number of uniform deviates to generate samples. If $\alpha_1 = \alpha_2$ and the same seed is used on the two runs, then both use the same pseudorandom number deviates, and the seeds for the next generation are identical. However, if $\alpha_1 \neq \alpha_2$, generation of the two runs may require different numbers of uniform deviates so that the seeds for the next generation differ. Then synchronization of pseudorandom numbers, so vital to inducing positive correlation, is lost. In practice, it is ill advised to use the same seed in this case, lest the practice unintentionally induce a negative correlation.

What about different logical operating rules? These include job routing through a system and job selection for service. For example, suppose one run takes jobs for service in the order of arrival, whereas the second run takes them in reverse order. As a consequence, we anticipate that the waiting times for customer i on the two runs are negatively correlated, so that sample average waiting times are negatively correlated, and therefore their difference has larger variance than would arise for independent runs. Interestingly, the long-run average waiting times are identical for both queueing disciplines.

Unless the implications for induced correlation are known and favorable, it is generally advisable to use independent runs, rather than cpn, when comparing different logical operating rules.

7.7.2 MULTIPLE COMPARISONS

What if we are comparing several sample averages as in Section 7.6? For pairwise comparisons, the discussion in these sections continues to apply with the variance of each pairwise difference being estimated as in Section 4.22 and simultaneity entering as in Section 7.1.

7.8 OPTIMAL SAMPLE-PATH LENGTHS

Other variance-reducing techniques exist that do not require correlation between sample averages. This section describes one for which variance reduction improves as $\max(h, 1/h)$ increases.

Let us examine the approach of Section 7.7 more closely for independent runs generating $\bar{X}$ and $\bar{Y}$ based on sample-path lengths t_1 and t_2 not necessarily equal but satisfying $t_1 + t_2 = 2t$. Then executing two independent runs with sample-path lengths

$$t_1 = \frac{2th}{1+h} \quad \text{and} \quad t_2 = \frac{2t}{1+h} \tag{7.16}$$

induces minimal variance

$$v_{\text{opt}} := \text{var}(\bar{X} - \bar{Y}) = [\sigma_\infty(X) + \sigma_\infty(Y)]^2 / 2t$$

and variance reduction

$$\text{VR}_{\text{opt}} := \frac{\text{var}(\bar{X} - \bar{Y})}{\text{var}\,\bar{X} + \text{var}\,\bar{Y}} = \frac{(1+h)^2}{2(1+h^2)}. \tag{7.17}$$

Table 7.3 shows the reduction in interval length that opt induces for selected $\max(h, 1/h)$. Clearly, there is a crossover point at which opt offers more of a reduction than cpn. In particular,

$$v_{\text{cpn}} \leq v_{\text{opt}} \quad \text{if } \max(h, 1/h) \leq 1 + 2\rho + 2\sqrt{\rho(1+\rho)},$$

$$v_{cpn} > v_{opt} \qquad \text{otherwise.}$$

As illustrations,

ρ	$1 + 2\rho + 2\sqrt{\rho(1 + \rho)}$
0	1
.25	2.618
.50	3.297
.75	4.791
.90	5.415
.95	5.622
1.0	5.828.

In particular, note that for all $\rho \in (0, 1)$ opt induces smaller variance than cpn if $\max(h, 1/h) > 5.828$. Also, $\frac{1}{2} \leq \text{VR}_{opt} \leq 1$ with

$$\text{VR}_{opt} \rightarrow \begin{cases} 1 & \text{as } \max(h, 1/h) \rightarrow 1, \\ \dfrac{1}{2} & \text{as } \max(h, 1/h) \rightarrow \infty, \end{cases}$$

whereas $0 \leq \text{VR}_{cpn} \leq 1$ with

$$\text{VR}_{cpn} \rightarrow \begin{cases} 1 - \rho & \text{as } \max(h, 1/h \rightarrow 1 \\ 0 & \text{as } \max(h, 1/h) \rightarrow \infty. \end{cases}$$

Clearly, cpn with $t_1 = t_2 = t$ has its greatest advantage when variances are equal, whereas opt, with $t_1 + t_2 = 2t$, achieves its best variance reduction of $\frac{1}{2}$ when $\sigma_\infty^2(X)$ and $\sigma_\infty^2(Y)$ differ substantially. The opt variance-reducing technique does have the limitation that h is unknown and, in principle, must be estimated in some way. We defer discussion of this issue until Section 7.10

Because of the independence of the sample paths $X_1, \ldots, X_{t_1}$ and $Y_1, \ldots, Y_{t_2}$, expression (7.13) can be used to estimate degrees of freedom for $(\bar{X} - \bar{Y})/\sqrt{\hat{\sigma}_\infty^2(X)/t_1 + \hat{\sigma}_\infty^2(Y)/t_2}$, and an asymptotically valid $100 \times (1 - \delta)$ percent confidence interval can be computed for $\mu_X - \mu_Y$.

7.9 UNEQUAL SAMPLING COSTS

The foregoing discussion assumes that generating costs per observation are identical, or close to identical, on each scenario. Cost is customarily measured in terms of cpu time expended per simulation observation. If these costs differ substantially, then we need to reformulate the opt approach and its comparison to cpn.

Table 7.3 Proportional reduction in confidence interval length for cpn and opt

$\max(h, 1/h)$	$1 - \sqrt{\mathrm{VR_{opt}}}$	ρ	$1 - \sqrt{\mathrm{VR_{cpn}}}$
1	0	0	0
		.25	.1340
		.50	.2929
		.75	.5000
		.90	.6838
		.95	.7764
		1.0	1.0
2	.05132	0	0
		.25	.1056
		.50	.2254
		.90	.4709
		.75	.3675
		.95	.5101
		1.0	.5528
4	.1425	0	0
		.25	.06066
		.50	.1255
		.75	.1956
		.90	.2407
		.95	.2564
		1.0	.2724
10	.2260	0	0
		.25	.02507
		.50	.05080
		.75	.07724
		.90	.09348
		.95	.09896
		1.0	.1045

Let b_1 and b_2 be the unit costs on runs 1 and 2, respectively, and suppose a total budget constraint

$$b = b_1 t_1 + b_2 t_2$$

exists. Then the variance-minimizing (t_1, t_2) is

$$t_1 = \frac{bh}{b_1^{1/2}\left(b_1^{1/2}h + b_2^{1/2}\right)} \quad \text{and} \quad t_2 = \frac{b}{b_2^{1/2}\left(b_1^{1/2}h + b_2^{1/2}\right)} \tag{7.18}$$

with

$$v_{\text{opt}} = \frac{1}{b}\left[b_1^{1/2}\sigma_\infty(X) + b_2^{1/2}\sigma_\infty(Y)\right]^2,$$

whereas cpn with $t_1 = t_2 = b/(b_1 + b_2)$ induces

$$v_{\text{cpn}} = \left[\sigma_\infty^2(X) - 2\,\rho\,\sigma_\infty(X)\sigma_\infty(Y) + \sigma_\infty^2(Y)\right](b_1 + b_2)/b.$$

As immediate consequences,

$$\frac{v_{\text{cpn}}}{v_{\text{opt}}} \to \begin{cases} 1 + \dfrac{b_1}{b_2} & \text{as} \quad h \to 0 & \text{i.} \\[2mm] 1 + \dfrac{b_2}{b_1} & \text{as} \quad h \to \infty & \text{ii.} \\[2mm] h^2 - 2\rho h + 1 & \text{as} \quad \dfrac{b_1}{b_2} \to 0 & \text{iii.} \\[2mm] 1 - 2\rho/h + 1/h^2 & \text{as} \quad \dfrac{b_1}{b_2} \to \infty & \text{iv.} \\[2mm] \dfrac{(h^2+1)(1+b_1/b_2)}{(hb_1^{1/2}/b_2^{1/2}+1)^2} & \text{as} \quad \rho \to 0 & \text{v.} \\[3mm] \dfrac{(h-1)^2(1+b_1/b_2)}{(hb_1^{1/2}/b_2^{1/2}+1)^2} & \text{as} \quad \rho \to 1. & \text{vi.} \end{cases}$$

Limits i, ii, and v indicate that opt induces smaller variance than cpn. Limit iii shows that opt leads to a smaller variance for $\rho < h/2$ and a larger variance for $\rho > h/2$. Therefore, limit iii favors opt for all $h > 2$. Conversely, limit iv favors opt for $\rho < \frac{1}{2}h$ and cpn for $\rho > \frac{1}{2}h$. This implies cpn for all $h < 2$.

Limit vi gives cpn its greatest potential advantage with regard to correlation. Nevertheless, opt induces a smaller variance if

$$\max(h/g, g/h) > 1 + g + 1/g + \sqrt{(g+1/g)^2 + 2(g+1/g)},$$

where

$$g := \left(\frac{b_1}{b_2}\right)^{1/2}.$$

More generally, opt induces smaller variance if

$$\max(h/g, g/h) > 1 + (g + 1/g)\rho + \sqrt{(g + 1/g)^2\rho^2 + 2(g + 1/g)\rho}, \qquad (7.19)$$

and cpn induces smaller variance otherwise.

As mentioned before, variances tend to increase as congestion increases. Also, the computing time to generate an observation tends to increase, but usually at a slower rate than variance. Hence, for $\sigma_\infty^2(X) > \sigma_\infty^2(Y)$ it is not unusual to expect $h \geq g$. Figure 7.3 displays the bound (7.19) for this case. If h^2 lies in the region above the line corresponding to a given ρ, then opt gives the smaller variance. Indeed, for $\rho = 1$ and $g^2 \leq 1$, we see that opt gives the smaller variance for all $h^2 \geq (3 + 2\sqrt{2})^2 = 33.97$. In summary,

- For fixed ρ and g, opt gives the greater variance reduction for sufficiently large h.

- For fixed h and ρ, cpn gives the greater variance reduction for sufficiently large g.

- For fixed h and g, cpn gives the greater variance reduction if $h > 1 + g + g^{1/2} + \sqrt{(1 + g)^2 + 2g^{1/2}(1 + g)}$ for all $\rho \in [0, 1]$.

- For fixed h and g, opt gives the greater variance reduction if $h > g^{1/2}$ as $\rho \to 0$.

7.10   ERRORS IN OPTIMALITY

To become operational, opt requires values for h and g. Since these are unknown, a simulationist customarily substitutes estimates for them. If preliminary runs make available rough estimates of $\sigma_\infty(X)$ and $\sigma_\infty(Y)$, then these can serve to estimate h. The question now becomes, How rough can these estimates be and yet lead to a variance reduction?

For the moment, assume that g is known and that αh is the estimate of h, where α is a positive quantity. If αh replaces h in computing t_1 and t_2 in expression (7.18), then n analogous to the ratio (7.17) we have

$$\mathrm{VR}_{\mathrm{opt}}(\alpha) := \frac{(gh/\alpha + 1)(gh\alpha + 1)}{(h^2 + 1)(g^2 + 1)}.$$

A variance reduction occurs (i.e., $\mathrm{VR}_{\mathrm{opt}}(\alpha) \leq 1$) for all

$$\max\left(\frac{h}{g}, \frac{g}{h}\right) \geq \max(\alpha, 1/\alpha). \qquad (7.20)$$

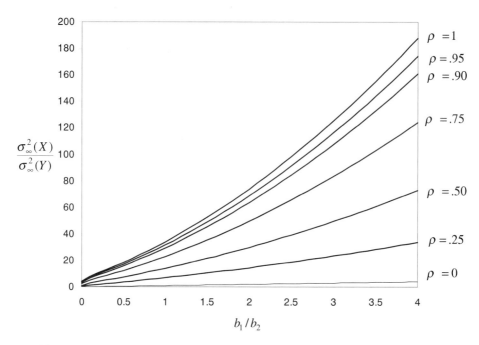

Figure 7.3   cpn vs. opt

For example, if in reality $h = 10$ and $r = 2$, then a variance reduction occurs if one overestimates h by a factor no greater than $\alpha = 5$ or if one underestimates h by a factor no smaller than $\alpha = \frac{1}{5}$. More generally, inequality (7.20) reveals that variance reduction is possible even with large errors of approximation.

Although the unit costs b_1 and b_2 rarely are known exactly, timings for preliminary debugging runs usually can provide good approximations. Therefore, degradation in variance reduction due to replacing b_1 and b_2 with estimates should be negligible.

7.11   OPT AND MULTIPLE REPLICATIONS

To extend opt to multiple comparisons, we focus on a special but commonly encountered case. Assume r scenarios with long-run averages $\mu_1, \ldots, \mu_r$. The principal objective is to compare μ_1 with each of the remaining μ_i. Let t_i denote sample-path length under Scenario i,

$$\sigma_{\infty i}^2 := \lim_{t_i \to \infty} t_i \operatorname{var} \hat{\mu}_i,$$

and

$$h_i := \frac{\sigma_{\infty 1}}{\sigma_{\infty i}}, \qquad\qquad i = 2, \ldots, r.$$

If cost per observation is the same on all r runs, then for fixed $rt = t_1 + \cdots + t_r$, choosing

$$t_1 = \frac{rt}{1 + 1/h_2 + \cdots + 1/h_r},$$

$$t_j = \frac{t_1}{h_j} \quad j = 2, \cdots, r,$$

minimizes each $\mathrm{var}(\hat{\mu}_1 - \hat{\mu}_i)$ for $i = 2, \ldots, r$. See expression (7.16).

Suppose unit costs $b_1, \ldots, b_r$ are unequal and let

$$g_j := \sqrt{\frac{b_1}{b_j}}, \quad j = 2, \ldots, r.$$

Then for a fixed cost $b = b_1 t_1 + \cdots + b_r t_r$, choosing

$$t_1 = \frac{b}{b_1 + \sum_{j-2}^{r}(\sqrt{b_1 b_{ij}}\,/h_j)}$$

and

$$t_j = \frac{t_1 g_1}{h_j}, \quad j = 2, \ldots, r,$$

minimizes all $\mathrm{var}(\hat{\mu}_1 - \hat{\mu}_i)$ for $i = 2, \ldots, r$. See expression (7.18). The commentary in Section 7.11 regarding estimation of g and h continues to apply here.

7.12   LESSONS LEARNED

- Bonferroni's inequality provides a means for computing a lower bound on the probability that several random events occur simultaneously.

- In computing a confidence interval for the difference of means based on two independent runs made with different input scenarios, Welch's approximation provides a means for computing degrees of freedom.

- Simultaneous confidence statements about pairwise differences for multiple runs, each with a diferent scenario, provide a basis for weaving a coherent story about what the data show.

- Common pseudorandom numbers (cpn) can reduce the variance of the difference of sample averages computed from simulation runs executed under different input scenarios.

- The potential variance reduction of cpn decreases as the variances of the individual sample averages diverge.

- cpn also apply to multiple comparisons of pairwise differences of means.

- For independent runs, opt is a variance-reducing technique that chooses sample-path lengths to minimize the variance of the difference of two sample averages.

- opt takes into consideration differences in sample-average variances and differences in unit data-collection costs on each run.

- opt also applies to multiple comparisons of pairwise differences in means where one mean is common to all differences.

7.13   EXERCISES

QUARRY-STONE CRUSHING (EXERCISE 6.2)

✐   **7.1** Exercise 6.2 provides estimates of long-run average profit as a function of λ, the arrival rate of external deliveries to the crusher. Presumably, these estimates have allowed you to estimate $\lambda_{\max}$, the rate that maximizes long-run average profit. For institutional reasons, management is inclined to choose $\lambda = .15$ as the permissible arrival rate, provided that this action results in no more than a 1.5 cent reduction in profit per ton.

a. Let $p(\lambda)$ denote long-run average profit at rate λ. Determine the credibility of the hypotheses

$$p(\lambda_{\max}) - p(.15) \leq d \text{ for all } d \in [1, 2].$$

b. Interpret the results of part a in light of management's qualification of no more than a 1.5 cent loss in profit per ton.

LOADING TANKERS (EXERCISE 6.4)

✐   **7.2** The port authority wants to reduce μ_r, the average rebate per tanker. One plan to do so replaces the tug by a new one that takes $\mathcal{E}(\frac{2}{3})$ time to berth an empty tanker, and $\mathcal{E}(\frac{5}{6})$ time to deberth a full tanker, and $\mathcal{E}(\frac{1}{2})$ to deadhead in either direction. As before, times are in hours. Let $\mu_r(\text{old})$ and $\mu_{r'}(\text{old})$ denote the quantities estimated in Exercise 6.4 with the old tug and let $\mu_r(\text{new})$ and $\mu_{r'}(\text{new})$ be the average rebates with the new tug.

a. Rerun Exercise 6.4 with these new times to estimate $\mu_r(\text{new})$ and $\mu_{r'}(\text{new})$. Since the system now has less congestion, the warm-up interval estimated in Exercise 6.3 should suffice.

b. Perform an analysis to determine the credibility of $\mu_r(\text{old}) - \mu_r(\text{new}) \leq d\mu_r(\text{old})$ for each $d \in [.2, .75]$.

c. Perform an analysis to estimate how large management can expect the reduction in $\mu_{r'}(\text{old}) - \mu_{r'}(\text{new})$ to be. What does this analysis reveal about waiting patterns for the old and new tugs?

TERMINAL PAD CONTROLLER (EXERCISE 6.6)

✎ **7.3** This exercise provides an opportunity to evaluate the benefits of common pseudorandom number streams for the case of $m = 8$ maximal characters in a packet. Part a establishes a baseline against which the results in part b, based on common pseudorandom numbers, can be compared. Let $\hat{\lambda}_{\min}$ denote your estimate of λ that minimizes mean delay $\mu(\lambda, 8)$ in Exercise 6.6.

a. Use the results of the runs in Exercise 6.6 to compute point and 99 percent interval estimates for the difference in mean delay $\mu(\lambda, 8) - \mu(\hat{\lambda}_{\min}, 8)$ for each $(\lambda \neq \hat{\lambda}_{\min})$ used for $m = 8$.

b. Repeat the simulation runs for all λ with the seeds used for the run with $\lambda = \hat{\lambda}_{\min}$ in Exercise 6.6 being the common pseudorandom numbers seeds on all runs. As a consequence, results for the different λ are correlated. Use these data to compute point and 99 percent interval estimates of $\mu(\lambda, 8) - \mu(\hat{\lambda}_{\min}, 8)$. Then for each fixed λ_i compare the statistical accuracies of those differences with those computed in part a.

MOBILE-TELEPHONE SERVICE (EXERCISE 6.8)

✎ **7.4** The set $\mathcal{C} = \{.1i; \ i = 1, \ldots, 9\}$ of cost evaluation points in Exercise 6.7 is intended merely to provide a perspective on the best choice of r for each m. In reality, cost c takes some value in $(0, 1)$ but is unknown. Suppose management has the prior Beta distribution $\mathcal{B}e(3, 1.11)$ (Section 8.14) for c.

a. For each (m, r) in Exercise 6.7, compute a point estimate $\hat{C}(m, r)$ of

$$C(m, r) := \text{expected long-run average cost}$$

for c distributed as $\mathcal{B}e(3, 1.11)$.

b. For each (m, r), compute an approximating 99 percent confidence interval for each $C(m, r)$ and for the function $\{C(m, r), \ r = 0, 1, \ldots, m - 2\}$.

c. For each m, estimate

$$C_{\min}(m) := \min_{0 \leq r \leq m-2} C(m, r)$$

and

$$r_{\min}(m) := \text{the minimizing } r.$$

One approach, called Alternative A, to system specification for fixed m chooses $r_{\min}(m)$. An alternative approach, called Alternative B, makes use of the piecewise constant function $\{r_{\min}(m, c), c \in \mathcal{C}\}$, where for each $r \in \{0, 1, \ldots, m - 2\}$

$$r_{\min}(m, c) := r \text{ that minimizes } C(m, r, c).$$

Since c changes randomly every 12 months, management is considering a policy that replaces $r_{\min}(m, c_{\text{old}})$ with $r_{\min}(m, c_{\text{new}})$ each time c changes, where c_{old} denotes the old cost and c_{new} the new.

d. For each m, compute the expected cost of Alternative B. Hint: Start by computing the probabilities of choosing each r for $r \in \{0, 1, \ldots, m - 2\}$.

e. Since there is an expense to changing r annually, management wants some assurance that any savings that Alternative B offers is at least 20 percent of $C_{\min}(m)$ for each m. Evaluate the credibility of this statement.

7.14 REFERENCES

Banks, J., J.S. Carson, III, and B. L. Nelson (1996). *Discrete-Event System Simulation*, second edition, Prentice-Hall, Upper Saddle River, New Jersey.

Bechhofer, R.E., T.J. Santner, and D. Goldman (1995). *Design and Analysis for Statistical Selection, Screening and Multiple Comparisons*, Wiley, New York.

Fishman, G.S. (1996). *Monte Carlo: Concepts, Algorithms, and Applications*, Springer-Verlag, New York.

Kendall, M.G., and A. Stuart (1961). *The Advanced Theory of Statistics*, **2**. Hafner, New York.

Law, A.M., and W. D. Kelton (1991). *Simulation Modeling and Analysis*, second edition, McGraw-Hill, New York.

Matejcik, F.J., and B.L. Nelson (1995a). Two-stage multiple comparisons with the best for computer simulation, *Operations Research*, **43**, 633–640.

Matejcik, F.J., and B.L. Nelson (1995b). Using common random numbers for indifference-zone selection and multiple comparisons in simulation, *Management Science*, **41**, 1935–1945.

Scheffé, H. (1970). Practical solutions of the Behrens–Fisher problem, *J. Amer. Statist. Assoc.*, **65**, 1501–1508.

CHAPTER 8

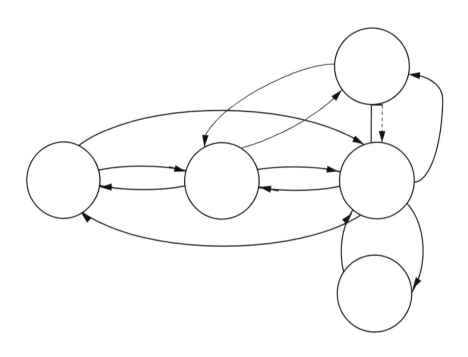

Sampling from Probability Distributions

Virtually every commercially available product for performing discrete-event simulation incorporates software for sampling from diverse probability distributions. Often, this incorporation is relatively seamless, requiring the user merely to pull down a menu of options, select a distribution, and specify its parameters. This major convenience relieves the user of the need to write her or his code to effect sampling.

Occasionally, a simulation requires samples from a distribution not available in the user's selected software. To remove this limitation, this chapter describes the theory underlying procedures and gives algorithms for generating samples from many commonly encountered distributions. Although several alternative algorithms may exist for producing a sample from a particular distribution, the chapter repeatedly focuses on those that guarantee a fixed mean computing time, or an upper bound on mean computing time that holds uniformly, regardless of the values that the parameters of the distribution assume. Moreover, for each distribution considered, the account concentrates on algorithms whose computing times compare favorably with the computing times of other proposed algorithms for generating a sample from the same distribution.

Expertise in generating samples from diverse distributions has grown constantly since the feasibility of performing Monte Carlo experiments became a reality during World War II.

Milestones in this development include the demonstrations in von Neumann (1951) of how the principles of conditional probability could be exploited for this purpose, in Marsaglia et al. (1964) of how a synthesis of probabilistic and computer-scientific considerations can lead to highly efficient generating procedures, and in Ahrens and Dieter (1974a,b) of how a bounded mean computing time could be realized for an arbitrary distribution.

Although the chapter describes procedures that are theoretically exact, the limitations of working on a finite word-size computer inevitably produce results that differ in practice from their theoretical ideals. The account notes where serious differences are most likely to occur and emphasizes the need to employ methods that keep potential errors within tolerable bounds.

Remarkably, all generating methods produce samples drawn from a specified distribution by transforming a sequence of independent *uniform deviates* each randomly drawn from $\mathcal{U}(0, 1)$. In practice, a sequence of (normalized) *pseudorandom numbers* (Chapter 9) takes the place of the idealized sequence of uniform deviates, and inevitably, this substitution induces an error of approximation. Chapter 9 describes in considerable detail how to control the extent to which the distribution of k-tuples of these normalized pseudorandom numbers departs from the uniform distribution on the k-dimensional unit hypercube for each $k \geq 1$. However, a more insidious source of error arises when the theoretically exact methods of this chapter use these pseudorandom numbers. Given several alternative methods for transforming uniform deviates into a sample drawn from a particular distribution, some are considerably more sensitive than others to the error of approximation that replacing the deviates by pseudorandom numbers induces. In particular, a growing body of evidence suggests that this sensitivity increases with the number of pseudorandom numbers that a method requires to generate a single value from a specified distribution (e.g., Afflerbach and Hörmann 1990 and Hörmann and Derflinger 1993).

In practice, this sobering discussion of error encourages two strategies. First, one should always employ a pseudorandom generator whose error of approximation is no more serious than the errors of approximation of available competing generators. Chapter 9 describes how to make this assessment and lists several generators that perform well according to it. Secondly, when selecting a method to generate samples from a particular distribution, one should prefer those that require few pseudorandom numbers per sample. This last stricture favors the inverse transform method (Section 8.1).

Today, many online computer program libraries provide programs for generating samples from a wide range of distributions. Using these programs can save considerable time and effort and, when appropriately implemented, can reduce the potential for incurring serious numerical errors. The conscientious user of any such program will want to assure herself or himself that the method on which it is based is technically sound and that its implementation keeps the inevitable error within an acceptable limit. This chapter aims at assisting the reader in making this evaluation by acquainting her or him with the exact methodologies for generating samples and with their potential sources of error. Also, it aims at providing the reader with the capacity to write a generating program, if necessary.

This chapter assumes the availability of a relatively inexhaustible source of pseudorandom numbers and Chapter 9 describes the detailed considerations that play roles in the design and implementation of a pseudorandom number generator. This particular organization gives more continuity to the topics that immediately concern the potential user of the Monte Carlo method. However, we emphasize that an understanding of issues related to pseudorandom number generation is essential for one knowledgeably to assess the suitability of computer library programs that claim to effect this generation.

Four principal methods exist for transforming uniform deviates drawn from $\mathcal{U}(0, 1)$ into samples drawn from a specified distribution. They are the inverse transform method, the composition method, the acceptance–rejection method, and the ratio-of-uniforms method. Sections 8.1 through 8.6 describe them in detail, emphasizing that sampling from particular distributions may require one to employ a combination of these methods. Of the remaining available methods, Section 8.7 describes the exact–approximate method, which illustrates how careful transformation of variates can reap benefits. Devroye (1986) describes many of the remaining methods.

8.1 INVERSE TRANSFORM METHOD

The inverse transform method provides the most direct route to generating a sample.

Theorem 8.1.

Let $\{F(z), a \leq z \leq b\}$ denote a distribution function with inverse distribution function

$$F^{-1}(u) := \inf\{z \in [a, b] : \quad F(z) \geq u, \ 0 \leq u \leq 1\}. \tag{8.1}$$

Let U denote a random variable from $\mathcal{U}(0, 1)$. Then $Z = F^{-1}(U)$ has the d.f. F.
The proof follows from the observation that $pr(Z \leq z) = pr[F^{-1}(U) \leq z] = pr[U \leq F(z)] = F(z)$.

This result immediately suggests the following generating scheme:

ALGORITHM ITM

Purpose: To generate Z randomly from $\{F(z), \ a \leq z \leq b\}$.
Input: Capability for evaluating $\{F^{-1}(u), \ 0 \leq u \leq 1\}$.
Output: Z.
Method:
    Randomly generate U from $\mathcal{U}(0, 1)$.
    $Z \leftarrow F^{-1}(U)$.
    Return Z.

8.1.1 CONTINUOUS DISTRIBUTIONS

Table 8.1 lists selected continuous distributions and their inverse d.f.s. Note that using the modified form allows one to gain computational efficiency while preserving the distribution of Z. For those continuous distributions whose inverse d.f.s have closed forms, the inverse transform method offers the most direct method of sampling. If no closed form is available but $F(z)$ can be easily computed, an iterative method can facilitate sampling. For finite a and b, Algorithm BISECTION with $k(z) = F(z) - U$, $z_1 = a$ and $z_2 = b$ determines $Z = z$.

If Z is unbounded with $f(z) > 0$, then the Newton–Raphson method, as described in Algorithm N-R with $k(z) = F(z) - U$, produces a solution. If $\{f(z)\}$ is bounded and unimodal, then the modal value provides a reasonable initial value z_0. In addition to the numerical error inherent in working on any finite word-size computer, the bisection and Newton–Raphson methods both induce an additional error based on the specified error tolerances. Since making ν smaller inevitably increases the number of iterations until success, one must inevitably balance cost and accuracy considerations here. See Devroye (1986) for a more complete discussion of this issue.

ALGORITHM BISECTION

Purpose: To compute the solution z to $k(z) = 0$ for $z \in [z_1, z_2]$.
Input: Function $\{k(z),\ z_1 \le z \le z_2\}$ and error tolerances $\nu_1, \nu_2 > 0$.
Output: z and $k(z)$.
Method: 1. Until $|z_L - z_U| \le \nu_1$ or $|k(z)| \le \nu_2$:
        a. $z = (z_L + z_U)/2$.
        b. Compute $k(z)$.
        c. If $k(z) < 0$, set $z_L \leftarrow z$; otherwise, set $z_U \leftarrow z$.
    2. Return with z and $k(z)$.

ALGORITHM N-R

Purpose: To compute the solution z to $k(z) = 0$.
Input: Functions $\{k(z), -\infty < z < \infty\}$ and $\{k'(z) = \partial k/\partial z, -\infty < z < \infty\}$, initial value z_0, and error tolerance $\nu > 0$.
Output: z.
Method: $y \leftarrow z_0$.
    Repeat
        $x \leftarrow y$
        compute $k(x)$ and $k'(x)$
        $y \leftarrow x - k(x)/k'(x)$
    until $|y - x| < \nu$.
    Return $z \leftarrow y$.

Table 8.1 Selected continuous distributions

distribution	p.d.f.		inverse d.f.	preferred form
uniform $\mathcal{U}(\alpha,\beta)$	$1/(\beta-\alpha)$	$\alpha \leq z \leq \beta$	$\alpha+(\beta-\alpha)u$	— —
beta $\mathcal{B}e(\alpha,1)$	$\alpha z^{\alpha-1}$	$\alpha>0,\ 0\leq z\leq 1$	$u^{1/\alpha}$	— —
beta $\mathcal{B}e(1,\beta)$	$\beta(1-z)^{\beta-1}$	$\beta>0,\ 0\leq z\leq 1$	$1-(1-u)^{1/\beta}$	$1-u^{1/\beta}$
exponential $\mathcal{E}(\beta)$	$\dfrac{1}{\beta}e^{-z/\beta}$	$\beta>0,\ z\geq 0$	$-\beta\ln(1-u)$	$-\beta\ln u$
logistic $\mathcal{L}(\alpha,\beta)$	$\dfrac{e^{-(z-\alpha)/\beta}}{\beta[1+e^{-(z-\alpha)/\beta}]^2}$	$\beta>0,\ -\infty<z<\infty$	$\alpha+\beta\ln[u/(1-u)]$	— —
non-central Cauchy $\mathcal{C}(\alpha,\beta)$	$\dfrac{\beta}{\pi[\beta^2+(z-\alpha)^2]}$	$\beta>0,\ -\infty<z<\infty$	$\alpha+\beta\tan\pi(u-1/2)$	$\alpha+\beta/\tan\pi u$
normal $\mathcal{N}(\mu,\sigma^2)$	$(2\pi\sigma^2)^{-1/2}e^{-(z-\mu)^2/2\sigma^2}$	$\sigma^2>0,\ -\infty<z<\infty$	Moro (1995)	
Pareto $\mathcal{P}a(\alpha,\beta)$	$\alpha\beta^\alpha/z^{\alpha+1}$	$z\geq\beta>0,\ \alpha>0$	$\beta/(1-u)^{1/\alpha}$	$\beta/u^{1/\alpha}$
Weibull $\mathcal{W}(\alpha,\beta)$	$(\alpha/\beta^\alpha)z^{\alpha-1}e^{-(z/\beta)^\alpha}$	$\alpha,\beta>0,\ z\geq 0$	$\beta[-\ln(1-u)]^{1/\alpha}$	$\beta(-\ln u)^{1/\alpha}$

Although ease of implementation gives the inverse transform method its appeal, it is not necessarily the most efficient method of sampling from a particular distribution. Its principal limitation is its heavy reliance on functional transformation. In particular, sampling from $\mathcal{B}e(\alpha, 1)$, $\mathcal{B}e(1, \beta)$, and $\mathcal{P}a(\alpha, \beta)$ call for two transformations per sample. For example, a sample from $\mathcal{B}e(\alpha, 1)$ has the form $\exp(\alpha^{-1} \ln u)$. Whereas computers that incorporate algorithms for evaluating these transformations into their hardware offer the fastest time implementation, the more common case is to have them evaluated by software with correspondingly longer times. Later sections describe alternative sampling methods designed to reduce the frequency of performing transformations and thus reducing computing time.

8.1.2 RESTRICTED SAMPLING

The inverse transform method offers one convenience that no other generating method has. It allows one to generate a sample from the modified distribution

$$\tilde{F}(z) := \frac{F(z) - F(a')}{F(b') - F(a')}, \quad a \leq a' \leq z \leq b' \leq b, \tag{8.2}$$

with the same ease as one generates a sample from the parent distribution $\{F(z), a \leq z \leq b\}$. We refer to this alternative as restricted sampling. Replacing U by $U' = F(a') + [F(b') - F(a')]U$ as the argument of the inverse d.f. accomplishes this result. Except for the additional computation of $F(a')$, $F(b')$, and U', the times to generate Z from $\{F(z)\}$ and $\{\tilde{F}(z)\}$ are the same.

8.1.3 DISCRETE DISTRIBUTIONS

The inverse transform method also applies to the case of discrete Z. Suppose that

$$F(z) = \begin{cases} q_z, & z = a, a + 1, \ldots, b, \\ 0, & \text{otherwise,} \end{cases} \tag{8.3}$$

where

$$q_{a-1} = 0,$$
$$q_z < q_{z+1}, \quad z = a, a + 1, \ldots, b - 1,$$
$$q_b = 1.$$

Then an immediate result of Theorem 8.1 is that

$$Z := \min\{z : q_{z-1} < U \leq q_z, U \sim \mathcal{U}(0, 1)\}$$

has the requisite distribution. Algorithm DI describes the steps.

ALGORITHM DI

Purpose: To generate Z randomly from the distribution $\{0 < q_a < q_{a+1} < \cdots < q_b = 1\}$.
Input: $\{q_a, \ldots, q_b\}$.
Output: Z.
Method:
    Randomly generate U from $\mathcal{U}(0, 1)$ and set $Z \leftarrow a$.
    While $U > q_Z : Z \leftarrow Z + 1$.
    Return Z.

Observe that the principal work here is a sequential search in which the mean number of comparisons until success is

$$q_a + 2(q_{a+1} - q_a) + 3(q_{a+2} - q_{a+1}) + \cdots + (b - a + 1)(q_b - q_{b-1}) = 1 - a + \mathrm{E}Z,$$

revealing that the mean generation time grows linearly with E Z.

ALGORITHM ITR

Purpose: To generate Z randomly from the p.m.f. $\{p_0, p_1, \ldots\}$.
Input: p_0 and $\{c(z + 1)\}$ from Table 8.2.
Output: Z.
Method:
    $p \leftarrow p_0, \ q \leftarrow p_0$ and $Z \leftarrow 0$.
    Randomly generate U from $\mathcal{U}(0, 1)$.
    While $U > q : Z \leftarrow Z + 1; p \leftarrow pc(Z); q \leftarrow q + p$.
    Return Z.

Implementation of this method can take at least two forms. Table 8.2 lists several commonly encountered discrete distributions and shows how to evaluate each successive probability p_{z+1} recursively from p_z given p_0. Then Algorithm ITR shows how to generate Z from any one of these distributions merely by specifying p_0 and $c(z+1)$. The discrete uniform distribution $\mathcal{U}_d(\alpha, \beta)$ is the exception here, merely requiring $Z \leftarrow \lfloor \alpha + (\beta - \alpha + 1)U \rfloor$. If repeated generation occurs from the same distribution with the same p_0, then computing this quantity before generation begins, storing it, and reusing it as input save a major part of the setup cost in computing p_0. For example, computing p_0 for the binomial distribution $\mathcal{B}(r, p)$ requires the operations $p_0 \leftarrow \exp[r \ln(1 - p)]$. As later sections show, more efficient generating procedures exist for each of these distributions in both the cases of varying and constant input parameter values.

The discrete inverse transform method also allows one to generate Z from the restricted distribution $\{(q_z - q_{a'-1})/(q_{b'} - q_{a'-1})\}; z = a', a'+1, \ldots, b'\}$, where $a \leq a' < b' \leq b$ merely by initially setting $p \leftarrow p_{a'}, q \leftarrow q_{a'}$, and $Z \leftarrow a'$, and replacing U by $q_{a'-1} + U(q_{b'} - q_{a'-1})$ in Algorithm ITR. Also, the method preserves the monotone relationship between U and Z, which allows one to induce negative correlation by using U to generate Z_1 and $1 - U$ to generate Z_2. Table 3.3 illustrates the effect for the Poisson distribution.

ACCUMULATED ROUND-OFF ERROR

In practice, the repeated updating $q \leftarrow q + p$ in Algorithm ITR creates a source of numerical error that increases with $b - a$. At each step, q represents the accumulation of all previously computed p's, each of which contains round-off error due to the finite word-size of the computer employed. The larger $b - a$ is, the greater is the potential number of additions that form q, and consequently, the greater is the accumulated round-off error it contains. If, for example, this round-off error produces a q that converges to $1 - \epsilon$, for $\epsilon > 0$, and one generates U such that $1 - \epsilon < U < 1$, then the sequential search for Z continues forever. While such an event is rare, it can occur in practice. Using extended-precision arithmetic to compute p and q reduces the frequency with which such an event occurs.

8.2   CUTPOINT METHOD

If repeated sampling is to occur from a specified discrete distribution with finite support, one can exploit other features of the inverse transform method to improve efficiency. Most notable is the cutpoint method as described in Chen and Asau (1974). The present discussion follows Fishman and Moore (1984), who give a detailed analysis of the efficiency of this method. The procedure uses sequential search to determine Z, but begins the search at a point in $\{a, a+1, \ldots, b\}$ considerably closer to the solution Z on average than the point used by Algorithms DI and ITR. It achieves this result by computing and storing m cutpoints or pointers

$$
I_j := \begin{cases} \min\{i : q_i > (j-1)/m, a \leq i \leq b\}, & j = 1, \ldots, m, \\ b, & j = m+1, \end{cases} \tag{8.4}
$$

beforehand. Then each generation randomly selects an integer $L = \lceil mU \rceil$ and starts the search at I_L. Algorithm CM describes the essential steps.

Since

$$
\mathrm{pr}(I_l \leq Z \leq I_{l+1} | L = l) = 1,
$$

Table 8.2   Selected discrete distributions $(0 < p < 1)$

distribution	p.m.f.		p_0	$c(z+1) = p_{z+1}/p_z$	mean no. of comparisons
binomial $\mathcal{B}(r,p)$	$\binom{r}{z}p^z(1-p)^{r-z}$	$z = 0,1,\ldots r$	$(1-p)^r$	$\dfrac{p(r-z)}{(1-p)(z+1)}$	$1+rp$
geometric $\mathcal{G}e(p)$	$(1-p)p^z$	$z = 0,1,\ldots$	$1-p$	p	$1/(1-p)$
hypergeometric $\mathcal{H}(\alpha,\beta,n)$	$\dfrac{\binom{\alpha}{z}\binom{\beta}{n-z}}{\binom{\alpha+\beta}{n}}$	α,β,n,z integer, $\alpha+\beta \geq n > 0$, $\max(0, n-\beta-1) \leq z \leq \min(\alpha,n)$	$\dfrac{\beta!(\alpha+\beta-n)!}{(\beta-n)!(\alpha+\beta)!}$	$\dfrac{(\alpha-z)(n-z)}{(z+1)(\beta-n+z+1)}$	$1+n\alpha/(\alpha+\beta)$
negative binomial $\mathcal{NB}(r,p)$	$\dfrac{\Gamma(r+z)}{\Gamma(r)\Gamma(z+1)}(1-p)^r p^z$	$r>0, z=0,1,\ldots$	$(1-p)^r$	$\dfrac{p(r+z)}{z+1}$	$1+rp/(1-p)$
Poisson $\mathcal{P}(\lambda)$	$\dfrac{\lambda^z e^{-\lambda}}{z!}$	$\lambda>0, z=0,1,\ldots$	$e^{-\lambda}$	$\lambda/(z+1)$	$1+\lambda$
uniform $\mathcal{U}_d(\alpha,\beta)$	$\dfrac{1}{\beta-\alpha}$	$\alpha<\beta$, $z=\alpha,\alpha+1,\ldots,\beta$, α and β integer	$z = \alpha + \lfloor(\beta-\alpha)\mu\rfloor$	—	—

the maximal number of comparisons needed to determine Z is $I_{L+1} - I_L + 1$, and the expected maximal number of comparisons has the upper bound

$$
\begin{aligned}
c_m &:= \frac{1}{m}(I_2 - I_1 + 1) + \frac{1}{m}(I_3 - I_2 + 1) + \cdots + \frac{1}{m}(I_{m+1} - I_m + 1) \\
&= (I_{m+1} - I_1 + m)/m.
\end{aligned}
\tag{8.5}
$$

ALGORITHM CM

Purpose: To generate Z randomly from the distribution
  $\{0 < q_a < q_{a+1} < \cdots < q_b = 1\}$.
Input: Distribution $\{q_a, \ldots, q_b\}$, and pointers $\{I_1, \ldots, I_m\}$ as in expression (8.4).
Output: Z.
Source: Fishman and Moore (1984).
Method: Randomly generate U from $\mathcal{U}(0, 1)$.
  $L \leftarrow \lceil mU \rceil$
  $Z \leftarrow I_L$.
  While $U > q_Z : Z \leftarrow Z + 1$.
  Return Z.

The algorithm requires a total of $b - a + m + 1$ locations: $b - a + 1$ to store $\{q_a, \ldots, q_b\}$ and m to store $I_1, \ldots, I_m$.

Observe that $c_m \leq 2$ if $m \geq b - a$, ensuring that no more than two comparisons are required on average to determine Z. Also, observe that

$$
J_m := 1 + Z - I_L
$$

gives the total number of comparisons, so that

$$
\mathrm{E}J_m = 1 + \mathrm{E}Z - \mathrm{E}I_L = 1 + \mathrm{E}Z - \frac{1}{m}\sum_{j=1}^{m} I_j.
\tag{8.6}
$$

Table 8.3 lists c_m and $\mathrm{E}J_m$ using the eight-point distribution in Fishman (1978, p. 459). As an illustration, $m = 4$ leads to $I_1 = 1$, $I_2 = 4$, $I_3 = 5$, and $I_4 = 7$. Also, $\mathrm{E}Z = 5.31$ and $\mathrm{E}I_L = 17/4 = 4.25$, so that $\mathrm{E}J_4 = 2.06$. If space is not an issue, one may regard $\mathrm{E}J_m - 1$ as the cost of sampling. Algorithm CMSET shows how to compute the points $I_1, \ldots, I_m$ in a presampling step.

ALGORITHM CMSET

Purpose: To compute pointers for use in Algorithm CM.
Input: Distribution $\{q_a, \ldots, q_b\}$ and number of pointers m.
Output: $I_1, \ldots, I_m$ as in expression (8.4).
Source: Fishman and Moore (1984).
Method:
> $i \leftarrow 0, j \leftarrow a - 1$ and $A \leftarrow 0$.
>   While $i < m$:
>>   While $A \leq i$:
>>> $j \leftarrow j + 1$.
>>> $A \leftarrow mq_j$.
>> $i \leftarrow i + 1$.
>> $I_i \leftarrow j$.
>   Return $I_1, \ldots, I_m$.

For repeated sampling from a continuous distribution, Ahrens (1993) describes a procedure that merges the cutpoint method with the acceptance–rejection method and whose computing time is negligibly influenced by the form of the distribution.

8.2.1   RESTRICTED SAMPLING

The cutpoint method offers several additional features that make it attractive in practice. Of these, restricted sampling is one of the most important. Suppose that one wants to generate a sample from the modified distribution $\{q_i = (q_i - q_{a'-1})/(q_{b'} - q_{a'-1})$; $a \leq a' \leq i \leq b' \leq b\}$. Generating U from $\mathcal{U}(q_{a'-1}, q_{b'})$ instead of from $\mathcal{U}(0, 1)$ in Algorithm CM achieves this result while preserving the efficiency of the method. In particular, once the original tables of $\{q_a, \ldots, q_b\}$ and $\{I_1, \ldots, I_m\}$ are computed and stored, one can sample from them with varying $a' < b'$ without incurring any new cost. No other method of sampling from tables offers this advantage.

Table 8.3   Example of cutpoint method

i	p_i	q_i	m	$m + b$	c_m	$\mathrm{E}J_m$
1	.01	.01	1	9	8.00	5.31
2	.04	.05	2	10	4.50	3.31
3	.07	.12	3	11	3.33	2.31
4	.15	.27	4	12	2.75	2.06
5	.28	.55	5	13	2.40	1.71
6	.19	.74	6	14	2.17	1.64
7	.21	.95	7	15	2.00	1.45
8	.05	1.00	8	18	1.88	1.44
			16	24	1.44	1.19

If the p.m.f. of interest changes after one generates each sample, then neither the cutpoint method nor the alias method (Section 8.4) can guarantee an O(1) generating time. For a commonly encountered form of changing k-cell p.m.f., Fishman and Yarberry (1997) describe a generating procedure that takes O($\log_2 k$) time. Also, see Fishman (1996, Section 3.31).

8.3   COMPOSITION METHOD

This approach to sampling relies on mixtures of random variables to achieve its effectiveness. Let Z denote a random variable on $[a, b]$ with d.f. F that has the decomposition

$$F(z) = \sum_{i=1}^{r} \omega_i F_i(z),$$

where

$$0 < \omega_i < 1, \quad i = 1, \dots, r,$$

$$\sum_{i=1}^{r} \omega_i = 1,$$

and $\{F_1(z), a \le a_1 \le z \le b_1 \le b\}, \dots, \{F_r(z), a \le a_r \le z \le b_r \le b\}$ also denote d.f.s. To generate Z, one performs the steps in Algorithm COMPM.

ALGORITHM COMPM

Purpose: To generate Z randomly from the distribution $\{F(z), a \le z \le b\}$.
Input: $\{\omega_i; i = 1, \dots, r\}$ and $\{F_i(z); a_i \le z \le b_i\}$ for $i = 1, \dots, r$.
Method:
   Randomly generate I from the mixing distribution $\{\omega_1, \dots, \omega_r\}$.
   Randomly generate Z from the distribution $\{F_I(z), a_I \le z \le b_I\}$.
   Return Z.

At least two motivations exist for using this method. First, generation directly from F may be difficult and expensive, whereas generation from $\{\omega_1, \dots, \omega_r\}$ and the chosen F_i may be simple and inexpensive. For example, suppose that

$$F(z) = \alpha(1 - e^{-z/\beta_1}) + (1 - \alpha)(1 - e^{-z/\beta_2})$$
$$= 1 - \alpha e^{-z/\beta_1} - (1 - \alpha)e^{-z/\beta_2}, \quad 0 < \alpha < 1, \quad \beta_1, \beta_2 > 0, \quad z \ge 0.$$

While the inverse transform method does not apply easily here, the composition method, with

$$\omega_1 := \alpha, \qquad\qquad \omega_2 := 1 - \alpha,$$
$$F_1(z) := 1 - e^{-z/\beta_1}, \qquad F_2(z) := 1 - e^{-z/\beta_2},$$

does.

The second motivation relates solely to the relative generating cost. A particularly good example arises in randomly sampling from the exponential distribution $\mathcal{E}(1)$. Recall that the inverse transform method (Table 8.1) requires the evaluation of a logarithm. If a computer contains hardware for effecting this evaluation, then the inverse transform method is usually the most efficient approach. If software, by way of a library subroutine, effects the evaluation, the computation time increases. In this case, the *rectangle–wedge–tail method* in MacLaren et al. (1964), based on the composition method, usually can generate an exponential sample in less time. Define $\lambda > 0, r = 2k + 1$, and

$$f(z) = e^{-z} = \omega_1 f_1(z) + \cdots + \omega_{2k+1} f_{2k+1}(z), \qquad z \geq 0,$$
$$\omega_j = \lambda e^{-j\lambda},$$

with

$$f_j(z) = 1/\lambda \quad (j-1)\lambda \leq z \leq j\lambda, \qquad j = 1, \ldots, k,$$
$$\omega_j = e^{-(j-k)\lambda}(e^{\lambda} - 1 - \lambda),$$
$$f_j(z) = (e^{-z+(j-k)\lambda} - 1)/(e^{\lambda} - 1 - \lambda) \quad (j - k - 1)\lambda \leq z \leq (j - k)\lambda,$$
$$j = k + 1, \ldots, 2k,$$
$$\omega_{2k+1} = e^{-k\lambda},$$
$$f_{2k+1}(z) = e^{-z+k\lambda}, \qquad z \geq k\lambda.$$

Figure 3.1 shows the decomposition. Here $\{f(z)\}$ has $2k + 1$ components of which the first k, $f_1, \ldots, f_k$, are rectangular (uniform) p.d.f.s on $[0, \lambda], [\lambda, 2\lambda], \ldots, [(k - 1)\lambda, k\lambda]$, respectively. Moreover, the probability of sampling from one of these is

$$\omega_1 + \cdots + \omega_k = \lambda e^{-\lambda}(1 - e^{-k\lambda})/(1 - e^{-\lambda}). \tag{8.7}$$

Components $k + 1, \ldots, 2k$ are the wedge-shaped p.d.f.s $f_{k+1}, \ldots, f_{2k}$, from which one can sample by the acceptance–rejection technique, which Section 8.5 describes in detail. Although the cost of this sampling is higher than the cost for $f_1, \ldots, f_k$, the frequency with

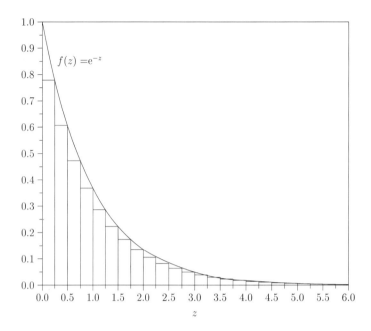

Figure 8.1 Exponential generation using the rectangle–wedge–tail method

which this more expensive sampling occurs is

$$\omega_{k+1} + \cdots + \omega_{2k} = (1 - e^{-k\lambda})[1 - \lambda e^{-\lambda}/(1 - e^{-\lambda})], \tag{8.8}$$

which can be made small relative to $\omega_1 + \cdots + \omega_k$ by an appropriate choice of λ and k.

If the tail component $2k + 1$ is chosen, then $Z \geq k\lambda$ with probability 1. Since

$$\mathrm{pr}(Z \leq z | Z > k\lambda) = [1 - e^{-z} - (1 - e^{-k\lambda})]/e^{-k\lambda} = 1 - e^{-z+k\lambda}, \quad z \geq k\lambda,$$

one can proceed as in Algorithm RWTE. In particular, one iterates until $J \leq 2k$ before generating Z. The mean number of iterations is $1/(1 - \omega_{2k+1})$. Actual implementation of this algorithm depends heavily on the features of the computer to be employed.

ALGORITHM RWTE

Purpose: To generate Z from randomly $\mathcal{E}(1)$ using the rectangle–wedge–tail method.
Input: $\lambda, \{\omega_1, \ldots, \omega_{2k+1}\}, \{f_{k+j}(z), (j-1)\lambda \leq z \leq j\lambda\}$ for $j = 1, \ldots, k$.
Method:

    $I \leftarrow -1$.
    Repeat
        randomly generate J from the distribution $\{\omega_1, \ldots, \omega_{2k+1}\}$

$$I \leftarrow I + 1$$

until $J \le 2k$.

If $J \le k$:

  Randomly generate U from $\mathcal{U}(0, 1)$.

  $Z \leftarrow [kI + (J - 1) + U]\lambda$.

  Return Z.

Otherwise:

  Randomly generate Y from the distribution $\{f_J(z), (J - k - 1)\lambda \le z \le (J - k)\lambda\}$.

  $Z \leftarrow kI\lambda + Y$.

  Return Z.

This technique requires one to set two parameters k and λ. As k increases for a fixed λ, the probability of iteration ω_{2k+1} decreases, but the size of the tables needed for choosing J from $\{\omega_1, \ldots, \omega_{2k+1}\}$ increases. As λ increases, $\omega_1 + \cdots + \omega_k$ in (8.7) eventually decreases, which is undesirable from the viewpoint of computational efficiency. As an illustration of the consequences of two choices, $k = 80$ and $\lambda = .05$ leads to

$$\omega_1 + \cdots + \omega_{80} = .9573,$$
$$\omega_{81} + \cdots + \omega_{160} = .0244,$$
$$\omega_{161} = .0183,$$

revealing the dominance of sampling from one of the 80 rectangular distributions. Also, note that the mean number of iterations is $1/(1 - e^{-80 \times .05}) = 1.0187$. The choice $k = 160$ and $\lambda = .05$ leads to

$$\omega_1 + \cdots + \omega_{160} = .9749,$$
$$\omega_{161} + \cdots + \omega_{320} = .0248,$$
$$\omega_{321} = .0003,$$

with a mean number of iterations $= 1/(1 - e^{-160 \times .05}) = 1.0003$. For a detailed account of how to prepare a highly efficient computer program to execute this algorithm, see Ahrens and Dieter (1974a).

8.4 ALIAS METHOD

The composition method of Section 8.3 also offers an alternative technique for generating a sample from a table. The procedure requires exactly one comparison, regardless of the size of the table (Walker 1974, 1977). This implausible, but nevertheless true, result makes this approach at least as appealing as the cutpoint method from the viewpoint of computational efficiency. Moreover, it is an application of the composition method of Section 8.3. Let Z denote a random variable that assumes value i in $\mathcal{T} = \{a, a + 1, \ldots, b\}$ with probability p_i. To generate a Z, the alias method first decomposes the support $\mathcal{T}$ into $b - a + 1$ 2-tuples $(A_a, B_a), (A_{a+1}, B_{a+1}), \ldots, (A_b, B_b)$ such that

$$\mathcal{T} = \cup_{i=a}^b (A_i, B_i)$$

and assigns probabilities $P(A_i) > 0$ to A_i and $P(B_i) = 1 - P(A_i)$ to B_i such that

$$p_i = \frac{1}{b - a + 1} \sum_{j=a}^b f_j(i), \tag{8.9}$$

where $\{f_j(i) \; i = a, a + 1, \ldots, b\}$ is a p.m.f. with

$$f_j(i) = \begin{cases} P(A_j) & \text{if } i = A_j, \\ P(B_j) & \text{if } i = B_j, \\ 0 & \text{otherwise.} \end{cases}$$

Expression (8.9) implies that selecting a 2-tuple with probability $1/(b - a + 1)$ and then choosing between A_I and B_I by Bernoulli sampling produces the desired result. Algorithm ALIAS describes the procedure.

ALGORITHM ALIAS

Purpose: To generate Z randomly from the p.m.f. $\{p_a, \ldots, p_b\}$.
Input: $\{A_a, \ldots, A_b\}, \{B_a, \ldots, B_b\}, \{P(A_a), \ldots, P(A_b)\}$.
Method:
  Randomly generate U_1 from $\mathcal{U}(0, 1)$.
  $Y \leftarrow a + \lfloor (b - a + 1)U_1 \rfloor$.
  Randomly generate U_2 from $\mathcal{U}(0, 1)$.
  If $U_2 \leq P(A_Y)$; return $Z \leftarrow A_Y$; otherwise, return $Z \leftarrow B_Y$.

There remains the issue of establishing that a decomposition satisfying expression (8.9) exists and of describing a procedure for computing this decomposition. Theorem 8.2 provides the proof for $a = 1$, which, without loss of generality, applies for $a > 1$ with n replaced by $b - a + 1$.

Theorem 8.2.

Any discrete distribution with a finite number n of mass points can be represented as an equiprobable mixture of n distributions, each of which has, at most, two mass points.

For the proof, see Kronmal and Peterson (1979), Dieter (1982), or Fishman (1996).

ALGORITHM ALSET

Purpose: To compute the setup for the alias method.
Input: $\{p_i; i = a, a + 1, \ldots, b\}$.
Output: $\{A_i, B_i, P(A_i); i = a, a + 1, \ldots, b\}$.
Source: Kronmal and Peterson (1979).
Method: $k \leftarrow (b - a + 1), \mathcal{S} \leftarrow \varnothing, \mathcal{G} \leftarrow \varnothing,$ and $i \leftarrow a$.

> While $i \leq b$:
>> $r_i \leftarrow k p_i$.
>> If $r_i < 1, \mathcal{S} \leftarrow \mathcal{S} + \{i\}$; otherwise, $\mathcal{G} \leftarrow \mathcal{G} + \{i\}$.
>> $i \leftarrow i + 1$.
>
> $i \leftarrow a$.
> While $i \leq b$:
>> Select and remove j from $\mathcal{S}$.
>> $A_i \leftarrow j; P(A_i) \leftarrow r_j$.
>> If $\mathcal{G} \neq \varnothing$:
>>> Select and remove l from $\mathcal{G}$.
>>> $B_i \leftarrow l; r_l \leftarrow r_l - (1 - r_j)$.
>>> If $r_l \leq 1, \mathcal{S} \leftarrow \mathcal{S} + \{l\}$; otherwise, $\mathcal{G} \leftarrow \mathcal{G} + \{l\}$.
>> $i \leftarrow i + 1$.

8.4.1 SETTING UP THE TABLES

Algorithm ALSET describes how to compute $\{A_i, B_i, P(A_i); i = 1, \ldots, b - a + 1\}$. To illustrate the procedure, consider the eight-point distribution in Table 8.3 for which Algorithm ALSET initially give:

$$r_1 \leftarrow .08,$$
$$r_2 \leftarrow .32,$$
$$r_3 \leftarrow .56,$$
$$r_4 \leftarrow 1.20,$$
$$r_5 \leftarrow 2.24,$$
$$r_6 \leftarrow 1.52,$$
$$r_7 \leftarrow 1.68,$$
$$r_8 \leftarrow .40,$$
$$\mathcal{S} \leftarrow \{1, 2, 3, 8\}, \qquad \mathcal{G} \leftarrow \{4, 5, 6, 7\}.$$

Then

Step 1. Pick 1 and 4,

$A_1 \leftarrow 1, P(A_1) \leftarrow .08, B_1 \leftarrow 4, r_4 \leftarrow 1.20 - (1 - .08) = .28,$
$S = \{2, 3, 4, 8\}, \mathcal{G} = \{5, 6, 7\}.$

Step 2. Pick 2 and 5,

$A_2 \leftarrow 2, P(A_2) \leftarrow .32, B_2 \leftarrow 5, r_5 \leftarrow 2.24 - (1 - .32) = 1.56,$
$S = \{3, 4, 8\}, \mathcal{G} = \{5, 6, 7\}.$

Step 3. Pick 3 and 5,

$A_3 \leftarrow 3, P(A_3) \leftarrow .56, B_3 \leftarrow 5, r_5 \leftarrow 1.56 - (1. - 56) = 1.12,$
$S = \{4, 8\}, \mathcal{G} = \{5, 6, 7\}.$

Step 4. Pick 4 and 5,

$A_4 \leftarrow 4, P(A_4) \leftarrow .28, B_4 \leftarrow 5, r_5 \leftarrow 1.12 - (1 - .28) = .40,$
$S = \{5, 8\}, \mathcal{G} = \{6, 7\}.$

Step 5. Pick 5 and 6,

$A_5 \leftarrow 5, P(A_5) \leftarrow .40, B_5 \leftarrow 6, r_6 \leftarrow 1.52 - (1 - .40) = .92,$
$S = \{6, 8\}, \mathcal{G} = \{7\}.$

Step 6. Pick 6 and 7,

$A_6 \leftarrow 6, P(A_6) \leftarrow .92, B_6 \leftarrow 7, r_7 \leftarrow 1.68 - (1 - .92) = 1.60,$
$S = \{8\}, \mathcal{G} = \{7\}.$

Step 7. Pick 8 and 7,

$A_7 \leftarrow 8, P(A_7) \leftarrow .40, B_7 \leftarrow 7, r_7 \leftarrow 1.60 - (1 - .40) = 1.00,$
$S = \{7\}, \mathcal{G} = \varnothing.$

Step 8. $A_8 \leftarrow 7, P(A_8) \leftarrow 1.00,$
$S = \varnothing, \mathcal{G} = \varnothing.$

Table 8.4 displays the corresponding probabilities where the row averages give the original p.m.f. $\{p_1, \ldots, p_8\}$ and the column entries sum to unity for each stratum $1, \ldots, 8$.

Table 8.4   Example of the alias method[†] $P(A_j)$ and $P(B_j)$

i/j	1	2	3	4	5	6	7	8	p_i
$A_1 = 1$	.08								.01
$A_2 = 3$		.32							.04
$A_3 = 3$			.56						.07
$B_1 = A_4 = 4$	.92			.28					.15
$B_2 = B_3 = B_4 = A_5 = 5$		.68	.44	.72	.40				.28
$B_5 = A_6 = 6$					.60	.92			.19
$B_6 = B_7 = A_8 = 7$						.08	.60	1.00	.21
$A_7 = 8$							.40		.05
$P(A_j) + P(B_j)$	1.00	1.00	1.00	1.00	1.00	1.00	1.00	1.00	1.00

[†] $P(A_j) =$ first entry in column j; $P(B_j) =$ second entry in column j; see expression (8.9) for computation of $\{p_1, \ldots, p_8\}$.

8.5 ACCEPTANCE–REJECTION METHOD

Although every d.f. has a unique inverse, this inverse may not exist in a form convenient for efficient computation. For example, the Erlang p.d.f.

$$f(y) = y^{\alpha-1}e^{-y}/(\alpha-1)!, \qquad \alpha = \text{integer} > 0, \ y \geq 0,$$

leads to the d.f. (Feller 1971, p. 11)

$$F(z) = \frac{1}{(\alpha-1)!} \int_0^z y^{\alpha-1}e^{-y}\mathrm{d}y = 1 - e^{-z} - e^{-z}\sum_{i=1}^{\alpha-1}(z^i/i!), \quad z \geq 0. \qquad (8.10a)$$

Although the bisection and Newton–Raphson methods can provide a solution, their computing times grow at least linearly with α. A more direct approach employs the expansion of the inverse d.f. (Fisher and Cornish 1960)

$$
\begin{aligned}
F^{-1}(u) = {} & 2\alpha + 2\zeta\alpha^{1/2} + 2(\zeta^2 - 1)/3 + (\zeta^3 - 7\zeta)/18\alpha^{1/2} \\
& - (6\zeta^4 + 14\zeta^2 - 32)/810\alpha + (9\zeta^5 + 256\zeta^3 - 433\zeta)/19440\alpha^{3/2} \\
& + (12\zeta^6 - 243\zeta^4 - 923\zeta^2 + 1472)/102060\alpha^2 \qquad (8.10b) \\
& - (3753\zeta^7 + 4353\zeta^5 - 289517\zeta^3 - 289717\zeta)/73483200\alpha^{5/2} \\
& + O(1/\alpha^3) \quad \text{as } \alpha \to \infty, \quad 0 \leq u \leq 1,
\end{aligned}
$$

where

$$\zeta = \Phi^{-1}(u).$$

Nevertheless, the time to compute a truncated version of expression (8.10b) and to discount negative values, which rarely occur, is not inconsequential. The acceptance-rejection method allows one to sidestep this issue by exploiting the properties of conditional probability.

Theorem 8.3 (von Neumann 1951).
  Let $\{f(z), a \leq z \leq b\}$ denote a p.d.f. with factorization

$$f(z) = cg(z)h(z),$$

where

$$h(z) \geq 0, \qquad \int_a^b h(z)\mathrm{d}z = 1, \qquad (8.11)$$

and

$$c = \sup_{z}[f(z)/h(z)] < \infty,$$

$$0 \le g(z) \le 1.$$

Let Z denote a random variable with p.d.f. $\{h(z)\}$ and let U be from $\mathcal{U}(0, 1)$. If $U \le g(Z)$, then Z has the p.d.f. $\{f(z)\}$.

PROOF.   The random variables U and Z have the joint p.d.f.

$$f_{U,Z}(u, z) = h(z), \qquad 0 \le u \le 1, \quad a \le z \le b.$$

Then Z has the conditional p.d.f.

$$h_Z(z|U \le g(Z)) = \frac{\int_0^{g(z)} f_{U,Z}(u, z)du}{\text{pr}[U \le g(Z)]}.$$

Observe that $c = \int_a^b g(z)h(z)dz$. Since

$$\int_0^{g(z)} f_{U,Z}(u, z)du = h(z)g(z)$$

and

$$\text{pr}[U \le g(Z)] = \int_a^b h(z)g(z)dz = 1/c,$$

it follows, that

$$h_Z(z|U \le g(Z)) = cg(z)h(z) = f(z),$$

which proves the theorem.   □

Since $g(z) = f(z)/c\, h(z)$, the assignment $c = \sup_z[f(z)/h(z)]$ maximizes the acceptance probability $g(z)$ for each $z \in [a, b]$. An analogous result holds for discrete distributions.

The significance of this result becomes most apparent when one can generate Z from $\{h(z)\}$ efficiently and c is close to unity. Algorithm AR describes how to implement the approach as a sequential sampling experiment that terminates with the first success ($U \le g(Z)$). Since $\text{pr}[U \le g(Z)] = 1/c$ is the success probability on any iteration or trial, and since all trials are independent, the number of trials until success is a geometric random variable with mean c, which explains why a small c is desirable. This procedure describes one of the first probabilistic algorithms used on a computer.

ALGORITHM AR

Purpose: To generate Z randomly from the distribution $\{f(z) = cg(z)h(z),\ a \leq z \leq b\}$.
Input: $\{g(z), h(z);\ a \leq z \leq b\}$.
Method:

> Repeat
>> randomly generate Z from the distribution $\{h(z)\}$
>> randomly generate U from $\mathcal{U}(0, 1)$
>> evaluate $g(Z)$
> until $U \leq g(Z)$.
> Return Z.

❏ EXAMPLE 8.1

To illustrate the method, consider generating Z from the half-normal distribution with p.d.f.

$$f(z) = \sqrt{2/\pi}\, \mathrm{e}^{-z^2/2}, \qquad 0 \leq z < \infty. \tag{8.12a}$$

Writing this density in the form

$$f(z) = \sqrt{2\mathrm{e}/\pi}\ \mathrm{e}^{-(z-1)^2/2}\mathrm{e}^{-z} \tag{8.12b}$$

leads to the factorization

$$h(z) = \mathrm{e}^{-z},$$
$$g(z) = \mathrm{e}^{-(z-1)^2/2}, \tag{8.12c}$$

and

$$c = \sqrt{2\mathrm{e}/\pi} = 1.3155.$$

Figure 8.2a shows the acceptance region $(u \leq g(-\ln v))$ and the rejection region $(u > g(-\ln v))$.

❏

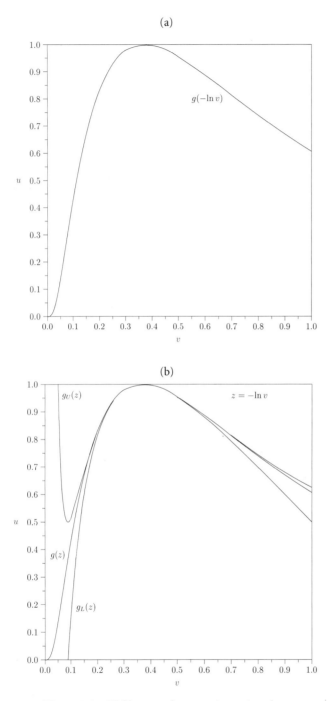

Figure 8.2 Half–normal generation using the acceptance–rejection method

At least two implementations are possible. Version 1, which uses the inverse transform method to generate the exponential deviate Z, requires a logarithmic evaluation for Z and an exponential evaluation for $g(Z)$. Alternatively, version 2 generates Y and Z from $\mathcal{E}(1)$ and avoids the exponentiation. If a highly efficient algorithm is available for generating exponential samples, then version 2 can be expected to be less costly.

VERSION 1

Repeat
        randomly generate X and V from $\mathcal{U}(0, 1)$
        $Z \leftarrow -\ln V$
        $g(Z) \leftarrow e^{-(Z-1)^2/2}$
until $X \leq g(Z)$.
Return Z.

VERSION 2

Repeat
        randomly generate Y and Z from $\mathcal{E}(1)$
until $Y \geq (Z - 1)^2/2$.
Return Z.

Half-normal generation provides one method for producing a sample from $\mathcal{N}(0, 1)$. Let W denote a random variable from $\mathcal{U}(0, 1)$. Then

$$
Z^* = \begin{cases} -Z & \text{if } W \leq \dfrac{1}{2}, \\ Z & \text{otherwise,} \end{cases}
$$

has the standard normal p.d.f. (Table 8.1).

8.5.1 SQUEEZE METHOD

On average, success in Version 1 of the half-normal example requires $2\sqrt{2e/\pi} = 2.6310$ function evaluation, one logarithmic evaluation, and one exponential evaluation per trial. Avoiding these can lead to a significant reduction in computing time. The squeeze method (Marsaglia 1977) allows one to do this. Suppose there exist simple functions $\{g_L(z), a \leq z \leq b\}$ and $\{g_U(z), a \leq z \leq b\}$ such that

$$
g_L(z) \leq g(z) \leq g_U(z) \quad a \leq z \leq b, \tag{8.13}
$$

where expression (8.11) defines $g(z)$. Here the modifier "simple" means that the time to evaluate $g_L(z)$ and $g_U(z)$ is less than the time to evaluate $g(z)$. Note that for some z, $g_L(z)$

may assume negative values or $g_U(z)$ may exceed unity. Algorithm ARS shows how to use these bounding functions to reduce computing time. On each trial, this algorithm leads to

$$\text{pr}[\text{evaluating } g(Z)] = \text{pr}[g_L(Z) < U \le g_U(Z)].$$

The checks $U \le g_L(Z)$ and $U > g_U(Z)$ are sometimes called preliminary tests or pretests.

ALGORITHM ARS

Purpose: To generate Z randomly from the p.d.f. $\{f(z), a \le z \le b\}$ using the acceptance–rejection and squeeze methods.

Input: p.d.f. $\{h(z), a \le z \le b\}$, functions $\{g_L(z), g(z), g_U(z); a \le z \le b\}$.

Method:
    Repeat
        randomly generate X from $\mathcal{U}(0, 1)$
        randomly generate Z from $\{h(z)\}$
        if $X \le g_L(Z)$, accept
        otherwise:
        if $X \le g_U(Z)$: if $X \le g(Z)$, accept
        otherwise: reject
    until accept.
    Return Z.

Let S denote the event $X \le g(Z)$, S_L the event $X \le g_L(Z)$, and S_U the event $X \le g_U(Z)$. Then

$$\text{pr}(S_L) = \int_a^b \max[0, g_L(z)]h(z)\mathrm{d}z,$$

$$\text{pr}(S_U) = \int_a^b \min[1, g_U(z)]h(z)\mathrm{d}z,$$

$$(8.14)$$

and, as before,

$$\text{pr}(S) = \int_a^b g(z)h(z)\mathrm{d}z = 1/c .$$

Then on a single trial

$$\text{pr}[\text{evaluating } g(Z)] = \text{pr}(S_U) - \text{pr}(S_L).$$

$$(8.15)$$

If N denotes the number of evaluations of $g(Z)$ until success, then

$$\mathrm{E}N = \text{pr}(\bar{S}_L, S_U|\bar{S})[\frac{1}{\text{pr}(S)} - 1] + \text{pr}(\bar{S}_L, S_U|S),$$

$$(8.16)$$

where $\bar{S}_L$ and $\bar{S}$ denote the events $X > g_L(Z)$ and $X > g(Z)$, respectively. Since

$$\mathrm{pr}(\bar{S}_L, S_U | \bar{S}) = \frac{\mathrm{pr}(S_U) - \mathrm{pr}(S)}{1 - \mathrm{pr}(S)}$$

and

$$\mathrm{pr}(\bar{S}_L, S_U | S) = \frac{\mathrm{pr}(S) - \mathrm{pr}(S_L)}{\mathrm{pr}(S)}, \tag{8.17}$$

it follows that

$$EN = \frac{\mathrm{pr}(S_U) - \mathrm{pr}(S_L)}{\mathrm{pr}(S)}.$$

To reduce the frequency of exponential evaluation, the squeeze method often employs the inequalities (Exercise 8.18)

$$1 - x \le e^{-x} \le 1 - x + x^2/2, \qquad x \ge 0. \tag{8.18}$$

To reduce the frequency of logarithmic evaluation, it often employs (Exercise 8.19)

$$\frac{x - 1}{x} \le \ln x \le -1 + x, \qquad x \ge 0, \tag{8.19}$$

or (Exercise 8.20)

$$x - 1/x \le 2 \ln x \le -3 + 4x - x^2, \qquad 0 < x \le 1. \tag{8.20}$$

Example 8.2 illustrates the benefit of using inequalities (8.18).

❏ EXAMPLE 8.2

For the half-normal distribution in Example 8.1 with $g(z) = e^{-(z-1)^2/2}$, inequalities (8.18) leads to the lower and upper bounds

$$g_L(z) = 1 - (z - 1)^2/2 \tag{8.21}$$

and

$$\begin{aligned} g_U(z) &= 1 - (z - 1)^2/2 + (z - 1)^4/8 \\ &= g_L(z) + (z - 1)^4/8, \end{aligned} \tag{8.22}$$

respectively. Moreover,

$$\text{pr}(S_L) = 1/2 + (1 + \sqrt{2})e^{-(1+\sqrt{2})} = .7159, \qquad (8.23a)$$

$$\text{pr}(S_U) = 13/8 - 16e^{-3} = .8284, \qquad (8.23b)$$

$$\text{pr}[\text{evaluating } g(Z) \text{ on a trial}] = \text{pr}(S_U) - \text{pr}(S_L) = .1125, \qquad (8.23c)$$

and the mean number of evaluations of $g(Z)$ until success is $EN = .1480$. Figure 8.2b shows $\{g_L(z)\}$, $\{g(z)\}$, and $\{g_U(z)\}$ for $z = -\ln v$ and $0 < v < 1$. In summary, incorporating the squeeze method into the acceptance–rejection method in this example reduces the mean number of evaluations of $g(Z)$ from 1.3155 to 0.1340. Of course, some of this saving is expended in the two additional pretests for S_L and $\bar{S}_U$.

8.5.2 THEORY AND PRACTICE

As with the inverse transform method, the acceptance–rejection method is a theoretically exact approach to sampling, provided that a source of i.i.d. samples are available from $\mathcal{U}(0, 1)$. Since every Monte Carlo experiment relies on a sequence of pseudorandom numbers to approximate that i.i.d. sequence from $\mathcal{U}(0, 1)$, an error of approximation always exists. This error has more serious implications for the acceptance–rejection method than for the inverse transform method. Indeed, it is also more serious for the ratio-of-uniforms method (Section 8.6) and any other procedure that demands more than one uniform deviate per draw. Since a careful choice of pseudorandom number generator (PNG) can reduce the severity of the problem, being familiar with the properties of one's chosen PNG is well advised. Fishman (1996, Section 7.6.1) illustrates the character of the error.

8.6 RATIO-OF-UNIFORMS METHOD

The acceptance–rejection method incurs four forms of work: generation of Z from $\{h(z)\}$, generation of U from $\mathcal{U}(0, 1)$, evaluation of $g(Z)$, and repeated generation until success occurs. The squeeze method reduces the frequency of evaluating $g(Z)$. The ratio-of-uniforms method replaces the generation of Z and U by the generation of $Z = a_1 + a_2 Y/X$, where (X, Y) is uniformly distributed over a bounded region where a_1, a_2 and the shape of the region depend on $\{f(z)\}$. In principle, the work to generate (X, Y) is considerably less, thus potentially reducing the computing time required to obtain a sample from $\{f(z)\}$. Theorem 8.4 represents a generalization of the original proposal of Kinderman and Monahan (1977) for the ratio-of-uniforms method and is used extensively by Ahrens and Dieter (1991) and Stadlober (1989, 1991) for generation from discrete distributions.

Theorem 8.4.

Let $\{r(z), -\infty < z < \infty\}$ denote a nonnegative integrable function and let $\mathcal{Q}$ denote the region

$$\mathcal{Q} = \mathcal{Q}(a_1, a_2) = \{(x, y) : 0 \le x \le r^{\frac{1}{2}}(a_1 + a_2 y/x), -\infty < a_1 < \infty, a_2 > 0\}. \quad (8.24)$$

Let (X, Y) be uniformly distributed on a bounded region $\mathcal{P} \supseteq \mathcal{Q}$ and set $W = X^2$ and $Z = a_1 + a_2 Y/X$. If $W \le r(Z)$, then Z has the distribution

$$f_Z(z) = r(z)/d, \quad -\infty < z < \infty, \quad (8.25)$$

where

$$d := 2a_2|\mathcal{Q}| = \int_{-\infty}^{\infty} r(z)\mathrm{d}z. \quad (8.26)$$

PROOF. If $(X, Y) \in \mathcal{Q}$, then X and Y have the joint p.d.f.

$$f_{X,Y}(x, y) = 1/|\mathcal{Q}|, \quad (x, y) \in \mathcal{Q},$$

and

$$f_{W,Z}(w, z) = J f_{X,Y}(\sqrt{w}, (z - a_1)\sqrt{w}/a_2), \quad 0 \le w \le r(z), \quad -\infty < z < \infty,$$

where J denotes the Jacobian

$$J = \left\| \begin{array}{cc} \dfrac{\partial x}{\partial w} & \dfrac{\partial x}{\partial z} \\[2mm] \dfrac{\partial y}{\partial w} & \dfrac{\partial y}{\partial z} \end{array} \right\| = \left\| \begin{array}{cc} \dfrac{1}{2\sqrt{w}} & 0 \\[2mm] \dfrac{z - a_1}{2a_2\sqrt{w}} & \dfrac{\sqrt{w}}{a_2} \end{array} \right\| = \dfrac{1}{2a_2},$$

so that

$$f_Z(z) = \int_0^{r(z)} f_{W,Z}(w, z)\mathrm{d}w = \frac{r(z)}{2a_2|\mathcal{Q}|}, \quad -\infty < z < \infty,$$

and

$$\int_{-\infty}^{\infty} f_Z(z)\mathrm{d}z = \frac{1}{2a_2|\mathcal{Q}|} \int_{-\infty}^{\infty} r(z)\mathrm{d}z = 1. \qquad \square$$

The equations

$$x(z) = r^{\frac{1}{2}}(z)$$

and

$$y(z) = (z - a_1)x(z)/a_2 = (z - a_1)r^{\frac{1}{2}}(z)/a_2$$

define the boundary of $\mathcal{Q}$, and for specified a_1 and a_2 every point in $\mathcal{Q}$ lies in the rectangle

$$\mathcal{P} := \{(x, y) : 0 \le x \le x^*, \ y_* \le y \le y^*\}, \tag{8.27a}$$

where

$$x^* := \sup_z r^{\frac{1}{2}}(z), \tag{8.27b}$$

$$y_* := \inf_z [(z - a_1)r^{\frac{1}{2}}(z)/a_2], \tag{8.27c}$$

and

$$y^* := \sup_z [(z - a_1)r^{\frac{1}{2}}(z)/a_2]. \tag{8.27d}$$

In order for y^* to be finite, $\{z^2 r(z)\}$ must be integrable.

The motivation behind the ratio-of-uniforms method is now clear. Suppose one sets $r(z) = df(z)$ and generates (X, Y) uniformly on $\mathcal{P}$. If $X \le r^{\frac{1}{2}}(a_1 + a_2 Y/X)$, then $(X, Y) \in \mathcal{Q}$, and $Z = a_1 + a_2 Y/X$ has the desired p.d.f. $\{f(z)\}$. Algorithm ROU describes the steps.

ALGORITHM ROU

Purpose: To generate Z randomly from the p.d.f. $\{f(z), -infty < z < \infty\}$ using the ratio-of-uniforms method.

Input: Function $\{r(z) = df(z), -\infty < z < \infty\}$, constants $a_1 \in (-\infty, \infty)$ and $a_2 \in (0, \infty)$.

Setup: $x^* \leftarrow \sup_z r^{\frac{1}{2}}(z), \ y_* \leftarrow \inf_z [(z - a_1)r^{\frac{1}{2}}(z)/a_2]$, and $y^* \leftarrow \sup_z [(z - a_1)r^{\frac{1}{2}}(z)/a_2].$

Method:
  Repeat
    randomly generate X from $\mathcal{U}(0, x^*)$
    randomly generate Y from $\mathcal{U}(y_*, y^*)$
    $Z \leftarrow a_1 + a_2 Y/X$
  until $X^2 \le r(Z)$.
  Return Z.

Since $\text{pr}(\text{accept } Z) = \text{pr}[X \le r^{\frac{1}{2}}(a_1 + a_2 Y/X)] = \text{pr}[(X, Y) \in \mathcal{Q}] = |\mathcal{Q}|/|\mathcal{P}|$, the mean number of trials until success has the form

$$|\mathcal{P}|/|\mathcal{Q}| = 2a_2 x^*(y^* - y_*)/d. \tag{8.28}$$

Generating (X, Y) from $\mathcal{P}$ takes little time compared to the potentially greater time to generate Z from $\{h(z)\}$ in expression (8.11) using the acceptance–rejection method. While the inclusive region $\mathcal{P}$ may assume a shape other than rectangular in order to make the ratio $|\mathcal{P}|/|\mathcal{Q}|$ closer to unity, one must balance this reduction against the potentially greater cost of generating X and Y from a less simple region. Section 8.13 describes how an alternatively shaped inclusive region does indeed pay off for Gamma generation. The original proposal in Kinderman and Monahan (1977) for this method chooses $a_1 = 0$ and $a_2 = 1$, and it is this assignment we use in the next several sections unless an alternative assignment is explicitly made. Section 8.15 shows how the judicious choice of a_1 and a_2 allows one to sample from a simple region to generate samples from discrete distributions with bounded mean computing times.

The acceptance check $X^2 \leq r(a_1 + a_2 Y/X)$ in Algorithm ROU can consume considerable time just as the acceptance check $U \leq g(Z)$ in Algorithm AR can. Again, appropriately chosen squeeze functions can reduce this time significantly.

❏ EXAMPLE 8.3

An example based on the half-normal distribution (8.12a) illustrates the features of the ratio-of-uniforms method. Here we take $a_1 = 0$ and $a_2 = 1$, so that

$$\mathcal{Q} = \{(x, y) : 0 \leq x \leq e^{-(y/2x)^2}\},$$
$$r(z) = e^{-z^2/2},$$
$$|\mathcal{Q}| = \frac{1}{2} \int_0^\infty e^{-z^2/2} dz = \frac{1}{2}\sqrt{\pi/2},$$
$$x^* = 1, y_* = 0, y^* = \sqrt{2/e},$$
$$|\mathcal{P}|/|\mathcal{Q}| = 4/\sqrt{\pi e} = 1.3688.$$

Figure 8.3a shows the acceptance region. The acceptance test in Algorithm ROU takes the form

$$X \leq e^{-(Y/2X)^2},$$

which is equivalent to the event T,

$$Y^2 \leq -4X^2 \ln X. \tag{8.29}$$

The acceptance probability

$$\text{pr}(T) = |\mathcal{Q}|/|\mathcal{P}| = \sqrt{\pi e}/4 = .7306 \tag{8.30}$$

is close to the acceptance probability $\mathrm{pr}(S) = \sqrt{\pi/2e} = .7602$ in Example 8.1, but the ratio-of-uniforms method requires no exponential deviate as the acceptance–rejection method in Example 8.1 does.

■

Pretesting based on the inequalities (8.19) again improves efficiency. Let T_L denote the event

$$Y^2 \leq 4X^2(1 - X) \tag{8.31}$$

and T_U the event

$$Y^2 \leq 4X(1 - X). \tag{8.32}$$

One accepts $Z = Y/X$ if T_L occurs on the first pretest, but rejects Z if $\bar{T}_U$ occurs on the second pretest, where

$$
\begin{aligned}
\mathrm{pr}(T_L) &= \frac{1}{\sqrt{2/e}} \int_0^1 \int_0^{2x\sqrt{1-x}} dy \, dx \\
&= \sqrt{2e}\, \Gamma\left(\tfrac{3}{2}\right) / \Gamma\left(\tfrac{7}{2}\right) = 4(2e)^{\frac{1}{2}}/15 \\
&= .6218
\end{aligned}
\tag{8.33a}
$$

and

$$
\begin{aligned}
\mathrm{pr}(T_U) &= \frac{1}{\sqrt{2/e}} \sqrt{\min[2/e, 4x(1 - x)]} dy \, dx \\
&= \frac{1}{2}\sqrt{1 - 2/e} + \frac{\sin^{-1}(-\sqrt{1 - 2/e}) + \pi/2}{2\sqrt{2/e}} \\
&= .8579.
\end{aligned}
\tag{8.33b}
$$

Then

$$\mathrm{pr}[\text{evaluating } r(Y/X)] = \mathrm{pr}(T_U) - \mathrm{pr}(T_L) = .2361.$$

If M denotes the number of evaluations of $r(Y/X)$ until success, then

$$
\begin{aligned}
\mathrm{E}M &= \frac{\mathrm{pr}(T_U) - \mathrm{pr}(T)}{1 - \mathrm{pr}(T)} \left[\frac{1}{\mathrm{pr}(T)} - 1\right] + \frac{\mathrm{pr}(T) - \mathrm{pr}(T_L)}{\mathrm{pr}(T)} \\
&= \frac{\mathrm{pr}(T_U) - \mathrm{pr}(T_L)}{\mathrm{pr}(T)} = .3232.
\end{aligned}
\tag{8.33c}
$$

(a)

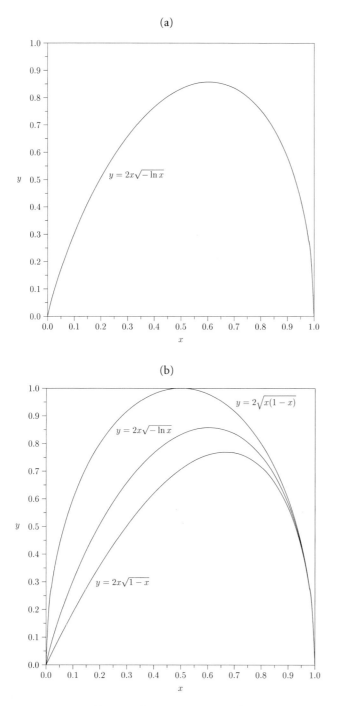

(b)

Figure 8.3 Half-normal generation using the ratio-of-uniforms method

This result follows in a manner analogous to that for EN in expression (8.16) for the acceptance–rejection method. Figure 8.3b shows the relevant curves.

In summary, incorporating the squeeze method into the ratio-of-uniforms method for generating a sample from the half-normal distribution reduces the mean number of evaluations of $r(Y/X)$ from 1.3688 to 0.3232. Again, the two pretests consume part of this time saving. However, one should note that $100 \times (1 - 0.3232/1.3688) = 76.39$ percent of the time no logarithmic or exponential evaluations are required, whereas the acceptance–rejection method in Example 8.1 requires one exponential deviate on every trial, regardless of whether or not the squeeze method is used.

8.7 EXACT-APPROXIMATION METHOD

The exact-approximation method in (Marsaglia 1984) demonstrates how for a specified p.d.f. one can identify a decomposition that allows generation from either a uniform or normal distribution most of the time and generation from a residual distribution for the remainder of the time. Since among all variate-generating procedures those for generating a uniform deviate are the least costly and those for generating a normal deviate are only marginally more costly, the substantial frequency with which the acceptance–rejection method makes direct use of uniform or normal generation gives the exact-approximation method its appeal.

Let Z have p.d.f. f and d.f. F. The exact-approximation method employs an approximation $\{m(x), 0 \leq x \leq 1\}$ to the corresponding inverse d.f. $\{F^{-1}(x), 0 \leq x \leq 1\}$ that allows one to generate a sample from the desired d.f. F. The result is exact. Assume that $\{m'(x) = dm/dx\}$ exists everywhere and is nonnegative. Let X denote a random variable with p.d.f. $\{f_X(x), 0 \leq x \leq 1\}$. Then $Z = m(X)$ has the p.d.f.

$$f_Z(z) = f_X(m^{-1}(z))/m'(m^{-1}(z)), \tag{8.34}$$

where m^{-1} denotes the inverse function of m. Since the objective is to achieve $f_Z(z) = f(z)$, one wants to determine the p.d.f. f_X that effects this result, namely

$$\begin{aligned} f_X(m^{-1}(z)) &= f_X(x) \\ &= f_Z(z)m'(m^{-1}(z)) \\ &= f(m(x))m'(x). \end{aligned} \tag{8.35}$$

If m closely approximates F^{-1}, then $\{f_X(x)\}$ is close to being uniform in appearance, suggesting the decomposition

$$f_X(x) = pI_{[0,1]}(x) + (1 - p)f^*(x), \qquad 0 < p \leq \min_{0 \leq x \leq 1} f_X(x), \tag{8.36}$$

where the upper bound on p ensures that f^* is a p.d.f. Clearly, assigning p the value of its upper bound offers the greatest benefit.

Algorithm EAM describes the method, making use of the fact that if $U \leq p$, then U/p is from $\mathcal{U}(0, 1)$. The motivation for large p is now clear. It reduces the frequency with which one must generate X from f^*, which Marsaglia recommends be done by the acceptance–rejection method.

ALGORITHM EAM

Purpose: To generate Z randomly from the p.d.f. $\{f(z)\}$.

Input: p, $\{m(x), 0 \leq x \leq 1\}$, and $\{f^*(x) = [f_X(x) - p]/(1 - p), 0 \leq x \leq 1\}$.

Method:

Randomly generate U from $\mathcal{U}(0, 1)$.

If $U \leq p$, $X \leftarrow U/p$. Otherwise: randomly generate X from the p.d.f. $\{f^*(x)\}$.

$Z \leftarrow m(X)$.

Return Z.

An example for the Cauchy p.d.f. (Section 8.12)

$$f_Z(z) = \frac{1}{\pi(1 + z^2)}, \quad -\infty < z < \infty, \tag{8.37}$$

denoted by $\mathcal{C}(0, 1)$, illustrates the method. From Table 8.1, the corresponding inverse of d.f. is $\{F^{-1}(x) = \tan \pi (x - 1/2), 0 \leq x \leq 1\}$. For $a = 2/\pi$ and $b = \pi - 8/\pi$, the function

$$m(x) = y \left(\frac{a}{1/4 - y^2} + b \right), \quad y := x - \frac{1}{2}, \quad 0 \leq x \leq 1, \tag{8.38}$$

approximates $\{F^{-1}(x)\}$ (Ahrens and Dieter 1988). Since

$$m'(x) = a \frac{1/4 + y^2}{(1/4 - y^2)^2} + b, \tag{8.39}$$

substitution into expression (8.35) gives

$$f_X(x) = \frac{1/\pi}{1 + m^2(x)} \left[a \frac{1/4 + y^2}{(1/4 - y^2)^2} + b \right].$$

The largest p that ensures that

$$f^*(x) = \frac{1}{1 - p} \left\{ \frac{1/\pi}{1 + m^2(x)} \left[a \frac{1/4 + y^2}{(1/4 - y^2)^2} + b \right] - p I_{[0,1]}(x) \right\} \geq 0 \tag{8.40}$$

is $p = .997738$.

Algorithm CA describes an implementation of the exact-approximation method for sampling Z from the Cauchy distribution $\mathcal{C}(0, 1)$ without the use of special functions. When sampling from f^*, it employs an acceptance–rejection procedure with a mean of 3.859819 additional samples from $\mathcal{U}(0, 1)$. However, Algorithm CA executes this branch with probability $1 - p = .002262$, resulting in an overall mean of 1.006469 samples from $\mathcal{U}(0, 1)$. Section 8.10 relies on Algorithm CA to provide a fast algorithm for generating samples from the normal distribution in Algorithm NA.

ALGORITHM CA

Purpose: To generate Z randomly from $\mathcal{C}(0, 1)$.
Output: Z.
Source: Ahrens and Dieter (1988).
Setup: Constants

$$
\begin{aligned}
a &= 0.6380\,6313\,6607\,7803, & b &= 0.5959\,4860\,6052\,9070, \\
q &= 0.9339\,9629\,5760\,3656, & W &= 0.2488\,7022\,8008\,3841, \\
A &= 0.6366\,1977\,2367\,5813, & B &= 0.5972\,9975\,9353\,9963, \\
H &= 0.0214\,9490\,0457\,0452, & P &= 4.9125\,0139\,5303\,3204.
\end{aligned}
$$

Method:

Randomly generate U from $\mathcal{U}(0, 1)$.
$T \leftarrow U - \frac{1}{2}$ and $S \leftarrow W - T^2$.
If $S > 0$, return $Z \leftarrow T(A/S + B)$.
Repeat:
> generate U and U' from $\mathcal{U}(0, 1)$
> $T \leftarrow U - \frac{1}{2}$, $S \leftarrow \frac{1}{4} - T^2$ and $Z \leftarrow T(a/S + b)$
until $S^2[(1 + Z^2)(HU' + P) - q] + S \leq \frac{1}{2}$.
Return Z.

8.8   ALGORITHMS FOR SELECTED DISTRIBUTIONS

The remainder of this chapter describes procedures for generating samples from selected continuous and discrete distributions. Throughout, the focus is on algorithms whose mean generating times are constant or bounded from above for all possible distributional parametric values. Section 3.14 illustrates the protection that using such algorithms offers in practice. The algorithms described here are not necessarily the computationally most efficient procedures known. However, they do compare favorably with the most efficient ones, which often depend on the preparation of extensive tables whose features are themselves machine dependent. The motivating factor in selecting algorithms for presentation here was their pedagogic value. Faced with the need to develop an algorithm for a highly specialized distribution, the reader of these sections should be able to craft a procedure that performs acceptably.

8.9 EXPONENTIAL DISTRIBUTION

If Y is drawn from $\mathcal{E}(1)$, then $Z = \lambda Y$, $\lambda > 0$, is from $\mathcal{E}(1/\lambda)$. Table 8.1 provides the transformation for the inverse transform approach that is the simplest and most direct method available. However, algorithms exist that when implemented in machine language generate an exponential sample in less time. These include the rectangle–wedge–tail approach in MacLaren et al. (1964) (Section 8.3) and Algorithms SA and EA in Ahrens and Dieter (1972, 1988). Of these, Algorithm EA is the fastest both in FORTRAN and assembler language versions (Ahrens and Dieter 1988). It generates a sample K from the geometric distribution $\mathcal{Ge}(1/2)$ (Section 8.21) and a sample V from the truncated exponential distribution

$$h(v) = 2e^{-v}, \qquad 0 \le v \le \ln 2. \tag{8.41}$$

Since

$$\mathrm{pr}(K = k) = (1/2)^{k+1}, \qquad k = 0, 1, \ldots, \tag{8.42}$$

$Z = K \ln 2 + V$ has the p.d.f.

$$f(z) = \mathrm{pr}(K = k)h(v) = e^{-z}, \quad z = k \ln 2 + v, \quad k = 0, 1, \ldots, \quad 0 \le v \le \ln 2, \tag{8.43}$$

so that Z has the desired distribution.

To sample V, Algorithm EA uses the exact-approximation method (Section 8.7) with

$$m(x) = a/(b - x) + c, \quad 0 \le x \le 1, \tag{8.44}$$

and constants $a = (4 + 3\sqrt{2}) \ln 2$, $b = 2 + \sqrt{2}$, and $c = -(1 + \sqrt{2}) \ln 2$ as an approximation to the inverse d.f.

$$F^{-1}(x) = -\ln(1 - x/2), \quad 0 \le x \le 1. \tag{8.45}$$

For the decomposition (8.36), expression (8.44) implies

$$f_X(x) = \frac{2e^{-m(x)}b(b - 1)\ln 2}{(b - x)^2}, \quad 0 \le x \le 1, \tag{8.46}$$

for which $p = \sqrt{2}\ln 2 = .980258$ is the largest probability that ensures

$$f^*(x) = \frac{1}{1-p}\left[\frac{2e^{-m(x)}b(b-1)\ln 2}{(b-x)^2} - pI_{[0,1]}(x)\right] \geq 0. \qquad (8.47)$$

To sample from f^*, Algorithm EA uses an acceptance–rejection method that requires $\alpha = 1.511076$ trials on average. Each trial uses two samples from $\mathcal{U}(0, 1)$ and an exponentiation. Therefore, the mean number of exponentiations is $100 \times \alpha(1-p) = 2.983144$ per 100 samples. Ahrens and Dieter (1988) describe a procedure that eliminates this relatively modest frequency.

ALGORITHM EA

Purpose: To generate a sample Z from $\mathcal{E}(1)$.

Output: Z.

Source: Ahrens and Dieter (1988).

Setup: Constants

$q = \ln 2 = 0.6931\ 4718\ 0559\ 9453,$ $a = 5.7133\ 6315\ 2645\ 4228,$
$b = 1.4142\ 1356\ 2373\ 0950,$ $c = -1.6734\ 0532\ 4028\ 4925,$
$p = 0.9802\ 5814\ 3468\ 5472,$ $A = 5.6005\ 7075\ 6973\ 8080,$
$B = 3.3468\ 1064\ 8056\ 9850,$ $h = 0.0026\ 1067\ 2360\ 2095,$
$D = 0.0857\ 8643\ 7626\ 9050.$

Method:

Randomly generate U from $\mathcal{U}(0, 1)$.
$K \leftarrow c$ and $U \leftarrow U + U$.
While $U < 1, U \leftarrow U + U$ and $K \leftarrow K + q$.
$U \leftarrow U - 1$.
If $U \leq p$, return $Z \leftarrow K + A/(B - U)$.
Repeat:
    randomly generate U and U' from $\mathcal{U}(0, 1)$
        $Y \leftarrow a/(b - U)$
until $(U'H + D)(b - U)^2 < e^{-(Y+c)}$.
Return $Z \leftarrow K + Y$.

8.9.1   WEIBULL DISTRIBUTION

Recall from Table 8.1 that the Weibull distribution has p.d.f.

$$f(z) = \frac{\alpha z^{\alpha-1}}{\beta^\alpha}\ e^{-(z/\beta)^\alpha},$$

d.f.

$$F(z) = 1 - e^{-(z/\beta)^{\alpha}}, \qquad \alpha, \beta > 0, \qquad z \geq 0,$$

and is denoted by $\mathcal{W}(\alpha, \beta)$. If Y is from $\mathcal{E}(\beta)$, then it is easily seen that $Z = Y^{1/\alpha}$ is from $\mathcal{W}(\alpha, \beta)$.

To sample Z from $\mathcal{W}(\alpha, \beta)$,

- Sample Y from $\mathcal{E}(\beta)$.

- Return $Z \leftarrow Y^{1/\alpha}$.

These steps are more general than those in Table 8.1, allowing for a faster method of exponential generation to be used if such an implementation is available.

8.9.2 INVERSE WEIBULL DISTRIBUTION

If $Y \sim \mathcal{W}(\alpha, \beta)$, then $Z = 1/Y$ has the p.d.f.

$$f(z) = \frac{\alpha}{\beta^{\alpha} z^{\alpha+1}} e^{-(1/\beta z)^{\alpha}}, \qquad \alpha, \beta > 0, \qquad z \geq 0,$$

which is called the inverse Weibull distribution and is denoted by $\mathcal{IW}(\alpha, \beta)$. Its mean, variance, and mode are, respectively,

$$\mathrm{E}Z = \Gamma(1 - 1/\alpha)/\beta,$$
$$\mathrm{var}Z = \left[\Gamma(1 - 2/\alpha) - \Gamma^2(1 - 1/\alpha)\right]/\beta^2,$$
$$\mathrm{mode} = \frac{1}{\beta}\left(\frac{\alpha}{1 + \alpha}\right)^{1/\alpha}.$$

The existence of $\mathrm{E}Z$ requires $\alpha \geq 1$, and of $\mathrm{var}Z$, $\alpha \geq 2$. Since $\mathrm{pr}(Y \leq a) = \mathrm{pr}(1/Z \leq a)$,

$$F_Z(z; \alpha, \beta) = 1 - F_Y(1/z; \alpha, \beta),$$

where F_Z and F_Y denote d.f.s of Z and Y, respectively.

Whereas the p.d.f. of Y has a right tail that decreases as $e^{-(y/\beta)^{\alpha}}$, the right tail of the p.d.f. of Z decreases polynomially as $1/z^{\alpha+1}$. Chapter 10 discusses this feature with regard to fitting data to a distribution.

To sample Z from $\mathcal{IW}(\alpha, \beta)$, first sample Y from $\mathcal{E}(\beta)$ and then take $Z = Y^{1/\alpha}$.

8.10 NORMAL DISTRIBUTION

Let Z denote a random variable with the normal p.d.f.

$$f(z) = \frac{1}{\sqrt{2\pi\sigma^2}}\, e^{-(z-\mu)^2/2\sigma^2}, \quad \sigma^2 > 0, \quad -\infty < z < \infty, \tag{8.48}$$

denoted by $\mathcal{N}(\mu, \sigma^2)$. Properties of the distribution include:

1. If Y is from $\mathcal{N}(0, 1)$, then $Z = \mu + \sigma Y$ is from $\mathcal{N}(\mu, \sigma^2)$.

2. If $Z_1, \ldots, Z_n$ are independent from $\mathcal{N}(\mu_1, \sigma_1^2), \ldots, \mathcal{N}(\mu_n, \sigma_n^2)$, respectively, then $Z = Z_1 + \cdots + Z_n$ is from $\mathcal{N}(\mu_1 + \cdots + \mu_n, \sigma_1^2 + \cdots + \sigma_n^2)$.

The first property makes clear that one need only concentrate on devising a procedure for generating a sample from $\mathcal{N}(0, 1)$, since the relatively inexpensive transformation $Z = \mu + \sigma Y$ generalizes its application. The second property shows that only one generation from $\mathcal{N}(0, 1)$ is necessary to generate $Z = Z_1 + \cdots + Z_n$.

Moro (1995) describes numerical evaluation by the inverse transform method with 10^{-9} accuracy. Examples 8.1 and 8.2 show how to generate a sample from the half-normal distribution from the acceptance–rejection and ratio-of-uniform methods, respectively. Multiplying the outcome by a randomly determined sign makes it a sample from $\mathcal{N}(0, 1)$.

The rectangle–wedge–tail method in Marsaglia et al. (1964) and its modification in Algorithm RT in Ahrens and Dieter (1972) provide the fastest algorithms for sampling from $\mathcal{N}(0, 1)$. Unfortunately, they rely on extensive tables the preparation of which may not appeal to a potential programmer. Shortly, we describe Algorithm NA in Ahrens and Dieter (1988), which is the fastest known table-free method in both FORTRAN and assembler language versions. Nevertheless, we first describe a commonly employed "slow" method that establishes the basis for Algorithm NA.

Theorem 8.5 (Box and Muller 1958).

Let U and V denote independent random variables from $\mathcal{U}(0, 1)$ and $\mathcal{E}(1)$, respectively. Then

$$X = (2V)^{\frac{1}{2}} \cos 2\pi U$$

and $\hspace{10cm}$ (8.49)

$$Y = (2V)^{\frac{1}{2}} \sin 2\pi U$$

are independent random variables from $\mathcal{N}(0, 1)$.

PROOF. Observe that $2V = X^2 + Y^2$, $\tan 2\pi U = Y/X$, and

$$f_{X,Y}(x, y) = f_{U,V}(u(x, y), v(x, y)) \left| \frac{\partial u}{\partial x} \frac{\partial v}{\partial y} - \frac{\partial u}{\partial y} \frac{\partial v}{\partial x} \right|$$

$$= \frac{1}{2\pi} e^{-(x^2+y^2)/2}, \qquad -\infty < x, y < \infty, \qquad (8.50)$$

since

$$\frac{\partial u}{\partial x} = \frac{-y \cos^2 2\pi u}{2\pi x^2} = \frac{-y}{4\pi v}, \qquad \frac{\partial u}{\partial y} = \frac{\cos^2 2\pi u}{2\pi x} = \frac{x}{4\pi v},$$

$$\frac{\partial v}{\partial x} = x, \qquad \frac{\partial v}{\partial y} = y.$$

Expression (8.50) gives the p.d.f. of two independent random variables, as required. □

The simplicity of implementation is most responsible for the common use of this approach. However, its reliance on square root, cosine, and sine evaluations makes it uncompetitive. However, it does establish a basis for Algorithm NA.

Observe that $\tan 2\pi U$ in Theorem 8.5 has the Cauchy distribution $\mathcal{C}(0, 1)$. Therefore, employing Algorithm EA to generate V and Algorithm CA to generate W and taking $X = \sqrt{2V/(1 + W^2)}$ and $Y = W\sqrt{2V/(1 + W^2)}$ accomplishes our purpose except for signs. Assigning a random sign to X and assigning the sign of W to Y solves this problem. Algorithm NA assumes that the value assigned to variable Y is saved between successive executions of the algorithm. As presented here, Algorithm NA is a slight variant of the one in Ahrens and Dieter (1988) that exploits the inevitable call to Algorithm CA.

ALGORITHM NA

Purpose: To generate Z from $\mathcal{N}(0, 1)$.
Output: Z.
Source: Ahrens and Dieter (1988).
Method:
    At compilation time, a flag F is set to 1.
    Execution
        $F \leftarrow -F$. If $F > 0$, return $Z \leftarrow Y$.
        Randomly generate U from $\mathcal{U}(0, 1)$.
        If $U < \frac{1}{2}$, $B \leftarrow 0$: otherwise, $B \leftarrow 1$.
        Randomly generate V from $\mathcal{E}(1)$ using Algorithm EA.
        $S \leftarrow V + V$.
        Randomly generate W from $\mathcal{C}(0, 1)$ using Algorithm CA (Section 8.12).
        $Z \leftarrow \sqrt{S/(1 + W^2)}$ and $Y \leftarrow WZ$.
        If $B = 0$, return Z; otherwise, return $-Z$.

8.11 LOGNORMAL DISTRIBUTION

Let Y be from $\mathcal{N}(\mu, \sigma^2)$. Then $Z = e^Y$ has the lognormal distribution with p.d.f.

$$f(z) = \frac{1}{\sqrt{2\pi}\,\sigma z}e^{-(\ln z - \mu)^2/2\sigma^2}, \qquad 0 \le z < \infty,$$

denoted by $\mathcal{LN}(\mu, \sigma^2)$. Occasionally, the distribution is described in terms of its moments

$$\mu_Z = EZ = e^{\mu + \sigma^2/2}$$

and

$$\sigma_Z^2 = \operatorname{var} Z = e^{2\mu + \sigma^2}(e^{\sigma^2} - 1).$$

Common practice generates X from $\mathcal{N}(0, 1)$, computes $Y = \mu + X\sigma$, and takes $Z = e^Y$ as the sample. The marginal increment in cost over normal generation is due to the exponential evaluation.

8.12 CAUCHY DISTRIBUTION

Suppose Z has the noncentral Cauchy p.d.f.

$$f(z) = \frac{\beta}{\pi[\beta^2 + (z - \alpha)^2]}, \qquad -\infty < \alpha < \infty, \quad \beta > 0, \quad -\infty < z < \infty,$$

denoted by $\mathcal{C}(\alpha, \beta)$. The distribution has the following properties:

1. If Y is from $\mathcal{C}(0, 1)$, then $Z = \alpha + \beta Y$ is from $\mathcal{C}(\alpha, \beta)$.

2. If $Z_1, \ldots, Z_n$ are independent from $\mathcal{C}(\alpha_1, \beta), \ldots, \mathcal{C}(\alpha_n, \beta)$, respectively, then $Z = Z_1 + \cdots + Z_n$ is from $\mathcal{C}(\alpha_1 + \cdots + \alpha_n, \beta)$.

3. If X_1 and X_2 are independent from $\mathcal{N}(0, 1)$, then $Z = X_1/X_2$ is from $\mathcal{C}(0, 1)$.

4. If (Y_1, Y_2) is uniformly distributed on the unit circle $y_1^2 + y_2^2 = \frac{1}{4}$, then $Z = Y_1/Y_2$ is from $\mathcal{C}(0, 1)$.

Without loss of generality, we take $\alpha = 0$ and $\beta = 1$. Algorithm CA in Section 8.7 generates Z from $\mathcal{C}(0, 1)$ using the exact-approximation method.

Table 8.1 gives the inverse d.f. for Z, which requires a tangent evaluation in practice and compares unfavorably in timing with Algorithm CA. To avoid this, one can generate X_1 and X_2 each from $\mathcal{N}(0, 1)$ and use $Z = X_1/X_2$. Alternatively, one can use the result based on uniform generation on the unit circle. In particular (Exercise 8.37), straightforward application of the acceptance–rejection technique gives

> Repeat
>> generate Y_1 and Y_2 each from $\mathcal{U}\left(-\frac{1}{2}, \frac{1}{2}\right)$ (8.51)
>> until $Y_1^2 + Y_2^2 \leq \frac{1}{4}$.
> Return Y_1/Y_2.

where the mean number of trials is $4/\pi = 1.2732$. Devroye (1986) suggests the pretest $|Y_1| + |Y_2| \leq 1$, which reduces the frequency of evaluating the two multiplications by half. However, one must add in the time required to evaluate the absolute value function twice.

8.13   GAMMA DISTRIBUTION

Let Z denote a random variable having the Gamma p.d.f.

$$f(z) = z^{\alpha-1} e^{-z/\beta} / \beta^\alpha \Gamma(\alpha), \quad \alpha, \beta > 0, \ z \geq 0,$$

denoted by $\mathcal{G}(\alpha, \beta)$. Then Z has the properties:

1. If $Z_1, \ldots, Z_n$ are independent from $\mathcal{G}(\alpha_1, \beta), \ldots, \mathcal{G}(\alpha_n, \beta)$, respectively, then $Z = Z_1 + \cdots + Z_n$ is from $\mathcal{G}(\alpha_1 + \cdots + \alpha_n, \beta)$.

2. If $\alpha = 1$, then Z has the exponential distribution $\mathcal{E}(\beta)$.

3. If α is an integer, then Z is the sum of α i.i.d. exponential random variables from $\mathcal{E}(\beta)$ and its distribution is called the Erlang distribution with α phases.

4. If α is an integer and $\beta = 2$, then Z has the chi-squared distribution with 2α degrees of freedom.

5. If Y is from $\mathcal{G}(\alpha, 1)$, then $Z = \beta Y$ is from $\mathcal{G}(\alpha, \beta)$.

6. If $\alpha < 1$, then $\{f(z)\}$ is unbounded at $z = 0$.

Without loss of generality, we hereafter take $\beta = 1$.

Unlike exponential, normal, and Cauchy generating times, Gamma generating time can vary with the shape parameter α. Erlang generation illustrates the potential danger. If α is integer, then Z can be viewed as the sum $Z = Z_1 + \cdots + Z_\alpha$, where $Z_1, \ldots, Z_\alpha$ are independent identically distributed random variables from $\mathcal{E}(1)$. This suggests

> $Z \leftarrow 0$ and $i \leftarrow 1$.
> Repeat

> randomly generate Z_i from $\mathcal{E}(1)$
> $Z \leftarrow Z + Z_i$
> $i \leftarrow i + 1$
> until $i > \alpha$.
> Return Z.

which takes $O(\alpha)$ time. If one uses the inverse transform method for the exponential deviates, then an opportunity to improve efficiency exists.
That is,

> $Y \leftarrow 1$ and $i \leftarrow 1$.
> Repeat
>> randomly generate U_i from $\mathcal{U}(0, 1)$
>> $Y \leftarrow Y U_i$
>> $i \leftarrow i + 1$
> until $i > \alpha$.
> Return $Z \leftarrow -\ln Y$.

Although only one logarithmic evaluation occurs, the generation of α uniform deviates retains the $O(\alpha)$ time. Also, the required integrality of α limits the applicability of this approach. The remainder of this section addresses the problem of Gamma generation for $\alpha \leq 1$ and the problem of Gamma generation with bounded mean computing time as $\alpha \to \infty$.

$\alpha \leq 1$

In their Algorithm GS, Ahrens and Dieter (1974b) generate Z from $\mathcal{G}(\alpha, 1)$ with $0 < \alpha \leq 1$, using a combination of the acceptance–rejection, composition, and inverse transform methods with

$$h(z) = \frac{e}{\alpha + e} I_{[0,1]}(z)\alpha z^{\alpha-1} + \frac{\alpha}{\alpha + e} I_{(1,\infty)}(z)e^{-z+1}, \tag{8.52a}$$

$$g(z) = I_{[0,1]}(z)e^{-z} + I_{(1,\infty)}(z)z^{\alpha-1}, \quad z \geq 0, \tag{8.52b}$$

and

$$c(\alpha) = (\alpha + e)/\alpha e \Gamma(\alpha), \tag{8.52c}$$

where $I_{\{A\}}(z)$ denotes the indicator function of the set $\mathcal{A}$ and $\{h(z)\}$, $\{g(z)\}$, and $c(\alpha)$ are defined in Theorem 8.3. Observe that $f(z) = c(\alpha)g(z)h(z)$ gives the Gamma p.d.f., sampling from the p.d.f. $\{h(z)\}$ uses the composition method and the mean number of trials until success $c(\alpha)$ has the upper bound $c \leq 1.39$. The composition method determines whether one generates Z in expression (8.52a) from the d.f. $\{z^\alpha, 0 \leq z \leq 1\}$ or from the d.f. $\{1 - e^{-z+1}, 1 < z < \infty\}$ by the inverse transform method. The acceptance–rejection method determines whether or not Z is from $\mathcal{G}(\alpha, 1)$.

Algorithm GS* describes the essential steps. It differs from those in Algorithm GS in Ahrens and Dieter (1974b) in minor ways. Observe the setup $b \leftarrow (\alpha + e)/e$. If repeated sampling is to occur for the same value of α, then it is advantageous to compute b once at the beginning of the experiment and use α and b, instead of α alone, as input, thus avoiding the repeated cost of computing b. Note that Algorithm GS* admits no simple pretest to exploit the squeeze method. Best (1983) suggests several modifications that reduce generating time by a relatively incidental factor for α close to zero and by one-half for α close to unity. As an extension of a method for $\alpha > 1$ to be described in the next section, Cheng and Feast (1980) describe a competitive method for $1/n \leq \alpha \leq 1$, where n is a fixed integer.

ALGORITHM GS*

Purpose: To generate Z randomly from $\mathcal{G}(\alpha, 1)$ with $0 < \alpha < 1$.
Input: α.
Setup: $b \leftarrow (\alpha + e)/e$.
Method:
    Repeat
        randomly generate U from $\mathcal{U}(0, 1)$; $Y \leftarrow bU$
        if $Y \leq 1$:
            $Z \leftarrow Y^{1/\alpha}$; randomly generate W from $\mathcal{E}(1)$
            if $W \geq Z$, success
        otherwise:
            $Z \leftarrow -\ln((b - Y)/\alpha)$
            randomly generate W_1 from $\mathcal{U}(0, 1)$; $W \leftarrow W_1^{1/(\alpha-1)}$
            if $W \leq Z$, success
        until success.
        Return Z.

$\alpha > 1$

Many alternative algorithms exist for this case of which those in Ahrens and Dieter (1974a, 1974b, 1982a), Cheng and Feast (1979), Marsaglia (1984), Minh (1988), and Schmeiser and Lal (1980) are demonstrably fast. The one due to Cheng and Feast uses the ratio-of-uniforms method (Section 8.6) and illustrates many of the nuances implicit in this approach.

The acceptance region is

$$\mathcal{Q}(\alpha) = \{0 \leq x \leq r^{\frac{1}{2}}(y/x) := [(y/x)^{\alpha-1}e^{-y/x}]^{\frac{1}{2}}\}, \tag{8.53}$$

implying that $\mathcal{Q}(\alpha)$ lies in the rectangle $\mathcal{P}(\alpha) := \{0 \leq x \leq x^* := [(\alpha - 1)/e]^{(\alpha-1)/2},$ $0 \leq y \leq y^* := [(\alpha + 1)/e]^{(\alpha+1)/2}\}$ (definitions 8.27b and 8.27d). Direct application of the

(a)

(b)

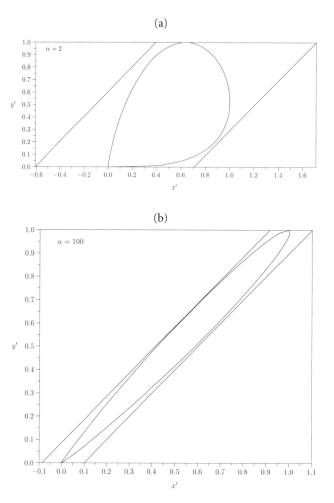

Figure 8.4 Gamma generation: ratio-of-uniforms method $x' := x/x^*$ and $y' := y/y^*$.

ratio-of-uniforms approach leads to mean number of trials until success

$$c(\alpha) = |\mathcal{P}(\alpha)|/|\mathcal{Q}(\alpha)| = 2(\alpha + 1)^{(\alpha+1)/2}(\alpha - 1)^{(\alpha-1)/2}/\Gamma(\alpha)e^{\alpha}, \tag{8.54a}$$

for which

$$c(\alpha)/\sqrt{\alpha} \to 2/(2\pi)^{\frac{1}{2}} \text{ as } \alpha \to \infty, \tag{8.54b}$$

revealing an $O(\alpha^{\frac{1}{2}})$ mean generating time. In addition, x^* and y^* need to be computed.

Table 8.5   Gamma variate generation: ratio-of-uniforms method[†]

		Mean number of trials until success	
α	$c(\alpha)$	$c'(\alpha)$	Ratio
1.5	1.331	2.019	.6592
2	1.406	1.848	.7608
2.5	1.499	1.761	.8561
3	1.593	1.709	.9321
4	1.773	1.647	1.077
5	1.941	1.612	1.204
6	2.096	1.590	1.318
7	2.241	1.574	1.424
8	2.378	1.562	1.522
9	2.507	1.553	1.614
10	2.631	1.545	1.703
15	3.177	1.524	2.085
20	3.643	1.514	2.406
25	4.056	1.507	2.691
30	4.431	1.503	2.948
35	4.777	1.500	3.185
40	5.099	1.498	3.404
45	5.402	1.496	3.611
50	5.689	1.495	3.805
∞	∞	1.482	∞

[†] Expression (8.54a) defines $c(\alpha)$; expression (8.55) defines $c'(\alpha)$.

To eliminate the asymptotic dependence on α, Cheng and Feast employ a parallelogram $\mathcal{P}'(\alpha)$, as in Figure 8.4, to enclose $\mathcal{Q}(\alpha)$. It leads to a mean number of trials until success

$$c'(\alpha) = \frac{|\mathcal{P}'(\alpha)|}{|\mathcal{Q}(\alpha)|} = \frac{2x^*y^*(1 + \sqrt{2/e})}{\Gamma(\alpha)\sqrt{\alpha}}, \qquad (8.55)$$

where

$$c'(\alpha) \to 1.482 \quad \text{as } \alpha \to \infty.$$

Table 8.5 reveals that the square $\mathcal{P}(\alpha)$ has an advantage for $\alpha \leq 3$, whereas for $\alpha \geq 4$, the parallelogram $\mathcal{P}'(\alpha)$ has an ever-increasing advantage as $\alpha \to \infty$. This suggests a procedure that uses the square for $\alpha \leq 3$ and the parallelogram for $\alpha \geq 4$. However, an additional issue plays a role.

Recall that $(X = x^*X', Y = y^*Y')$ is uniformly distributed on $\mathcal{P}(\alpha)$ if (X', Y') is uniformly distributed in the unit square $[0, 1]^2$. If generation from the same $\mathcal{G}(\alpha, 1)$ is to occur repeatedly, then one can compute x^* and y^* once at the beginning, store them, and use α, x^*, and y^* as input to the algorithm on each call. If many distinct α's are used, this approach may not be feasible, and without modification one would repeatedly incur the relatively high setup cost of computing x^* and y^*. A modification of the acceptance region reduces this cost substantially at a relatively small increase in the mean number of trials until success for small α.

Let (X', Y') be uniformly distributed on $[0, 1]^2$. Let $X = x^* X'$ and $Y = y^{**} Y'$, where $y^{**} \geq y^*$. Then (X, Y) is uniformly distributed on the rectangle

$$\mathcal{P}_1(\alpha) := \{(x, y) : 0 \leq x \leq x^*, \ 0 \leq y \leq y^{**}\}, \tag{8.56}$$

where $\mathcal{Q}(\alpha) \subseteq \mathcal{P}(\alpha) \subseteq \mathcal{P}_1(\alpha)$. Let $W = X^2$ and $Z = Y/X$. If $W \leq r(Z)$, then Z is from $\mathcal{G}(\alpha, 1)$. Observe that this inequality implies

$$\frac{2}{\alpha - 1} \ln X' \leq \ln(bY'/X') - bY'/X' + 1, \tag{8.57}$$

where $b := y^{**}/x^*(\alpha - 1)$, suggesting that one merely has to take $b \geq y^*/x^*(\alpha - 1) = (\alpha + 1)/(\alpha - 1)^{(\alpha+1)/2}/e$ and $Z = (\alpha - 1)bY'/X'$. Cheng and Feast suggest

$$b := (\alpha - 1/6\alpha)/(\alpha - 1), \tag{8.58}$$

which ensures $b \geq y^*/x^*(\alpha - 1)$ and eliminates the need to evaluate x^*, y^*, and y^{**}.

Since W and Z have the joint p.d.f.

$$f_{W,Z}(w, z) = w/b(\alpha - 1), \qquad 0 \leq z \leq b(\alpha - 1)/w, \quad 0 \leq w \leq 1,$$

it follows that

$$\begin{aligned}
\mathrm{pr}[W \leq r(Z)] &= |\mathcal{Q}(\alpha)|/|\mathcal{P}_1(\alpha)| = \Gamma(\alpha)e^{\alpha-1}/2b(\alpha - 1)^\alpha \tag{8.59} \\
&= \Gamma(\alpha)e^{\alpha-1}/2(\alpha - 1/6\alpha)(\alpha - 1)^{\alpha-1},
\end{aligned}$$

and Z given $W \leq r(Z)$ has the desired Gamma p.d.f. Since X' and Y' are uniformly distributed over $[0, 1]^2$, the mean number of trials is $2(\alpha/2\pi)^{1/2}$ as $\alpha \to \infty$, which agrees with the limit (8.54b).

Since $\mathcal{P}'(\alpha) \cap [0, 1]^2 \subset \mathcal{P}(\alpha)$, this alteration also applies to sampling from $\mathcal{P}'(\alpha)$. While preserving the validity of the ratio-of-uniforms approach to Gamma generation for $\alpha > 1$, the change merely scales $c(\alpha)$ and $c'(\alpha)$ by $b(\alpha - 1)x^*/y^*$, which decreases from 1.133 at $\alpha = 1$ to unity as $\alpha \to \infty$.

Algorithm GKM1 describes the generating method when (X', Y') are randomly generated on the unit square $\mathcal{P}(\alpha)$. Note the pretest $mX' - d + V + V^{-1} \leq 0$, which is based on the inequality $\ln w \leq w - 1$ for $w > 0$. This pretest reduces the frequency of logarithmic evaluation (Exercise 8.42). Whenever possible, evaluating a, b, m, and d beforehand, storing them, and using them as input in place of α on repeated generations for the same α saves time.

ALGORITHM GKM1

Purpose: To sample Z randomly from the Gamma distribution $\mathcal{G}(\alpha, 1)$ with $\alpha > 1$.
Input: α.
Setup: $a \leftarrow \alpha - 1, b \leftarrow (\alpha - 1/6\alpha)/a, m \leftarrow 2/a, d \leftarrow m + 2$.
Source: Cheng and Feast (1979).
Method:
    Repeat
        randomly generate X' and Y' from $\mathcal{U}(0, 1)$
        $V \leftarrow bY'/X'$
        if $mX' - d + V + V^{-1} \leq 0$, accept   (fast acceptance)
        if $m \ln X' - \ln V + V - 1 \leq 0$, accept
    until accept.
    Return $Z \leftarrow aV$.

Algorithm GKM2 describes the steps when sampling (X', Y') from the parallelogram $\mathcal{P}'(\alpha)$. Whenever possible, evaluating the setup, a, b, m, d and especially f, beforehand, storing them and using them as input on repeated generations for the same α saves time. Note that $\sqrt{2/e} = .857764$. It is left to the reader (Exercise 33) to show that if $0 < X' < 1$, then (X', Y') is uniformly distributed on $\mathcal{P}'(\alpha) \cap [0, 1]^2$ and that on any trial

$$\text{pr}(0 < X' < 1) = 1 - \frac{e \mid 2}{2\sqrt{\alpha e}(\sqrt{e} + \sqrt{2})} \geq .5328. \tag{8.60}$$

ALGORITHM GKM2

Purpose: To generate Z randomly from $\mathcal{G}(\alpha, 1)$ for $\alpha > 1$.
Input: α.
Setup: $a \leftarrow \alpha - 1, b \leftarrow (\alpha - 1/6\alpha)/a, m, \leftarrow 2/a, d \leftarrow m + 2, f \leftarrow \sqrt{\alpha}$.
Source: Cheng and Feast (1979).
Method:
    Repeat
        Repeat
        randomly generate X and Y' from $\mathcal{U}(0, 1)$; $X' \leftarrow Y' + (1 - 1.857764X)/f$
        until $0 < X' < 1$.
        $V \leftarrow bY'/X'$
        if $mX' - d + V + V^{-1} \leq 0$, accept   (fast acceptance)
        if $m \ln X' - \ln V + V - 1 \leq 0$, accept
    until accept.
    Return $Z \leftarrow aV$.

The step $f \leftarrow \sqrt{\alpha}$ keeps Algorithm GKM2 from completely dominating Algorithm GKM1 timewise. However, the associated $O(1)$ evaluation time increasingly favors Algorithm GKM2 when compared to the $O(\alpha^{\frac{1}{2}})$ mean generating time for Algorithm GKM1 as $\alpha \to \infty$. Algorithm GKM3 reconciles these two options. Cheng and Feast recommend $\alpha_0 = 2.5$.

ALGORITHM GKM3

Purpose: To generate Z randomly from $\mathcal{G}(\alpha, 1)$ for $\alpha > 1$.
Input: α.
Source: Cheng and Feast (1979).
Method:
   If $\alpha \leq \alpha_0$: use Algorithm GKM1.
   Otherwise: use Algorithm GKM2.

8.13.1   INVERTED GAMMA DISTRIBUTION

If $Y \sim \mathcal{G}(\alpha, \beta)$, then $Z = 1/Y$ has the p.d.f.

$$f(z) = \frac{1}{\Gamma(\alpha)\beta^\alpha z^{\alpha+1}}\, e^{-1/\beta z}, \quad \alpha, \beta > 0, \quad z \geq 0,$$

which is called the inverted Gamma distribution and denoted by $\mathcal{IG}(\alpha, \beta)$. Its mean, variance, and mode are, respectively,

$$EZ = \frac{1}{(\alpha - 1)\beta},$$

$$\mathrm{var}\, Z = \frac{1}{(\alpha - 2)(\alpha - 1)^2 \beta^2},$$

and

$$\mathrm{mode} = \frac{1}{(\alpha + 1)\beta}.$$

The existence of EZ requires $\alpha > 1$, and of $\mathrm{var}\, Z$, $\alpha > 2$. Since $\mathrm{pr}(Y \leq a) = \mathrm{pr}(1/Z \leq a)$,

$$F_Z(z; \alpha, \beta) = 1 - F_Y(1/z; \alpha, \beta),$$

where F_Z and F_Y denote the d.f.s of Z and Y, respectively.

   Whereas the p.d.f. of Y has a right tail that decreases exponentially, the p.d.f. of Z decreases polynomially as $1/z^{\alpha+1}$. Chapter 10 discusses this feature with regard to fitting data to a distribution.

   To sample Z from $\mathcal{IG}(\alpha, \beta)$, first sample Y from $\mathcal{G}(\alpha, \beta)$ and then take $Z = 1/Y$.

8.14 BETA DISTRIBUTION

The Beta distribution has p.d.f.

$$f(z) = \frac{1}{B(\alpha, \beta)} z^{\alpha-1}(1-z)^{\beta-1}, \quad \alpha, \beta > 0, \quad 0 \le z \le 1, \tag{8.61}$$

where B denotes the beta function, and is denoted by $\mathcal{B}e(\alpha, \beta)$. The distribution has the following properties:

1. If X and Y are independent random variables from $\mathcal{G}(\alpha, 1)$ and $\mathcal{G}(\beta, 1)$, respectively, then $Z = X/(X + Y)$ has the Beta distribution $\mathcal{B}e(\alpha, \beta)$.

2. Let $U_1, \ldots, U_n$ denote independent random variables each from $\mathcal{U}(0, 1)$ and let $U_{(1)} < \cdots < U_{(n)}$ denote the corresponding order statistics. Then $U_{(k)}$ has the Beta distribution $\mathcal{B}e(k, nk + 1)$ for $k = 1, \ldots, n$.

3. If Z is from $\mathcal{B}e(\alpha, \beta)$, then $1 - Z$ is from $\mathcal{B}e(\beta, \alpha)$.

Property 1 provides an immediate way to generate a Beta variate given the availability of a procedure for gamma generation. However, faster Beta generation methods exist. Property 2 reveals how to generate order statistics for any parent distribution for which generation via the inverse transform method applies. See Fishman (1996, Section 3.26). The third property shows that any development for $\alpha \ge \beta$ applies equally to $\beta \le \alpha$ by simple transformation.

Table 8.1 gives the inverse d.f.s that allow Beta generation in O(1) time for $\alpha = 1$ or $\beta = 1$. However, the derivation of an algorithm with bounded mean computing time for arbitrary α and β calls for considerably more attention. We first consider the cases $\max(\alpha, \beta) < 1$ and $\min(\alpha, \beta) > 1$ separately and then consider the case of $\min(\alpha, \beta) < 1$ and $\max(\alpha, \beta) > 1$.

$\max(\alpha, \beta) < 1$

Atkinson and Whittaker (1976, 1979) describe an especially fast Beta generating procedure for this case. It combines the composition, inverse transform, and acceptance–rejection methods in a manner similar to that in Section 8.13 (Ahrens and Dieter 1974b) for Gamma generation with $\alpha < 1$. For $\max(\alpha, \beta) < 1$ and $0 \le t \le 1$, let

$$h(z) = pI_{[0,t]}(z)\frac{\alpha z^{\alpha-1}}{t^\alpha} + (1-p)I_{(t,1]}(z)\frac{\beta(1-z)^{\beta-1}}{(1-t)^\beta}, \tag{8.62a}$$

$$g(z) = I_{[0,t]}(z)\left(\frac{1-z}{1-t}\right)^{\beta-1} + I_{(t,1]}(z)\left(\frac{z}{t}\right)^{\alpha-1}, \quad 0 \le z \le 1,$$

$$p = \beta t/[\beta t + \alpha(1-t)],$$

and

$$c(\alpha, \beta, t) = \Gamma(\alpha + \beta + 1)[\beta t + \alpha(1 - t)]/\Gamma(\alpha)\Gamma(\beta)(\alpha + \beta)t^{1-\alpha}(1 - t)^{1-\beta}. \quad (8.62b)$$

Here $\{h(z)\}$ is a p.d.f., $0 \leq g(z) \leq 1$,

$$c(\alpha, \beta, t) = 1/\int_0^1 g(z)h(z)\mathrm{d}z = \sup_{0 \leq z \leq 1} [f(z)/h(z)],$$

and $c(\alpha, \beta, t)g(z)h(z)$ is identical with $f(z)$ in expression (8.61). To generate a Beta variate, one first determines whether $0 \leq Z \leq t$ with probability p or $t < Z \leq 1$ with probability $1 - p$. If $0 \leq Z \leq t$, then generation occurs from $\{\alpha z^{\alpha-1}/t^{\alpha}, 0 \leq z \leq t\}$ using the inverse transform method. If $(t < Z \leq 1)$, then generation occurs from $\{\beta(1 - z)^{\beta-1}/(1 - t)^{\beta}, t < z \leq 1\}$, again using the inverse transform method. In both cases, the acceptance–rejection method determines whether or not Z is from $\mathcal{B}e(\alpha, \beta)$. The assignment $t = t^*$, where

$$t^* = \left[1 + \sqrt{\frac{\beta(1 - \beta)}{\alpha(1 - \alpha)}}\right]^{-1}, \quad (8.63)$$

minimizes the mean number of trials $c(\alpha, \beta, t)$. We leave to the reader the proof (Exercise 8.45) that $c(\alpha, \beta, t)$ is bounded above for all $0 \leq \alpha, \beta \leq 1$ and all $0 \leq t \leq 1$.

Algorithm AW describes the procedure with two preliminary tests that use the inequality $e^x > x + 1$, which holds for all $x > 0$. If repeated generation for the same α and β occurs, then evaluating t prior to the first generation, storing it, and using it as input with α and β on each execution of the algorithm eliminates the repeated computation of the square root. For repeated generation with many different α's and β's, storing all the computed optimal t's may not be a viable option. An alternative is to set $t = \frac{1}{2}$ as a parameter, which eliminates the square root evaluation but increases the mean number of trials $c(\alpha, \beta, \frac{1}{2})$. However, $\sup_{0 < \alpha, \beta \leq 1} c(\alpha, \beta, \frac{1}{2}) < \infty$. Note that if $U \leq p$, then U/p is from $\mathcal{U}(0, 1)$.

ALGORITHM AW

Purpose: To generate Z randomly from $\mathcal{B}e(\alpha, \beta)$ for $\max(\alpha, \beta) < 1$.

Input: α and β.

Setup: $t \leftarrow \left[1 + \sqrt{\frac{\beta(1-\beta)}{\alpha(1-\alpha)}}\right]^{-1}$ and $p \leftarrow \beta t/[\beta t + \alpha(1 - t)]$.

Method:

    Repeat

        randomly generate U from $\mathcal{U}(0, 1)$ and Y from $\mathcal{E}(1)$

        if $U \leq p$:

            $Z \leftarrow t(U/p)^{1/\alpha}$

if $Y \geq (1 - \beta)(t - Z)/(1 - t)$, accept   (fast acceptance)

if $Y \geq (1 - \beta) \ln \left(\frac{1-Z}{1-t} \right)$, accept

otherwise:

$$Z \leftarrow 1 - (1 - t) \left(\frac{1-U}{1-p} \right)^{1/\beta}$$

if $Y \geq (1 - \alpha)(Z/t - 1)$, accept   (fast acceptance)

if $Y \geq (1 - \alpha) \ln(Z/t)$, accept

until accept.

Return Z.

$\min(\alpha, \beta) > 1$

Cheng (1978) describes Algorithm BB, which has bounded mean computing time and is known to be fast. Schmeiser and Babu (1980) and Zechner and Stadlober (1993) describe alternative fast algorithms. The Cheng procedure relies on the acceptance–rejection method and exploits the observation that if Y is a random variable with p.d.f.

$$f_Y(y) = y^{\alpha-1}/B(\alpha, \beta)(1 + y)^{\alpha+\beta}, \qquad y \geq 0, \tag{8.64}$$

then $Z = Y/(1 + Y)$ is from $\mathcal{B}e(\alpha, \beta)$. Therefore, one need only concentrate on sampling Y using the acceptance–rejection approach with the factorization

$$f_Y(y) = c(\alpha, \beta)g(y)h(y),$$

where

$$h(y) = \lambda \mu y^{\lambda-1}/(\mu + y^\lambda)^2, \qquad y \geq 0,$$
$$g(y) = y^{\alpha-\lambda}/(\mu + y^\lambda)^2(\alpha + \beta)^{\alpha+\beta}/4(1 + y)^{\alpha+\beta}\alpha^{\alpha+\lambda}\beta^{\beta-\lambda},$$
$$c(\alpha, \beta) = 4\alpha^\alpha \beta^\beta/\lambda B(\alpha, \beta)(\alpha + \beta)^{\alpha+\beta}, \tag{8.65}$$
$$\lambda = \begin{cases} \min(\alpha, \beta) & \text{if } \min(\alpha, \beta) \leq 1, \\ [(2\alpha\beta - \alpha - \beta)/(\alpha + \beta - 2)]^{\frac{1}{2}} & \text{otherwise}, \end{cases}$$

and

$$\mu = (\alpha/\beta)^\lambda.$$

For α and $\beta \geq 1$ $c(\alpha, \beta) \leq 4/e = 1.47$; for either $\alpha \leq 1$ and $\beta \to 0$ or $\beta \leq 1$ and $\alpha \to 0$, $c(\alpha, \beta) = 4$, which gives the relatively poor upper bound for α and $\beta \leq 1$.

Since

$$H(x) = \int_0^x h(y)dy = x^\lambda/(\mu + x^\lambda), \qquad x \geq 0,$$

the inverse transform method allows one to generate a sample from $\{h(y)\}$ in constant time.

Algorithm BB* corresponds to Algorithm BB in Cheng (1978) in content. It contains several pretests that reduce the required number of logarithmic and exponential transformations. If repeated samples are to be taken with the same α and β, then evaluating the quantities in the setup beforehand, storing them, and using them as input to successive executions reduces computing time. There is no way to avoid the logarithmic and exponential evaluation in evaluating V and W. Note in Algorithm BB* that we use $U_1 U_1$ in place of U_1^2. Since some programming languages compute U_1^2 as $\exp(2 \ln U_1)$, it is desirable to avoid this expensive evaluation method wherever possible.

ALGORITHM BB[*]

Purpose: To generate Z randomly from $\mathcal{B}e(\alpha, \beta)$ for $\min(\alpha, \beta) > 1$.
Input: α and β.
Setup: $d_1 = \min(\alpha, \beta), d_2 = \max(\alpha, \beta), d_3 = d_1 + d_2,$
  $d_4 = [(d_3 - 2)/(2d_1 d_2 - d_3)]^{\frac{1}{2}}, d_5 = d_1 + 1/d_4.$
Method:

   Repeat

   randomly generate U_1 and U_2 from $\mathcal{U}(0, 1)$
   $V \leftarrow d_4 \ln[U_1/(1 - U_1)];\ W \leftarrow d_1 e^V;\ Z_1 \leftarrow U_1 U_1 U_2$
   $R \leftarrow d_5 V - 1.38629436;\ S \leftarrow d_1 + R - W$
   if $S + 2.60943791 > 5Z_1$, accept   (preliminary test 1 with $\ln 4 =$
                    1.38629436 and $1 + \ln 5 = 2.60943791$)

   otherwise:
      $T \leftarrow \ln Z_1$
      if $S \geq T$, accept   (preliminary test 2)
      otherwise: if $R + d_3 \ln[d_3/(d_2 + W)] \geq T$, accept
   until accept.
   If $d_1 = \alpha$, return $Z \leftarrow W/(d_2 + W)$; otherwise, return $Z \leftarrow d_2/(d_2 + W)$.

For $\min(\alpha, \beta) < 1$ and $\max(\alpha, \beta) > 1$, see Fishman (1996, Section 3.15).

8.14.1 BETA-PRIME DISTRIBUTION

If $Y \sim \mathcal{B}e(\alpha, \beta)$ then $Z = Y^{-1} - 1$ has the p.d.f.

$$f(z) = \frac{\Gamma(\alpha)\Gamma(\beta)}{\Gamma(\alpha + \beta)} \frac{z^{\beta-1}}{(1 + z)^{\alpha+\beta}}, \qquad \alpha, \beta > 0, \quad z \geq 0,$$

sometimes called the Beta-prime distribution (Johnson et al. 1994). Note that Z has infinite support and that f decreases as $O(z^{-(1+\alpha)})$ as $z \to \infty$.

Observe that

$$\mathrm{E} Z^l = \frac{\Pi_{i=1}^l (l + \beta - i)}{\Pi_{i=1}^l (\alpha - i)}, \qquad l < \alpha.$$

Moreover, Z has infinite mean for $\alpha \leq 1$. For $1 < \alpha \leq 2$, it has infinite variance.

To sample from this distribution, first sample Y from $\mathcal{Be}(\alpha, \beta)$ and then return $Z = Y^{-1} - 1$.

8.15 REVISITING THE RATIO-OF-UNIFORMS METHOD

Section 8.13 shows how $a_1 = 0, a_2 = 1$, and a bounded region $\mathcal{P}$ other than rectangular create a bounded-computing-time algorithm for Gamma generation using the ratio-of-uniforms method. This section describes another variant of the method that is particularly well adapted for efficient sampling from Poisson, binomial, and hypergeometric distributions.

As in Section 8.13, assume that we wish to generate samples from a bounded p.d.f. $\{f(z), -\infty < z < \infty\}$. Let

$$r(z) := \frac{f(z)}{\sup_t f(t)} \tag{8.66}$$

and recall that

$$\mathcal{Q}(a_1, a_2) := \{(x, y) : \; x \leq r^{1/2}(a_1 + a_2 y / x), \; -\infty < a, < \infty, a_2 > 0\}.$$

Theorem 8.4 asserts that if (X, Y) is uniformly distributed on $\mathcal{Q}(a_1, a_2)$, then $Z = a_1 + a_2 Y / X$ has the p.d.f. f.

Let $\mathcal{P}$ be the rectangle

$$\mathcal{P} := \{(x, y) : \; 0 \leq x \leq 1, \quad -1 \leq y \leq 1\}. \tag{8.67}$$

Expressions (8.27b) through (8.27d) imply $\mathcal{P} \supseteq \mathcal{Q}(a_1, a_2)$ if and only if

$$0 \leq r(z) \leq 1, \qquad -\infty < z < \infty, \tag{8.68}$$

and

$$-1 \leq \frac{z - a_1}{a_2} r^{1/2}(z) \leq 1, \qquad -\infty < z < \infty. \tag{8.69}$$

Definition (8.66) ensures inequality (8.68). However, two cases need to be considered for inequality (8.69). In the first, $| y | /x = | z - a_1 | /a_2 \leq 1$, for which inequality (8.68) implies inequality (8.69). In the second, $| y | /x = | z - a_1 | /a_2 \geq 1$, which requires

$$r(z) \leq \frac{a_2^2}{(z - a_1)^2} \quad \forall z \in (-\infty, a_1 - a_2) \cup (a_1 + a_2, \infty). \qquad (8.70)$$

Recall that

$$| \mathcal{Q}(a_1, a_2) | = \frac{1}{2a_2} \int_{-\infty}^{\infty} r(z) \mathrm{d}z = \frac{1}{2a_2 \sup_t f(t)}. \qquad (8.71)$$

Then one chooses a_1 and a_2 to minimize the mean number of trials until success

$$\frac{| \mathcal{P} |}{| \mathcal{Q}(a_1, a_2) |} = 4a_2 \sup_t f(t) \qquad (8.72)$$

subject to $-\infty < a_1 < \infty$, $a_2 > 0$, and inequalities (8.68) and (8.70).

The application of the ratio-of-uniforms method to generation from a discrete distribution is now relatively straightforward. Suppose Z has the p.m.f. $\{p_i; i = a, a + 1, \ldots, b\}$. Then one takes

$$f(z) = \begin{cases} p_i & \text{if } i \leq z < i + 1 \quad \text{for } a \leq i \leq b, \\ 0 & \text{otherwise,} \end{cases} \qquad (8.73)$$

and applies the theory of this section to find desirable values of a_1 and a_2. The next several sections illustrate the generality that the method offers in practice.

8.16   POISSON DISTRIBUTION

Let Z have the Poisson p.m.f.

$$p_i = e^{-\lambda} \lambda^i / i!, \qquad \lambda > 0, \qquad i = 0, 1, \ldots, \qquad (8.74)$$

denoted by $\mathcal{P}(\lambda)$. This distribution has the following properties:

1. If $X_1, X_2, \ldots$ are i.i.d. random variables from $\mathcal{E}(1)$ and Z is the smallest nonnegative integer such that $X_1 + \cdots + X_{Z+1} > \lambda$, then Z has the Poisson distribution $\mathcal{P}(\lambda)$.

2. If $U_1, U_2, \ldots$ are i.i.d. random variables from $\mathcal{U}(0, 1)$ and Z is the smallest nonnegative integer such that $\Pi_{i=1}^{Z+1} U_i > \lambda$, then Z has the Poisson distribution $\mathcal{P}(\lambda)$.

3. If Z is from $\mathcal{P}(\lambda)$, then $(Z - \lambda)/\sqrt{\lambda}$ has a distribution that converges to $\mathcal{N}(0, 1)$ as $\lambda \to \infty$.

Algorithm ITR applies the inverse transform method with $p_0 = e^{-\lambda}$ and $c(z) = \lambda/z$, $z > 0$. Although Section 8.1 shows the mean computing time as $O(\lambda)$, its relatively small setup cost allows this procedure to compare favorably timewise with more sophisticated algorithms with bounded mean computing time for small λ. An alternative algorithm, based on property 2, is often encountered in practice, but timing studies in Atkinson (1979) and Schmeiser and Kachitvichyanukul (1981) show it to be more time-consuming that Algorithm ITR. No doubt, the multiple generation of $U_1, U_2, \ldots$ accounts for this excess time consumption.

Let Y be from $\mathcal{N}(0, 1)$. Property 3 implies that treating $\max(0, \lfloor \lambda + 1/2 + Y\sqrt{\lambda} \rfloor)$ as a Poisson variate from $\mathcal{P}(\lambda)$ incurs an error that diminishes as λ increases. The factor $\frac{1}{2}$ is a continuity correction that reduces the error of approximation for every fixed λ. Although this constant-time approach is acceptable in problems in which this error is small relative to errors from other sources, it is unacceptable in problems where results are sensitive to a departure from the Poisson distribution. This may be especially so in applying the Monte Carlo method to estimate the distributional character of a statistic that is a function of Poisson random variables.

Atkinson (1979), Ahrens and Dieter (1982b), Devroye (1981), Schmeiser and Kachitvichyanukul (1981), and Ahrens and Dieter (1991) all propose Poisson generating algorithms with bounded mean computing times. Among these, Algorithm PD in Ahrens and Dieter (1982) appears least time-consuming, but not simple. Since the 1991 proposal of Ahrens and Dieter for $\lambda \geq 1$, which exploits the ratio-of-uniforms method, is considerably simpler and not too much more time-consuming, we describe it here.

We take

$$r(z) := f(z)/\max_x f(x), \qquad -\infty < z < \infty, \tag{8.75}$$

where expression (8.73) defines $\{f(z)\}$ for discrete distributions. The Poisson distribution has its mode at $\lfloor \lambda \rfloor$, so that

$$r(z) = \frac{p_{\lfloor z \rfloor}}{\max_t p_{\lfloor t \rfloor}} = \lambda^{\lfloor z \rfloor - \lfloor \lambda \rfloor} \lfloor \lambda \rfloor! / \lfloor z \rfloor! \tag{8.76}$$

and the mean number of trials is

$$|\mathcal{P}|/|\mathcal{Q}(a_1, a_2)| = 4a_2 e^{-\lambda} \lambda^{\lfloor \lambda \rfloor} / \lfloor \lambda \rfloor!. \tag{8.77}$$

To simplify the minimization problem, Ahrens and Dieter recommend

$$\hat{a}_1 := \hat{a}_1(\lambda) := \lambda + \frac{1}{2} \tag{8.78}$$

for a_1. For this choice, the a_2 that minimizes the ratio (8.77) subject to inequality (8.71b), $\lambda \geq 1$, and $a_2 > 0$ is

$$a_2^* := a_2^*(\lambda) := (\hat{a}_1 - k^*)\sqrt{\lambda^{k^* - \lfloor \lambda \rfloor} \lfloor \lambda \rfloor! / k^*!}, \qquad (8.79a)$$

where k^* takes one of the two values $k^* = \lfloor \hat{a}_1 - \sqrt{2\hat{a}_1} \rfloor$ and $k^* = \lceil \hat{a}_1 - \sqrt{2\hat{a}_1} \rceil$. Also,

$$a_2^*(1) = \frac{3}{2} \qquad (8.79b)$$

and

$$\lim_{\lambda \to \infty} \lambda^{-1/2} a_2^*(\lambda) = \sqrt{2/e} = 0.8578. \qquad (8.79c)$$

Since computing a_2^* in line is relatively time consuming, Ahrens and Dieter recommend the less costly

$$\hat{a}_2 := \hat{a}_2(\lambda) := \sqrt{2\hat{a}_1/e} + \frac{3}{2} - \sqrt{3/e}, \qquad (8.79d)$$

which manifests the same limits (8.79c) as the optimal a_2. Then

$$|\mathcal{P}|/|\mathcal{Q}(\hat{a}_1, \hat{a}_2)| = 4[\sqrt{(2\lambda + 1)/e} + \frac{3}{2} - \sqrt{3/e}]e^{-\lambda} \lambda^{\lfloor \lambda \rfloor} / \lfloor \lambda \rfloor! \qquad (8.80a)$$

and, using Stirling's formula $j! \approx \sqrt{2\pi} j^{j+1/2} \exp(-j + \theta/12j)$ for integer $j \geq 1$ and $0 < \theta < 1$,

$$\lim_{\lambda \to \infty} |\mathcal{P}|/|\mathcal{Q}(\hat{a}_1, \hat{a}_2)| = 4/\sqrt{\pi e} = 1.3688. \qquad (8.80b)$$

Computations in Stadlober (1989) illustrate how well $\hat{a}_1$ and $\hat{a}_2$ fare with respect to the optimal a_1 and a_2. For $\lambda = 5$, $a_1 = 5.277$ and $a_2 = 2.271$ minimize the mean number of trials (8.77). For the recommended $\hat{a}_1 = 5.5$ in expression (8.78), expression (8.79a) gives $a_2^* = 2.425$ as best possible, and expression (8.79d) gives $\hat{a}_2 = 2.461$, the more easily computed bound. Therefore, the mean number of trials based on $\hat{a}_1$ and $\hat{a}_2$ is within 8.37 percent of the achievable minimum, and Stadlober shows that this percentage progressively decreases to zero as λ increases.

Algorithm PRUA* describes a modification of Algorithm PRUA' in Stadlober (1989), which is a modification of Algorithm PRUA in Ahrens and Dieter (1991). Note that the setup calls for a precomputed table of $\gamma_k = \ln k!$ for $k = 0, 1, \ldots, k'$. Moreover, for $k > k'$, $\ln k!$ can be computed by using a truncated form of the expansion (Hastings 1955)

$$\gamma_k := \ln \sqrt{2\pi} + (k + 1/2) \ln k - k + 1/12k - 1/360k^3 + 1/1260k^5 - 1/1680k^7 \text{ as } k \to \infty. \qquad (8.81)$$

Stadlober reports an absolute truncation error less than 7.9×10^{-9} when terms up to $O(1/k^3)$ are retained. The choice of k' is at the user's discretion and, in principle, depends on space availability. Making k' large reduces the need to compute d_7 and γ_k in line and considerably improves the efficiency of the procedure. If Poisson generation with $\lfloor \lambda \rfloor \leq k'$ is contemplated, one can omit the test in the setup entirely. The rejection test $Z \geq d_8$, based on the precision factor d_0, allows the algorithm to avoid computing insignificant probabilities. As before, if repeated sampling is to occur with the same λ, then one can compute the setup prior to any generation, store $d_3, d_4, d_5, d_6, d_7, \gamma_{d_7}, d_8$, and d_9, and use these as input on successive executions.

ALGORITHM PRUA*

Purpose: To generate K randomly from $\mathcal{P}(\lambda)$ with $\lambda \geq 1$.

Input: λ and in-line function for computing $\gamma_k = \ln k!$.

Setup: Precomputed table of $\gamma_k = \ln k!$ for $k = 0, 1, \ldots, k'$.

Constants $d_0 = 7$ (for nine-decimal-digit precision), $d_1 = 2\sqrt{2/e} = 1.715527770$, $d_2 = 3 - 2\sqrt{3/e} = 0.898916162$.

$d_3 \leftarrow \lambda + 1/2$, $d_4 \leftarrow \sqrt{d_3}$, $d_5 \leftarrow d_1 d_4 + d_2$, $d_6 \leftarrow \ln \lambda$, $d_7 \leftarrow \lfloor \lambda \rfloor$, $d_8 \leftarrow \lfloor d_3 + d_0(d_4 + 1) \rfloor$.

If $d_7 > k'$, compute γ_{d_7} in line.

$d_9 \leftarrow d_6 d_7 - \gamma_{d_7}$.

Method:

Repeat

    randomly generate X and Y from $\mathcal{U}(0, 1)$.

    $Z \leftarrow d_3 + d_5(Y - 1/2)/X$

    if $Z < 0$ or $Z \geq d_8$,   (preliminary test 1 fast rejection)

  otherwise:

    $K \leftarrow \lfloor Z \rfloor$; if $K > k'$, compute γ_K in line

    $T \leftarrow K d_6 - \gamma_K - d_9$

    if $T \geq X(4 - X) - 3$, accept   (preliminary test 2 fast acceptance)

    otherwise:

      if $X(X - T) \geq 1$, reject   (preliminary test 3 fast rejection)

      otherwise:

        if $2 \ln X \leq T$, accept

until accept.

Return K.

Preliminary test 1 accounts for the possibility of $Z < 0$. Preliminary tests 2 and 3 use the squeeze method with the inequalities $x - 1/x \leq 2 \ln x \leq -3 + 4x - x^2$, $0 <$

$x \leq 1$, (Section 8.5.1) to reduce the frequency of evaluating $\ln X$ by 86 percent according to Stadlober (1989), who also indicates that nine-digit precision in computation ensures no loss of accuracy in computing T. For γ_k taken exclusively from a stored table, Stadlober also shows that this approach compares well with all other known bounded mean computing time Poisson generation algorithms for $\lambda \geq 5$, except when compared to Algorithm PD of Ahrens and Dieter (1982b). However, the present approach offers considerably greater simplicity in description and implementation than that algorithm does.

Stadlober (1989) shows that, with the exception of Algorithm PD in Ahrens and Dieter (1982b), all bounded mean computing time algorithms for Poisson generation consume more time than Algorithm ITR does for small λ. This remains true whether or not one includes the setup times. Also, using the relatively simple Algorithm ITR for Poisson generation, for say $\lambda \leq 10$, and Algorithm PRUA* otherwise, appears to be a judicious compromise. If speed is the overriding consideration, then the less simple Algorithm PD deserves attention.

8.17 BINOMIAL DISTRIBUTION

Let Z have the binomial distribution

$$p_i = \binom{n}{i} p^i (1-p)^{n-i}, \quad 0 < p < 1, \quad n \text{ integer} > 0, \quad i = 0, 1, \ldots, n, \quad (8.82)$$

denoted by $\mathcal{B}(n, p)$. The distribution has the following properties:

1. If $Z_1, Z_2, \ldots, Z_n$ are independent Bernoulli random variables with parameter p, then $Z = Z_1 + \cdots + Z_n$ is from $\mathcal{B}(n, p)$.

2. Let $\mu = np$. For fixed μ, the distribution of Z converges to $\mathcal{P}(\mu)$ as $n \to \infty$ and $p \to 0$ simultaneously.

3. If Z is from $\mathcal{B}(n, p)$, then $(Z - np)/\sqrt{np(1-p)}$ has a distribution that converges to $\mathcal{N}(0, 1)$ as $n \to \infty$.

Algorithm ITR applies the inverse transform method with $p_0 = (1-p)^n$ and $c(z) = (n - z + 1)p/z(1-p)$, which consumes $O(np)$ time. If one takes $p_0 = (1 - \bar{p})^n$ and $c(z) = (n - z + 1)\bar{p}/z(1 - \bar{p})$, where $\bar{p} = \min(p, 1 - p)$, and returns Z if $p \leq \frac{1}{2}$ and $n - Z$ if $p > \frac{1}{2}$, then the algorithm has $O(n \min(p, 1 - p))$ time. Algorithm BER provides an alternative approach that merely generates n independent Bernoulli samples and takes their total as Z. Although this avoids the need for the logarithmic and exponential evaluations implicit in $(1 - p)^n$, it takes $O(n)$ time. Kachitvichyanukul and Schmeiser (1988) report that the inverse transform approach performs faster than the Bernoulli approach does.

ALGORITHM BER

Purpose: To generate Z randomly from $\mathcal{B}(n, p)$.
Input: n and p.
Method:
$Z \leftarrow 0$ and $i \leftarrow 1$.
While $i \leq n$:
      Randomly generate U from $\mathcal{U}(0, 1)$.
      $Z \leftarrow Z + \lfloor U + p \rfloor$.
      $i \leftarrow i + 1$.
Return Z.

Although asymptotic properties 2 and 3 may tempt one to employ Poisson or normal generation to produce an approximate binomial Z, the resulting error may lead to seriously misleading results in the Monte Carlo experiment.

Fishman (1979), Ahrens and Dieter (1980), Devroye and Naderasamani (1980), Kachitvichyanukul and Schmeiser (1988), and Stadlober (1991) all describe binomial generating algorithms with bounded mean computing times. We describe Stadlober's approach based on the ratio-of-uniforms method for $p \leq \frac{1}{2}, np \geq 1$, and

$$r(z) := \frac{p_{\lfloor z \rfloor}}{\max_t p_{\lfloor t \rfloor}} = \frac{\lfloor (n+1)p \rfloor ! (n - \lfloor (n+1)p \rfloor)!}{\lfloor z \rfloor ! (n - \lfloor z \rfloor)!} [p/(1-p)]^{\lfloor z \rfloor - \lfloor (n+1)p \rfloor}. \qquad (8.83)$$

The mean number of trials is

$$|\mathcal{P}|/|\mathcal{Q}(a_1, a_2)| = 4a_2 \binom{n}{\lfloor (n+1)p \rfloor} p^{\lfloor (n+1)p \rfloor} (1 - p)^{n - \lfloor (n+1)p \rfloor}. \qquad (8.84a)$$

For a_1, Stadlober recommends

$$\hat{a}_1 := np + \frac{1}{2}, \qquad (8.84b)$$

which is optimal for $p = \frac{1}{2}$. For a_2,

$$a_2^* = (\hat{a}_1 - k^*)\sqrt{\frac{m!(n-m)!}{k^*!(n-k^*)!} \left(\frac{p}{1-p}\right)^{k^*-m}}, \quad m := \lfloor (n+1)p \rfloor, \qquad (8.85)$$

where $k^* = \left\lfloor \hat{a}_1 - \sqrt{2\hat{a}_1(1-p)} \right\rfloor$ or $k^* = \left\lceil \hat{a}_1 - \sqrt{2\hat{a}_1(1-p)} \right\rceil$ maximizes $|\mathcal{Q}(\hat{a}_1, a_2)|/|\mathcal{P}|$. Also, for $np = 1$,

$$\lim_{\substack{n \to \infty \\ p \to 0}} a_2^*(n, p) = 1.5, \qquad (8.86a)$$

and for fixed p,

$$\lim_{n \to \infty} \frac{a_2^*(n, p)}{\sqrt{np(1 - p)}} = \sqrt{2/e} = 0.8578. \tag{8.86b}$$

For convenience of computation, Stadlober recommends

$$a_2 = \hat{a}_2 = \sqrt{2 \left[np(1 - p) + \frac{1}{2} \right] / e} + \frac{3}{2} - \sqrt{\frac{3}{e}}, \tag{8.87}$$

which achieves the limits (8.86a) and (8.86b). Also,

$$\lim_{n \to \infty} |\mathcal{P}| / |\mathcal{Q}(\hat{a}_1, a_2^*)| = 4/\sqrt{\pi e} = 1.3688 \tag{8.88a}$$

and

$$\lim_{n \to \infty} |\mathcal{P}| / |\mathcal{Q}(\hat{a}_1, \hat{a}_2)| = 4/\sqrt{\pi e} = 1.3688. \tag{8.88b}$$

Algorithm BRUA* is a modification of Algorithm BRUA in Stadlober (1991). It allows for a precomputed table of $\{\gamma_k = \ln k\,!; \ k = 1, \ldots, k'\}$, and clearly, if $k' \leq n$, then no in-line function evaluation of γ_k occurs. As with Algorithm PRUA*, a large table improves the computing time considerably. If repeated binomial generation with the same n and p is to occur, then computing the setup beforehand, storing it, and using it as input on each successive execution reduces computing time. Proofs of the fast acceptance and fast rejection tests are left to the reader (Exercises 8.51 and 8.52).

ALGORITHM BRUA*

Purpose: To generate Z randomly from $\mathcal{B}(n, p)$ with $n \ \min(p, 1 - p) \geq 1$.

Input: n, p and in-line function for computing $\gamma_k = \ln k!$.

Setup: Precomputed table of $\{\gamma_k = \ln k!; k = 0, 1, \ldots, k'\}$.

Constants $d_0 = 7$ (for nine decimal digit precision), $d_1 = 2\sqrt{\frac{2}{e}} = 1.715527770$, $d_2 = 3 - 2\sqrt{\frac{3}{e}} = 0.898916162$.

$d_3 \leftarrow \min(p, 1 - p), d_4 \leftarrow nd_3 + \frac{1}{2}, d_5 \leftarrow 1 - d_3, d_6 \leftarrow \sqrt{nd_3d_5 + \frac{1}{2}}$

$d_7 \leftarrow d_1d_6 + d_2, d_8 \leftarrow d_3/d_5, d_9 \leftarrow \ln d_8, d_{10} \leftarrow \lfloor (n + 1)d_3 \rfloor,$

$d_{11} \leftarrow \min(n + 1, \lfloor d_4 + d_0d_6 \rfloor).$

If $d_{10} > k'$, compute $\gamma_{d_{10}}$ in line.

If $n - d_{10} > k'$, compute $\gamma_{n - d_{10}}$ in line.

$d_{12} \leftarrow \gamma_{d_{10}} + \gamma_{n - d_{10}}.$

Method:

    Repeat

        randomly generate X and Y from $\mathcal{U}(0, 1)$

        $W \leftarrow d_4 + d_7(Y - \frac{1}{2})/X$

        if $W < 0$ or $W \geq d_{11}$, reject   (fast rejection)

        otherwise:

                $Z \leftarrow \lfloor W \rfloor$

                if $Z > k'$, compute γ_z in line

                if $n - Z > k'$, compute γ_{n-z} in line

                $T \leftarrow (Z - d_{10})d_9 + d_{12} - \gamma_z - \gamma_{n-z}$

                if $X(4 - X) - 3 \leq T$, accept   (fast acceptance)

                otherwise:

                      if $X(X - T) \geq 1$, reject   (fast rejection)

                      otherwise:

                            if $2 \ln X \leq T$, accept

    until accept.

    If $p \leq \frac{1}{2}$, return Z; otherwise, return $n - Z$.

For small $n \min(p, 1 - p)$, Algorithm ITR is less time-consuming than virtually all bounded-time algorithms are for binomial generation. Accordingly, using Algorithm ITR for $n \min(p, 1 - p) \leq 10$, and Algorithm BRUA* otherwise provides a reasonable compromise.

8.18 HYPERGEOMETRIC DISTRIBUTION

The hypergeometric distribution has the form

$$p_z = \binom{\alpha}{z}\binom{\beta}{n-z} \Big/ \binom{\alpha+\beta}{n}, \quad \alpha, \beta \geq 0, \quad \alpha + \beta \geq n \geq 1,$$

$$\max(0, n - \beta) \leq z \leq \min(n, \alpha), \tag{8.89}$$

and is denoted by $\mathcal{H}(\alpha, \beta, n)$. In the study of production and quality control problems, it applies to a finite population of items of size $\alpha + \beta$ of which α have a particular characteristic and β do not. If one randomly draws n items, $n \leq (\alpha + \beta)$, from the population without replacement and Z is the number of the drawn items that have the characteristic, then Z has the p.m.f. (8.89).

Algorithm ITR with $p_0 = \beta!(\alpha + \beta - n)!/(\beta - n)!(\alpha + \beta)!$ and $c(z) = (\alpha - z)(n - z)/(z+1)(\beta - n + z + 1), z \geq \max(0, n - \beta)$, provides the most direct generating method and takes $O\left(\frac{n\alpha}{\alpha+\beta}\right)$ time. Replacing α by $\max(\alpha, \beta)$ and β by $\min(\alpha, \beta)$ everywhere in p_0 and $c(z)$ and returning Z if $\alpha \leq \beta$ and $n - Z$ otherwise reduces the time to $O(n \min(\alpha, \beta)/(\alpha + \beta))$. In the present case, the setup cost for p_0 is considerably larger than it is when Algorithm

ITR is applied to Poisson and binomial generation. Algorithm HYP provides an alternative approach based on direct sampling without replacement and takes $O(n)$ time, but avoids the need to compute p_0. However, it can require n uniform deviates, in contrast to one deviate for Algorithm ITR. If repeated generation with the same α, β, and n is to occur, then precomputing p_0, storing it, and using it on successive executions of Algorithm ITR makes this approach considerably more efficient than sampling without replacement as in Algorithm HYP (Kachitvichyanukul and Schmeiser 1985).

ALGORITHM HYP

Purpose: To generate Z randomly from $\mathcal{H}(\alpha, \beta, n)$.
Input: α, β, and n where $\alpha \geq 1$, $\beta \geq 1$, and $\alpha + \beta \geq n \geq 1$.
Method:
    $d_1 \leftarrow \alpha + \beta - n, d_2 \leftarrow \min(\alpha, \beta), Y \leftarrow d_2$ and $i \leftarrow n$.
    While $Y \times i > 0$:
        Randomly generate U from $\mathcal{U}(0, 1)$.
        $Y \leftarrow Y - \lfloor U + Y/(d_1 + i) \rfloor$.
        $i \leftarrow i - 1$.
    $Z \leftarrow d_2 - Y$.
    If $\alpha \leq \beta$, return Z; otherwise, return $Z \leftarrow n - Z$.

Kachitvichyanukul and Schmeiser (1985) describe a bounded mean computing time algorithm for generating a hypergeometric sample that employs the acceptance–rejection method. Stadlober (1989) describes a bounded mean computing time algorithm that employs the ratio-of-uniforms method of Section 8.15. We describe this last method here.

We have

$$r(z) := \frac{p_{\lfloor z \rfloor}}{\max_t \ p_{\lfloor t \rfloor}} = \frac{m!(\alpha - m)!(n - m)!(\beta - n + m)!}{\lfloor z \rfloor!(\alpha - \lfloor z \rfloor)!(n - \lfloor z \rfloor)!(\beta - n + \lfloor z \rfloor)!} \tag{8.90}$$

and

$$m := \lfloor (n + 1)(\alpha + 1)/(\alpha + \beta + 2) \rfloor$$

with the mean number of trials

$$|\mathcal{P}|/|\mathcal{Q}(a_1, a_2)| = 4a_2 \binom{\alpha}{m}\binom{\beta}{n - m} \Big/ \binom{\alpha + \beta}{n}. \tag{8.91}$$

For a_1 and a_2, Stadlober recommends

$$\hat{a}_1 := n\alpha/(\alpha + \beta) + \frac{1}{2} \tag{8.92a}$$

and

$$\hat{a}_2 := \sqrt{2\left(\sigma^2 + \frac{1}{2}\right)\bigg/ e} + \frac{3}{2} - \sqrt{\frac{3}{e}}, \tag{8.92b}$$

where

$$\sigma^2 := n \cdot \frac{\alpha + \beta - n}{\alpha + \beta - 1} \cdot \frac{\alpha}{\alpha + \beta} \cdot \frac{\beta}{\alpha + \beta}.$$

An extensive numerical study in Stadlober (1989) shows that for $a_1 = \hat{a}_1$ in expression (8.92a), the choice of $a_2 = \hat{a}_2$ in expression (8.92b) was always greater than the a_2 that minimizes the ratio (8.91), namely,

$$a_2^*(\alpha, \beta, n) := (\hat{a}_1 - k^*) e^{(\delta_m - \delta_{k^*})/2}, \tag{8.93}$$

where

$$\delta_j := \gamma_j + \gamma_{\alpha+j} + \gamma_{n-j} + \gamma_{\beta-n+j}$$

and

$$\gamma_j := \ln j!,$$

and k^* assumes the value

$$\left\lfloor a_1 - \sqrt{2a_1 \frac{\beta}{\alpha + \beta}[1 - n/(\alpha + \beta)]} \right\rfloor \quad \text{or} \quad \left\lceil a_1 - \sqrt{2a_1 \frac{\beta}{\alpha + \beta}[1 - n/(\alpha + \beta)]} \right\rceil.$$

Again, computing $\hat{a}_2$ instead of a_2^* requires considerably less time.

Algorithm HRUA* is a modification of Algorithms HRUA and HRUA' in Stadlober (1989). Note that the setup cost is now considerable, and the admonition to precompute the quantities in the setup, if possible, is now strongest. For $k = Z, \alpha + Z, n - Z$, and $\beta - n + Z$, the in-line function for δ_Z takes γ_k from the table if $k \le k^*$ and computes γ_k as in (105) otherwise. Proof of the fast acceptance and rejection tests again are left to the reader (Exercises 8.55 and 8.56).

Let $\mu = n\alpha/(\alpha + \beta)$. For small μ Algorithm HYP is considerably less time-consuming than Algorithm HRUA* would be. Based on the timing study in Stadlober (1989) for algorithms close to HRUA*, using Algorithms HYP for $\mu \le 3$ and HRUA* for $\mu > 0$ provides a reasonable compromise.

ALGORITHM HRUA[*]

Purpose: To generate Z randomly from $\mathcal{H}(\alpha, \beta, n)$.

Input: α, β, n with $1 \leq n \leq (\alpha + \beta)/2$; in-line function for $\gamma_k = \ln k!$; in-line function for $\delta_k = \gamma_k + \gamma_{\min(\alpha,\beta)-k} + \gamma_{m-k} + \gamma_{\max(\alpha,\beta)-m+k}$.

Setup: Precomputed table of $\{\gamma_k = \ln k!; k = 0, 1, \ldots, k'\}$.

Constants $m \leftarrow \min(n, \alpha + \beta - n)$, $d_0 = 7$ (for nine-decimal-digit precision), $d_1 = 2\sqrt{2/e} = 1.715527770, d_2 = 3 - 2\sqrt{3/e} = 0.898916162$.

$d_3 \leftarrow \min(\alpha, \beta)$, $d_4 \leftarrow d_3/(\alpha + \beta)$, $d_5 \leftarrow 1 - d_4$, $d_6 \leftarrow md_4 + \frac{1}{2}$,

$d_7 \leftarrow \sqrt{\frac{\alpha+\beta-m}{\alpha+\beta-1}nd_4d_5} + \frac{1}{2}$, $d_8 \leftarrow d_1d_7 + d_2$,

$d_9 \leftarrow \lfloor (m + 1)(d_3 + 1)/(\alpha + \beta + 2) \rfloor$,

$d_{10} \leftarrow \delta_{d_9}$, $d_{11} \leftarrow \min(\min(m, d_3) + 1, \lfloor d_6 + d_0d_7 \rfloor)$.

Method:

    Repeat

        randomly generate X and Y from $\mathcal{U}(0, 1)$

        $W \leftarrow d_6 + d_8(Y - 1/2)/X$

        if $W < 0$ or $W \geq d_{11}$, reject   (fast rejection)

        $Z \leftarrow \lfloor W \rfloor$.

        otherwise:

            $T \leftarrow d_{10} - \delta_Z$

            if $X(4 - X) - 3 \leq T$, accept   (fast acceptance)

            otherwise:

                if $X(X - T) \geq 1$, reject   (fast rejection)

        otherwise:

            if $2 \ln X \leq T$, accept

    until accept.

        If $\alpha \leq \beta$, $Z \leftarrow m - Z$.

        If $m \geq n$, return Z; otherwise, return $Z \leftarrow \alpha - Z$.

8.19 GEOMETRIC DISTRIBUTION

Let Z have the geometric p.m.f.

$$p_i = (1 - p)p^i, \quad 0 < p < 1, \quad i = 0, 1, \ldots, \qquad (8.94)$$

denoted by $\mathcal{G}e(p)$. This distribution has the following properties:

1. Let $Y_1, Y_2, \ldots$ be a sequence of i.i.d. Bernoulli distributed random variables with success probability $1 - p$. Then

$$Z = \min(i \geq 1 : Y_1 + \cdots + Y_i = 1) - 1.$$

2. Let $\beta = -1/\ln p$ and let X denote a random variable from $\mathcal{E}(\beta)$. Then $\lfloor X \rfloor$ has the p.m.f. (8.94).

Algorithm ITR with $p_0 = 1 - p$ and $c(z) = p, z > 0$, provides a straightforward generating method with an $O(1/(1 - p))$ mean computing time. Observe that the setup cost for p_0 is relatively incidental when compared to the corresponding costs for Poisson, binomial, and hypergeometric generation using Algorithm ITR. However, the $O(1/(1 - p))$ computing time limits the appeal of this approach as p increases. For example, $p = .50$ gives $1/(1 - p) = 2$, whereas $p = .90$ gives $1/(1 - p) = 10$.

Algorithm GEO provides a bounded mean computing time algorithm by exploiting property 2 relating to the exponential and geometric distributions. If repeated generation is to occur with the same p, then computing $\beta = -1/\ln p$ beforehand, storing it, and using it as input to successive executions reduces the unit generation cost. Also, the availability of an efficient source of exponential samples benefits this approach. As p increases, one comes to favor Algorithm GEO over Algorithm ITR. However, the exact p at which one is indifferent as to which of these procedures one uses remains a function of the particular computer and programming language being employed and the form of exponential generation.

ALGORITHM GEO

Purpose: To generate Z randomly from $\mathcal{G}e(p)$.
Input: p.
Setup: $\beta = -1/\ln p$.
Method:

Randomly generate Y from $\mathcal{E}(1)$.
$Z \leftarrow \lfloor \beta Y \rfloor$.
Return Z.

8.20 NEGATIVE BINOMIAL DISTRIBUTION

Let Z have the negative binomial p.m.f.

$$p_i = \frac{\Gamma(r + i)}{\Gamma(i + 1)\Gamma(r)}(1 - p)^r p^i, \qquad 0 < p < 1, \quad r > 0, i = 0, 1, \ldots, \qquad (8.95)$$

denoted by $\mathcal{NB}(r, p)$. This distribution has the following properties:

1. Let $Y_1, Y_2, \ldots$ be an infinite sequence of i.i.d. Bernoulli distributed random variables with success probability $1 - p$. For r integer,

$$Z = \min(i \geq 1 : Y_1 + \cdots + Y_i = r) - r$$

has the p.m.f. (8.95).

2. For r integer, let $Z_1, \ldots, Z_r$ denote i.i.d. random variables from $\mathcal{G}e(p)$. Then $Z = Z_1 + \cdots + Z_r$ has the p.m.f. (8.95).

3. Let X be from $\mathcal{G}(r, 1)$ and let Y given X be from $\mathcal{P}(Xp/(1 - p))$. Then Y has the p.m.f. (8.95).

In the case of r integer, $\{p_i\}$ is sometimes called the Pascal distribution.

Algorithm ITR with $p_0 = (1 - p)^r$ and $c(z) = p(r + z - 1)/z, z \geq 1$, provides a generating method that takes $O(r/(1 - p))$ mean computing time. If repeated generating with the same r and p is to occur, then computing p_0 beforehand, storing it, and using it as input to successive executions spreads the one-time setup cost over all generations, thus reducing the unit cost. For r integer, applying Algorithm ITR with $p_0 = 1 - p$ and $c(z) = p$, $z \geq 1, r$ times and summing the results provides an alternative with $O(r/(1 - p))$ mean computing time that eliminates the setup cost entirely.

Algorithm NEGBIN describes a procedure that exploits property 3. If bounded mean computing time algorithms such as GKM3 and PRUA* are employed for Gamma and Poisson generation, respectively, and $r \geq 1$, then Algorithm NEGBIN, using those algorithms, has a bounded mean computing time.

ALGORITHM NEGBIN

Purpose: To generate Z randomly from $\mathcal{N}\mathcal{B}(r, p)$.
Input: r and p.
Method:
    Randomly generate X from $\mathcal{G}(r, 1)$.
    Randomly generate Z from $\mathcal{P}(Xp/(1 - p))$.
    Return Z.

8.21 MULTINOMIAL DISTRIBUTION

Let $\mathbf{Z} = (Z_1, \ldots, Z_r)^{\mathrm{T}}$ have the multinomial distribution

$$f(z_1, \ldots, z_r) = n! \prod_{i=1}^{r} \left(\frac{p_i^{z_i}}{z_i!} \right), \tag{8.96}$$

where

$$0 < p_i < 1,$$
$$p_1 + \cdots + p_r = 1,$$

$$z_i \text{ integer } \geq 0,$$
$$z_1 + \cdots + z_r = n,$$

and denoted by $\mathcal{M}(n, p_1, \ldots, p_r)$. Observe that Z_j has the binomial p.m.f.

$$f_j(z_j) = \binom{n}{z_j} p_j^{z_j} (1 - p_j)^{n-z_j},$$

and more importantly, Z_j given $Z_1 = z_1, \ldots, Z_{j-1} = z_{j-1}$ has the binomial p.m.f.

$$f_j(z_j | Z_1 = z_1, \ldots, Z_{j-1} = z_{j-1}) = \binom{n - z_1 - \cdots - z_{j-1}}{z_j} w_j^{z_j} (1 - w_j)^{n-z_1-\cdots-z_{j-1}-z_j},$$

where

$$w_j := p_j/(1 - p_1 - \cdots - p_{j-1}), \qquad 2 \leq j \leq r. \tag{8.97}$$

Algorithm MULTN exploits this property. If used together with a bounded mean computing time algorithm such as BRUA*, the worst-case mean computing time is $O(r)$. However, an opportunity does exist to improve efficiency further.

Let

$$S_j := Z_1 + \cdots + Z_j, \qquad j = 1, \ldots, r,$$
$$S_0 := 0,$$

and let N denote the mean number of required binomial generations. Then

$$
\begin{aligned}
\mathrm{E}N &= \sum_{j=1}^{r} j \operatorname{pr}(S_{j-1} < n, S_j = n) \\
&= \sum_{j=1}^{r} [\operatorname{pr}(S_j = n) - \operatorname{pr}(S_{j-1} = n)] \\
&= \sum_{j=1}^{r} j \left[\left(\sum_{i=1}^{j} p_i \right)^n - \left(\sum_{i=1}^{j-1} p_i \right)^n \right] \\
&= r - \sum_{i=1}^{r-1} \left(\sum_{k=1}^{i} p_k \right)^n .
\end{aligned}
\tag{8.98}
$$

Clearly, $\mathrm{E}N$ is minimized if $p_1 \geq p_2 \geq \cdots \geq p_r$, suggesting that one rearrange cell probabilities into descending order before executing Algorithm MULTN. The saving is especially substantial if a single p_j is considerably larger than the remaining ones.

ALGORITHM MULTN

Purpose: To generate $\mathbf{Z} = (Z_1, \ldots, Z_r)^{\mathrm{T}}$ randomly from the multinomial distribution $\mathcal{M}(n, p_1, \ldots, p_r)$.

Input: $n, r, p_1, \ldots, p_r$.

Method:

   $q \leftarrow 1; m \leftarrow n; j \leftarrow 1$.
   Until $m \cdot (r - j + 1) = 0$:
         Randomly generate Z_j from $\mathcal{B}(m, p_j/q)$.
         $m \leftarrow m - Z_j; q \leftarrow q - p_i; j \leftarrow j + 1$.
   Return $\mathbf{Z} = (Z_1, \ldots, Z_{j-1}, 0, \ldots, 0)^{\mathrm{T}}$.

8.22   BERNOULLI TRIALS

In principle, generating a sequence of n independent Bernoulli trials with success probability p is straightforward. One generates $U_1, \ldots, U_n$ from $\mathcal{U}(0, 1)$ and forms $Z_i = \lfloor U_i + p \rfloor$ for $i = 1, \ldots, n$. This takes $\mathrm{O}(n)$ time. If np is small, few successes occur, making an alternative worth considering. Let X be from $\mathcal{G}e(1 - p)$, where one can regard $T = 1 + X$ as the time to the first success in an infinite Bernoulli sequence. Then Algorithm SKIP generates the set of indices or times $T_1, \ldots, T_{j-1}$ at which successive successes occur. The computing time for Algorithm ITR with $p_0 = p$ and $c(z) = 1 - p, z > 0$, is $\mathrm{O}(np/(1 - p))$, which, when p is small, more than compensates for the added cost of generating X from $\mathcal{G}e(1 - p)$.

ALGORITHM SKIP

Purpose: To generate random times $T_1, \ldots, T_{j-1}$ at which Bernoulli trial successes occur.

Input: Success probability p and number of successes n.

Method:

   $T_0 \leftarrow 0; j \leftarrow 0$.
   Until $T_j > n$:
         $j \leftarrow j + 1$.
         Randomly generate X from $\mathcal{G}e(1 - p)$.
         $T_j \leftarrow T_{j-1} + X + 1$.
   If $j = 1$, return indicating no successes.
   Otherwise, return $T_1, \ldots, T_{j-1}$.

8.23 LESSONS LEARNED

- All algorithms for generating samples from probability distributions rely on the availability of a procedure for generating a sequence of numbers with values in $(0, 1)$ that resembles a sequence of i.i.d. random variables from the uniform distribution on this interval. Chapter 9 describes *pseudorandom number generation*, an essential ingredient in creating these sequences.

- The principal algorithms for generating these samples are the *inverse-transform, composition, acceptance–rejection, ratio-of-uniforms*, and *exact-approximation* methods. All are theoretically correct and apply to both continuous and discrete distributions.

- Random variate generating algorithms fall into three categories: those with computing times that are independent of parameter values (1), those that tend to grow as functions of the parameters of the distribution (2), and those that have an upper bound on computing time (3). Whenever possible, it is preferable to use an algorithm of type (1) or type (3).

- For continuous distributions, the inverse-transform method usually executes in time independent of parameter values. Examples are exponential and normal variate generation

- The *cutpoint method*, based on the inverse-transform approach, and the *alias method*, based on the composition approach, allow sampling from a table in time independent of its size.

- Virtually all competitive acceptance–rejection and ratio-of-uniforms algorithms have upper bounds on computing time.

- The *squeeze method* (Section 8.5.1) provides a means of further reducing computing times for the acceptance–rejection and ratio-of-uniforms techniques.

- Virtually all simulation languages provide library routines for sampling from commonly encountered distributions. These include the exponential, Gamma, Erlang, normal, uniform, Weibull, binomial, Poisson, geometric, and negative binomial. Some routines may be based on approximation rather than theoretically exact algorithms. Also, some may have parameter-dependent computing times.

8.24 EXERCISES

INVERSE TRANSFORM METHOD

8.1 Prove that if U is from $\mathcal{U}(0, 1)$, then $V = 1 - U$ is from $\mathcal{U}(0, 1)$.

8.2 Show that $X = -\beta \ln U$ is from $\mathcal{E}(\beta)$.

8.3 For the trapezoidal p.d.f.

$$f(z) = \begin{cases} \dfrac{z - c}{ab} & \text{if } c \leq z \leq a + c, \\[2mm] \dfrac{1}{b} & \text{if } a + c \leq z \leq b + c, \\[2mm] \dfrac{a + b + c - z}{ab} & \text{if } b + c \leq z \leq a + b + c, \quad 0 < a \leq b: \end{cases} \tag{8.99}$$

a. Devise an algorithm for generating samples based on the inverse transform method.

b. Show that for appropriately chosen a_1, a_2, and a_3, and U_1 and U_2 independent of $\mathcal{U}(0, 1)$, $Z = a_1 U_1 + a_2 U_2 + a_3$ has the trapezoidal p.d.f. in equation (8.1).

8.4 Let $V_1, \ldots, V_n$ be i.i.d. from $\mathcal{U}(a, b)$ and let $V_{(1)}, V_{(2)}, \ldots, V_{(n)}$ denote the order statistics of $V_1, \ldots, V_n$. That is, $V_{(1)} \leq V_{(2)} \leq \cdots \leq V_{(n)}$. It is known that $V_{(j)}$ has the p.d.f.

$$f_j(v) = \frac{\Gamma(n + 1)}{\Gamma(j)\Gamma(n - j + 1)}(v - a)^{j-1}(b - v)^{n-j}(b - a)^{-n} \quad 0 \leq a \leq v \leq b$$

where $\Gamma(k) = (k - 1)!$ for k integer ≥ 1. Let $X_1, \ldots, X_n$ denote i.i.d. random variables that have the non-central Cauchy distribution

$$F(x) = \frac{1}{2} + \frac{1}{\pi} \tan^{-1}\left(\frac{x - c}{d}\right) \quad d > 0, -\infty < c < \infty, -\infty < x < \infty.$$

a. Describe an algorithm for randomly generating $Y = \max(X_1, \ldots, X_n)$ using exactly one uniform deviate U from $\mathcal{U}(0, 1)$.

b. Describe an algorithm for randomly generating $Y = \max(X_1, \ldots, X_n)$ and $Z = \min(X_1, \ldots, X_n)$ using exactly two independent uniform deviates U and V from $\mathcal{U}(0, 1)$. **Hint**: Conditioning plays a role.

8.5 Apply the inverse transform method for generating a sample, X, from the d.f.

$$F(x) = \left(1 - \frac{2 \cos^{-1} x}{\pi}\right)^2 \quad 0 \leq x \leq 1.$$

8.6 Let U and V be independent samples, each drawn from $\mathcal{U}(0, 1)$. Use U and V together with the inverse transform and composition methods to generate a sample X from the d.f.

$$F(x) = \int_0^\infty h(y)G(x, y)dy,$$

where

$$h(y) = \frac{1}{\beta}e^{-y/\beta} \quad \beta > 0, \quad y \geq 0$$

and

$$G(x, y) = \frac{1}{1 + e^{-(x-y)/c}} c > 0, -\infty < x < \infty.$$

8.7 For the triangular p.d.f.

$$f(z) = \begin{cases} \dfrac{2(z - m_1)}{(m_3 - m_1)(m_2 - m_1)} & \text{if } m_1 \leq z \leq m_2, \\ \dfrac{2(m_3 - z)}{(m_3 - m_1)(m_3 - m_2)} & \text{if } m_2 \leq z \leq m_3 : \end{cases}$$

a. Devise an algorithm for generating random samples based on the inverse transform method.

b. For the special case of $m_2 - m_1 = m_3 - m_2$, devise a generating algorithm that relies on a linear combination of independent U_1 and U_2 each from $\mathcal{U}(0, 1)$.

CUTPOINT METHOD

8.8 Derive the mean generating cost $d_0 \mathrm{E}J_m^* + (d_1 - d_0)\mathrm{E}(J_m^*|Z > b*)(1 - q_{b*})$ for Algorithm CM.

ALIAS METHOD

8.9 Apply cutpoint method 2 for large $b - a$ to the Poisson distribution with mean $\lambda = 100$ and using $m = 25, 50, 100, 200$ cutpoints. For each m, compute:

a. Cutpoints $I_1, \ldots, I_m$

b. Mean number of comparisons

c. Total space requirements.

✏ **8.10** Prove that every p.m.f. $\{p_1, \ldots, p_n\}$ can be expressed as an equiprobable mixture of $n - 1$ two-point p.m.f.s.

✏ **8.11** Define $k = b - a + 1$ and suppose that $p_a < 1/k$ and $p_i \geq 1/k, a + 1 \leq i \leq b$. Show that $r_l \leq 1, \forall l \in \mathcal{G}$, in step $*$ in Algorithm ALSET. This property ensures that the algorithm continues to iterate until all aliases are assigned.

ACCEPTANCE–REJECTION METHOD

✏ **8.12** Consider the discrete p.m.f.

$$f(z) = cg(z)h(z), \quad z = a, a + 1, \ldots, b,$$

where

$$h(z) \geq 0, \quad \sum_{z=a}^{b} h(z) = 1,$$
$$c = \max_z [f(z)/h(z)], \quad 0 \leq g(z) \leq 1.$$

Suppose one samples Z from $\{h(z)\}$ and U from $\mathcal{U}(0, 1)$. If $U \leq g(Z)$, show that Z has the p.m.f. $\{f(z)\}$.

✏ **8.13** Let X denote a random variable with distribution function

$$F(x) = \left(\frac{x - a}{b - a}\right)^{\alpha} \quad \alpha > 0, \quad 0 \leq a \leq x \leq b. \tag{8.100}$$

Give an algorithm for sampling from (1) based on the acceptance-rejection method. Identify any additional restrictions on α, a, and b and give the mean number of trials until success.

✏ **8.14** Let X have bounded p.d.f. f over the finite domain $[a, b]$ and define $\lambda = \max_x[f(x)]$. Let U and Y be independent and from $\mathcal{U}(0, 1)$ and $\mathcal{U}(a, b)$, respectively, and let S denote the event $U \leq f(Y)/\lambda$.

a. Show that if S holds, then Y has the p.d.f. f.

b. Determine the success probability $\text{pr}(S)$.

c. Suppose one randomly generates X from the p.d.f.

$$f(x) = \begin{cases} (e^{-x+1} - 1)/(e - 2), & 0 \leq x \leq 1, \\ 0, & \text{otherwise,} \end{cases}$$

using the aforementioned acceptance–rejection scheme. Determine the corresponding $\text{pr}(S)$.

d. Describe an upper bound pretest that reduces the mean number of exponentiations until success. Do not address the issue of a lower bound pretest. **Hint:** A picture may be helpful.

e. Modify the generating method in part c to increase $\text{pr}(S)$ by $\frac{1}{2}$ without introducing any new nonlinear transformations to generate U and Y. Hint: The picture in part d may again be helpful.

8.15 (Fishman 1976). Using the p.d.f.

$$h(z, v) = v^{-1} e^{-z/v}, \qquad v > 0, \ z \geq 0 :$$

a. Devise a factorization $c(\alpha, v) g(z, \alpha, v) h(z, v)$ for the Gamma p.d.f.

$$f(z) = \frac{z^{\alpha-1} e^{-z}}{\Gamma(\alpha)}, \qquad \alpha \geq 1, \ z \geq 0,$$

for implementing the acceptance–rejection method.

b. Find the v that minimizes the mean number of trials.

c. Show analytically how the mean number of trials varies as a function of increasing α.

d. Prepare an algorithm for Gamma generation that minimizes the mean number of exponential and logarithmic function evaluations on each trial. Assume no pretest.

8.16 (Cheng 1977). Using the p.d.f.

$$h(z, \alpha) = \lambda \mu z^{\lambda-1}/(\mu + z^\lambda)^2, \qquad z \geq 0,$$

where

$$\mu = \alpha^\lambda,$$
$$\lambda = (2\alpha - 1)^{1/2}, \qquad \alpha \geq 1 :$$

a. Devise a factorization $c(\alpha)g(z, \alpha)h(z, \alpha)$ for the Gamma p.d.f.

$$f(z) = \frac{z^{\alpha-1}e^{-z}}{\Gamma(\alpha)}, \qquad \alpha \geq 1, \; z \geq 0,$$

for implementing the acceptance–rejection method.

b. Show that $\lim_{\alpha\to\infty} c(\alpha) = 2/\sqrt{\pi}$.

c. Other choices of μ and λ are possible. Explain the appeal of the choices made here.

d. Prepare an algorithm for Gamma generation that minimizes the number of exponential and logarithmic function evaluations per trial. Assume no pretest.

8.17 (Ahrens and Dieter 1974b). Using the truncated Cauchy p.d.f.

$$h(z, \beta, \eta) = \frac{\beta}{\theta\pi[\beta^2 + (z-\eta)^2]}, \qquad \beta > 0, \; -\infty < \eta < \infty, \; z \geq 0,$$

where

$$\theta = \frac{1}{2} - \frac{1}{\pi}\tan^{-1}(-\eta/\beta):$$

a. Derive a factorization $c(\alpha, \beta, \eta)g(z, \alpha, \beta, \eta)h(z, \beta, \eta)$ for the Gamma p.d.f.

$$f(z) = \frac{z^{\alpha-1}e^{-z}}{\Gamma(\alpha)}, \qquad \alpha \geq 1, \; z \geq 0,$$

for implementing the acceptance–rejection method.

b. Show that $\eta = \alpha - 1$ and $\beta = (2\alpha - 1)^{1/2}$ are acceptable choices and give a rationale for using them.

c. With the choice of β and η in part b, show that $\lim_{\alpha\to\infty} c(\alpha, \beta, \delta) = \pi/2$.

d. Prepare an algorithm for Gamma generation that minimizes the number of exponential and logarithmic function evaluations per trial. Assume no pretest.

SQUEEZE METHOD

8.18 Prove the inequalities

$$1 - x \leq e^{-x} \leq 1 - x + x^2/2, \qquad x \geq 0.$$

✎ **8.19** Prove the inequalities

$$1 - 1/x \leq \ln x \leq -1 + x, \qquad x \geq 0.$$

✎ **8.20** Prove the inequalities

$$x - 1/x \leq 2 \ln x \leq -3 + 4x - x^2, \qquad 0 < x \leq 1.$$

✎ **8.21** Prove the results for $\text{pr}(S_L)$ and $\text{pr}(S_U)$ in expressions (8.23a) and (8.23b), respectively.

✎ **8.22** Prove the result in expression (8.23c).

✎ **8.23** For the algorithm developed in Exercise 8.15:

a. Devise a pretest that reduces the mean number of exponential and logarithmic function evaluations.

b. Derive an expression for the reduced mean number of function evaluations in using the pretest and compare it with the corresponding mean in Exercise 8.15 as α increases.

✎ **8.24** Consider the inequality $\phi z - \ln \phi - 1 \geq \ln z$, $\forall \phi$ and $z > 0$. For the algorithm in Exercise 8.16:

a. Devise a pretest that reduces the number of exponential and logarithmic function evaluations.

b. Derive an expression for the mean number of function evaluations in using the pretest and compare it with the corresponding mean in Exercise 8.16 as α increases.

✎ **8.25** For the algorithm in Exercise 8.17:

a. Devise a pretest that reduces the number of exponential and logarithmic function evaluations.

b. Derive an expression for the reduced mean number of function evaluations and compare it with the corresponding mean in Exercise 8.17 as α increases.

✎   **8.26** Let X denote a continous random variable with p.d.f.

$$f(x) = \frac{1}{2} \sin x \quad 0 \le x \le \pi. \tag{8.101}$$

a. Devise a factorization for applying the acceptance-rejection method to sample X from $\{f(x)\}$.

b. Based on this factorization, describe an algorithm for generating a sample.

c. Give the probability of success on a trial and the mean number of trials until success.

d. Give the mean number of sine evaluations until success.

e. Using the squeeze method, devise a piecewise linear lower bound, $\{g_L(x)\}$, for $\{g(x)\}$ that reduces the mean number of sine evaluations until success.

f. Describe an algorithm that incorporates this lower bound into the acceptance–rejection method.

g. Give the mean number of sine evaluations until success for the algorithm in part f.

RATIO-OF-UNIFORMS METHOD

✎   **8.27** For the p.d.f. in Exercise 8.14:

a. Describe a procedure based on the ratio-of-uniforms method for randomly generating a sample.

b. Determine the acceptance probability $\text{pr}(T)$.

c. Devise a pretest that reduces the frequency of exponential and logarithmic function evaluations.

d. Determine the mean number of exponential and logarithmic function evaluations until success.

✎   **8.28** Prove $\text{pr}(T_L)$ and $\text{pr}(T_U)$ in expressions (8.33a) and (8.33b), respectively.

✎   **8.29** (Kinderman and Monahan 1980). Consider generating samples from the Gamma p.d.f.

$$f(z) = z^{\alpha-1} e^{-z} / \Gamma(\alpha), \qquad > 1, \ z \ge 0,$$

✎ **8.30** Using the ratio-of-uniforms method with $a_1 = \alpha - 1$ and $a_2 = 1$.

a. Give $\{r(z)\}$ and devise the bounds x^*, y_*, and y^* in expression (8.27a).

b. Prepare an algorithm for generating random samples from $\mathcal{G}(\alpha, 1)$.

c. Derive an expression for the mean number of trials until success and characterize its behavior as $\alpha \to \infty$.

d. Define $d = \ln[\alpha/(\alpha - 1)]$ and $W = (Z + 2 \ln X)/(\alpha - 1)$. Let T_L denote the event $W \leq 2 - \alpha/Z + d$. Show that T_L provides an acceptance pretest that reduces the mean number of exponential and logarithmic function evaluations.

e. Derive an expression for $\text{pr}(T_L)$ and characterize its behavior as $\alpha \to \infty$.

f. For d and W as defined in part d, let T_U denote the event $W \geq Z/\alpha + d$. Show that T_U provides a rejection pretest that reduces the mean number of exponential and logarithmic function evaluations.

g. Derive an expression for $\text{pr}(T_U)$ and characterize its behavior as $\alpha \to \infty$.

h. Amend the algorithm of part b to incorporate the pretests of parts d and f.

i. Characterize the mean number of exponential and logarithmic function evaluations as $\alpha \to \infty$.

✎ **8.31** Give an algorithm for sampling from (8.100) based on the ratio-of uniforms method. Identify any additional restrictions on α, a, and b and give the mean number of trials until success.

✎ **8.32** For the p.d.f in Exercise 8.26,

a) Devise a formulation for sampling X from $\{f(x)\}$ using the ratio-of-uniforms method with the enclosing $\mathcal{P}$ being the smallest rectangle containing $\mathcal{Q}$.

b. Describe an algorithm for implementing the procedure.

c. Give the mean number of trials until success.

d. Devise a pretest for reducing the mean number of sine evaluations until success.

e. Amend the algorithm of part a to incorporate the pretest.

f. For the amended algorithm of part e, give the mean number of sine evaluations until success.

✏ **8.33** For the p.d.f in Exercise 8.26,

a. Devise a formulation for sampling X from $\{f(x)\}$ using the ratio-of-uniforms method with the $\mathcal{P}$ being the smallest right triangle enclosing $\mathcal{Q}$. For this alternative approach:

b. Describe an algorithm for implementing this procedure.

c. Give the mean number of trials until success.

d. Devise a pretest for reducing the mean number of sine evaluations until success.

e. Amend the algorithm of part a to incorporate the pretest.

f. For the amended algorithm of part e, give the mean number of sine evaluations until success.

EXACT-APPROXIMATION METHOD

✏ **8.34** (Marsaglia 1984). For the half-normal d.f. $F(z) = \sqrt{2/\pi} \int_0^z e^{-w^2/2}\,dw$, consider the approximation

$$M(z) := (1 - e^{-z^2/c})^{1/2}, \qquad c > 0,\ z \geq 0.$$

a. Find $\{m(x), 0 \leq x \leq 1\}$, the inverse of $\{M(z)\}$.

b. Show that $\{f_X(x)\}$ in expression (8.36) takes the form

$$f_X(x) = cx(1-x^2)^{c/2-1}/\sqrt{\pi/2}\,m(x), \qquad 0 \leq x \leq 1.$$

c. For the corresponding $\{f^*(x)\}$ in expression (8.36) to be a p.d.f., one requires

$$\min_{0 \leq x \leq 1} cx(1-x^2)^{c/2-1}/\sqrt{\pi/2}\,m(x) \geq p, \qquad 0 \leq p \leq 1.$$

Determine the c that maximizes the left side of this inequality and then determine the largest possible value for p.

EXPONENTIAL GENERATION

✏ **8.35** Let X and Y be independent random variables from $\mathcal{U}(0, 1)$ and $\mathcal{U}(0, 2/e)$, respectively.

a. Using the ratio-of-uniforms approach, describe an algorithm for exponential generation.

b. Let W denote the event $Y \le -2X \ln X$. Show that $\mathrm{pr}(W) = e/4 = 0.6796$.

c. Using the inequalities (8.19), find a function $\{h_*(x)\}$ that minorizes $-2x \ln x$ and a function $\{h^*(x)\}$ that majorizes $-2x \ln x$.

d. Let W_L denote the event $Y \le h_*(X)$ and W_U the event $Y \le h^*(X)$. Derive expressions for $\mathrm{pr}(W_L)$ and $\mathrm{pr}(W_U)$ and evaluate them.

e. Determine the mean required number of evaluations of $\ln X$.

NORMAL GENERATION

✏ **8.36** Let U_1 and U_2 be independent random variables from $\mathcal{U}(-\frac{1}{2}, \frac{1}{2})$. Show that if $U_1^2 + U_2^2 \le \frac{1}{4}$, then $Y = U_1/(U_1^2 + U_2^2)^{1/2}$ has the same distribution as $\cos 2\pi U$ and $Z = U_2/(U_1^2 + U_2^2)^{1/2}$ has the same distribution as $\sin 2\pi U$, where U is from $\mathcal{U}(0, 1)$. See Theorem 8.5.

CAUCHY GENERATION

✏ **8.37** Let X_1 and X_2 be independent random variables from $\mathcal{U}(-\frac{1}{2}, \frac{1}{2})$.

a. Show that if $X_1^2 + X_2^2 \le 1$, then $Z = X_1/X_2$ has the Cauchy distribution $\mathcal{C}(0, 1)$.

b. Determine $\mathrm{pr}(\text{acceptance})$.

GAMMA GENERATION

✏ **8.38** Show that if X is from $\mathcal{G}(\alpha, 1)$, then $Z = \beta X$ is from $\mathcal{G}(\alpha, \beta)$.

✏ **8.39** Prove the correctness of Algorithm GS*.

✏ **8.40** Prove the result in expression (8.54b).

✏ **8.41** Prove the result in expression (8.57).

✏ **8.42** Determine the mean number of logarithmic function evaluations for Algorithm GKM1.

✏ **8.43** For Algorithm GKM2:

a. Show that if $0 < X' < 1$, then (X', Y') is uniformly distributed on $\mathcal{P}'(\alpha) \cap \mathcal{I}^2$, and on any trial

$$\mathrm{pr}(0 < X' < 1) = 1 - \frac{e + 2}{2\sqrt{\alpha e}(\sqrt{e} + \sqrt{2})}.$$

b. Determine the mean number of logarithmic function evaluations.

✏ **8.44** Determine the feasibility of employing the acceptance–rejection method for sampling Z from $\mathcal{G}(\alpha, 1), \alpha \geq 1$, for $\{h(z) = \sqrt{2/\pi}e^{-z^2/2}, z \geq 0\}$.

BETA GENERATION

✏ **8.45** Show that $\{c(\alpha, \beta, t), 0 \leq t \leq 1\}$ in expression (8.62b) is bounded for all $0 < \alpha$, $\beta < 1$. Find $\sup_{0 < \alpha, \beta \leq 1} c(\alpha, \beta, t^*)$ for t^* in expression (8.63).

✏ **8.46** For $c(\alpha, \beta)$ as in expression (8.65):

a. Show that $c(\alpha, \beta) \leq 4/e$ for all $\alpha, \beta > 0$.

b. Show that $\lim_{\beta \to 0} c(\alpha, \beta) = 4$ for $0 < \alpha \leq 1$.

POISSON GENERATION

✏ **8.47** Derive an expression for the mean number of evaluations of $\ln X$ in Algorithm PRUA*.

✏ **8.48** For $\lambda > 1.75$ and $\hat{a}_1 = \lambda + \frac{1}{2}$, show that a_2^* in definition (8.79a) with $k^* = \lceil \hat{a}_1 - \sqrt{2\hat{a}_1} \rceil$ minimizes the mean number of trials (8.77). Hint: First, show that the continuous version of the problem has its optimal solution at $z_0 > \lceil \hat{a}_1 - \sqrt{2\hat{a}_1} \rceil$.

✏ **8.49** Prove the validity of preliminary test 2 in Algorithm PRUA*.

✏ **8.50** Prove the validity of preliminary test 3 given that preliminary test 2 fails in Algorithm PRUA*.

BINOMIAL GENERATION

✏ **8.51** Prove the validity of the fast acceptance test in Algorithm BRUA*.

✏ **8.52** Prove the validity of the second fast rejection test given the failure of the fast acceptance test in Algorithm BRUA*.

8.53 Derive an expression for the mean number of evaluations of $\ln X$ in Algorithm BRUA*.

8.54 Derive an expression for the mean number of logarithmic evaluations in Algorithm BRUA*.

HYPERGEOMETRIC GENERATION

8.55 Prove the validity of the fast acceptance test in Algorithm HRUA*.

8.56 Prove the validity of the second fast rejection test given the failure of the fast acceptance test in Algorithm HRUA*.

8.57 Derive an expression for the mean number of logarithmic evaluations in Algorithm HRUA*.

NEGATIVE BINOMIAL GENERATION

8.58 Prove that if X is from $\mathcal{G}(r, 1)$ and Y given X is from $\mathcal{P}(Xp/(1-p)), 0 < p < 1$, then Y has the negative binomial distribution $\mathcal{NB}(r, p)$.

HANDS-ON-EXERCISES

These exercises provide opportunities to prepare and execute computer programs for sampling on a computer and to analyze the results. Each requires a source of pseudorandom numbers as input.

CUTPOINT METHOD

H.1 For the distribution

i	p_i	i	p_i
1	.001	11	.040
2	.008	12	.108
3	.010	13	.162
4	.011	14	.037
5	.081	15	.012
6	.049	16	.028
7	.048	17	.041
8	.059	18	.029
9	.034	19	.100
10	.142		

a. Use Algorithm CMSET to determine $m = 19$ cutpoints.

b. Using these cutpoints as input to a computer program based on Algorithm CM, randomly generate 10,000 samples from the distribution. Create a plot that compares the histogram of these data with the distribution $\{p_i\}$ and record the computing time per sample (for comparison with the corresponding time for the alias method in Algorithm ALIAS in Exercise 8.H.2).

ALIAS METHOD

H.2 For the distribution in Exercise 8.H.1:

a. Use Algorithm ALSET to determine the aliases $B_1, \ldots, B_{19}$ and the probabilities $P(A_1), \ldots, P(A_{19})$.

b. Using the output of Algorithm ALSET as input to a computer program based on Algorithm ALIAS, randomly generate 10,000 samples from the distribution. Create a plot that compares the histogram of the data with the distribution $\{p_i\}$ and record the computing time per sample.

c. Compare the computing time per sample here with that in Exercise 8.H.1

GAMMA GENERATION

H.3 This exercise compares the performances of the acceptance–rejection and ratio-of-uniforms methods for generating random samples from $\mathcal{G}(\alpha, 1)$, $1 \le \alpha \le 2$.

a. Using the algorithm with the optimal choice of ν and pretest devised in Exercises 3 and 18, prepare a computer program to generate samples from $\mathcal{G}(\alpha, 1)$, $\alpha \ge 1$.

b. Prepare a computer program based on Algorithm GKM1.

c. For each program and each $\alpha \in \{1.2, 1.4, 1.6, 1.8, 2\}$, generate 10,000 samples from $\mathcal{G}(\alpha, 1)$ and record the computing time.

d. Compare the 10 recorded times and determine which algorithm is preferred for $\alpha \in [1, 2]$.

e. Based on analytical considerations, give an argument as to why you would or would not have anticipated these relative performances.

POISSON GENERATION

H.4 This exercise applies the ratio-of-uniform method to sampling from the Poisson distribution. Section 8.15 contains the supplementary theory on which the method is based.

You are encouraged to study it and Section 8.16 carefully beore performing the computing assignment. For $\lambda = 1, 2, 4, 8, 16, 32$, generate independent samples of size $n = 100,000$.

a. Using a class-adopted simulation programming language, write a program to generating samples from the Poisson distribution using its built-in-routine for generating Poisson variates and then execute it. For each λ, record the CPU generating time and compute the sample average and sample variance of the n outcomes.

b. Write a program in a class-adopted programming language to generate samples from the Poisson distribution using Algorithm PRUA* in Section 8.16 and execute it. For each λ, compute $d_3, \ldots, d_9$ within the program on each of the n calls to PRUA*. For each λ, record the CPU generating time and compute the sample average and sample variance of the n outcomes. Also, for each λ create and display the histogram of number of trials until success, its sample average, and sample variane. In programming these computations, avoid storing data for all n trials by computing the sample average and sample variance recursively (e.g., Section 6.4. Recursive Computation) and, similarly, a histogram with ordinates $\{1, 2, \ldots, 100\}$. The time to compute these quantities should be relatively negligble.

c. For each λ, compute $d_3, \ldots, d_9$ and install the computed values in the amended program of part b. Execute the program and again record the CPU generating time and compute the sample average and sample variance of the n outcomes.

d. How do the three programs compare with regard to CPU generating time?

e. For each λ, how do the sample averages of the number of trials compare with theory?

Note: If the simulation programming language selected for part a allows for programming as in parts b and c and is also used there, the comparison will be more meaningful. If $n = 100,000$ does not suffice to get meaningful estimates of average generation time, increase n.

NEGATIVE BINOMIAL GENERATION

H.5 This exercise applies the ration-of-uniform method to sampling from the negative binomial distribution. Section 8.15 contains the supplementary theory on which the method is based. You are encouraged to study it study it and Sections 8.13 ($\alpha > 1$), 8.16, and 8.20 care fully before performing the computing assignment. For each $\lambda := \frac{rp}{1-p} \in \{1, 2, 4, 8, 16, 32\}$ and $r \in \{1, 5, 2.5, 5, 10\}$, generate samples of size $n = 100,000$.

a. Using a class-adopted simulation programming language, write a program to generate samples from the negative binomial distribution using its built-in routine for generating negative binomial variates and then execute it. If note exists, use Algorithms NEGBIN with the language's built-in Poisson and Gamma routines. For each λ and r, record CPU generating time and compute the sample average and sample variance of the n outcomes.

b. Write and execute a program in a class-adopted programming language to generate samples from the negative binomial distribution using Property 1 or 2 of the Poisson distribution (Section 8.16) and Algorithms NEGBIN, GKM3, and PRUA*. For each λ and r, compute $d_3, \ldots, d_9$ within the program on each of the n calls to PRUA*. For each λ and r, record the CPU generating thime excute and comput the sample average variance of the n outcomes. Also, for each λ and r, create and display the histogram of number of trials until success, its sample average, and its sample variance. In programming these computations, avoid storing data for all n trials by computing the sample mean and variance recursively (e.g., Section 6.4 Recursive Computation) and, similarly, a histogram with ordinates $\{1, 2, \ldots, 100\}$. The time to compute these quantities should be relatively negligible.

c. How do the two programs compare with regard to CPU generating time?

d. For each λ and r, how do the sample averages of number of trials compare?

Note: If the simulation programming language selected for part a allows for programming as in part b and is also used there, the comparsion will be more meaningul. If $n = 100,000$ does not suffice to get meninful estimates of average generation time, increase n.

8.25 REFERENCES

Abramowitz, M., and I. Stegun (1964). *Handbook of Mathematical Functions*, Applied Mathematics Series 55, National Bureau of Standards, Washington, D.C.

Afflerbach, L., and W. Hörmann (1990). Nonuniform random numbers: a sensitivity analysis for tranformation methods, *Lecture Notes in Economics and Mathematical Systems*, G. Plug and U. Dieter, eds., Springer-Verlag, New York.

Agrawal, A., and A. Satyanarayana (1984). An $O(|\mathcal{E}|)$ time algorithm for computing the reliability of a class of directed networks, *Oper. Res.*, **32**, 493–515.

Aho, A.V., J.E. Hopcroft, and J.D. Ullman (1974). *The Design and Analysis of Computer Algorithms*, Addison-Wesley, Reading, MA.

Ahrens, J.H. (1989). How to avoid logarithms in comparisons with uniform random variables, *Computing*, **41**, 163–166.

Ahrens, J.H. (1993). Sampling from general distributions by suboptimal division of domains, *Grazer Math. Berichte* **319**, Graz, Austria.

Ahrens, J.H., and U. Dieter (1972). Computer methods for sampling from the exponential and normal distributions, *Comm. ACM*, **15**, 873–882.

Ahrens, J.H., and U. Dieter (1974a). *Non-Uniform Random-Numbers*, Institut für Math. Statistik, Technische Hochschule in Graz, Austria.

Ahrens, J.H., and U. Dieter (1974b). Computer methods for sampling from Gamma Beta, Poisson and binomial distributions, *Computing*, **12**, 223–246.

Ahrens, J.H., and U. Dieter (1980a). Sampling from the binomial and Poisson distributions: a method with bounded computation times, *Computing*, **25**, 193–208.

Ahrens, J.H., and U. Dieter (1982a). Generating Gamma variates by a modified rejection technique, *Comm. ACM*, **25**, 47–53.

Ahrens, J.H., and U. Dieter (1982b). Computer generation of Poisson deviates from modified normal distributions, *ACM Trans. Math. Software*, **8**, 163–179.

Ahrens, J.H., and U. Dieter (1985). Sequential random sampling, *ACM Transactions on Mathematical Software*, **11**, 157–169.

Ahrens, J.H., and U. Dieter (1988). Efficient tablefree sampling methods for the exponential, Cauchy and normal distributions, *Comm. ACM*, **31**, 1330–1337.

Ahrens, J.H., and U. Dieter (1991). A convenient sampling method with bounded computation times for Poisson distributions, *The Frontiers of Statistical Computation, Simulation and Modeling*. P.R. Nelson, E.J. Dudewicz, A.Öztürk, and E.C. van der Meulen, editors American Science Press, Syracuse, NY, 137–149.

Anderson, T.W. (1958). *An Introduction to Multivariate Statistical Analysis*, John Wiley and Sons.

Atkinson, A.C. (1979). The computer generation of Poisson random variables, *Appl. Statist.* **28**, 29–35.

Atkinson, A.C., and J. Whittaker (1976). A switching algorithm for the generation of beta random variables with at least one parameter less than one, *J. Roy. Statist. Soc.*, Series A, **139**, 462–467.

Best, D.J. (1978). A simple algorithm for the computer generation of random samples from a Student's *t* or symmetric beta distribution, Proceedings of the 1978 COMPSTAT Conference, Leiden, 341–347.

Best, D.J. (1983). A note on gamma variate generators with shape parameter less than unity, *Computing*, **30**, 185–188.

Box, G.E.P., and M.E. Muller (1958). A note on the generation of random normal deviates, *Ann. Math. Statist*, **29**, 610–611.

Broder, A.Z. (1989). Generating random spanning trees, *Thirtieth Annual Symposium on Foundations of Computer Science*, 442–447.

Broder, A.Z., and A.R. Karlin (1989). Bounds on the cover time, *J. Theoretical Probability*, **2**, 101–120.

Chen, H-C., and Y. Asau (1974). On generating random variates from an empirical distribution, *AIIE Trans.*, No. 2, **6**, 163–166.

Cheng, R.C.H. (1977). The generation of Gamma variables with nonintegral shape parameter, *Appl. Stat.*, **26**, 71–75.

Cheng, R.C.H. (1978). Generating beta variates with nonintegral shape parameters, *Comm. ACM*, **21**, 317–322.

Cheng, R.C.H., and G.M. Feast (1979). Some simple Gamma variate generators, *Appl. Statist.*, **28**, 290–295.

Cheng, R.C.H., and G.M. Feast (1980). Gamma variate generators with increased shape parameter range, *Comm. ACM*, **23**, 389–394.

Devroye, L. (1981). The computer generation of Poisson random variables, *Computing*, **26**, 197–207.

Devroye, L. (1986). *Non-Uniform Random Variate Generation*, Springer-Verlag, New York.

Devroye, L. and A. Naderisamani (1980). A binomial variate generator, Tech. Rep., School of Computer Science, McGill University, Montreal.

Dieter, U. (1982). An alternate proof for the representation of discrete distributions by equiprobable mixtures, *J. Appl. Prob.*, **19**, 869–872.

Feller, W. (1971). *An Introduction to Probability Theory and Its Applications*, Volume II, second ed., Wiley, New York.

Fisher, R.A., and E.A. Cornish (1960). The percentile points of distributions having known cumulants, *Technometrics*, **2**, 209–225.

Fishman, G.S. (1976). Sampling from the gamma distribution on a computer, *Comm. ACM*, **19**, 407–409.

Fishman, G.S. (1978). *Principles of Discrete Event Simulation*, Wiley, New York.

Fishman, G.S. (1979). Sampling from the binomial distribution on a computer, *J. Amer. Statist. Assoc.*, **74**, 418–423.

Fishman, G.S. (1996). *Monte Carlo: concepts, Algorithms and Applications*, Springer-Verlag, New York.

Fishman, G.S., and L.R. Moore (1984). Sampling from a discrete distribution while preserving monotonicity, *Amer. Statist.*, **38**, 219–223.

Fishman, G.S., and L.R. Moore (1986). An exhaustive analysis of multiplicative congruential random number generators with modulus $2^{31} - 1$, *SIAM J. Sci. Stat. Comput.*, **7**, 24–45.

Fishman, G.S., and L.S. Yarberry (1993). Generating a sample from a k-cell table with changing probabilities in $O(\log_2 k)$ time, *ACM Trans. Math. Software*, **19**, 257–261.

Gerontidis, I., and R.L. Smith (1982). Monte Carlo generation of order statistics from general distributions, *Appl. Statist*, **31**, 238–243.

Hastings, C., Jr. (1955). *Approximations for Digital Computers*, Princeton University Press, Princeton, NJ.

Hoeffding, W. (1940). Masstabinvariante Korrelationstheorie, *Schriften des Mathematischen Instituts und des Instituts für Angewandte Mathematik der Universität Berlin*, **5**, 197–233.

Hoeffding, W. (1963). Probability inequalities for sums of bounded random variables, *J. Amer. Statist. Assoc.*, **58**, 13–29.

Hörmann, W., and G. Derflinger (1993). A portable random number generator well suited for the rejection method, *ACM Transactions on Mathematical Software*, **19**, 489–495.

Johnson, N.L., S. Kotz, and N. Balakrishnan (1994). *Continuous Univariate Distributions, Volume II*, second edition, Wiley, New York.

Kachitvichyanukul, V., and B. Schmeiser (1985). Computer generation of hypergeometric random variates, *J. Statist. Comp. Simul.*, **22**, 127–145.

Kachitvichyanukul, V., and B. Schmeiser (1988). Binomial random variate generation, *Comm. ACM*, **31**, 216–222.

Kinderman, A.J., and J.F. Monahan (1977). Computer generation of random variables using the ratio of uniform deviates, *ACM Trans. Math. Software*, **3**, 257–260.

Kinderman, A.J., and J.F. Monahan (1980). New methods for generating Student's t and Gamma variables, *Computing*, **25**, 369–377.

Knuth, D. (1973). *The Art of Computer Programming: Sorting and Searching*, Addison-Wesley, Reading, MA.

Kronmal, Richard A., and A.V. Peterson, Jr. (1979). On the alias method for generating random variables from a discrete distribution, *The Amer. Statist.*, **4**, 214–218.

Kulkarni, V.G. (1990). Generating random combinatorial objects, *J. Algorithms*, **11**, 185–207.

Lurie, D., and H.O. Hartley (1972). Machine-generation of order statistics for Monte Carlo computations, *The Amer. Statist.*, **26**, 26–27.

MacLaren, M.D., G. Marsaglia, and T.A. Bray (1964). A fast procedure for generating exponential random variables, *Comm. ACM*, **7**, 298–300.

Marsaglia, G. (1984). The exact-approximation method for generating random variables in a computer, *J. Amer. Statist.*, **79**, 218–221.

Marsaglia, G., K. Ananthanarayanan, and N.J. Paul (1976). Improvements on fast methods for generating normal random variables, *Inf. Proc. Letters*, **5**, 27–30.

Marsaglia, G., M.D. MacLaren, and T.A. Bray (1964). A fast procedure for generating normal random variables, *Comm. ACM*, **7**, 4–10.

Marsaglis, G., A. Zaman, and J. Marsaglia (1994) Rapid evaluation of the inverse of the normal distribution function, *Statistics and Probability Letters*, **19**, 259–266.

Minh, D.L. (1988). Generating gamma variates, *ACM Trans. on Math. Software*, **14**, 261–266.

Moro, B. (1995). The full Monte, *Risk*, **8**, 57–58.

Ross, S. and Z. Schechner (1986). Simulation uses of the exponential distribution, *Stochastic Programming*, M. Lucertini, ed., in *Lecture Notes in Control and Information Sciences*, Springer-Verlag, New York, **76**, 41–52.

Schmeiser, B.W., and A.J.G. Babu (1980). Beta variate generation via exponential majorizing functions, *Oper. Res.*, **28**, 917–926. Errata: *Oper. Res.*, **31**, 1983, 802.

Schmeiser, B.W., and R. Lal (1980). Squeeze methods for generating gamma variates, *J. Amer. Statist. Assoc.* **75**, 679–682.

Schmeiser, B.W., and V. Kachitvichyanukul (1981). Poisson random variate generation, Res. Memo. 8184, School of Industrial Engineering, Purdue University.

Schreider, Y.A. (1964). *Method of Statistical Testing*, Elsevier, Amsterdam.

Schucany, W.R. (1972). Order statistics in simulation, *J. Statist. Comput. and Simul.*, **1**, 281–286.

Stadlober, E. (1982). Generating Student's *t* variates by a modified rejection method, *Probability and Statistical Inference*, W. Grossmann, G.Ch. Plug, and W. Wertz, eds., Reidel, Dordrecht, Holland, 349–360.

Stadlober, E. (1989). Sampling from Poisson, binomial and hypergeometric distributions: ratio of uniforms as a simple and fast alternative, *Math. Statist. Sektion 303*, Forschungsgesellschaft Joanneum, Graz, Austria.

Stadlober, E. (1991). Binomial random variate generation: a method based on ratio of uniforms, *The Frontiers of Statistical Computation, Simulation, and Modeling*, volume 1; P.R. Nelson, E.J. Dudewicz, A. Öztürk, E.C. van der Meulen, editors, American Sciences Press, Syracuse, NY, 93–112.

von Neumann, J. (1951). Various techniques used in connection with random digits, *Monte Carlo Method*, Applied Mathematics Series 12, National Bureau of Standards, Washington, D.C.

Walker, A.J. (1974). New fast method for generating discrete random numbers with arbitrary frequency distributions, *Electronic Letters*, **10**, 127–128.

Walker, A.J. (1977). An efficient method for generating discrete random variables with general distributions, *ACM Trans. on Math. Software*, **3**, 253–256.

Wilks, S.S. (1962). *Mathematical Statistics*, Wiley, New York.

Wong, C.K. and M.C. Easton (1980). An efficient method for weighted sampling without replacement, *SIAM J. of Comput.*, **9**, 111–113.

Zechner, H., and E. Stadlober (1993). Generating Beta variates via patchwork rejection, *Computing*, **50**, 1–18.

CHAPTER **9**

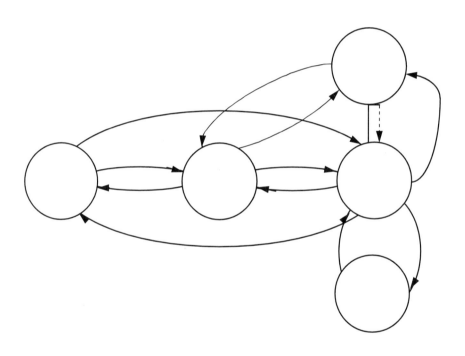

Pseudorandom Number Generation

Chapter 8 reveals that every algorithm that generates a sequence of i.i.d. random samples from a probability distribution as output requires a sequence of i.i.d. random samples from $\mathcal{U}(0, 1)$ as input. To meet this need, every discrete-event simulation programming language provides a *pseudorandom number generator* that produces a sequence of nonnegative integers $Z_1, Z_2, \ldots$ with integer upper bound $M > Z_i \ \forall i$ and then uses $U_1, U_2, \ldots$, where $U_i := Z_i/M$, as an approximation to an i.i.d. sequence from $\mathcal{U}(0, 1)$.

Since $U_i \in \mathcal{M} \subseteq \mathcal{M}_0 := \{0, \frac{1}{M}, \frac{2}{M}, \ldots, \frac{M-1}{M}\}$, this generating procedure incurs a discretization error that depends on how large M is and on how closely $\mathcal{M}$ resembles $\mathcal{M}_0$. Whereas an i.i.d. sequence from $\mathcal{U}(0, 1)$ is continuous, $U_1, U_2, \ldots$ is a discrete sequence. Moreover, the selected generating procedure determines how well the distribution of any subset of $U_1, U_2, \ldots$, namely $U_{j_1}, U_{j_2}, \ldots, U_{j_k}$ for arbitrary positive integers $k, j, \ldots, j_k$, approximates the distribution of a k-tuple of i.i.d. uniform random variables on $[0, 1)$. Most importantly, there is the issue of how these discretization and distributional errors in the generation of $U_1, U_2, \ldots$ propagate themselves in the sequence generated by any one of the algorithms in Chapter 8.

All generators described in this chapter exploit the concept of *congruence*. We say that *a is congruent to b modulo M* if M divides $a - b$. In symbols, one writes $a \equiv b \pmod{M}$, where M is called the *modulus* and $\equiv$ is called the congruence symbol. As examples, $7 \equiv 1$ (mod 6), $11 \equiv 2$ (mod 3), $338 \equiv 8$ (mod 11), and $-5 \equiv 1$ (mod 6). Also, $X = (B + C)$ (mod M) denotes that X equals the quantity $B + C$ reduced modulo M.

9.1 LINEAR CONGRUENTIAL GENERATORS

To generate $Z_1, Z_2, \ldots$ virtually all discrete-event simulation programming languages employ *multiplicative congruential pseudorandom number generators* of the form

$$Z_i = AZ_{i-1} \ (\text{mod } M), \qquad i \geq 1, \tag{9.1}$$

where Z_0 denotes the *seed*, A is a positive integer, called the *multiplier*, and M is the *modulus*. Expression (9.1) is algebraically equivalent to

$$Z_i = AZ_{i-1} - MK_i, \qquad i \geq 1, \tag{9.2}$$

where

$$K_i := \lfloor AZ_{i-1}/M \rfloor.$$

To generate numbers from expression (9.1), one needs to *prime* the generator with a seed. For example, to generate Z_1, Z_0 acts as the seed; to generate Z_2, Z_1 acts as the seed; to generate Z_n, Z_{n-1} acts as the seed. The simplicity of priming this generator represents one of its attractive features.

Reproducibility is another important feature. In designing a discrete-event simulation experiment, it is not unusual to conduct a preliminary run to identify errors in program logic and previously unexpected sample-path behavior. Such an observation often prompts one to amend the program. By recording the seed for the first run, one can make a new run with the modified program but using the same seed as input. Provided that the modified program leaves the logic for assigning pseudorandom numbers to each source of variation untouched, the sample-path behavior of the new run reflects only the code change and contains no new variation due to sample generation. Therefore, the ability to reproduce the exact same sampling variation on successive runs is a valuable tool in judging sample-path changes that result from modified programs. All pseudorandom number generators described in this chapter offer this feature of reproducibility.

Expression (9.1) makes it clear that the number of distinct values the Z_i assume cannot exceed $M - 1$. Just how many distinct values are possible depends on the relationship between

A and M. To see how this occurs, first note that

$$
\begin{aligned}
Z_1 &= AZ_0 - MK_1, \\
Z_2 &= A^2 Z_0 - MK_2 - MK_1 A, \\
Z_3 &= A^3 Z_0 - MK_3 - MK_2 A - MK_1 A^2, \\
&\vdots \quad \vdots \\
Z_n &= A^n Z_0 - M(K_{n-1}A + \cdots + K_1 A^{n-1}).
\end{aligned}
\tag{9.3}
$$

Since $1 \le Z_i < M$ for $1 \le i \le n$, the quantity $K_n + K_{n-1}A + \cdots + K_1 A^{n-1}$ must be the largest integer in $A^n Z_0 / M$, so that expression (9.1) has the alternative representation

$$
Z_n = A^n Z_0 \ (\text{mod } M).
\tag{9.4}
$$

Observe that Z_0 completely determines the $\{Z_i, i \ge 1\}$ integer sequence once A and M are fixed. Therefore, our ability to make $U_1, U_2, \ldots$ resemble a continuous-valued i.i.d. sequence with elements from $\mathcal{U}(0, 1)$ can, at best, depend on the choice of A and M.

9.1.1 PERIOD

Perhaps the most obvious property of the generator (9.1) is that once $Z_n = Z_0$ for some integer $n > 0$, then $Z_{n+i} = Z_i$ for all $i = 1, 2, \ldots$. We call

$$
P(Z_0, A, M) := \min(n \ge 1 : Z_n = Z_0)
\tag{9.5}
$$

the *period* of the generator for given Z_0, A, M. Clearly, for given Z_0 and M we would like to choose the multiplier A to maximize the period. In particular, we would like $P(Z_0, A, M) = M - 1$, which would make the period identical for all seeds $Z_0 \in \mathcal{M}_0$, and thus independent of the starting seed. Recall from expressions (9.4) and (9.5) that $Z_n = Z_0$ implies

$$
(A^n - 1)Z_0 \ (\text{mod } M) = 0.
\tag{9.6}
$$

If $n = M - 1$ is the smallest integer satisfying the congruence (9.6), then it must hold for all $Z_0 \in \mathcal{M}_0$, so that

$$
(A^{M-1} - 1) \ (\text{mod } M) = 0.
\tag{9.7}
$$

Table 9.1 Sequences for $Z_i = AZ_{i-1}(\mathrm{mod}\ 11), i > 1; Z_0 = 1$

					Z_i					
$i \setminus A$	1	2	3	4	5	6	7	8	9	10
0	1	1	1	1	1	1	1	1	1	1
1	1	2	3	4	5	6	7	8	9	10
2		4	9	5	3	3	5	9	4	1
3		8	5	9	4	7	2	6	3	
4		5	4	3	9	9	3	4	5	
5		10	1	1	1	10	10	10	1	
6		9				5	4	3		
7		7				8	6	2		
8		3				4	9	5		
9		6				2	8	7		

9.1.2 PRIME MODULUS

If M is prime, there exist $\phi(M - 1)$ integers $A \in \mathcal{M}_0$ for which $n = M - 1$ is the smallest integer satisfying the congruence (9.7), where

$$\phi(n) := \Pi_{p|n}\left(1 - \frac{1}{p}\right) \tag{9.8}$$

and the product is taken over all distinct primes p dividing n. An A that satisfies expression (9.7) for M prime is called a *primitive root* of M. Expression (9.8) defines the *Euler totient function*.

As illustration, the largest integer that a 32-bit register in a computer can hold is $M = 2^{31} - 1$, which happens to be prime. Since $M - 1$ has the prime factorization $M - 1 = 2^{31} - 2 = 2 \times 3^2 \times 7 \times 11 \times 31 \times 151 \times 331$, these exist $\phi(M - 1) = 534{,}600{,}000$ primitive roots of $2^{31} - 1$, each one of which guarantees a full period $P = M - 1$ for the multiplicative congruential generator (9.1).

To illustrate the relationship between A and the period, Table 9.1 lists the sequences that result for the prime-modulus generator $Z_i = AZ_{i-1}$ (mod 11). The table reveals $\phi(11 - 1) = 10 \times (1 - \frac{1}{2})(1 - \frac{1}{5}) = 4$ primitive roots $A = 2, 6, 7,$ and 8. It also reveals that $A = 6$ generates the same full period as $A = 2$, but in reverse order. Moreover, $A = 7$ and $A = 8$ have a similar relationship. We next explain these relationships.

9.1.3 MULTIPLICATIVE INVERSE

Consider two generators

$$X_i = AX_{i-1} \pmod{M}$$

and

$$Y_{i-1} = BY_i \pmod{M}$$

with multipliers A and B and modulus M. We want to establish the condition on A and B that guarantees $X_i = Y_i$ given $X_{i-1} = Y_{i-1}$ for all $i \geq 1$. Since

$$X_i = AX_{i-1} - KM,$$
$$Y_{i-1} = BY_i - K'M,$$

for $K := \lfloor A\,X_{i-1}/M \rfloor$ and $K' := \lfloor B\,Y_i/M \rfloor$, $X_{i-1} = Y_{i-1}$ is equivalent to

$$X_i = A(BY_i - K'M) - KM,$$

or, equivalently,

$$(ABY_i - X_i) = (K + AK')M \equiv 0 \pmod{M}.$$

Therefore, $X_i = Y_i$ given $X_{i-1} = Y_{i-1}$ if and only if

$$(AB - 1)X_i \equiv 0 \pmod{M},$$

or, equivalently, for prime M,

$$AB - 1 \equiv 0 \pmod{M}. \tag{9.9}$$

If B satisfies this congruence for given M and A, we say that B is the *multiplicative inverse of A modulo M.*

9.1.4 FINDING PRIMITIVE ROOTS

If A is a primitive root of M, then $B := A^K \pmod{M}$ is also a primitive root of M if and only if the greatest common divisor of K and $M - 1$ is unity. Notationally, $\gcd(K, M-1) = 1$.

This property implies that once a primitive root is known, others can be computed by exponentiation and modulo reduction subject to the restriction on the exponent. As illustration, 2 is the smallest (base) primitive root of 11, so that $2^3 \pmod{11} = 8$, $2^7 \pmod{11} = 7$, and $2^9 \pmod{11} = 6$. For $M = 2^{31} - 1$, 7 is the base primitive root, so that any $B = 7^K \pmod{2^{31}-1}$ is a primitive root, provided that K has no factors 2, 3, 7, 31, 151, and 331. For example, $7^5 \pmod{2^{31} - 1} = 16807$ is the multiplier found in the multiplicative congruential generators in SLAM and SIMAN. Alternatively, GPSS/H uses $B = 7^K \pmod{2^{31} - 1} = 742938295$, where $K{=}981902737$ can easily be shown to have no factors 2, 3, 7, 11, 31, 151, and 331. SIMSCRIPT II.5 uses $B = 7^K \pmod{2^{31}-1} = 630360016$, where $K = 1177652351$

contains none of these factors. Section 9.5 rates these alternative multipliers with regard to their randomness properties.

9.2   MODULUS 2^β

Multiplicative congruential generators with $M = 2^\beta$ for $\beta \geq 3$ are also in common use, although less so than prime modulus generators. The appeal of "powers-of-two" generators stems from the computational efficiency they allow.

All modern-day computers perform computations in base 2 using either integer or floating-point arithmetic. Integer arithmetic takes less computing time, and because of the relatively slow speeds that characterized computers in the earlier days a natural preference arose for using integer rather than floating-point arithmetic, whenever possible.

To evaluate Z_i in expression (9.1) takes multiplication (AZ_{i-1}), division and truncation ($K_i = \lfloor AZ_{i-1}/M \rfloor$), multiplication ($MK_i$), and subtraction ($AZ_{i-1} - MK_i$). Of these, multiplication and division are the more time-consuming operations. By choosing a modulus $M = 2^\beta$, a pseudorandom-number-generating program can be written in the machine language of the chosen computer that would use integer arithmetic and replace all but the first multiplication (AZ_{i-1}) by bit inspection and replacement.

As illustration, suppose that AZ_{i-1} has the binary representation $B_k B_{k-1}, \ldots, B_1$ for some positive integer k, where

$$B_j \in \{0, 1\}, \qquad j = 1, \ldots, k,$$
$$B_k = 1,$$

so that

$$AZ_{i-1} = \sum_{j=1}^{k} 2^{j-1} B_j.$$

Then

$$2^\beta \lfloor AZ_{i-1}/2^\beta \rfloor = \begin{cases} 0 & \text{if } k \leq \beta, \\ \displaystyle\sum_{j=1+\beta}^{k} 2^{j-1} B_j & \text{if } k > \beta, \end{cases}$$

revealing that division, truncation, multiplication, and subtraction incurred in computing

$$Z_i = AZ_{i-1} \ (\mathrm{mod} \ 2^\beta) = \sum_{j=1}^{\min(k,\beta)} 2^{j-1} B_j$$

can be replaced by the less costly steps of determining k by bit inspection and then replacing bits $\min(k, \beta) + 1, \ldots, k$ by zeros.

The desire to eliminate the multiplication AZ_{i-1} as well has led to a recommendation for the form of A. However, Section 9.53 shows that this form introduces a disastrous departure from randomness, especially in three dimensions.

Although integer arithmetic is faster, it imposes restrictions on word size that impose other design constraints on pseudorandom number generation (png). Moreover, the speeds of current-day computers are so much greater than in the past as to make the penalty incurred in using floating-point arithmetic considerably less of a contentious issue. As a consequence, many png programs rely on extended-precision floating-point arithmetic to perform all computations.

Without its computational edge, one might have expected $M = 2^\beta$, as a multiplier, to give way to M prime. Since this has not happened in total, we need to familiarize the simulationists with the properties of powers-of-two moduli.

What does $M = 2^\beta$ imply about the sequences that the generator (9.1) produces? If the seed Z_{i-1} is an even integer, then, for any positive integer A, AZ_{i-1} is even and so is $Z_i = AZ_{i-1}(\bmod\ 2^\beta) = AZ_{i-1} - K_i 2^\beta$, where $K_i = \lfloor AZ_{i-1}/2^\beta \rfloor$. Suppose that $Z_{i-1} = Y_{i-1} \times 2^\gamma$ for some odd integer Y_{i-1} and integer $\gamma < \beta$. Then

$$Z_i = AZ_{i-1}\ (\bmod\ 2^\beta) = (AY_{i-1} \times 2^\gamma)\ (\bmod\ 2^\beta) = Y_i \times 2^\gamma,$$

where

$$Y_i := AY_{i-1}\ (\bmod\ 2^{\beta-\gamma}).$$

Therefore, an even seed $Z_{i-1} = Y_{i-1} \times 2^\gamma$ induces a sequence $\{Z_j,\ j \geq i\}$ of even integers with period no larger than $2^{\beta-\gamma} - 1$.

If A and Z_{i-1} are odd integers, then $\{Z_j = AZ_{j-1}(\bmod\ 2^\beta), j \geq 1\}$ is a sequence of odd integers with period no greater than $2^{\beta-1}$. However, the upper bound is not achievable. For $M = 2^\beta$, there exists no multiplier A that admits a full period $P(A, 2^\beta, Z_0) = 2^\beta - 1$. Indeed, the maximal achievable period is $2^{\beta-2}$, which occurs for Z_0 odd and

$$A \equiv 3 \pmod 8$$

and

$$A \equiv 5 \pmod 8.$$

Equivalently, $A \in \{3+8j;\ j = 0, 1, \ldots, 2^{\beta-3}\} \cup \{5+8j;\ j = 0, 1, \ldots, 2^{\beta-3}\}$. For $M = 2^{32}$, this implies a period of $2^{30} = 1{,}073{,}741{,}824$ odd numbers on each of two disjoint cycles. Moreover, this collection of values that A can assume implies that there are 2^{30} maximal period generators for $M = 2^{32}$.

Table 9.2 $Z_i = AZ_{i-1} (\text{mod } 64)^a$

| | A = 13 | | | | A = 11 | | | |
i	Z_i	$Z_{(i)}$	Z_i	$Z_{(i)}$	Z_i	$Z_{(i)}$	Z_i	$Z_{(i)}$
0	1	1	3	3	1	1	7	5
1	13	5	39	7	11	3	13	7
2	41	9	59	11	57	9	15	13
3	21	13	63	15	51	11	37	15
4	17	17	51	19	49	17	23	21
5	29	21	23	23	27	19	61	23
6	57	25	43	27	41	25	31	29
7	37	29	47	31	3	27	21	31
8	33	33	35	35	33	33	39	37
9	45	37	7	39	43	35	45	39
10	9	41	27	43	25	41	47	45
11	53	45	31	47	19	43	5	47
12	49	49	19	51	17	49	55	53
13	61	53	55	55	59	51	29	55
14	25	57	11	59	9	57	63	61
15	5	61	15	63	35	59	53	63
16	1		3		1		7	

$^a\{Z_{(i)}; i = 0, 1, \ldots, 16\}$ is the ordered sequence of $\{Z_i; i = 0, 1, \ldots, 16\}$.

Table 9.2 illustrates the sequences that these generators can produce. Here $A = 13 \equiv 5$ (mod 8), whereas $A = 11 \equiv 3$ (mod 8). Most notably, each A leads to two disjoint odd sequences, depending on the starting Z_0. Also, $Z_{(i)} - Z_{(i-1)} = 4$ for $A = 13$, whereas $Z_{(i)} - Z_{(i-1)} \in \{2, 6\}$ for $A = 11$, implying that the latter sequences have less uniformity than the former. More generally, maximal-period multiplicative congruential generators with $M = 2^\beta$ and $\beta \geq 3$ have the following properties (Jannson 1966):

Property M1. For $A \equiv 5$ (mod 8) and $Z_0 \equiv 1$ (mod 4), $\{Z_i, 1 \leq i \leq P\}$ is a permutation of the integers $4j + 1$; $j = 0, 1, \ldots, 2^{\beta-2} - 1$.

Property M2. For $A \equiv 5$ (mod 8) and $Z_0 \equiv 3$ (mod 4), $\{Z_i, 1 \leq i \leq P\}$ is a permutation of the integers $4j + 3$; $j = 0, 1, \ldots, 2^{\beta-2} - 1$.

Property M3. For $A \equiv 3$ (mod 8) and $Z_0 \equiv 1$ or 3 (mod 4), $\{Z_i, 1 \leq i \leq P\}$ is a permutation of the integers $8j + 1$ and $8j + 3$; $j = 0, 1, \ldots, 2^{\beta-3} - 1$.

Property M4. For $A \equiv 3$ (mod 8) and $Z_0 \equiv 5$ or 7 (mod 4), $\{Z_i, 1 \leq i \leq P\}$ is a permutation of the integers $8j + 5$ and $8j + 7$; $j = 0, 1, \ldots, 2^{\beta+3} - 1$.

Here $P := P(A, M, Z_0)$. Because of the inherent nonuniformity that $A \equiv 3$ (mod 8) induces, we hereafter focus exclusively on $A \equiv 5$ (mod 8) for $M = 2^\beta$.

9.2.1 MIXED CONGRUENTIAL GENERATORS

Consider the mixed generator

$$Z_i = (AZ_{i-1} + C) \pmod{M}, \qquad i \geq 1, \tag{9.10}$$

which has the additional parameter C. For $M = 2^\beta$, C an odd integer, $A \equiv 1 \pmod 4$, and $\gcd(M, C) = 1$, this generator has full period M consisting of all integers $\{0\} \cup \mathcal{M}_0 = \{0, 1, \ldots, M - 1\}$. Note that:

  i. Z_i assumes all values in $\{0, 1, \ldots, M - 1\}$ once per cycle.

 ii. Any odd C satisfies $\gcd(M, C) = 1$.

iii. Bit shifting suffices to implement modulo reduction.

In the early days of discrete-event simulation, the computational convenience of bit shifting while preserving the full period $P = M$ gave this generator considerable appeal. With substantially enhanced speeds, computational convenience has become less of an issue.

While the full-period linear congruential generator (9.10) with $M = 2^\beta$ may appear to be different from the multiplicative congruential generator (9.1), elementary analysis reveals the former merely to be an affine transformation of the latter (e.g., Overstreet 1974). Assume $\{Z_i = AZ_{i-1} \pmod{2^\beta}, \ i \geq 1\}$ with $A \equiv 5 \pmod 8$ and $Z_0 \equiv 1 \pmod 4$, and let $Y_i := \lfloor Z_i/4 \rfloor$ and $A' := \lfloor A/4 \rfloor$. Since $Z_i \in \{4\nu + 1; \nu = 0, 1, \ldots, 2^{\beta-2}\}$, $Z_i = 4Y_i + 1$. Since $A = 1 + 4(2K + 1)$ for some integer K, $A = 4A' + 1$. Therefore,

$$Z_i = AZ_{i-1} \qquad (\text{mod } 2^\beta), \tag{9.11}$$

$$4Y_i + 1 = A(4Y_{i-1} + 1) \qquad (\text{mod } 2^\beta)$$

$$= 4AY_{i-1} + 4A' + 1 \qquad (\text{mod } 2^\beta),$$

so that

$$Y_i = AY_{i-1} + A' \qquad (\text{mod } 2^{\beta-2}),$$

where $Y_i \in \{0, 1, \ldots, 2^{\beta-2}\}$, thus proving our assertion. Since $A \equiv 5 \pmod 8$, $A = 1 + 4(1 + 2^K)$ for some integer K. Therefore, $A \equiv 1 \pmod 4$ and $A' = \lfloor 1/4 + 1 + 2K \rfloor$ is odd, so that $\{Y_i\}$ has "full period" $2^{\beta-2}$. Analogous results obtain for $A \equiv 5 \pmod 8$ and $Z_0 \equiv 3 \pmod 4$ and for $A \equiv 3 \pmod 8$ and $Z_0 \equiv 1 \pmod 2$.

In summary, we see that an analysis of the randomness properties of maximal-period generators (9.1) with modulus $2^{\beta+2}$ suffices to characterize the randomness properties of full-period generators (9.10) with modulus 2^{β}.

9.3 IMPLEMENTATION AND PORTABILITY

As with any proposed numerical technique, implementing a pseudorandom number generator on a computer calls for considerable care. Moreover, an implementation appropriate for one computer may produce a different (incorrect) sequence for the same seed when employed on another computer with different internal architecture. Paying attention to certain features of a generator before attempting implementation can significantly reduce the potential for such error.

All binary computers with a word size of $\alpha + 1$ bits can perform integer arithmetic and store the exact result, provided that it does not exceed $2^{\alpha} - 1$ in absolute value. This assumes that the leftmost bit contains the sign of the number. Likewise, an α-bit mantissa in floating-point words ensures this exact storage in performing floating-point arithmetic. In what follows we assume that $2^{\alpha} - 1$ is the largest storable integer.

Consider an implementation of $\{Z_i = AZ_{i-1} \, (\mathrm{mod}\, M), i \geq 1\}$ that executes the steps

$$S \leftarrow AZ_{i-1}, \qquad\qquad (9.13)$$
$$K \leftarrow \lfloor S/M \rfloor,$$
$$Z_i \leftarrow S - KM.$$

Computing S exactly and directly requires that

$$A(M - 1) \leq 2^{\alpha} - 1$$

for a prime modulus generator (with period $P = M - 1 < 2^{\alpha} - 1$), and requires $A \times (2^{\beta} - 1) \leq 2^{\alpha} - 1$ (Property M2) for a multiplicative congruential generator with modulus $M = 2^{\beta}$ and $A \equiv 5 \, (\mathrm{mod}\, 8)$.

To illustrate the limitation that these bounds impose, consider the primitive root $A = 950706376$ for $M = 2^{31} - 1$, which leads to $2^{60} < A(M - 1) < 2^{61}$. Therefore, directly performing integer arithmetic to compute AZ_{i-1} would inevitably lead to an overflow error for some i on 32-bit computers. Since floating-point representations on these computers have 24-bit mantissas in single precision and at most 56-bit mantissas in double precision, directly performing floating-point arithmetic on AZ_{i-1} also leads to an error for some i.

At least two solutions to this problem exist. The first uses extended-precision arithmetic software and by specifying an adequate level of precision, eliminates the overflow problem. However, this type of software is inevitably machine-dependent, thereby limiting the *portability* of any pseudorandom number generation program that relies on it. A portable computer program produces identical output on different computing platforms when given identical input. In the context of discrete-event simulation experiments, portability is a highly desirable feature.

The second alternative partitions arithmetic operations in a way that eliminates the overflow problem on a wide range of computing platforms and therefore ensures portability. For example, let $A_1 := A \pmod{2^{16}}$, $A_2 := A - A_1$, $X_i = A_1 Z_{i-1} \pmod{2^{31}}$, and $Y_i := A_2 Z_{i-1} \pmod{2^{31}}$. Then expression (9.1) has the equivalent form

$$Z_i = (X_i + Y_i + \lfloor A_1 Z_{i-1}/2^{31} \rfloor + \lfloor A_2 Z_{i-1}/2^{31} \rfloor) \pmod{2^{31} - 1},$$

$A_1 Z_{i-1} \le 2^{47} - 1$, and $A_2 Z_{i-1} \le 2^{62} - 1$. However, the least significant 16 bits of $A_2 Z_{i-1}$ are identically zero. Therefore, this partition of $A Z_{i-1}$ allows one to compute Z_i with no overflow, provided that a 47-bit mantissa is available.

Hörmann and Derflinger (1993) give algorithms for effecting this portability for $M = 2^{31} - 1$ and $A < 2^{30}$. Implementations of these algorithms are demonstrably faster than that in Fishman (1996, Section 7.5).

For $M = 2^{31} - 1$, a third alternative for ensuring portability across platforms restricts the choice of the multiplier A to a subset of the primitive roots of M that satisfy specified conditions (Schrage 1979, Park and Miller 1988, and Fishman 1996). Since a major objective is to choose an A that makes $U_1, U_2, \ldots$ resemble an i.i.d. uniform sequence as closely as possible, these additional restrictions can in no way improve the resemblance and are likely to reduce it. Therefore, we do not encourage this approach to portability.

9.4 HOW RANDOM ARE PSEUDORANDOM NUMBERS?

How well does $U_1, U_2, \ldots$ resemble a sequence of i.i.d. random variables from $\mathcal{U}(0, 1)$? We have already indicated that prime M and A a primitive root of M establish one-dimensional uniformity over the integers $\{1, \ldots, M-1\}$, whereas $M := 2^\beta$ $(\beta > 3)$ and $A = 5 \pmod 8$ do likewise over the integers $\{3 + 8j; j = 0, 1, \ldots, 2^{\beta-3}\}$ for $Z_0 = 3$ and $\{5 + 8j; j = 0, 1, \ldots, 2^{\beta-3}\}$ for $Z_0 = 5$. But what about the independence property? Do large values of U_{i+1} follow large (small) values of U_i, indicating dependence; or is U_{i+1} independent of the value of U_i?

Example 8.1 illustrates why independence within 2-tuples is necessary. If U_i and U_{i+1} are independent and each from $\mathcal{U}(0, 1)$ and $U_i \le e^{-(U_{i+1}-1)^2/2}$, then $-\ln U_{i+1}$ has the half normal p.d.f. (8.12a). If U_l and U_{i+1} are not independent then $-\ln U_{i+1}$ does not have the desired p.d.f.

Because of the deterministic way in which the generator (9.1) produces Z_i, some dependence between Z_i and Z_{i+1} is inevitable. Presumably, we want to choose A for a given M in a way that keeps this dependence to an acceptable minimum. But what about the joint dependence of Z_i, Z_{i+1}, Z_{i+2} or, more generally, of $Z_i, \ldots, Z_{i+k-1}$? Regrettably, analysis

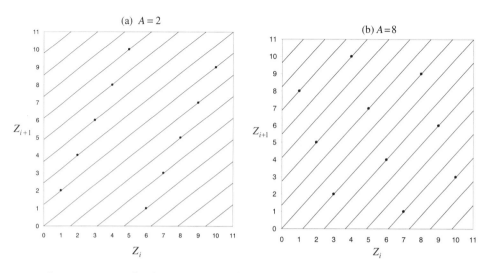

Figure 9.1 2-tuples for $Z_i = AZ_{i-1} \pmod{11}$

shows that no A performs equally well in all dimensions 1 through k for arbitrary k with respect to given error norms. However, for a given M we can identify those A whose worst errors across the k dimensions are smallest.

If U_i and U_{i+1} are independent from $\mathcal{U}(0, 1)$, then the sequence $\{(U_i, U_{i+1}); \; i = 1, 2, \ldots\}$ would have its elements (U_i, U_{i+1}) uniformly distributed in the unit square $[0, 1]^2$. Since the generator (9.1) induces discrete (U_i, U_{i+1}), the closest that we can come to this idealization is to have 2-tuples uniformly distributed on the grid $\{x_j := \frac{j}{M}; j = 1, \ldots, M-1$ and $y_j = \frac{j}{M}; j = 1, \ldots, M - 1\}$.

Figure 9.1 gives us a preview of what we can expect. It shows overlapping 2-tuples for the generator in Table 9.1 with $A = 2$ and 8. Observe that ten 2-tuples appear in each figure, whereas one expects one hundred 2-tuples, one at each vertex of the 2-dimensional 10×10 grid. Also, the points for $A = 2$ appear less random than those for $A = 8$. A second illustration makes the issue of nonuniformity more salient.

Recall that the most commonly employed multipliers for $M = 2^{31} - 1$ in discrete-event simulation are 16807 (e.g., SIMAN), 630360016 (e.g., SIMSCRIPT II.5), and 742938285 (e.g., GPSS/H). To reveal the discrete patterns they induce, Figure 9.2 shows their overlapping 2-tuples that fall in the square $[0, .001]^2$. Since any square of width and length .001 in the unit square exhibits approximately the same appearance, $[0, .001]^2$ contains about $(.001)^2 \times M \approx 2147$ 2-tuples. Figure 9.2 reveals that for $A = 16807$ these 2-tuples are packed together tightly on 17 parallel hyperplanes. By contrast, Figure 9.2b shows that the 2-tuples for $A = 630360016$ are considerably more uniformly distributed on 29 parallel lines, whereas Figure 9.2c shows 50 parallel lines for $A = 742938285$.

These numbers arise from counting the number of parallel lines within the set of parallel lines that have maximal distance between successive lines. For $(0, 1)^2$, these numbers

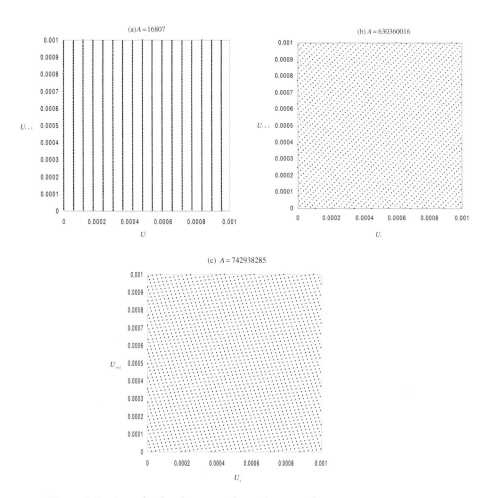

Figure 9.2 2-tuples for three pseudorandom number generators: $Z_i = A Z_{i-1} \pmod{2^{31} - 1}$

suggest that 2-tuples for $A = 16807$ are distributed on about 17,000 parallel lines in the unit square, 2-tuples for $A = 630360016$ are distributed on about 29,000 parallel lines, and 2-tuples for $A = 74938285$ are distributed on about 50,000 parallel lines.

Since the number of these lines increases with the distance between successive parallel lines and since for a given multiplier the distribution of 2-tuples becomes less uniform as the distance between parallel lines increases because of the increasing area void of 2-tuples, it follows that $A = 16807$ offers less-uniform 2-tuples than $A = 630360016$, which offers less-uniform 2-tuples than $A = 742938285$.

These observations for 2-tuples extend more generally to k-tuples. In particular, all prime modulus generators of the form induce $M - 1$ k-tuples $\{\mathbf{Z}_i = (Z_i, \ldots, Z_{i+k-1}); 1 \leq i \leq M - 1\}$ on the vertices of the k-dimensional grid $\mathcal{G} = \{\mathbf{z} = (z_1, \ldots, z_k) : z_i = j, 1 \leq$

$j \leq M - 1, 1 \leq i \leq k\}$, thereby leaving $(M - 2)[(M - 1)^k - 1]$ vertices unoccupied. Hence, the density of points diminishes with increasing dimension k. Since all vertices are not occupied, the second observation regarding uniformity becomes increasingly important as k increases. That is, some arrangments of the $M - 1$ k-tuples appear closer to uniformity than others. Since uniformity is a principal consideration, one prefers the primitive root A, for a specified prime modulus M, whose k-tuples come closest to uniformity as measured by an objective criterion for each dimension $k = 1, 2, \ldots$. Theorem 9.1 provides a basis for making this choice.

Theorem 9.1.

For each $k \geq 1$, let $\mathbf{q} := (q_0, q_1, \ldots, q_{k-1})$ denote points in the set

$$\mathcal{Q}_k(M) := \{-M, -M + 1, \ldots, -1, 0, 1, \ldots, M - 1\}^k \setminus \underbrace{\{0, \ldots, 0\}}_{k \; times}$$

and let

$$Z_i(\mathbf{q}, k) := \sum_{s=0}^{k-1} q_s Z_{i+s}, \tag{9.14}$$

where

$$Z_i = AZ_{i-1} \pmod{M},$$

M is prime, and A is a primitive root of M. Consider the subset of $\mathbf{q}$ vectors

$$\mathcal{Q}_k(A, M) := \left\{ \mathbf{q} \in \mathcal{Q}_k(M) : \sum_{s=0}^{k-1} q_s A^s \equiv 0 \pmod{M} \right\}. \tag{9.15}$$

i. For each $\mathbf{q} \in \mathcal{Q}_k(A, M)$ the k-tuples $\{\mathbf{U}_{ik} := (U_i, U_{i+1}, \ldots, U_{i+k-1});$ $i = 1, \ldots, M - 1\}$, where $U_i := Z_i/M$, lie on parallel hyperplanes in $(0, 1)^k$.

ii. The total number of these parallel hyperplanes does not exceed $\sum_{s=0}^{k-1} | q_s | - 1$.

iii. The distance between two parallel hyperplanes is proportional to $\left(\sum_{s=0}^{k-1} q_s^2 \right)^{-1/2}$.

PROOF OF PART I. Since $Z_{i+s} = (A^s Z_i) \pmod{M}$,

$$Z_i(\mathbf{q}, k) \pmod{M} = \left(Z_i \sum_{s=0}^{k-1} q_s A^s \right) \pmod{M}$$

for each $\mathbf{q} \in \mathcal{Q}_k(M)$. Therefore, for each $\mathbf{q} \in \mathcal{Q}_k(A, M)$

$$Z_i(\mathbf{q}, k) \ (\mathrm{mod}\ M) = 0,$$

or, equivalently,

$$Z_i(\mathbf{q}, k) \equiv 0 \pmod{M}.$$

This implies that for $i = 1, \ldots, M - 1$ there exist integers $c_i(\mathbf{q}, k) \in \{-M, -M + 1, \ldots, -1, 0, 1, \ldots, M - 1\}$ such that

$$Z_i(\mathbf{q}, k) = c_i(\mathbf{q}, k)M,$$

or, equivalently,

$$\sum_{s=0}^{k-1} q_s U_{i+s} = c_i(\mathbf{q}, k). \tag{9.16}$$

Therefore, all k-tuples $\{\mathbf{U}_i; i = 1, \cdots, M - 1\}$ lie on hyperplanes $\{q_0 u_0 + \ldots + q_{k-1} u_{k-1} = c: 0 < u_j < 1; j = 0, \ldots, k - 1\}$ indexed by integers $c \in \mathcal{C}_k(\mathbf{q}) := \cup_{i=1}^{M-1}\{c_i(\mathbf{q}, k)\}$. Moreover, since all hyperplanes have the same coefficients $q_0, q_1, \ldots, q_{k-1}$, they are parallel, and this parallelism is preserved when equation (9.15) is reduced modulo 1 onto the k-dimensional unit hypercube. $\qquad\square$

Proof of Part II.   Since $1 \le Z_i < M$,

$$Z_i(\mathbf{q}, k) < M \sum_{s=0}^{k-1} \max(0, q_s),$$

so that

$$c_i(\mathbf{q}, k) \le \sum_{s=0}^{k-1} \max(0, q_s).$$

Conversely,

$$Z_i(\mathbf{q}, k) > M \sum_{s=0}^{k-1} \min(0, q_s),$$

so that

$$c_i(\mathbf{q}, k) \ge \sum_{s=0}^{k-1} \min(0, q_s) + 1.$$

Let $c_{\min}$ and $c_{\max}$ denote the largest and smallest integers in $\mathcal{C}_k(\mathbf{q})$. Since $\mid \mathcal{C}_k(\mathbf{q}) \mid$ denotes the number of parallel hyperplanes,

$$\mid \mathcal{C}_k(\mathbf{q}) \mid \le c_{\max} - c_{\min} + 1 \le \sum_{s=0}^{k-1} \max(0, q_s) - 1 - \sum_{s=0}^{k-1} \min(0, q_s) - 1 + 1 = \sum_{s=0}^{k-1} \mid q_s \mid -1.$$

$\square$

PROOF OF PART III.. Let $\mathbf{x} := (x_0, \ldots, x_{k-1})$ and $\mathbf{y} := (y_0, \ldots, y_{k-1})$ denote points on the parallel hyperplanes $\{q_0 u_0 + \cdots + q_{k-1} u_{k-1} = c : 0 < u_j < 1; j = 0, \cdots, k-1\}$ and $\{q_0 u_0 + \cdots + q_{k-1} u_{k-1} = c' : 0 < u_j < 1; j = 0, \ldots, k-1\}$, respectively, for c and $c' \in \mathcal{C}_k(\mathbf{q})$. To find the distance between these hyperplanes, we fix $\mathbf{x}$ and find the solution to

$$\min_{\mathbf{y}} \sum_{s=0}^{k-1} (x_s - y_s)^2, \qquad \mathbf{y} := (y_0, y_1, \ldots, y_{k-1}),$$

subject to

$$\sum_{s=0}^{k-1} q_s x_s = c$$

and

$$\sum_{s=0}^{k-1} q_s y_s = c'.$$

Let $z_s = x_s - y_s$. Then the problem is equivalent to

$$\min_{\mathbf{z}} \left[\sum_{s=0}^{k-1} z_s^2 + \lambda \left(c - c' - \sum_{s=0}^{k-1} q_s z_s \right) \right], \qquad \mathbf{z} := (z_0, z_1, \ldots, z_{k-1}),$$

where λ is a Lagrange multiplier. The solution is

$$2z_s = \lambda q_s, \qquad s = 0, \ldots, k-1,$$

so that

$$\lambda = \frac{2(c - c')}{\sum_{s=0}^{k-1} q_s^2},$$

and the resulting distance between hyperplanes is

$$\left(\sum_{s=0}^{k-1} z_s^2\right)^{1/2} = \frac{|c - c'|}{(\sum_{s=0}^{k} q_s^2)^{1/2}}. \qquad \square$$

Corollary 9.2.

    For $\{Z_i = AZ_{i-1} \ (mod \ 2^\beta); \beta > 3, i \geq 1\}$ *with* $A \equiv 5 \ (mod \ 8)$, *Theorem 9.2 continues to hold with* $\sum_{s=0}^{k-1} q_s A^s \equiv 0 \ (mod \ 2^{\beta-2})$.

This implies that all k-tuples for generators with $M = 2^\beta$ and $A \equiv 5$ (mod 8) lie on parallel hyperplanes, and thus any methods for discerning departures from uniformity apply to these as well as to prime modulus generators.

9.5   CHOOSING AMONG MULTIPLIERS

To predict the seriousness of error and thereby to identify perferred multipliers, we need to understand more about the k-tuples in expression (9.15). Reducing both sides of the expression modulo 1 leads to

$$\sum_{s=0}^{k-1} q_s U_{i+s} \equiv 0 \ (\text{mod } 1) \qquad (9.17)$$

for all $2^k M^{k-1} - 1$ vectors $\mathbf{q} \in \mathcal{Q}_k(A, M)$. Therefore, for each $\mathbf{q}$, all k-tuples $\{\mathbf{U}_{jk}, 1 \leq j \leq P\}$ lie on the intersection of k-dimensional parallel hyperplanes and $[0,1)^k$. We now use this representation to study the randomness properties of k-tuples. As one basis for assessment, consider the Euclidean distance between adjacent parallel hyperplanes,

$$d_k(\mathbf{q}, A, M) = \Delta_k(\mathbf{q})/(\mathbf{q}\mathbf{q}^\mathrm{T})^{1/2}, \qquad (9.18)$$

where

$$\Delta_k(\mathbf{q}) := \min_{c \neq c' \in \mathcal{C}_k(\mathbf{q})} \lceil c - c' \rceil. \qquad (9.19)$$

In particular,

$$d_k(A, M) = \max_{\mathbf{q} \in \mathcal{Q}_k(A,M)} d_k(\mathbf{q}, A, M) \qquad (9.20)$$

denotes the maximal distance between two adjacent hyperplanes over all families of parallel hyperplanes that the multiplier A and modulus M induce. Since one can show that for every

$\bar{\mathbf{q}} \in \mathcal{Q}_k(A, M)$ with $\Delta_k(\bar{\mathbf{q}}) > 1$ there exists a $\mathbf{q} \in \mathcal{Q}_k(A, M)$ with $\Delta_k(\mathbf{q}) = 1$ such that

$$d_k(\bar{\mathbf{q}}, A, M) = d_k(\mathbf{q}, A, M),$$

we hereafter restrict attention to the $\mathbf{q}$ with $\Delta_k(\mathbf{q}) = 1$.

If two multipliers A_1 and A_2 have $d_k(A_1, M) > d_k(A_2, M)$, then A_1 indicates a more sparse and therefore less uniform arrangement for k-tuples in k-dimensional space than A_2 implies. More generally, for multipliers $A_1, \ldots, A_r$ one in principle prefers the A_i for which $d_k(A_i, M)$ is minimal. This method of analyzing multipliers is called the *spectral test*.

As examples, Figures 9.1a and 9.1b give $d_2(2, 11) = 1/(2^2 + 1)^{1/2} = .4472$ and $d_2(8, 11) = 1/(3^2 + 1)^{1/2} = .3162$, respectively, which supports the earlier assertion based on visual evidence that multiplier $A = 8$ gives more uniformly distributed 2-tuples than multiplier $A = 2$.

A limit exists on how small the maximal distance (9.19) can be; in particular, it is known that (Cassels 1959, p. 332)

$$[t(M)]^{1/k} d_k(A, M) \geq \gamma_k := \begin{cases} \left(\dfrac{3}{4}\right)^{1/4} & k = 2, \\ 2^{-1/6} & k = 3, \\ 2^{-1/4} & k = 4, \\ 2^{-3/10} & k = 5, \\ \left(\dfrac{3}{64}\right)^{1/12} & k = 6, \\ 2^{-3/7} & k = 7, \\ 2^{-1/2} & k = 8, \end{cases} \qquad (9.21)$$

where

$$t(M) := \begin{cases} M & \text{if } M \text{ is prime,} \\ 2^{\beta-2} & \text{if } M = 2^\beta. \end{cases}$$

For $M = 11$, $d_2(A, 11) \geq .2806$, revealing that $d_2(8, 11)$ is within 13 percent of the best achievable bound for 2-tuples. For $M = 2^{31} - 1$,

$$d_k(A, 2^{31} - 1) \geq \begin{cases} .2008 \times 10^{-4} & k = 2, \\ .6905 \times 10^{-3} & k = 3, \\ .3906 \times 10^{-2} & k = 4, \\ .1105 \times 10^{-1} & k = 5, \\ .2157 \times 10^{-1} & k = 6, \end{cases} \qquad (9.22)$$

and for $M = 2^{48}$,

$$d_k(A, 2^{48}) \geq \begin{cases} .1109 \times 10^{-6} & k = 2, \\ .2158 \times 10^{-4} & k = 3, \\ .2903 \times 10^{-3} & k = 4, \\ .1381 \times 10^{-2} & k = 5, \\ .3814 \times 10^{-2} & k = 6, \end{cases} \qquad (9.23)$$

indicating the relative coarseness of the grids in as few as four dimensions. This coarseness enhances the importance of choosing an A that makes $d_k(A, M)$ as small as possible across dimensions.

Table 9.3 gives results for high-performing multipliers, with respect to $d_k(A, M)$ for $M = 2^{31} - 1$, 2^{32}, and 2^{48}. For modulus M, the quantity $S_k(A, M)$ denotes the ratio of the minimal achievable distance $\gamma_k / t^{1/k}(M)$ to the worst-case distance $d_k(A, M)$ for the multiplier A. For $M = 2^{31} - 1$, the first five entries are the best among the 534,600,000 primitive roots of M. For $M = 2^{32}$, the entries are best among the $2^{\beta-3} = 2^{29} = 536,870,912$ possible candidates, and for $M = 2^{48}$ the five entries are the best among 60 million multipliers tested. For $M = 2^{32}$, the second entry under each multiplier is its multiplicative inverse modulo M. For $M = 2^{31} - 1$, the table also shows results for the two multipliers $A = 630360016$ in SIMSCRIPT II.5 and $A - 16807$ in SIMAN and SLAM. As anticipated from Figure 9.2, $d_2(742938285, 2^{31} - 1) < d_2(630360016, 2^{31} - 1) < d_2(16807, 2^{31} - 1)$. More generally, each of the best five dominates 630360016 and 16807 for $2 \leq k \leq 6$. Recall that GPSS/H uses 742938285. Clearly, there is room for improvement in the pseudorandom number generators in SIMAN and SLAM.

9.5.1 NUMBER OF PARALLEL HYPERPLANES

Other mathematical norms exist for choosing among multipliers, and they generally lead to the same, or close to the same, ordering among multipliers in each dimension k. See Fishman (1996, Chapter 7). These alternatives include a bound on the number of hyperplanes (Theorem 9.2 ii),

$$N_k(A, M) := \sum_{s=0}^{k-1} |q_s| - 1,$$

for each $\mathbf{q} \in \mathcal{Q}_k(A, M)$.

Although $N_k(A, M)$ tends to decrease as the distance between successive parallel hyperplanes $d_k(A, M)$ decreases, $N_k(A, M)$ is not as discerning a measure as $d_k(A, M)$. First, it is an upper bound, whereas $d_k(A, M)$ is the actual minimal distance. Second, the number of hyperplanes depends on the orientation of the hyperplanes in $[0,1]^k$, whereas $d_k(A, M)$ is rotation invariant.

Table 9.3   Distance between parallel hyperplanes for $\{Z_i = AZ_{i-1} \pmod{M}, i \geq 1\}$

$$S_k(A, M) := \gamma_k / d_k(A, M) t^{1/k}(M)$$

Modulus M	Multiplier A	Dimension (k)				
		2	3	4	5	6
$2^{31} - 1$	742938285[b]	.8673	.8607	.8627	.8320	.8342
	950706376[b]	.8574	.8985	.8692	.8337	.8274
	1226874159[b]	.8411	.8787	.8255	.8378	.8441
	62089911[b]	.8930	.8903	.8575	.8630	.8249
	1343714438[b]	.8237	.8324	.8245	.8262	.8255
	630360016	.8212	.4317	.7832	.8021	.5700
	16807	.3375	.4412	.5752	.7361	.6454[a]
2^{32}	1099087573[c]	.8920	.8563	.8604	.8420	.8325
	4028795517					
	2396548189[c]	.8571	.9238	.8316	.8248	.8248
	3203713013					
	2824527309[c]	.9220	.8235	.8501	.8451	.8332
	1732073221					
	3934873077[c]	.8675	.8287	.8278	.8361	.8212
	1749966429					
	392314069[c]	.9095	.8292	.8536	.8489	.8198
	2304580733					
2^{48}	68909602460261[c]	.8253	.8579	.8222	.8492	.8230
	33952834046453[c]	.9282	.8476	.8575	.8353	.8215
	43272750451645[c]	.8368	.8262	.8230	.8400	.8213
	127107890972165[c]	.8531	.8193	.8216	.8495	.8224
	55151000561141[c]	.9246	.8170	.9240	.8278	.8394

[a] Entries are $S_k(A, M) \times 1000$.
[b] Best among all primitive roots of $2^{31} - 1$. Source: Fishman and Moore (1986).
[c] Best among all multipliers $A \equiv 5^j \pmod{2^\beta}$, $j \geq 1$. Source: Fishman (1990).

To illustrate the problem, Figure 9.3a contains seven parallel lines with incremental distance d when the lines are oriented parallel to the 45-degree line; but Figure 9.3b shows only five lines with incremental distance d when the lines are oriented parallel to the y-axis. While this example places more value on $d_k(A, M)$ than on $N_k(A, M)$ as an indicator of uniformity, $N_k(A, M)$ nevertheless can provide useful information, as Section 9.5.3 shows.

9.5.2   SHOULD THE SIMULATIONIST BE CONCERNED ABOUT A THEORETICALLY POOR MULTIPLIER?

Since Figure 9.2 reveals that $A = 16807$ induces larger null regions in $[0, 1)^2$ than $A = 360360016$ and $A = 742938285$, its use in practice is more suspect than the others. However, some say that even with these large null regions $A = 16807$ continues to be an acceptable generator "to the level of approximation" one aims for when using simulation

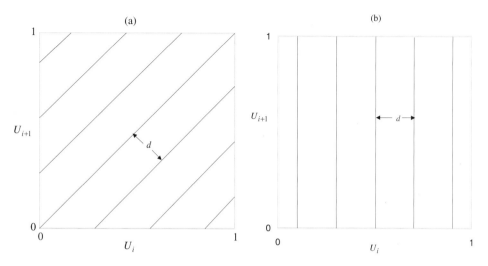

Figure 9.3 Number of parallel hyperplanes

output statistics to approximate unknown system quantities. Regrettably, both the criticism and defense of $A = 16807$ are far too abstract for a reader to assess the strength of each argument. To move the debate closer to reality, we return to sampling from the half-normal p.d.f. (8.12a) and corresponding d.f. $\{2\Phi(x) - 1,\ 0 \leq x < \infty\}$. In particular, we compare the discretization errors that each of the three multipliers induces for half-normal generation via the acceptance–rejection method (Section 8.5). To benchmark these errors, we first study discretization error for the inverse transform method (Section 8.1.3).

Applying the inverse transform method leads to the sequence

$$X_i = \Phi^{-1}\left(\frac{1 + U_i}{2}\right), \quad i = 1, 2, \ldots. \tag{9.24}$$

Since a prime-modulus generator produces all fractions $\{j/M;\ j = 1, \ldots, M - 1\}$ per period,

$$x_{(j)} := \Phi^{-1}\left(\frac{1 + j/M}{2}\right), \qquad j = 1, \ldots, M - 1,$$

defines, for all prime-modulus generators, the ordered sequence of approximating half-normal values that expression (9.23) produces. Moreover, the discretization errors between successively ordered values have bounds

$$
x_{(j+1)} - x_{(j)} \leq \begin{cases} 9.622 \times 10^{-10} & \text{if } 0 < x_{(j)} \leq 1, \\ 4.312 \times 10^{-9} & \text{if } 1 < x_{(j)} \leq 2, \\ 5.254 \times 10^{-8} & \text{if } 2 < x_{(j)} \leq 3, \\ 1.740 \times 10^{-6} & \text{if } 3 < x_{(j)} \leq 4. \end{cases}
$$

Since $2\Phi(4) - 1 = 1 - .6334 \times 10^{-4} \geq .99993$, we conclude that discretization error is at most 1.740×10^{-6} for all numbers except those greater than 4, which are rarely encountered.

These favorable results apply for any full-period generator. However, no closed form exists for the inverse Φ^{-1}, so that this method of half-normal generation must rely on time-consuming methods of numerical approximation. Nevertheless, the bounds on discretization error provide a basis for judging the discretization errors induced by the acceptance–rejection method based on Algorithm AR for $A = 16807, 630360016$, and 742938285. In the present notation:

ALGORITHM H-N

Given: Z_0 (seed).
Method: $i \leftarrow -1$.
        Repeat
                $i \leftarrow i + 2$
                $Z_i \leftarrow (AZ_{i-1}) \,(\bmod\, M)$
                $Z_{i+1} \leftarrow (AZ_i) \,(\bmod\, M)$
                $U_i \leftarrow Z_i / M$
                $U_{i+1} \leftarrow Z_{i+1} / M$
                $Y \leftarrow -\ln(1 - U_i)$
        Until $U_{i+1} \leq e^{-(Y_i - 1)^2/2}$.
        Return Y.

Clearly, discretization error depends on the properties of the 2-tuples $\{(U_i, U_{i+1})\}$ that lie in

$$
\mathcal{L} := \left\{ (u_1, u_2) : \quad -u_2 + Au_1 \equiv 0 \,(\bmod\, 1),\, u_1 \in \left\{ \frac{1}{M}, \dots, \frac{M-1}{M} \right\} \right\}.
$$

Observe that the points fall on A parallel lines with slope A, whereas the accepted points are in

$$
\mathcal{L}_{AR} := \left\{ (u_1, u_2) \in \mathcal{L} : \quad u_2 \leq e^{-[\ln(1-u_1)-1]^2/2} \right\}.
$$

Let $j = 1, \ldots, A$ index the parallel lines (from left to right). Then $u_1 = k(j)/M$, where

$$k(j) := \lceil M(j-1)/A \rceil,$$

denotes the smallest u_1 on line j and $l(j)/M$, where

$$l(j) := \min \left\{ l \in \mathcal{M} \, : \, Al/M - (j-1) > e^{-[\ln(1-l/M-1)]^2/2} \right\}$$

denotes the smallest u_1 on line j that is rejected.

Let

$$y_{(l)} := -\ln(1 - l/M), \qquad l = 1, \ldots, M-1,$$

where $y_{(l)}$ is strictly increasing in l. Then Algorithm H-M produces no values on the positive axis in the intervals

$$\bigcup_{j=1}^{A-1} \left(y_{(l(j)-1)}, \, y_{(k(j+1))} \right).$$

Since $y_{(l(j))}$ is rejected, no value occurs in the intervals

$$\left(y_{(\max[k(j), l(j)-1])}, \, y_{(l(j))} \right] \qquad \text{and} \qquad \left(y_{(l(j))}, \, y_{(k(j+1))} \right).$$

Let Δy denote the length of an excluded interval, which we hereafter call a gap. Table 9.4 displays bounds on gaps for the three multipliers of interest. For $A = 16807$, the gaps are considerably larger than for the other multipliers. They are also substantially larger than those that the inverse transform distribution induces. Although Algorithm H-N is not the commonly encountered method for generating half-normal samples, the gaps for $A = 16807$ represent what can happen when a multiplier that performs relatively poorly on the spectral test is used. At a minimum, Table 9.4 should encourage potential users of this multiplier to switch to any of the first five multipliers in Table 9.3 that perform best on the spectral test in dimensions 2 through 6.

9.5.3 DANGEROUS MULTIPLIERS

Recall that $A \equiv \pm 3 \pmod 8$ and $M = 2^\beta$ ($\beta \geq 3$) yield a generator with maximal period $2^{\beta-2}$. Consider a multiplier $A \in \{2^\nu - 3; \, \nu \geq \beta/2\} \cup \{2^\nu + 3; \, \nu \geq \beta/2\}$ that ensures

Table 9.4 Bounds on lengths of excluded intervals, $1 \le y < 4$

A		y		Δy	
16807	$1 \le$	< 2	$1.27 \times 10^{-9} \le$		$\le 1.73 \times 10^{-4}$
	$2 \le$	< 3	$1.73 \times 10^{-4} \le$		$\le 1.03 \times 10^{-3}$
	$3 \le$	< 4	$1.03 \times 10^{-3} \le$		$\le 3.22 \times 10^{-3}$
6300360016	$1 \le$	< 2	$1.27 \times 10^{-9} \le$		$\le 4.61 \times 10^{-9}$
	$2 \le$	< 3	$4.61 \times 10^{-9} \le$		$\le 2.76 \times 10^{-8}$
	$3 \le$	< 4	$2.76 \times 10^{-8} \le$		$\le 8.57 \times 10^{-8}$
742938285	$1 \le$	< 2	$1.27 \times 10^{-9} \le$		$\le 3.91 \times 10^{-9}$
	$2 \le$	< 3	$3.91 \times 10^{-9} \le$		$\le 2.34 \times 10^{-8}$
	$3 \le$	< 4	$2.34 \times 10^{-8} \le$		$\le 7.27 \times 10^{-8}$

this maximal period and leads to

$$Z_{i+1} = (2^{\nu} \pm 3)Z_i \qquad (\text{mod } 2^{\beta}),$$

$$Z_{i+2} = (\pm 3 \times 2^{\nu+1} + 9)Z_i \quad (\text{mod } 2^{\beta}),$$

$$Z_{i+2} = \pm 6\, Z_{i+1} \mp 9\, Z_i \qquad (\text{mod } 2^{\beta}). \tag{9.25}$$

In the early days of simulation, multipliers of this form offered the appeal of allowing shift and add operations to replace the more costly direct multiplication $A Z_i$. For example, $A = 2^{\nu} + 3$ allows:

- Shift Z_i ν bits to the left and call this number B.

- Shift Z_i one bit to the left and add it to B.

- Add Z_i to B.

- $Z_{i+1} = B$.

Normalizing both sides of equation (9.25) and rearranging terms leads to

$$U_{i+2} \mp 6\,U_{i+1} \pm 9\,U_i \equiv 0 \pmod{1},$$

revealing that the triplets $\{(U_i, U_{i+1}, U_{i+2}); i \geq 1\}$ all lie on no more than $N_3(A, M) = 15$ parallel hyperplanes.

Regrettably, a multiplier $A = 2^{16} + 3$ in this class was used for many years in the pseudorandom number generator RANDU in IBM's Scientific Subroutine package. Interestingly, the bad performance in three dimensions was discovered empirically.

This last example demonstrates that although $d_k(A, M)$ is a better measure of k-dimensional nonuniformity than $N_k(A, M)$, a bound on the number of hyperplanes may be more easily computed and thus be of more immediate use.

9.6 MULTIPLE REPLICATIONS

As discussed in Section 4.20, using separate pseudorandom number streams, one for each source of stochastic variation, creates an opportunity for variance reduction. Suppose there are r sources of variation and the jth requires k_j pseudorandom numbers to generate a sample from its corresponding probability distribution. This focuses our interest on the $(k_1 + \cdots + k_r)$-tuples

$$
\begin{aligned}
& Z_i, Z_{i+1}, \ldots, Z_{i+k_1-1}, \ldots, Z_{i+l}, Z_{i+l+1}, \ldots, Z_{i+l+k_2-1}, \ldots, \\
& \quad Z_{i+(r-1)l}, Z_{i+(r-1)l+1}, \ldots, Z_{i+(r-1)l+k_r-1}, \qquad i \geq 1,
\end{aligned}
\tag{9.26}
$$

where $l > \max_j k_j$ denotes the offset in the pseudorandom number sequence between successive streams.

It is easily seen that these points lie on $(k_1 + \cdots + k_r)$-dimensional hyperplanes

$$\sum_{j=1}^{r} \sum_{s=0}^{k_j-1} q_{(j-1)l+s} Z_{i+(j-1)l+s} \equiv 0 \pmod{M}, \tag{9.27}$$

where $\mathbf{q} := (q_0, q_1, \ldots, q_{(r-1)l+k_r-1})$ is an element of

$$\mathcal{Q}'_{(r-1)l+k_r}(A, M) := \left\{ \mathbf{q} \in \mathcal{Q}'_k(M) : \sum_{l=1}^{r} \sum_{s=0}^{k_j-1} q_{(j-1)l+s} A^{(j-1)l+s} \equiv 0 \pmod{t(M)} \right\},$$

$t(M)$ is defined in Section 9.5, and the maximal distance between successive parallel hyperplanes is

$$d'_{k_1+\cdots+k_r}(A, M) := \max_{\mathbf{q} \in \mathcal{Q}'(r-1)\mathbf{l}+\mathbf{k_r}(\mathbf{A},\mathbf{M})} \left(\sum_{j=1}^{r} \sum_{s=0}^{k_j-1} q^2_{(j-1)l+s} \right)^{-1/2}. \tag{9.28}$$

This motivates our interest in identifying the multiplier A that minimizes this maximal distance.

Since inequality (9.21) has already revealed the grid coarseness for the generator (9.1) for dimensions as small as 4, identifying the best A in $k_1 + \cdots + k_r > 6$ dimensions would serve little purpose. This motivates us to defer discussion of multiple streams until Section 9.7.1 and to turn to an alternative form of pseudorandom number generation that offers an expeditious way of reducing this coarseness up to significantly higher dimensions. While not currently implemented in any simulation language, its availability in C code makes it a prime candidate for future implementation.

9.7   COMBINED GENERATORS

Consider J multiplicative pseudorandom number generators (L'Ecuyer and Tezuka 1991)

$$Z_{ji} = A_j Z_{j,i-1} \pmod{M_j}, \qquad j = 1, \ldots, J, \tag{9.29}$$

where $M_1, \ldots, M_J$ are distinct primes and A_j is a primitive root of M_j. For example, for $J = 4$,

$$
\begin{aligned}
M_1 &= 2^{31} - 1 = 2147483647, & A_1 &= 45991, \\
M_2 &= 2^{31} - 105 = 2147483543, & A_2 &= 207707, \\
M_3 &= 2^{31} - 225 = 2147483423, & A_3 &= 138556, \\
M_4 &= 2^{31} - 325 = 2147483323, & A_4 &= 49689.
\end{aligned}
\tag{9.30}
$$

Let $\delta_1, \ldots, \delta_J$ denote arbitrary integers such that for each j, δ_j and M_j have no common factors. Consider the combined generator

$$\tilde{Z}_i := \left(\sum_{j=1}^{J} \delta_j Z_{ji} \right) \pmod{M_1}, \quad i \geq 1, \tag{9.31}$$

and the normalized sequence $\{\tilde{U}_i := \tilde{Z}_i / M_1, i \geq 1\}$. What can we say about its properties?

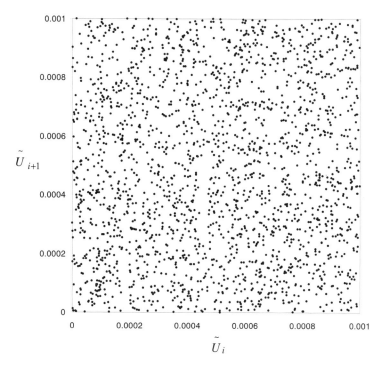

Figure 9.4 2-tuples for combined generator

To answer this question, we need to consider an ancillary generator. Let

$$U_{ji} := Z_{ji}/M_i,$$
$$V_i := \left(\sum_{j=1}^{J} \delta_j U_{ji} \right) (\text{mod } 1), \quad i \geq 1, \tag{9.32}$$

and

$$\tilde{M} := \Pi_{j=1}^{J} M_j.$$

Then

- $\{V_i, i \geq 1\}$ obeys the recurrence

$$Y_i = \tilde{A} \, Y_{i-1} \pmod{\tilde{M}}, \tag{9.33}$$

where $Y_i = \tilde{M} \, V_i$ and $\tilde{A}$ is computable (L'Ecuyer and Tezuka 1991) and does not depend on $\{\delta_j\}$.

- $\{V_i\}$ and therefore $\{Y_i\}$ have period

$$P = \mathrm{lcm}(M_1 - 1, \ldots, M_J - 1),$$

the least common multiple of $M_1 - 1, \ldots, M_J - 1$, where

$$P \le (M_1 - 1) \times \cdots \times (M_J - 1)/2^{J-1}.$$

- $\tilde{U}_i = V_i + \epsilon_i$, where tight lower and upper bounds can be computed on ϵ_i.

These are important properties. They imply that the period is at least $\min_{1 \le j \le J} (M_j - 1)$ and that judiciously chosen $M_1, \ldots, M_J$ can induce a period as large as $(M_1 - 1) \times \cdots \times (M_J - 1)/2^{J-1}$. In particular, the maximum is achieved if $(M_1 - 1)/2, \ldots, (M_J - 1)/2$ have no factors in common, as is the case for $M_1, \ldots, M_4$ in expression (9.29), for which $P \approx 2.13 \times 10^{37}$, in contrast to a period of $2^{31} - 2 \approx 2.14 \times 10^9$ for the generator (9.1) with $M = 2^{31} - 1$. Indeed, the density of points in $[0, 1)$ increases by a factor of 10^{27}.

The first property tells us that the normalized $\{V_i,\ i \ge 1\}$ satisfy a one-step recursion $V_i = \tilde{A}\, V_{i-1} \pmod 1$. The third reveals that $\tilde{U}_i$, based on the generator (9.30), differs from V_i by a negligible amount. In particular, the assignment (9.29) leads to

$$\tilde{M} = 21267641435849934371830464348413044909 \approx 2 \times 10^{37},$$
$$\tilde{A} = 5494569482908719143153333426731027229,$$

and induces the bounds $-1.04308128 \times 10^{-7} \le \epsilon_i \le 1.99303031 \times 10^{-7}$ (L'Ecuyer 1999). Therefore, it is not unreasonable to characterize the properties of $\{U_i\}$ by those of $\{V_i\}$, which, fortunately, we can determine via the spectral test.

Table 9.5 lists the distances $\{d_k(\tilde{A}, \tilde{M}),\ k = 2, \ldots, 10\}$ between successive hyperplanes for the generator based on the assignment (9.29). For comparison, it also lists the distances for $A = 16807, 630360016$, and 742938285 and $M = 2^{31} - 1$. The unequivocal superiority of the generator (9.30) is clear. It induces a considerably denser grid than the other generators do.

Recall that Figures 9.2a, 9.2b, and 9.2c each show regular patterns of about 2147 2-tuples in $[0, .001) \times [0, .001)$ over the period $M - 1 = 2^{31} - 2$ of the three prime-modulus multiplicative congruential generators of the basic form (9.1). For the combined generator defined in expressions (9.28), (9.29), and (9.30), run with arbitrarily selected seeds for each of its component generators, Figure 9.5 displays the subset of 2147 2-tuples of $\{(\tilde{U}_i, \tilde{U}_{i+1});\ i = 1, \ldots, 2^{31} - 2\}$ that fall into $[0, .001) \times [0, .001)$. The choice of run length was made to allow comparison between Figures 9.2 and 9.5 with density held relatively fixed. Most notably, the absence of any regularity in Figure 9.5 allows us to conclude that for sequences of length $2^{31} - 2$ the combined generator succeeds in eliminating the apparent lattice structure inherent in generators of the form (9.1) and depicted in Figures 9.2a, 9.2b, and 9.2c.

Table 9.5 Distance between parallel hyperplanes

	$d_k(A, 2^{31} - 1)$			$d_k(\tilde{A}, \tilde{M})$
k	$A = 16807$	$A = 630360016$	$A = 742938285$	
2	5.95×10^{-5}	2.42×10^{-5}	2.32×10^{-5}	2.38×10^{-19}
3	1.57×10^{-3}	1.60×10^{-3}	8.02×10^{-4}	4.18×10^{-13}
4	6.79×10^{-3}	4.99×10^{-3}	4.53×10^{-3}	5.24×10^{-10}
5	1.50×10^{-2}	1.38×10^{-2}	1.33×10^{-2}	3.41×10^{-8}
6	3.34×10^{-2}	3.34×10^{-2}	2.59×10^{-2}	6.31×10^{-7}
7	6.04×10^{-2}		5.53×10^{-2}	4.73×10^{-6}
8	7.91×10^{-2}		6.82×10^{-2}	2.09×10^{-5}
9	.113		.106	7.78×10^{-5}
10	.125		.108	2.30×10^{-4}

9.7.1 MULTIPLE PSEUDORANDOM NUMBER STREAMS

Recall from Section 9.6 that it is the properties of the $(k_1 + \cdots + k_r)$-tuples that ultimately interest us. Moreover, we cannot expect these to be particularly good for $M = 2^{31} - 1$ and 2^{32}. To simplify the exposition in the current context, we take $k_1 = \cdots = k_r = k$ and focus on the kr-tuples

$$\tilde{\mathbf{Z}}_i := (\tilde{Z}_i, \tilde{Z}_{i+1}, \ldots, \tilde{Z}_{i+k-1}, \ldots, \tilde{Z}_{i+l}, \tilde{Z}_{i+l+1}, \ldots, \tilde{Z}_{i+l+k-1}, \ldots,$$
$$\tilde{Z}_{i+(r-1)l}, \tilde{Z}_{i+(r-1)l+1}, \ldots, \tilde{Z}_{i+(r-1)l+k-1}),$$

which, because of the third property of generator (9.31), we assume are close to those for the kr-tuples

$$\mathbf{Y}_i := (Y_i, Y_{i+1}, \ldots, Y_{i+k-1}, Y_{i+l}, Y_{i+l+1}, \ldots, Y_{i+l+k-1}, \ldots,$$
$$Y_{i+(r-1)l}, Y_{i+(r-1)l+1}, \ldots, Y_{i+(r-1)l+k-1}).$$

For assignment (9.29), it is of interest to determine $d'_{klr}(\tilde{A}, \tilde{M})$, the maximal distance between parallel hyperplanes, for r pseudorandom number streams with displacement l between successive streams for k dimensions. An empirical analysis in L'Ecuyer and Andres (1997a) reveals that

$$d'_{klr}(\tilde{A}, \tilde{M}) \leq \begin{cases} 2.02 \times 10^{-19} & r = 2, \\ 3.22 \times 10^{-13} & r = 3, \\ 3.92 \times 10^{-10} & r = 4, \\ 2.78 \times 10^{-8} & r = 5, \\ 4.66 \times 10^{-7} & r = 6, \\ 3.46 \times 10^{-6} & r = 7, \\ 1.53 \times 10^{-5} & r = 8, \end{cases}$$

for $k = 1, \ldots, 20$, and $l = 2^{41}$. This choice of displacement l induces the least upper bound on $d'_{klr}(\tilde{A}, \tilde{M})$ for all $2 \leq r \leq 8$ and $1 \leq k \leq 20$. Since these are considerably smaller than one could expect to achieve for generators of the form (9.1) for k-tuples taken on r streams for these ranges of k and r and any displacement l, we recommend the L'Ecuyer and Andres combined generator (9.31) with the assignment (9.30) for replacing currently implemented pseudorandom number generators. Although the L'Ecuyer–Andres implementation takes $r = 2^{31}$ streams each of length $l = 2^{41}$ by default, it allows the user to change these values.

9.7.2 COST OF IMPROVED QUALITY

These improved properties come at the expense of increased computing time per pseudorandom number generation. The increase is between three and four times the computing time for an efficient implementation of the generator (9.1). The upper bound of four arises because the function call to the generator occurs once regardless of whether generator (9.1) or (9.29) is used.

As already mentioned in Section 9.2, careful attention to multiplication modulo $\tilde{M}$ can save time. The C programs clcg4.c and clcg4.h and the Modula-2 code clcg4.mod and clcg4.def, available at http://www.iro.umontreal.ca/~lecuyer/papers.html, take full advantage of the option to conserve time. A test program, prg 2.c, is also available there.

For the practitioner, the issue now becomes one of quality of results versus computing time. If a simulation using generator (9.1) devotes $100 \times a$ percent of its time to generating pseudorandom numbers, then a switch to generator (9.30) induces a relative increase in computing time of at most $4a + (1 - a) = 3a + 1$. For example, $a = .05, .10$, and $.20$ induce $3a + 1 = 1.15, 1.30$, and 1.60, respectively.

In Monte Carlo experiments that focus narrowly on numerical integration, pseudorandom number generation can be the principal time consumer. If, for example, $a = .80$, then the switch from generator (9.1) to generator (9.30) implies as much as a $3a + 1 = 3.4$ time increase in computing time. No doubt, this substantial increment in cost is what has retarded the implementation of better pseudorandom number generators in this type of experiment.

By contrast, pseudorandom number generation in discrete-event simulation rarely consumes more than a marginal amount of computing time, most of which is devoted to memory management and, in particular, list processing. It is not unusual to have $a < .20$. In this case, the penalty of increased computing time seems worth the benefit of higher-quality pseudorandom numbers.

9.8 LESSONS LEARNED

- Every simulation language provides a routine for generating *pseudorandom* integers $Z_i, Z_{i+1}, \ldots$, where Z_i is less than some positive integer M $\forall i$. The integers are normalized by mapping them ($U_i = Z_i/M$) into points in $(0, 1)$, which are then used as input for generating samples from diverse distributions.

- *Multiplicative congruential pseudorandom generators* are the most common mechanism for generating these integers. Each has *modulus M* and integer *multiplier A*. The generators execute rapidly and need little memory. Each require a *seed Z_0* to *prime* the generator. The simultaneous choice of M and A determines the density of integers generated from $\{1, \ldots, M - 1\}$. For given M, the choice of A determines how well k-tuples generated on $(0, 1)^k$ resemble random k-tuples uniformly distributed on $(0, 1)^k$.

- If M is prime, then choosing A to be a *primitive root* of M ensures that the sequence runs through a complete permutation of the integers $\{1, \ldots, M - 1\}$ before repeating itself.

- If $M = 2^\beta$ for $\beta \geq 3$, then choosing $A = \pm 3 \pmod 8$ ensures a *maximal period $M/4$* on a subset of the $M/2$ odd integers in $\{1, 3, \ldots, M/2 - 1\}$. The period denotes the number of iterations of the generator before it repeats itself.

- k-tuples $(U_i, U_{i+1}, \ldots, U_{i+k-1})$ for $i \geq 1$ lie on parallel hyperplanes in $(0, 1)^k$. The choice of M and A dictates the number of parallel hyperplanes and the maximal distance between successive parallel hyperplanes. A small number of widely spaced hyperplanes signals a poor approximation to randomness for the k-tuples. Among all multipliers A that maximize the period for a given M, one prefers the one that minimizes the maximal distance between successive hyperplanes in k dimensions. This optimal A usually varies with k.

- Table 9.3 gives good choices of multipliers for $M = 2^{31} - 1$, 2^{32}, and 2^{48}.

- Since $U_i, U_{i+1}, \ldots$ are the input to a procedure for generating samples $X_j, X_{j+1}, \ldots$ from a particular distribution F, departures from uniformity in the U_i induce a distribution $\tilde{F}$ for the X_j that only approximates F. Most importantly, a given sequence $U_i, U_{i+1}, \ldots$ used as input to different procedures for generating samples from F induces different $\tilde{F}$'s and therefore different approximations to the desired distribution F. The acceptance–rejection and ratio-of-uniforms methods can induce substantial errors in this way.

- By combining several multiplicative pseudorandom number generators according to specified rules, generators can be created with exceptionally large periods. For example, combining four generators with individual periods of about 2.14×10^9 leads to a combined generator with period of about 2×10^{37}.

- In dimensions $k = 2, \ldots, 8$, the maximal distances between successive hyperplanes in $(0, 1)^k$ for the combined generator are substantially smaller than for the more conventional multiplicative congruential generators.

- Implementations of the combined generator are available via the Internet. See Section 9.7 for the address.

9.9 EXERCISES

9.1 Show that $A = 742938295$ is a primitive root of $2^{31} - 1$.

9.2 Show that $A = 630360016$ is a primitive root of $2^{31} - 1$.

9.3 Show that $A = 746061395$ is the multiplicative inverse of $630360016 \pmod{2^{31}-1}$.

9.4 Determine whether $A = 48271$ and $M = 2^{31} - 1$ gives a full period $M - 1$.

9.5 Determine whether $A = 343$ and $M = 2^{31} - 1$ gives a full period $M - 1$.

9.6 For the multiplicative congruential generator

$$Z_i = AZ_{i-1} \pmod{2731},$$

a. Determine the number of values $\pmod{2731}$ for A that give a full period 2730.

b. Enumerate all these values.

c. Enumerate all pairs of values of A_1 and A_2 for which the generator

$$Z_i = A_1 Z_{i-1} \pmod{2731}$$

gives the same sequence

$$Z_i = A_z Z_{i-1} \pmod{2731},$$

but in reverse order, and for which both generators have full period. Show your work in detail.

9.7 Consider the full-period pseudorandom number generator

$$Z_i = AZ_{i-1} \pmod{M},$$

where $M = 2^\beta - \gamma, \gamma > 0, \beta > \log \gamma / \log 2$, and M is prime. Construct a portable pseudorandom number generator that will produce the identical sequence of numbers on all computers whose floating-point representations have at least a $\lceil 3\beta/2 \rceil$-bit mantissa.

✐   **9.8** A simulator wants to generate all the integers $\{0, 1, \ldots, 4997\}$ in a pseudorandom order. He plans to use the sequence $\{Y_i = Z_i - 1, i = 1, \ldots, 4998\}$, where

$$Z_i = A_1 Z_{i-1} \ (\text{mod } M)$$

with $M = 4999$ prime, $A_1 = 2187$, and $Z_0 \in \{1, 2, \ldots, 4998\}$. A second simulator suggests an alternative multiplier $A_2 = 2182$. Determine which, if either, of the multipliers generates all the desired integers. Give complete details.

✐   **9.9** Consider the pseudorandom number generator

$$Z_i = 7340035 Z_{i-1} \ (\text{mod}(65536 \times 524288)).$$

a. Determine the period of the generator.

b. Determine an upper bound on the maximal number of parallel hyperplanes on which the 2-tuples $\{(Z_i, Z_{i+1}), i \geq 1\}$ lie.

c. Show that the maximal number of parallel hyperplanes on which the 3-tuples $\{(Z_i, Z_{i+1}, Z_{i+2}), i \geq 1\}$ lie cannot exceed 15.

✐   **9.10** The programming language Prime Sheffield Pascal provides a library function RANDOM that uses a multiplicative congruential generator $Z_i = A Z_{i-1} \ (\text{mod } M)$ with $A = 16807$ and $M = 2^{31}$ to produce random numbers. Determine whether this generator has the maximal achievable period for this modulus. Display all steps in your proof.

✐   **9.11** Because of the small word size available on her computer, a simulator decides to employ the random number generator

$$Z_i = 20397 Z_{i-1} \ (\text{mod } 2^{15}).$$

Her experiment calls for generating exponential samples with p.d.f.

$$f(x) = e^{-x}, \quad x \geq 0.$$

She decides to apply the inverse transform method to the normalized Z_i to generate these exponential samples. Let X be a resulting sample. Find the smallest and largest values that X can assume before repeating itself (a) if $Z_0 = 1$ and (b) if $Z_0 = 7$. Give their numerical values to six significant digits.

✐   **9.12** Consider the multiplicative congruential generator

$$Z_i = 1048579 Z_{i-1} \ (\text{mod } 536870912).$$

a. Determine the period.

b. State the conditions under which the numbers 2875, 4876321, and 578766 would each appear in a sampling experiment that uses P pseudorandom numbers, where P denotes the maximal period.

✐ **9.13** For $A = 65541$ and $M = 2^{32}$, show that the period of generator (9.1) is 2^{30}.

✐ **9.14** For $A = 16777221$ and $M = 2^{48}$, show that the period of generator (9.1) is 2^{46}.

9.10 REFERENCES

Cassels, J.W.S. (1959). *An Introduction to the Geometry of Numbers*, Springer-Verlag, Berlin.

Fishman, G.S., and L. Moore (1986). An exhaustive analysis of multiplicative congruential random number generators with modulus $2^{31} - 1$, *SIAM J. Sci. and Statist. Comput.*, **7**, 24–45.

Fishman, G.S. (1990). Multplicative congruential random number generators with modulus 2^{β}: an exhaustive analysis for $\beta = 32$ and a partial analysis for $\beta = 48$, *Math. Comp.*, **54**, 331–334.

Fishman, G.S. (1996). *Monte Carlo: Concepts, Algorithms, and Applications*, Springer-Verlag, New York.

Hörmann, W., and G. Derflinger (1993). A portable random number generator well suited for the rejection method, *ACM Trans. Math. Software*, **19**, 489–495.

Jannson, B. (1966). *Random Number Generators*, Almquist and Wiksell, Stockholm.

L'Ecuyer, P., and S. Tezuka (1991). Structural properties for two classes of combined random number generators, *Mathematics of Computation*, **57**, 735–746.

L'Ecuyer, P., and T.H. Andres (1997a). A random number generator based on the combination of four LCGs, version: November 25, 1997, University of Montreal, available at http://www.iro.umontreal.ca/~lecuyer.

L'Ecuyer, P., and T.H. Andres (1997b). A random number generator based on the combination of four LCGs, *Mathematics and Computers in Simulation*, **44**, 99–107.

L'Ecuyer, P. (1999). Personal communication.

Overstreet, Jr., C.L. (1974). The relationship between the multiplicative and mixed generators modulo 2^{b}, Computer Science Department, Virginia Polytechnic Institute and State University, Blacksburg.

Park, S.K., and K.W. Miller (1988). Random number generators: good ones are hard to find. *Comm. ACM*, **31**, 1192–1201.

Payne, W.H., J.R. Rabung, and T.P. Bogyo (1969). Coding the Lehmer pseudorandom number generator, *Comm. ACM*, **12**, 85–86.

Schrage, L. (1979). A more portable FORTRAN random number generator, *ACM Trans. Math. Software*, **5**, 132–138.

Chapter 10

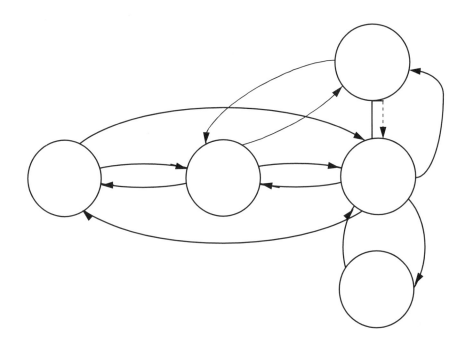

Preparing the Input

The modeling representations of Chapter 2 and the programming concepts of Chapter 3 provide us with the ability to create relatively sophisticated simulation models and executable programs, but only after we have identified all sources of stochastic variation, specified sampling distributions that characterize each source, and assign numerical values to the parameters of the distributions. This chapter addresses these issues.

In practice, specifying distributions and assigning numerical value go hand in hand. Collectively, we refer to them as *input preparation*. If no historical numerical data are available, then a simulationist customarily relies on *plausibility arguments*. These arise from at least three sources:

- Theoretical deductions that follow from an input process specified properties

- Expert opinion based on knowledge of how the particular input process behaves in the real world

- Expert opinion about the behavior of input processes in closely related systems.

For example, if an arrival process can be viewed as the *superposition* of a very large number of independent renewal processes each with relatively small arrival rate, then theoretical deduction can show that interarrival times in the arrival process of interest have a distribution well approximated by the exponential (Section 10.4.1).

If, in addition, expert opinion based on historical recollection indicates that the arrival rate λ lies in $[\lambda_1, \lambda_2]$ for $0 < \lambda_1 < \lambda_2 < \infty$, then the simulationist has a candidate distribution and an interval for the value of its parameter. Using the lower bound λ_1 in the simulation would induce lower congestion than expected in the real system, whereas using λ_2 would induce the opposite behavior. This may encourage the simulationist to compare the output of, say, two runs, one with $\lambda = \lambda_1$ and the other with $\lambda = \lambda_2$, to see whether the uncertainty about λ induces variation in the output that is within tolerable bounds. If it does, then the actual value of λ is not a major issue. If variation is not within tolerable bounds, then attention must be devoted to finding a better way to estimate λ.

Naturally, the availability of empirical data on the input process of interest or on one closely resembling it is of great value. It creates an opportunity for input preparation that enables the executing simulation program to generate stochastic sample paths that mimic those observed in either the real or related system. How well these match depends on the amount of historical data available and on the techniques employed in analyzing data.

Some input preparation depends exclusively on either expert opinion or on empirical data analysis. However, a blend of both can endow a simulation with a considerably more representative input than either alone. To effect this blend often calls for *hypothesis testing* as well as estimation. Several factors militate against embracing this joint approach. First, it requires a deeper knowledge of statistical theory. Second, it inevitably takes more of the simulationist's time and effort. Third, many theoretical results in the study of delay systems have their greatest value in facilitating analytical solution. Since a simulation model provides a numerical solution, why burden the analysis with specification aimed at analytical solution?

While the first two factors need to be evaluated in light of the simulationist's skills and time constraint, the third needs to be tempered by recognizing the limitations of relying on a data analysis devoid of plausibility arguments. Using a sample record of finite length to fit an input distribution inevitably invites errors both in selecting the distribution and estimating its parameters. By combining this analysis with plausibility arguments, a simulationist prevents the sampling errors in the distribution fitting process from having undue influence in the final form of the input.

10.1 ARRIVAL PROCESSESS

Interarrival times, service times, and batch sizes (number of items demanding service in an arrival) are principal sources of stochastic variation in a delay system. Our discussion focuses on these with the admonition that other sources of stochastic variation peculiar to particular systems can often be analyzed using techniques applicable to these three.

From the perspective of modeling and programming it is convenient to regard interarrival times as a sequence of i.i.d. random variables. This applies equally well to service

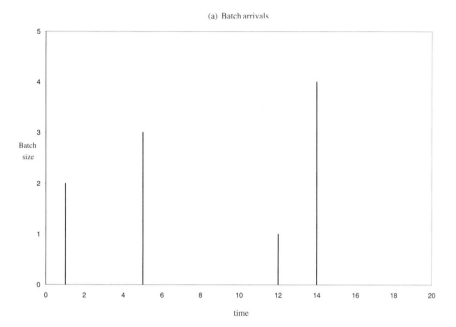

(a) Batch arrivals

Figure 10.1 Arrival processes

times and batch sizes. Then one need only specify a single distribution for each source from which to generate samples when executing a simulation run. In reality, the i.i.d. assumption does not always apply, and input preparation needs to acknowledge that and accommodate dependence. Section 10.11 does this. For now, we retain the assumption merely as a convenient backdrop to addressing issues relevant to the objectives of input preparation. The issues raised here apply equally well when the i.i.d. assumption does not hold.

Discrete-event simulation methodology often characterizes the arrival process in one of two ways. One views it as a sequence of 2-tuples $(T_1, D_1), (T_2, D_2), \ldots$, where $T_1, T_2, \ldots$ is a sequence of i.i.d. interarrival times and $D_1, D_2, \ldots$ is a sequence of batch sizes. If $D_i = 1 \; \forall i$, then we are back to the classical single-item arrival queueing model. If $D_1, D_2, \ldots$ are i.i.d. random variables, then we have a random batch-size arrival process. Figure 10.1a illustrates the concept.

The second characterization consists of a sequence of i.i.d. 3-tuples $(T_1, T_1', D_1), (T_2, T_2', D_2), \ldots$ where $T_1, T_2, \ldots$ and $T_1', T_2', \ldots$ are independent sequences of i.i.d. nonnegative random variables, and D_i items in the ith $(i > 1)$ batch begin arriving at time $T_1 + \sum_{l=1}^{i-1} (T_l + T_l')$ and end arrival at time $\sum_{l=1}^{i} (T_l + T_l')$.

The transmission of records in a communication network often fits this second characterization, where T_i' denotes the time required to transmit the ith record and $T_i + T_i'$ denotes the time that elapses between the end of transmission of the $(i-1)$st record and the end of transmission of the ith record. Occasionally, it is convenient to think of this as a fluid

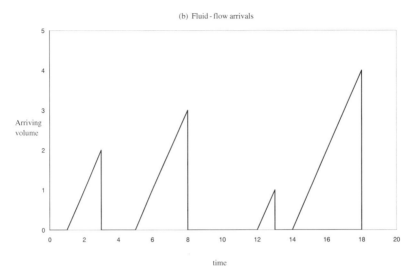

(b) Fluid-flow arrivals

Figure 10.1   (continued)

flow into a storage tank rather than as a discrete count on items (Section 2.6). Figure 10.1b gives one possible conceptualization of this process.

10.2   EMPIRICAL AND FITTED DISTRIBUTIONS

When empirical data are available on input processes, a simulationist can use them in at least two ways to prepare input. One constructs empirical distribution functions (e.d.f.s), based exclusively on the data, and generates samples from them during the simulation. The other fits the data to theoretical distributions and then generates samples from the fitted distributions.

The e.d.f. approach induces stochastic variation in a simulated sequence that exactly mimics that in the empirical data. Some of these fluctuations arise from the intrinsic properties of the underlying stochastic process. These are desirable. However, others arise from the less-than-complete stochastic representation on which the e.d.f. is based. These are undesirable. Therefore, simulated sample sequences generated by an e.d.f. show desirable variation due to the intrinsic character of the processes and undesirable variation due to finite sample size.

As the length n of the empirical record increases, the unwanted fluctuations diminish in relative importance when compared to the wanted fluctuations, so that the e.d.f. approach generates samples that are increasingly (as n increases) representative of sample paths on the underlying stochastic process. When very long empirical records are available, the e.d.f.

approach has much to commend it as a means of generating sample sequences. Section 10.3 describes its essential features.

For small and moderate n, the fitted d.f. approach has appeal. Fitting a theoretical distribution to data smoothes out the variations in the empirical distribution due to finite sample size. It also enlarges the domain of values for sampling. For example, given a sample record of service times $X_{(1)} \leq X_{(2)} < \cdots \leq X_{(n)}$, the e.d.f. approach generates values on $[X_{(1)}, X_{(n)}]$ at best. However, if expert opinion tells us that in reality service times take values in the *support* $[x_1, x_2]$, where $x_1 \leq X_{(1)} < \cdots \leq X_{(n)} < x_2$, then fitting a distribution with this support allows this expert opinion to be merged with the sample record to generate samples on $[x_1, x_2]$.

Smoothing and increasing the domain for sampling are the desirable features of the fitted d.f. approach. As n grows, these benefits diminish in importance when compared to the e.d.f. approach. First, the e.d.f. becomes smooth; second, its domain for sampling increases, and third, because the fitted d.f. is inevitably an approximation, tests of its adequacy begin to detect statistically significant discrepancies between fitted and empirical distributions. Section 10.12 discusses distribution fitting in several different contexts.

Although our discussion reveals a tendency to favor the e.d.f. approach for large sample sizes and the fitted d.f. approach for small and moderate sizes, following this inclination would be too simplistic in practice. Circumstances, as in Section 10.13, exist where the e.d.f. approach offers more convenience and more accuracy for relatively small sample sizes. Moreover, the characterization of arrival processes in Section 10.4.1 encourages the reader to focus on the (theoretical) exponential distribution as a candidate for fitting, even in large samples. Therefore, we encourage the reader to recognize that circumstances for her/his particular study must be the determinant of which approach to follow. More often than not, both play roles in a single simulation study.

10.3 EMPIRICAL D.F. APPROACH

Let X denote a random variable with d.f. $\{F(x), x \epsilon X\}$ and p.d.f. or p.m.f. $\{f(x), x \in X\}$. If X denotes interarrival or service time, then F is usually continuous with support $X \in [0, \infty)$, $[a, \infty)$ for some fixed $a > 0$, or $[a, b)$ for fixed $0 \leq a < b < \infty$. For the moment, assume that X is continuous on $[0, \infty)$.

Let $X_1, \ldots, X_n$ denote n i.i.d. observations from F. Then the e.d.f.

$$F_n(x) := \frac{1}{n} \sum_{i=1}^{n} I_{[X_i, \infty)}(x), \quad x \geq 0, \tag{10.1}$$

provides an estimator of $\{F(x)\}$ and the empirical histogram

$$H_n(y_j) := F_n(y_j + \Delta/2) - F_n(y_j - \Delta/2), \tag{10.2}$$

(a)

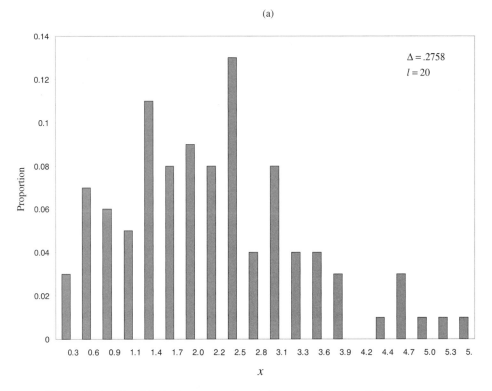

Figure 10.2 Empirical histograms for service-time data, $n = 100$

where for some fixed $1 < l < n$,

$$y_1 := \frac{\Delta}{2} + \min X_i,$$
$$y_j := y_{j-1} + \Delta \qquad j = 2, \ldots, l,$$
$$\Delta := \frac{\max X_i - \min X_i}{l},$$

approximates $f(y)\Delta$ evaluated at $y \in \{y_1, \ldots, y_l\}$. Figure 10.2a shows an empirical histogram for $n = 101$ i.i.d. service times based on an $l = 20$ cell partition of the data.

The graph of a histogram can provide insight into which theoretical distributions are good candidates for fitting. For example, the Gamma and the Wishart distributions with their skewed right tails are good prospects for Figure 10.2. However, the ability to discern shape in a histogram depends crucially on l. Customarily, l is chosen to balance the need to discern the underlying shape of this frequency chart against the need to smooth out fluctuations that obscure shape. Large l improves resolution but admits large fluctuations. Small l smooths out fluctuations but reduces resolution. Inevitably, the choice of l is a compromise. A comparision between Figures 10.2a and 10.2b shows the tradeoff.

(b)

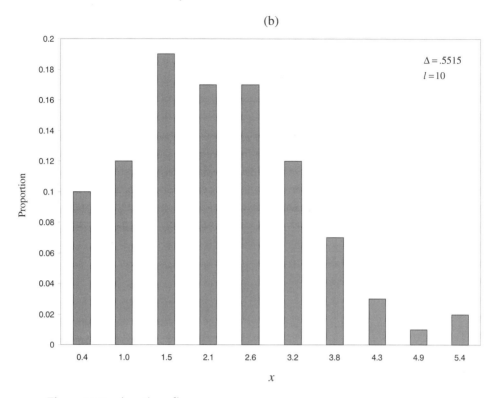

$\Delta = .5515$
$l = 10$

Figure 10.2 (continued)

10.3.1 SAMPLING PROPERTIES

To assess how well $\{F_n(x)\}$ approximates $\{F(x)\}$, we turn to its sampling properties. For each $x \in \mathcal{X}$ they are:

- $F_n(x) \rightarrow F(x)$ w.p.1 as $n \rightarrow \infty$.

- $\text{s.e.}(F_n(x)) := \sqrt{\text{var} F_n(x)} = \sqrt{F(x)[1 - F(x)]/n}$ (standard error).

- $\text{r.e.}(F_n(x)) := \sqrt{\dfrac{\text{var} F_n(x)}{\text{E} F_n(x)}} = \sqrt{[1 - F(x)]/F(x)n}$ (relative error).

- $\text{r.e.}(1 - F_n(x)) := \sqrt{\dfrac{\text{var} F_n(x)}{\text{E}[1 - F_n(x)]}} = \sqrt{F(x)/[1 - F(x)]n}$.

- $\text{corr}[F_n(x), F_n(y)] := \dfrac{\text{corr}[F_n(x), F_n(y)]}{\sqrt{\text{var} F_n(x) \text{var} F_n(y)}} = \sqrt{\dfrac{F(y)[1 - F(x)]}{F(x)[1 - F(y)]}}, \qquad y \le x.$

- $\sup_{x \in \mathcal{X}} | F_n(x) - F(x) |$ has the Kolmogorov–Smirnov (K-S) distribution for each fixed positive integer n. Therefore, $\{[F_n(x) - d_n(\delta), F_n(x) + d_n(\delta)], \forall x \in \mathcal{X}\}$ covers the true d.f. $\{F(x), x \in \mathcal{X}\}$ w.p. $1 - \delta$, where $d_n(\delta)$ denotes the $1 - \delta$ quantile of the Kolmogorov–Smirnov distribution. As $n \rightarrow \infty$, $n^{1/2} d_n(.05) \rightarrow 1.3581$

and $n^{1/2}d_n(.01) \to 1.6276$. For $n \geq 100$, $1.3581/n^{1/2}$ and $1.6276/n^{1/2}$ approximate $d_n(.05)$ and $d_n(.01)$, respectively, with negligible error.

- $\sqrt{n}F_n(x) \xrightarrow{d} \mathcal{N}(F(x), F(x)[1 - F(x)])$ as $n \to \infty$. Therefore, for $x \in \mathcal{X}$, $\{nF_n(x) + \phi^2/2 \mp \phi\{\phi^2/4 + nF_n(x)[1 - F_n(x)]\}^{1/2}\}/(n + \phi^2)$, where $\phi := \Phi^{-1}(1 - \delta/2)$, covers $F(x)$ with probability that converges to $1 - \delta$ as $n \to \infty$.

These properties also hold if X is continuous over an interval of finite length $\mathcal{X} := [x_1, x_2]$. If X is discrete, the properties hold except that the confidence region based on the K-S distribution corresponds to a confidence level greater than or equal to $1 - \delta$.

10.3.2 ASSESSING THE APPROXIMATION

The standard error, relative error, K-S distribution, and normal limit all provide ways of assessing how well the e.d.f. approximates the true, but unknown, d.f. Replacing $F(x)$ by $F_n(x)$ in s.e. $(F_n(x))$ and r.e. $(F_n(x))$ provides estimates of these errors. The K-S and limiting normal properties provide more incisive measures.

The K-S distribution leads to a confidence region for the entire d.f. $\{F(x)\}$ at the $100 \times (1 - \delta)$ percent confidence level for each positive integer n. By contrast, the limiting normal result provides a $100 \times (1 - \delta)$ percent confidence interval for a selected x as n becomes large. Both results have benefits and limitations. For example, the K-S result gives a half-length of $1.6276/n^{1/2}$ for $\delta = .01$ and $n \geq 100$ for all x, leading to $\{F(x) \in [\max[0, F_n(x) - .1628], \min[1, F_n(x) + .1628]], \forall x \in \mathcal{X}\}$ for the e.d.f. with $n = 100$ on which the empirical histogram in Figure 10.2 is based.

Figure 10.3 shows $\{\tilde{F}_n(x)\}$ and its 99 percent K-S confidence region where $\{\tilde{F}_n(x)\}$, as defined in expression (10.3), differs from $\{F_n(x)\}$ in a relatively minor way. The width of the region is far too large to regard the e.d.f. as a relatively accurate approximation of $\{F(x)\}$. However, the limiting normal result reveals a better pointwise accuracy. Table 10.1 shows individual and simultaneous confidence intervals (Section 7.1) for five values of x.

These reveal shorter inteval lengths than the K-S approach, suggesting in general, that for moderate and large n, computing confidence intevals for selected x provides a more informative picture of statistical accuracy than the K-S approach. Nevertheless, in the present

Table 10.1 99% individual and simultaneous confidence intervals, $n = 100$

x	$F_n(x)$	Individual Lower	Individual Upper	Simultaneous Lower	Simultaneous Upper
1	.17	.09461	.2865	.08419	.3133
2	.47	.3474	.5964	.3252	.6200
3	.78	.6578	.8674	.6309	.8803
4	.93	.8342	.9723	.8084	.9767
5	.98	.9042	.9961	.8793	1.000

case, the lengths of the normal intervals also imply that $\{F_n(x)\}$ is not a particularly accurate approximation to $\{F(x)\}$.

10.3.3   SAMPLING FROM $\{F_n(x)\}$

To generate a random sample X' from $\{F_n(x)\}$, we proceed as follows. Let $\{x_1 < x_2 < \cdots < x_{k(n)}\}$ denote the ordered collection of the $k(n)$ distinct values in $\{X_1, \dots, X_n\}$,

$$a := 1,$$
$$b := k(n),$$
$$p_1 := F_n(x_1),$$
$$q_1 := p_1,$$
$$p_i := F_n(x_i) - F_n(x_{i-1}),$$

and

$$q_i := q_{i-1} + p_i, \qquad i = 2, \dots, k(n).$$

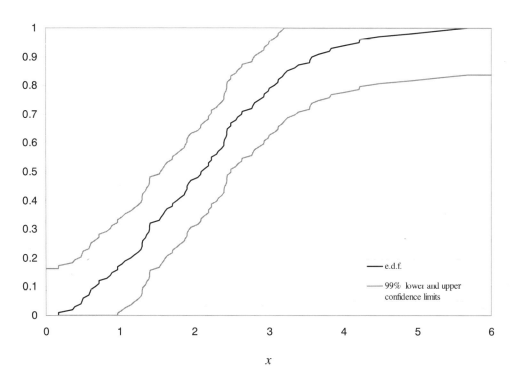

Figure 10.3   Empiral distribution function $\{\tilde{F}_n(x), x \in X\}$ and .99 K-S limits for service-time data, $n = 100$

Then the following steps produce a sample X' from $\{F_n(x)\}$:

- Sample Z using the cutpoint method (Section 8.2) with $\{q_a, \ldots, q_b\}$ as input or using the alias method (Section 8.4) with $\{p_a, \ldots, p_b\}$ as input.

- $X' = x_Z$.

Repeated application of these steps generates a sequence of i.i.d. random samples from the restricted support $\{x_1 < \cdots < x_{k(n)}\}$ rather than from the infinite support $\mathcal{X}$. To increase the set of values that can be sampled from $\{x_1 < \ldots < x_{k(n)}\}$ to $[0, x_{k(n)}]$, let

$$x_{a-1} := 0,$$
$$F_n(0) := 0,$$
$$j'(x) := \min(i \geq a : x_i > x),$$

and

$$\tilde{F}_n(x) := F_n(x_{j(x)-1}) + \frac{x - x_{j(x)-1}}{x_{j(x)} - x_{j(x)-1}} \left[F_n(x_{j(x)}) - F_n(x_{j(x)-1}) \right]. \tag{10.3}$$

Then repeated execution of the steps

- Sample Z using the cutpoint method with $\{q_a, \ldots, q_b\}$ as input and U from $\mathcal{U}(0, 1)$;

- $X' \leftarrow x_{Z-1} + (x_Z - x_{Z-1}) \frac{U - F_n(x_{Z-1})}{F_n(x_Z) - F_n(x_{Z-1})}$;

produces a sequence of i.i.d. samples from $\{\tilde{F}_n(x), 0 < x \leq x_{k(n)}\}$. For an alternative method of interpolation that relies on the alias method, see Ahrens and Kohrt (1981).

10.3.4 THE QUANTILE METHOD

As n increases, so does $k(n)$ and, therefore, the memory space requirements for both the cutpoint and alias methods. Indeed, each grows proportionately to $k(n)$. If space is an issue, one can resort to a space-conserving algorithm that, in effect, relies on an estimator of the inverse d.f.

$$Q(\beta) = \inf_{x \in \mathcal{X}} [F(x) \geq \beta], \qquad 0 \leq \beta \leq 1.$$

Let $X_{(1)} \leq X_{(2)} \leq \cdots \leq X_{(n)}$ denote the order statistics of the sample $X_1, \ldots, X_n$. For some positive integer $k \leq n$,

$$Q_n(0) := 0,$$
$$Q_n(j/k) := X_{(\lceil jn/k \rceil)}, \qquad j = 1, \ldots, k,$$

is a k-point estimate of $\{Q(\beta), 0 \leq \beta \leq 1\}$. Then repeated execution of the step

- Sample U from $\mathcal{U}(0, 1)$.

- $J \leftarrow \lfloor ku \rfloor$.

- $Z \leftarrow Q_n(J/k) + k(U - J/k)\left[Q_n((J + 1)/k) - Q_n(J/k)\right]$.

- $X'' \leftarrow Z$;

produces a sequence of i.i.d. samples from

$$\ddot{F}_n(x) = \frac{1}{k}\sum_{j=0}^{k-1}\left[\frac{x - Q_n(j/k)}{Q_n((j + 1)/k) - Q_n(j/k)} + j\right] I_{[Q_n(j/k), Q_n((j+1)/k))}(x),$$

$$0 \leq x \leq Q_n(1),$$

which is an approximation to $\{F(x), x \in \mathcal{X}\}$ based on linear interpolation between sample quantiles. If X is discrete on the integers $\{0, 1, 2, \ldots\}$, then $\lfloor Z \rfloor$ replaces Z in the assignment.

As in the cutpoint and alias methods, the time to generate a sample via this quantile approach is constant and independent of n and k. However, its unique appeal is that it partitions the unit interval into k subintervals, where k remains fixed as $n \to \infty$. Unlike the cutpoint and alias methods, whose memory requirements grow linearly in n, the quantile approach has a memory requirement proportional to fixed k but independent of n.

How statistically accurate is this quantile approach to estimation? Unlike the e.d.f. the sample quantiles $\{Q_n(j/k), j = 1, k - 1\}$ are not algebraic functions of the observations, and thus its sampling properties are not available in algebraic form for finite sample size n. However, limiting results exist. Suppose F has a continuous and positive p.d.f. $\{f(x), x \in \mathcal{X}\}$. Then for each $j = 1, \ldots, k - 1$ with jn/k integer:

- $Q_n(j/k) \to Q(j/k)$ w.p.1   as $n \to \infty$

- $n^{1/2}Q_n(j/k) \xrightarrow{d} \mathcal{N}\left(Q(j/k), \frac{(j/k)(1-j/k)}{f(Q(j/k))}\right)$ as $n \to \infty$.

With minor adjustments, these properties continue to hold for nonintegral jn/k as n becomes large. As a consequence of the consistency property, we have

$$\tilde{F}_n(Q_n(j/k) \to j/k, \quad \text{w.p.1} \quad n \to \infty,$$

ensuring that $F_n(x)$ converges to $F(x)$, for each $x \in \mathcal{Q}_k := \{Q(1/k), \ldots, Q((k - 1)/k)\}$, as desired.

However, no such convergence occurs for $x \in \mathcal{X} \setminus \mathcal{Q}_k$. How serious is this absence of convergence? The answer depends on k and n. If $n = 100$ and $k = 10$, the quantile method is an imprecise method at best. For $n = 10^6$, as may occur when large data sets are available, choosing $k = 1000$ makes $\{Q_n(j/k)\}$ a sequence of $.001j$ quantiles for $j = 1, \ldots, 1000$, so that linear interpolation in the quantile method is over relatively small probability intervals where the error of approximation is far less than in the $k = 10$ and $n = 100$ case. Since $n = 10^6$ would require space of order four (single precision) or eight (double precision) megabytes for the cutpoint and alias methods, a simulationist inevitability has to make a choice between a large space commitment and an error of interpolation at the .001-increment level.

10.4   CHARACTERIZATIONS FROM THEORY

A considerable body of theory exists about delay systems. Although much of it relates to those with special properties, some theory applies under relatively weak assumptions about the system under study. For the simulationist, these properties can guide the choice of sampling distributions and the estimation of their parameters. We discuss two widely applicable results: the characterizations of interarrival times (Section 10.2.1) and of delay times (Section 10.2.2).

10.4.1   AN EXPONENTIAL LIMIT

Definition 10.1.

Let $X_1, X_2, \ldots$ denote a sequence of i.i.d. nonnegative random variables and let

$$S_0 := 0,$$
$$S_n := X_1 + \ldots + X_n, \qquad n = 1, 2, \ldots.$$

Then $\{N(t) := \sup(n : S_n \leq t), \ t \geq 0\}$ is a called a renewal process. In particular, for $t > 0$, $N(t)$ denotes the number of events that occur in the time interval $(0, t]$.

Definition 10.2.

Let $\{N_1(t), t \geq 0\}, \ldots, \{N_m(t), t \geq 0\}$ denote m independent renewal processes and let $\{N_{0m}(t) := N_1(t) + \cdots + N_m(t), t \geq 0\}$ denote their superposition. That is, $N_i(t)$ denotes the number of events of type i that occur in $(0, t]$, whereas $N_{0m}(t)$ denotes the number of events of all types that occur. In general, $\{N_{0m}(t)\}$ is not a renewal process. However, in at least one special case, it exhibits that property in the limit.

For each $i = 1, \ldots, m$, let $X_{i1}, X_{i2}, \ldots$ denote a sequence of i.i.d. nonnegative random variables with d.f. $\{F_{im}(t), t \geq 0\}$, and let

$$S_{in} := X_{i1} + \cdots + X_{in},$$
$$N_i(t) := \sup(n : S_{in} \leq t),$$
$$\lambda_{im} := 1/EX_{i1},$$

and

$$\lambda := \lambda_{1m} + \cdots + \lambda_{mm}.$$

If $\lambda \to$ constant as $m \to \infty$ and if for each m there exist $\epsilon > 0$ and $t' > 0$ such that

$$F_{im}(t') \leq \epsilon,$$

then (e.g., Heyman and Sobel 1982, p. 156)

$$\text{pr}\,[N_{0m}(t) = 0] \to e^{-\lambda t}, \qquad \text{as } m \to \infty.$$

Since by definition

$$\text{pr}\,[N_{0m}(t) = 0] = \text{pr}(S_1 > t),$$
$$= \text{probability that the next event occurs at time greater than } t,$$

the limiting distribution (as $m \to \infty$) of the time to the next event is the exponential $\mathcal{E}(1/\lambda)$. Gnedenko and Kovalenko (1989) attribute this result to Grigelionis (1962).

For many open-loop delay systems, it is not untenable to characterize the arrival process as the limit of a large number of independent renewal processes each with relatively small arrival rate. As a consequence, the exponential distribution becomes a candidate for data analysis aimed at fitting an interarrival time distribution. This does not mean that we should consider only this distribution. It does mean that within the framework of data analysis a test for the suitability of the exponential distribution offers us some protection from allowing the empirical data, based on a finite sample size, to dictate exclusively the estimated values of the parameters. Estimates based on theory together with data analysis inevitably result in more thoughtfully fitted distributions.

10.4.2 HEAVY TRAFFIC

It is of interest to know how sensitive simulated output is to the choice of input distributions as a function of the level of congestion (traffic) in a simulated system. A result in Köllerstrom (1974) provides a partial answer in the case of *heavy traffic*. A system with traffic intensity ρ close to unity is said to be in heavy traffic. Let $T_1, T_2, \ldots$ denote i.i.d. interarrival times with mean μ_T and variance σ_T^2 to an m-server queue using a first-come-first-served discipline. Let $S_1, S_2, \ldots$ denote i.i.d. service times with mean μ_S and variance σ_S^2, $\rho := \mu_S/m\,\mu_T$,

$$\omega := \frac{\mu_T\left(\rho^2 \gamma_S^2 + \gamma_T^2\right) + (1 - \rho)^2}{2(1 - \rho)},$$

$$\gamma_S = \frac{\sigma_S}{\mu_S}, \text{ and } \gamma_T := \frac{\sigma_T}{\mu_T}.$$

If W denotes waiting time, then under relatively mild conditions on S_i, T_i, and $S_i - kT_i$, W/ω converges to a random variable from $\mathcal{E}(1)$ as $\rho \uparrow 1$ (Köllerstrom 1974).

For ρ close to unity, this implies that the distribution of W is approximately $\mathcal{E}(1/\omega)$, where W depends only on the traffic intensity ρ, the coefficients of variation γ_S and γ_T, and is independent of the parent distributions of the interarrival times $T_1, T_2, \ldots$ and the service times $S_1, S_2, \ldots$.

For this commonly encountered delay system, this result implies that as the system approaches saturation, the dependence of at least one of its output distributions becomes less sensitive to the form of its input distributions and more sensitive to their first two moments. For input data analysis for *heavy-traffic systems*, the result lessens our concern about input distributional shapes but encourages us to obtain estimates for their means and variances that are as accurate as possible.

10.5   EXPERT OPINION

Just as the characterization of a systems's logic flow benefits from the expert opinion of those most familiar with the real system, so can the specification of input characteristics benefit from their knowledge. Combining that knowledge with statistical inferences drawn from an analysis of historical data is the most likely recipe for successful input modeling. When no data are available, we need to rely exclusively on expert opinion. An example shows how this knowledge can be converted into an input distribution.

Recall the example of Section 1.5 where drums were delivered daily to the incinerator complex. No historical data were available for delivery times. However, incinerator operators revealed that drums never arrived before 11 a.m. nor after 4 p.m. When pressed by the simulationist as to the most likely time of arrival, they indicated 2 p.m.

With only these pieces of information, distribution fitting had limited possibilities. The simulationist chose to fit the triangular distribution (Exercise 8.7). In particular, $m_1 = 11$, $m_2 = 14$, and $m_3 = 16$ after the times were converted to a 24-hour basis. As a consequence,

$$f(t) = \begin{cases} 2(t-11)/15, & 11 \leq t \leq 14, \\ (16-t)/5, & 14 \leq t \leq 16, \end{cases}$$

$$F(t) = \begin{cases} (t-11)^2/15, & 11 \leq t \leq 14, \\ 1 - (16-t)^2/10, & 14 < t \leq 16. \end{cases}$$

In particular,

$$
F(t) - F(t-1) = \begin{cases}
\dfrac{2}{30}, & t = 12, \\[2mm]
\dfrac{6}{30}, & t = 13, \\[2mm]
\dfrac{10}{30}, & t = 14, \\[2mm]
\dfrac{9}{30}, & t = 15, \\[2mm]
\dfrac{3}{30}, & t = 16.
\end{cases}
$$

Sampling delivery times from this $\{F(t)\}$ would result in almost two-thirds of delivery occurring between 1 and 3 p.m. Since the incinerator operators were more certain about the endpoints m_1 and m_3 than the modal value m_2, the simulationist chose to run at least two additional simulations, one with $m_2 = 13$ and the other with $m_2 = 15$, and to compare the output with that for $m_2 = 14$. Little variation in output would indicate that system response was relatively insensitive to the choice of most likely delivery time.

10.6 CHOOSING A DISTRIBUTION

By fitting a distribution, we mean selecting a theoretical p.m.f. or p.d.f., estimating the values of its parameters from data or expert judgment, and, in the data case, assessing how well the resulting fitted d.f. approximates the corresponding e.d.f. Graphical comparison provides one means of assessment; applying one or more statistical tests provides a second.

Commercial software is available for fitting. For example, BestFit (www.lithec. sk/uk/dts_bestfit.htm), ExpertFit (www.does.org/masterli/s3.htm), and Stat::Fit (www. geerms.com/default.htm) each use empirical data to estimate the values of parameters for a host of candidate distributions and provide graphical displays and test statistics to guide the choice among them. Also, some simulation systems such as Arena provide a capability for fitting distributions. Using this software saves the simulationist's time; but he/she must ultimately make the choice among the candidates.

Unfortunately, goodness-of-fit tests for distributions tend to be considerably less discriminating than desired. In particular, the standard approach tests H_0: a particular distribution fits the data, against the omnibus alternative H_1: one of the many nonspecified alternatives fits the data. This setup is known to have low *power*, meaning that the test accepts H_0 more often than it should, thus being less discerning than an analyst would like. Moreover, some of these software applications rank the desirability of the host of candidate distributions by the values of their test statistics on a particular test. This practice encourages users to choose the best-performing distribution according to the ranking. However, little basis exists in the statistical literature to support these ranking schemes. In the remainder of this chapter we concentrate more on graphical analysis to assess the adequacy of fitted distributions. These are usually available from the software fitting applications or are easily computed by more general-purpose applications such as Microsoft Excel.

For nonnegative data, the most difficult part of a distribution to fit is its right tail. Graphical analysis may reveal discrepancies between the tails of empirical and fitted d.f.s, but except in the case of an exceptionally large sample size, the ability of statistical tests to reject based on these discrepancies is relatively low.

If a simulationist knows something about alternative tail behaviors, then her/his capacity to choose among alternative fits can be greatly enhanced. This section provides an overview of these alternatives.

10.6.1 EXPONENTIAL TAILS

The Gamma distribution (Section 8.13 and Table 10.3) is a commonly fitted distribution that has an exponentially decreasing right tail. It often provides a basis for fitting interarrival and service time data. It has mean $\mu = \alpha\beta$, variance $\sigma^2 = \alpha\beta^2$, mode at $x = \max[0, \beta(\alpha - 1)]$. It also facilitates hypothesis testing for exponentiality ($\alpha = 1$) within the context of standard statistical theory. Section 10.12.2 illustrates this feature. For $\alpha = 1$ its (finite) mode is at $x = 0$. For $\alpha < 1$, its (infinite) mode is also at zero. For $\alpha > 1$, its (finite) mode is at some positive x. Table 10.2 lists the maximum likelihood equations that need to be solved iteratively for this distribution.

10.6.2 SUB- AND HYPEREXPONENTIAL TAILS

The Weibull distribution (§8.9.1 and Table 10.3) provides an alternative basis for fitting with mean $\beta\Gamma(1 + 1/\alpha)$, variance $\beta^2[\Gamma(1 + 2/\alpha) - \Gamma^2(1 + 1/\alpha)]$, mode at $x = \beta \max[0, (1 - /\alpha)^{1/\alpha}]$, and right tail that decreases as $\exp[-(x/\beta)^\alpha]$. This last property implies that the data, more than model selection, dictate tail behavior, being more rapid than exponential if $\hat{\alpha} > 1$ and less if $\hat{\alpha} < 1$. The Weibull also facilitates hypothesis testing for exponentiality ($\alpha = 1$) within the context of the likelihood ratio test (Section 10.12.2). The location of its mode follows analogously to that for the Gamma. Table 10.3 lists the maximum likelihood estimates that need to be solved iteratively for the Weibull.

10.6.3 LOGNORMAL DISTRIBUTION

Fitting data to an exponential, Gamma, or Weibull p.d.f. implicitly restricts its mode and right tail. For exponential fitting, the mode is at zero and the tail exponential. A Gamma fit with $\alpha < 1$ puts the (infinite) mode at zero and retains the exponential tail. A Weibull fit with $\alpha < 1$ again puts the (infinite) mode at zero but makes the tail diminish subexponentially. A Gamma fit with $\alpha > 1$ puts the mode at a positive value and continues to give an exponential tail. A Weibull fit with $\alpha > 1$ puts the mode at a positive value but makes the right tail decrease hyperexponentially.

Occasionally, the empirical histogram reveals tail behavior that does not fit one of these profiles. This is especially true when the empirical histogram shows the mode at a positive value and a right tail that decreases at less than an exponential rate. If a simulationist were to ignore this discrepancy between the e.d.f. and a fitted exponential, Gamma, or Weibull d.f. the resulting sampling sequence during a simulation experiment would show proportionately fewer large right tail values than the empirical evidence suggests. Accordingly, we need to look at other options. This section discusses the lognormal distribution and §10.6.4 the inverted Gamma and inverse Weibull distributions.

From Section 8.13, a random variable from the lognormal distribution $\mathcal{LN}(\mu, \sigma^2)$ has mean $\exp[-(\mu+\sigma^2/2)]$, variance $\exp(2\mu+\sigma^2)\times[\exp(\sigma^2)-1]$, mode at $x = \exp(\mu+\sigma^2/2)$, and right tail whose dominant terms decreases as $\exp[-(\ln x)^2/2\sigma^2]$ as x increases. Most notably, tail behavior is slower than exponential in x, thus providing a candidate for data with a "heavy" tail.

10.6.4 INVERTED GAMMA AND INVERSE WEIBULL DISTRIBUTIONS

While an alternative to subexponential, exponential, and hyperexponential decrease, the lognormal distribution can place so much probability mass in the right tail that the fitted distribution compares poorly with the e.d.f. in other regions. The inverted Gamma distribution (Section 8.13.1) and the inverse Weibull distribution (Section 8.9.2) provide alternatives with nonexponential tails that may allow for a better fit.

For $\alpha > 2$, the inverted Gamma has finite mean $1/(\alpha - 1)\beta$ and variance $1/(\alpha - 2)(\alpha - 1)^2\beta^2$. It also has positive mode at $1/(\alpha + 1)\beta$, right tail that decreases as $x^{-(\alpha+1)}$ as x increases, and a flatter left tail than the Gamma.

For $\alpha > 2$, the inverse Weibull distribution has finite mean $\Gamma(1-1/\alpha)/\beta$ and variance $[\Gamma(1 - 2/\alpha) - \Gamma^2(1 - 1/\alpha)]\beta^2$. It also has positive mode at $1/(1 + 1/\alpha)^{1/\alpha}\beta$, right tail that decreases as $x^{-(\alpha+1)}$ as x increases, and left tail that is flatter than that for the Weibull.

The inverted Gamma and inverse Weibull distributions can provide better fits than the Gamma and Weibull distributions for data whose empirical histogram tends to decrease at a polynomial rate. Figure 10.4 shows the difference among these distributions for a common mean $\mu = 2.6$ and variance $\sigma^2 = 3.38$. In particular, the p.d.f.s in Figure 10.4a reveal the differences in left tail and modal behavior. Figure 10.4b shows the exceedance probabilities. It reveals that the y tend to be larger for the inverted Gamma and inverse Weibull for both small and large x but smaller (in this example) for $x \in [2, 3]$.

Figure 10.4c shows the left-tail exceedance probabilities

$$\mathrm{pr}[X > x \mid X \geq x(.99)],$$

where

$$x(\eta) := \inf[x : F(x) \geq \eta], \quad 0 < \eta < 1,$$

and F is the d.f. In the present case with $\eta = .99$,

$$
x(.99) = \begin{cases}
8.6 & \text{for} \quad \mathcal{G}\ (2, 1.300), \\
9.5 & \text{for} \quad \mathcal{IG}\ (2.4, .1282), \\
9.0 & \text{for} \quad \mathcal{W}\ (2.953, 2.914), \\
4.9 & \text{for} \quad \mathcal{IW}\ (2.938, .5257).
\end{cases}
$$

In addition to revealing the wide disparity among the .99 quantiles, Figure 10.4c shows that although samples are generated from this part of each distribution no more than once per hundred trials on average, when such an outcome X does occur the values of X tend to be considerably larger for the inverted Gamma and inverse Weibull distributions than for the Gamma and Weibull. We can easily recognize the significance of this property for service times. In some repair processes, the vast majority of items require moderate repair times, but occasionally an item requires an excessively large repair time.

For completeness, recall that the Beta-prime distribution of Section 8.14.1 also has a right tail that decreases as a polynomial in x. We leave to the reader the exercise of showing that this distribution has $\mu = \beta/(\alpha - 1)$, $\sigma^2 + \mu^2 = \beta(1 + \beta)/(\alpha - 1)(\alpha - 2)$, and, for the particular $\mu = 2.6$ and $\sigma^2 = 3.38$, $\alpha = 4.7692$ and $\beta = 9.800$. A graph of the corresponding p.d.f. is of interest when compared to those in Figure 10.4.

10.7   SUPPORT

Let X denote a continuous random variable with p.d.f. or p.m.f. f_X. The support of X, supp(X), denotes all points in $\Re$ for which $f_X(x)$ is positive. For example, $Z \sim \mathcal{G}\ (\alpha, \beta)$ has supp$(Z) = [0, \infty)$, whereas $X := a + bZ$, for $b > 0$, has supp$(X) = [a, \infty)$. If $Z \sim \mathcal{Be}(\alpha, \beta)$, then supp$(Z) = [0, 1]$, whereas $X := a + (b - a)Z$, for $b > a$, has supp$(X) = [a, b]$.

Given a sample sequence $X_1, \ldots, X_n$ for which supp$(X_i) = [a, \infty)$ for known $a \neq 0$, one customarily works with the transformed data, $X_i' := X_i - a$, $i = 1, \ldots, n$, when fitting a distribution. If supp$(X_i) = [a, b]$ for known $b > a$, then $X_i' := (X_i - a)/(b - a)$, $i = 1, \ldots, n$, has support $[0, 1]$. If unknown, the support must be estimated prior to estimating, or along with, the values of the parameters of the selected distribution. Clearly, $a \leq \min_{1 \leq i \leq n} X_i$ and $b \geq \max_{1 \leq i \leq n} X_i$. When expert opinion indicates finite support $[a, b]$ and gives estimates $\tilde{a}$ and $\tilde{b}$ satisfying $\tilde{a} \leq \min_{1 \leq i \leq n} X_i$ and $\tilde{b} > \max_{1 \leq i \leq n} X_i$, it is not unusual to fit $X_i' := (X_i - \tilde{a})/(\tilde{b} - \tilde{a})$, $i = 1, \ldots, n$, to the beta distribution $\mathcal{Be}(\alpha, \beta)$. An approach, more firmly grounded in theory, simultaneously fits a, b, α, β using the maximum likelihood method (e.g., Johnson et al. 1994a and 1994b).

When b is assumed infinite, similar approaches are used with $X_i' = X_i - \tilde{a}$, $i = 1, \ldots, n$, and the exponential, Gamma, Weibull, and other distributions. As an example,

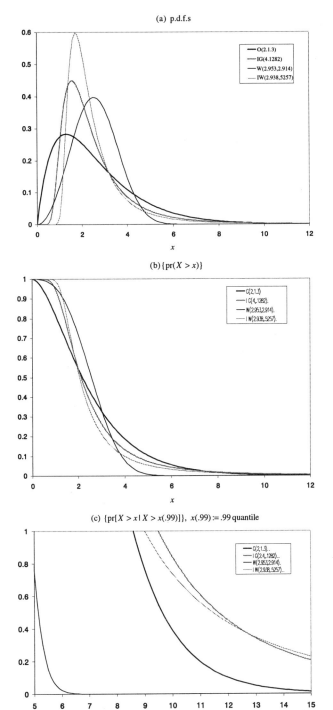

(a) p.d.f.s

(b) {pr(X > x)}

(c) {pr[X > x | X > x(.99)]}, x(.99) := .99 quantile

Figure 10.4
Selected distributions
with $\mu = 2.6$ and
$\sigma^2 = 3.38$

suppose that the X_i have the p.d.f.

$$f_X(x) = \lambda e^{-\lambda(x-a)}, \quad \lambda > 0, \quad x \geq a \geq 0.$$

Then it is easily seen that the m.l.e. of a and λ are, respectively,

$$\hat{a} = \min_{1 \leq i \leq n} X_i$$

and

$$\hat{\lambda} = \frac{1}{n} \sum_{i=1}^{n} (X_i - \hat{a}).$$

For discrete data, comparable techniques can be used with the binomial distribution if the support is finite or with the Poisson, negative binomial, or other distributions, if the support is infinite.

10.8 MAXIMUM LIKELIHOOD ESTIMATION

Many alternative methods exist for estimating the values of the parameters of a distribution, each having its own virtues and limitations. The present account focuses on the *method of maximum likelihood* principally because of the optimal large-sample statistical properties with which it endows its estimators, its well-developed theory for testing hypotheses about parameter values, and in today's computing environment the relative ease of computing by exact or iterative methods the estimated values of parameters of distributions of interest from sample data.

Let $\{f(x; \theta_1, \ldots, \theta_q), x \in \mathcal{X}\}$ denote a p.d.f. or p.m.f. with x taking values in the support $\mathcal{X}$ and with unknown parameter vector $\boldsymbol{\theta} := (\theta_1, \ldots, \theta_q)$ taking a value in some set Θ. Let $\mathbf{X} := (X_1, \ldots, X_n)$ denote a vector of n i.i.d. observations from f. Then

$$L(\boldsymbol{\theta}, \mathbf{X}) := \Pi_{i=1}^{q} f(X_i, \theta_1, \ldots, \theta_q), \quad \boldsymbol{\theta} \in \Theta,$$

denotes the corresponding *likelihood function*, and $\hat{\theta}$, an element of Θ satisfying

$$L(\hat{\boldsymbol{\theta}}, \mathbf{X}) = \sup_{\boldsymbol{\theta} \in \Theta} L(\boldsymbol{\theta}, \mathbf{X}).$$

By way of interpretation, $\hat{\theta}$ is the value that maximizes the likelihood of observing the given sample $\mathbf{X}$.

For the distributions we consider here, all of which are described in Chapter 8, the maximum likelihood estimator (m.l.e.) $\hat{\boldsymbol{\theta}}$ of $\boldsymbol{\theta}$, has the folloing important properties:

- **Uniqueness.** $\hat{\theta}$ is unique

- **Strong Consistency.** $\hat{\theta} \to \theta$ as $n \to \infty$ w.p.1

- $n^{1/2} \hat{\theta}$ has covariance matrix

$$
\Sigma_n(\boldsymbol{\theta}) = \text{E}
\begin{bmatrix}
L_{11} & L_{12} & \dots & L_{1q} \\
& L_{22} & & \vdots \\
\vdots & & \ddots & \\
L_{q1} & & \dots & L_{qq}
\end{bmatrix}^{-1},
$$

where

$$
L_{ij} := \left. \frac{\partial \ln L(\boldsymbol{\theta}, \mathbf{X}) \big|_{\theta_i = \hat{\theta}_i}}{\partial \theta_i \partial \theta_j} \right|_{\theta_j = \hat{\theta}_j}, \quad i, j = 1, \dots, q
$$

- **Effeciency.** If $\tilde{\boldsymbol{\theta}} := (\tilde{\theta}_1, \dots, \tilde{\theta}_q)$ is an alternative estimate of $\boldsymbol{\theta}$, then

$$
\lim_{n \to \infty} \frac{\text{var} \hat{\theta}_i}{\text{var} \tilde{\theta}_i} \le 1, \quad i = 1, \dots, q,
$$

$n \text{ var } \hat{\theta}_i$ being the ith diagonal element of $\Sigma(\boldsymbol{\theta})/n$, where

$$
\Sigma(\boldsymbol{\theta}) := \lim_{n \to \infty} \Sigma_n(\boldsymbol{\theta})
$$

- **Asymptotic Normality.** $n^{1/2}(\hat{\boldsymbol{\theta}} - \boldsymbol{\theta}) \xrightarrow{\text{d}} \mathcal{N}(0, \Sigma(\boldsymbol{\theta}))$ as $n \to \infty$

- **Invariance.** For any function, $\{h(\boldsymbol{\theta})\}$, $\{h(\hat{\boldsymbol{\theta}})\}$ is its m.l.e., thus inheriting the properties of strong consistency, efficiency, and asymptotic normality.

In large samples, efficiency implies that no alternative estimator of θ_i has smaller variance than the m.l.e. Asymptotic normality provides a distribution theory for assessing how well $\hat{\boldsymbol{\theta}}$ approximates $\boldsymbol{\theta}$. In particular, for each i $\left[\hat{\theta}_i \pm \Phi^{-1}(1 - \delta/2)\text{s.e.}(\hat{\theta}_i) \right]$ with

$$
\text{s.e.}(\hat{\theta}_i) := (\text{var } \hat{\theta}_i)^{1/2}
$$

provides a $100 \times (1 - \delta)$ percent confidence inteval for θ_i, if var $\hat{\theta}_i$ were known. Since it is not, we replace it with the ith diagonal element of $\Sigma(\hat{\boldsymbol{\theta}})/n$, which by the invariance property is the m.l.e. of var $\hat{\theta}_i$ and therefore inherits the property of efficiency. That is, for large n no alternative estimator of var $\hat{\theta}_i$ has smaller variance.

The relative error

$$\text{r.e.}(\hat{\theta}_i) := (\text{var } \hat{\theta}_i)^{1/2}/\theta_i$$

provides an additional measure of how well $\hat{\theta}_i$ approximates θ_i.

These m.l.e. properties hold for any distribution representable in the form

$$f(x, \boldsymbol{\theta}) = \exp\left[A(x)B(\boldsymbol{\theta}) + C(x) + D(\boldsymbol{\theta})\right],$$

where A, B, C, and D are arbitrary functions. Collectively, they define the class of Darmois–Koopman distributions (e.g., Johnson et al. 1994a). Virtually all distributions discussed in this chapter admit representations of this form.

❏ EXAMPLE 10.1

Let $X_1, \ldots, X_n$ denote i.i.d. observations from $\mathcal{E}(\beta)$ with p.d.f.

$$f(x) = \beta^{-1}e^{-x/\beta}, \qquad \beta > 0, \quad x \geq 0.$$

Then $q = 1$ and $\theta_1 = \beta$, so that

$$\ln L(\boldsymbol{\theta}, \mathbf{X}) = -n \ln \beta + \beta^{-1} \sum_{i=1}^{n} X_i,$$

$$\hat{\theta}_1 = \hat{\beta} = \frac{1}{n} \sum_{i=1}^{n} X_i,$$

$$\text{var}\hat{\beta} = \beta^2/n,$$

$$\text{s.e.}(\hat{\beta}) = \beta/n^{1/2},$$

and

$$\text{r.e.} (\hat{\beta}) := \frac{\text{s.e.}(\hat{\beta})}{\beta} = \frac{1}{n^{1/2}}.$$

By the invariance property, $\hat{\beta}/n^{1/2}$ is the m.l.e. of s.e.$(\hat{\beta})$.

❏

Table 10.2 lists selected continuous and discrete distributions whose m.l.e.s are available in closed form. Table 10.3 lists selected continuous and discrete distributions whose m.l.e.s are computed by iterative methods. In particular, Algorithm N-R in Section 8.1.1 describes the Newton–Raphson method for finding the m.l.e. for one parameter using $\{k(z)\}$ and $\{k'(z)\}$ in Table 10.3 for the Gamma, Weibull, and negative binomial distributions. Then the rightmost column gives the form of the second parameter estimator. For example, Algorithm N-R determines $\hat{\alpha}$ for the Gamma distribution and $\hat{\beta} = A/\hat{\alpha}$. An error bound $\nu \leq 10^{-6}$ is recommended for error tolerance in the algorithm.

Tables 10.3 and 10.4 also list the large-sample form for var $\hat{\theta}$. In the remainder of this chapter these quantities are evaluated with var$\hat{\theta}$ replaced by its m.l.e.

All iterative computations rely on the availability of procedures for computing the digamma function $\{\psi(z) := \frac{d \ln \Gamma(z)}{dz}\}$ and the polygamma function $\{\psi^{(1)}(z) := \frac{d\psi(z)}{dz}\}$. These are available in Mathematica and Mathcad. Since these are slowly converging infinite series, we encourage a user of the iterative procedures in Table 10.3 to employ one of these packages rather than to write his or her own procedures for these functions from first principles.

❏ EXAMPLE 10.2

In fitting the Gamma distribution to the service-time data with empirical histogram in Figure 10.2, the m.l.e. are

$$\hat{\alpha} = 3.028, \qquad \hat{\beta} = .7138,$$
$$\text{s.e.}(\hat{\alpha}) = .4069, \qquad \text{s.e.}(\hat{\beta}) = .1043,$$
$$\text{r.e.}(\hat{\alpha}) = .1344, \qquad \text{r.e.}(\hat{\beta}) = .1461,$$
$$\text{corr}(\hat{\alpha}, \hat{\beta}) = -.9194.$$

❏

Figure 10.5a compares the exceedance probabilities for the e.d.f. in Figure 10.2 with those for the fitted d.f. Figure 10.5b shows their difference, and Figure 10.5c compares the conditional right tail exceedance probabilities. The first two graphs reveal a relatively good fit for just $n = 100$ observations. In particular, Figure 10.5a shows that the fitted d.f. introduces smoothness that is notably absent in the e.d.f.

Because $X_{(1)}$ and $X_{(n)}$ are finite, sampling from an e.d.f. limits its sample values to $[X_{(1)}, X_{(n)}]$. This can be a serious limitation when X is believed to have infinite support on the nonnegative line or at least an upper bound much larger than $X_{(n)}$. Figure 10.5c shows how the fitted d.f. overcomes this limitation of the e.d.f.

Table 10.2 Selected distributions whose m.l.e. are available in closed form
$(A := n^{-1}\sum_{i=1}^n X_i,\ B := n^{-1}\sum_{i=1}^n X_i^2,\ A' := n^{-1}\sum_{i=1}^n \ln X_i,\ B' := n^{-1}\sum_{i=1}^n (\ln X_i)^2)$

distribution	mean	variance	mode	m.l.e.	properties		
normal $\mathcal{N}(\mu,\sigma^2)$ $-\infty<\mu<\infty$ $\sigma^2>0$ $-\infty<x<\infty$	μ	σ^2	μ	$\hat{\mu}=A$ $\hat{\sigma}^2=B-A^2$	$\mathrm{E}\hat{\mu}=\mu$ $\mathrm{E}\hat{\sigma}^2=\frac{n-1}{n}\sigma^2$	$\mathrm{var}\,\hat{\mu}=\frac{\sigma^2}{n}$ $\mathrm{var}\,\hat{\sigma}^2=\frac{2\sigma^4}{n}$	$\mathrm{corr}(\hat{\mu},\hat{\sigma}^2)=0$
uniform $\mathcal{U}(a,b)$ $-\infty<a<b<\infty$ $a\le x\le b$	$\frac{a+b}{2}$	$\frac{(b-a)^2}{12}$		$\hat{a}=\min X_i$ $\hat{b}=\max X_i$	$\mathrm{E}\hat{a}=a+\frac{b-a}{n+1}$ $\mathrm{E}\hat{b}=b-\frac{b-a}{n+1}$	$\mathrm{var}\,\hat{a}=\frac{n(b-a)^2}{(n+1)^2(n+2)}$ $\mathrm{var}\,\hat{b}=\frac{n(b-a)^2}{(n+1)^2(n+2)}$	$\mathrm{corr}(\hat{a},\hat{b})=\frac{1}{n}$
exponential $\mathcal{E}(\beta)$ $\beta>0$ $x\ge 0$	β	β^2	0	$\hat{\beta}=A$	$\mathrm{E}\hat{\beta}=\beta$	$\mathrm{var}\,\hat{\beta}=\frac{\beta^2}{n}$	
lognormal $\mathcal{LN}(\mu,\sigma^2)$ $-\infty<\mu<\infty$ $\sigma^2>0$ $x\ge 0$	$e^{\mu+\sigma^2/2}$	$e^{2\mu+\sigma^2}(e^{\sigma^2}-1)$	$e^{\mu+\sigma^2/2}$	$\hat{\mu}=A'$ $\hat{\sigma}^2=B'-(A')^2$	$\mathrm{E}\hat{\mu}=\mu$ $\mathrm{E}\hat{\sigma}^2=\frac{n-1}{n}\sigma^2$	$\mathrm{var}\,\hat{\mu}=\frac{\sigma^2}{n}$ $\mathrm{var}\,\hat{\sigma}^2=\frac{2\sigma^4}{n}$	$\mathrm{corr}(\hat{\mu},\hat{\sigma}^2)=0$

Table 10.2 (continued)

distribution	mean	variance	mode	m.l.e.	properties
Bernoulli $Ber(p)$ $0 \le p \le 1$ $x = 0,1$	p	$p(1-p)$	0 if $p < \frac{1}{2}$ 1 if $p \ge \frac{1}{2}$	$\hat{p} = A$	$E\hat{p} = p$ $\operatorname{var}\hat{p} = \frac{p(1-p)}{n}$
Poisson $P(\lambda)$ $\lambda > 0$ $x = 0,1,\dots$	λ	λ	$\lambda - 1$ and λ, if $[\lambda] = \lambda$, $[\lambda]$ otherwise	$\hat{\lambda} = A$	$E\hat{\lambda} = \lambda$ $\operatorname{var}\hat{\lambda} = \frac{\lambda}{n}$
geometric $Ge(p)$ $0 \le p \le 1$ $x = 0,1,\dots$	$\frac{p}{1-p}$	$\frac{p}{(1-p)^2}$	0	$\hat{p} = \frac{A}{1+A}$	$E\hat{p} \to p$ as $n \to \infty$ $n\operatorname{var}\hat{p} \to p(1-p)^2$ as $n \to \infty$

Table 10.3   Selected distributions whose m.l.e.s require iteration $A := n^{-1}\sum_{i=1}^n X_i$, $B := n^{-1}\sum_{i=1}^n X_i^2$, $C := n^{-1}\sum_{i=1}^n \ln X_i$, $D := n^{-1}\sum_{i=1}^n \ln(1-X_i)$, $E(z) := n^{-1}\sum_{i=1}^n X_i^z$, $G(z) := n^{-1}\sum_{i=1}^n X_i^z \ln X_i$, $H(z) := n^{-1}\sum_{i=1}^n X_i^z(\ln X_i)^2$, $I(z) := n^{-1}\sum_{i=1}^n \psi(z+X_i)$, $J(z) := n^{-1}\sum_{i=1}^n \psi^{(1)}(z+X_i)$

distribution	mean	variance	mode
Gamma $\mathcal{G}(\alpha,\beta)$ $\alpha,\beta>0$ $x\geq 0$	$\alpha\beta$	$\alpha\beta^2$	$\max[0,(\alpha-1)\beta]$
Weibull $\mathcal{W}(\alpha,\beta)$ $\alpha,\beta>0$ $x\geq 0$	$\beta\Gamma(1+1/\alpha)$	$\beta^2[\Gamma(2+1/\alpha)-\Gamma^2(1+1/\alpha)]$	$\max[0,\beta(1-1/\alpha)^{1/\alpha}]$
negative binomial $\mathcal{NB}(r,p)$ $r>0,\ 0\leq p\leq 1$ $x=0,1,\ldots$	$\frac{rp}{1-p}$	$\frac{rp}{(1-p)^2}$	w and $w+1$ if $\lfloor w\rfloor = w$ $\lfloor w\rfloor$ otherwise $w := (rp-1)/(1-p)$

asymptotic properties

distribution	variances		correlation
Gamma	$\mathrm{var}\,\hat\alpha \approx \frac{\alpha}{[\alpha\psi^{(1)}(\alpha)-1]n}$	$\mathrm{var}\,\hat\beta \approx \frac{\beta^2}{\psi^{(1)}(\alpha)[\alpha\psi^{(1)}(\alpha)-1]n}$	$\approx -\frac{1}{\alpha\psi^{(1)}(\alpha)}$
Weibull	$\mathrm{var}\,\hat\alpha \approx 1.109(\beta/\alpha)^2/n$	$\mathrm{var}\,\hat\beta \approx .608\alpha^2/n$	$\approx .3812$
negative binomial[a]	$\mathrm{var}\,\hat r \approx \frac{1}{M(r,p)n}$	$\mathrm{var}\,\hat p \approx \frac{1}{n}\left[\frac{p(1+p)}{r}+\frac{p^2}{r^2 M(r,p)}\right]$	$\approx \frac{-p}{M(r,p)r}\left[\mathrm{var}\,\hat r\ \mathrm{var}\,\hat p\right]^{1/2}$

distribution	iterate on $z=$	$k(z)$	$k'(z)$	second estimator
Gamma	α	$\psi(z)-\ln z+\ln A-C$	$\psi^{(1)}(z)-\frac{1}{z}$	$\hat\beta = \frac{A}{\hat\alpha}$
Weibull	α	$\frac{1}{z}-\frac{G(z)}{E(z)}+C$	$-\frac{1}{z^2}-\frac{H(z)}{E(z)}+\left[\frac{G(z)}{E(z)}\right]^2$	$\hat\beta = [E(\hat\alpha)]^{1/\hat\alpha}$
negative binomial	r	$\ln(\frac{z}{z+A})+I(z)-\psi(z)$	$\frac{1}{z}-\frac{1}{z+A}+J(z)-\psi^{(1)}(z)$	$\hat p = \frac{A}{\hat r+A}$

[a] $M(x,y):=\sum_{j=0}^\infty \frac{B(x,y)}{j+2}\left(\frac{y}{1+y}\right)^{j+2}$, where B denotes the beta function.

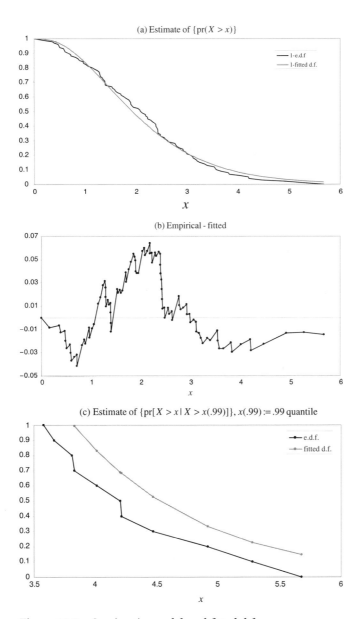

Figure 10.5 Service-time e.d.f. and fitted d.f.

10.8.1 BETA DISTRIBUTION

For n observations $X_1, \ldots, X_n$ from the Beta distribution $\mathcal{B}e(\alpha, \beta)$ (Section 8.14), the logarithm of the likelihood function is

$$\ln L(\alpha, \beta, \mathbf{X}) = n\left[\ln \Gamma(\alpha + \beta) - \ln \Gamma(\alpha) - \ln \Gamma(\beta) + (\alpha - 1)C + (\beta - 1)D\right],$$

where Table 10.3 defines C and D. The challenge is to solve

$$\psi(\hat{\alpha} + \hat{\beta}) - \psi(\hat{\alpha}) + \hat{\alpha} \, C = 0,$$
$$\psi(\hat{\alpha} + \hat{\beta}) - \psi(\hat{\beta}) + \hat{\beta} \, D = 0,$$

simultaneously for the m.l.e. $\hat{\alpha}$ and $\hat{\beta}$, where ψ denotes the digamma function. This presents more formidable problems than solving for the other m.l.e.s in Table 10.3.

Bivariate search provides one approach, starting with the initial values

$$\tilde{\alpha} := (A - B)A/(B - A^2)$$

and

$$\tilde{\beta} := (A - B)(1 - A/(B - A^2)),$$

the moment estimators of α and β (Fielitz and Myers 1975). Familiarity with bivariate search techniques is necessary.

A second approach reduces the problem to a univariate search for $\hat{\alpha}$ by combining the likelihood equations to obtain (Beckman and Tietjen 1978)

$$\psi(\hat{\alpha}) - \psi(\psi^{-1}(D - C + \psi(\alpha)) + \hat{\alpha}) - C = 0,$$

where ψ^{-1} denotes the inverse of ψ, and then computes

$$\hat{\beta} = \psi^{-1}(D - C + \psi(\hat{\alpha})).$$

Since mathematical-applications software rarely provides for the direct computation of $\psi^{-1}(\cdot)$, this approach also has a limitation.

A third approach approximates $\hat{\alpha}$ and $\hat{\beta}$ by relatively gross interpolation in a table of selected values of C and D (Gnanadesikan et al. 1967). The table is reproduced in Fishman (1973, p. 359). Law and Kelton (1991, pp. 411–413) contains a more extensive table.

A fourth approach again resorts to bivariate search using entries in the aforementioned tables to localize the region to be searched.

A fifth alternative uses the moment estimators $\tilde{\alpha}$ and $\tilde{\beta}$ in place of the m.l.e. Since var $\tilde{\alpha} \geq$ var$\hat{\alpha}$ and var $\tilde{\beta} \geq$ var $\hat{\beta}$ for large n, this approach produces less accurate estimates. However, it does offer the convenience of being easy to compute.

For large n, $\hat{\alpha}$, and $\hat{\beta}$ have

$$\text{var } \hat{\alpha} \approx \frac{1}{\psi^{(1)}(\alpha + \beta)[1/Q(\beta, \alpha) - Q(\alpha, \beta)]n},$$

$$\text{var } \hat{\beta} \approx \frac{1}{\psi^{(1)}(\alpha + \beta)[1/Q(\alpha, \beta) - Q(\beta, \alpha)]n},$$

$$\text{corr}(\hat{\alpha}, \hat{\beta}) \approx \frac{-1}{[Q(\alpha, \beta)Q(\beta, \alpha)]^{1/2}},$$

where

$$Q(\alpha, \beta) := \frac{\psi^{(1)}(\alpha)}{\psi^{(1)}(\alpha + \beta)}.$$

10.8.2 GENERALIZED BETA DISTRIBUTION

The Beta distribution of Section 8.14 is a special case of the generalized Beta distribution

$$f(z) = \frac{\Gamma(\alpha + \beta)}{\Gamma(\alpha)\Gamma(\beta)} \cdot \frac{(z - z_1)^{\alpha-1}(z_2 - z)^{\beta-1}}{(z_2 - z_1)^{\alpha+\beta-1}}, \quad z_1 \leq z \leq z_2.$$

If z_1 and z_2 are known exactly, then one transforms $X_1, \ldots, X_n$ to $\{(X_i - z_1)/(z_2 - z_1); i = 1, \ldots, n\}$ and proceeds as above. If z_1 and z_2 are unknown, estimation becomes more problematic. In particular, the m.l.e.s for α and β are sensitive to values assumed for z_1 and z_2. AbouRizk et al. (1994) provide software called BetaFit for estimating α, β, z_1, and z_2 by several different methods. The reference shows that least-squares estimation generally gives as good and often a better fit than the m.l.e. VIBES, accessible on the Internet at http://www.ie.ncsu.edu/jwilson/page3.html, provides the software for BetaFit.

10.8.3 BETA-PRIME DISTRIBUTION

To estimate the m.l.e. of the Beta-prime distribution, we make use of its relationship to the Beta distribution. Recall from Section 8.14.1 that if Z is from $\mathcal{B}e(\alpha, \beta)$, then $X = Z^{-1} - 1$ is from the Beta-prime distribution with support $[0, \infty)$ and right tail that decreases polynomially. To estimate α and β for this distribution for sample data $X_1, \ldots, X_n$, one replaces X_i by $1/(1 + X_i)$ for $i = 1, \ldots, n$ everywhere in the adopted m.l.e. procedure for $\mathcal{B}e(\alpha, \beta)$.

10.8.4 INVERTED GAMMA AND INVERSE WEIBULL M.L.E.S

Maximum likelihood estimators are easily computed for the inverted Gamma distribution (Section 8.13.1) and the inverse Weibull distribution (Section 8.9.2) by exploiting their relationships with the Gamma and Weibull distributions respectively. Recall that if X is from $\mathcal{IG}(\alpha, \beta)$, then $1/X$ is from $\mathcal{G}(\alpha, \beta)$. Accordingly, the m.l.e.s $\hat{\alpha}$ and $\hat{\beta}$ for $\mathcal{IG}(\alpha, \beta)$ are obtainable using $\{k(z)\}$ and $\{k'(z)\}$ for the Gamma distribution in Table 10.3 with the revised definitions

$$A := n^{-1} \sum_{i=1}^{n} X_i^{-1}$$

and

$$B := n^{-1} \sum_{i=1}^{n} X_i^{-2}.$$

In an analogous manner, the m.l.e.s $\hat{\alpha}$ and $\hat{\beta}$ for $\mathcal{IW}(\alpha, \beta)$ are obtainable using $\{k(z)\}$ and $\{k'(z)\}$ for the Weibull distribution with

$$E(z) := n^{-1} \sum_{i=1}^{n} X_i^{-z},$$

$$G(z) := n^{-1} \sum_{i=1}^{n} X_i^{-z} \ln(X_i^{-1}),$$

and

$$H(z) := n^{-1} \sum_{i=1}^{n} X_i^{-z} [\ln(X_i^{-1})]^2.$$

10.9   ADEQUACY OF FIT

Previous sections encourage the reader to rely on graphical comparison to assess the adequacy of fit of a particular distribution. The approach allows the simulationist to spot discrepancies between fitted and empirical curves, be they d.f.s or densities. The magnitude of these discrepancies, their locations, and the use to be made of a curve provide a basis for deciding adequacy. For example, a particular fitted distribution may show substantial deviations from the e.d.f. in a region. If, when used to generate samples, this part of the fitted curve comes into play a relatively small percentage of the time, the error may be tolerable.

While certainly an informal approach, graphical comparisons are easy to effect via commonly available spreadsheet software such as Microsoft Excel. Ease of use gives the approach its appeal. A more formal assessment uses *goodness-of-fit tests*. For a sample i.i.d. sequence $X_1, \ldots, X_n$ and selected distribution F, one tests the hypothesis H′: The data $X_1, \ldots, X_n$ come from F. If $T(X_1, \ldots, X_n)$, the test statistic for a particular goodness-of-fit test, has d.f. G_n under H′, then H′ is accepted at the user-specified significance level $\delta \in (0, 1)$ if $1 - G_n(T(X_1, \ldots, X_n)) \leq \delta$. Otherwise, it is rejected.

Conover (1999) describes goodness-of-fit tests, including chi-squared, Kolmogorov–Smirnov, and Anderson–Darling. Spreadsheet and distribution-fitting software packages commonly offer one or several of these tests.

A goodness-of-fit test provides a single number on the basis of which one accepts or rejects H′. While this yes–no approach has a natural appeal, it can at best be used as a supplement but not as a substitute for graphical comparison.

Several d.f. and density-fitting software packages recommend or at least encourage the use of a fitted distribution whose test-statistic value on a particular test implies a higher rank than the values of the corresponding statistic for, say, $N - 1$ other fitted distributions, where N varies with the package. Experience indicates that reliance on this comparison of test-statistic values in no way replaces the benefit to be gained from graphical comparison of a fitted d.f. with the corresponding empirical d.f.

10.10 ATTRIBUTES AND FITTING DISTRIBUTIONS

Recall that Chapter 2 focuses on entities of delay systems and on their attributes. The values that the attributes assume frequently influence the dynamics of the simulation. For example, they can affect decision-flow logic for:

- Routing that temporary entities follow as they flow through the system

- Sequencing of jobs (entities) for service

- Matching resources with workload

- Scheduling of resources.

They can also affect sample averages collected during a simulation. For example, management decision-making may require estimates of long-run averages and exceedance probabilities for waiting time and queue length for jobs partitioned by attribute values.

Whenever any of these issues arise in a simulation study and attribute values affect the values that interarrival and service times assume, relying on distributional fitting as in Section 10.8 runs the risk of seriously misrepresenting system dynamics within the corresponding simulation model. Section 10.10.1 addresses this problem. When none of these issues arise but attribute values nevertheless affect interarrival and service times, the fitting of unimodal distributions as in Section 10.8 may again be inappropriate. Section 10.10.2 addresses this topic.

10.10.1 ATTRIBUTES AFFECT DECISION FLOW

If an attribute affects decision flow and also interarrival and service times, then, to ensure the fidelity of the system's logic, the corresponding simulation program must incorporate procedures for explicitly sampling these quantities from distributions conditioned on attribute values. For example, if customers in the airline reservation problem of Chapter 2 were selected for service based on type, then this mandate would apply. We use the service time distributions for that problem to describe the issues that arise in estimating parameter values.

In the airline reservation problem, recall that $100p$ percent of callers want multidestination reservations and are assumed to have service times from the Erlang distribution

$\mathcal{G}(2, 1/\omega)$. Also, $100(1 - p)$ percent of callers want a single-destination reservation and have service times from $\mathcal{E}(1/\omega)$. Suppose that the service-time distributions are unknown and need to be estimated.

Three data-analysis scenarios are possible:

1. The proportion p is known and service-time data are available by type of customer.

2. The proportion p is known and service-time data are not identified by type.

3. The proportion p is unknown and service-time data are available by type.

4. The proportion p is unknown and service-time data are not identified by type.

When Scenario 1 obtains, a service-time distribution is fitted separately for each type of data. Given the value of p and the two fitted distributions, a simulationist now has the essentials for generating service times by type in the simulation.

If the historical data in Scenario 3 contain n_1 service times for single-destination callers and n_2 service times for multiple-destination callers, then one estimates p by $\hat{p} = n_2/(n_1 + n_2)$ and fits the two service-time distributions as with Scenario 1.

If p is known but the historical data do not reveal type (Scenario 2), estimation becomes much harder. Suppose we want to fit $\mathcal{G}(\alpha_1, \beta_1)$ to service times for single-destination callers and $\mathcal{G}(\alpha_2, \beta_2)$ to service times for multiple-destination callers. That is, we want to estimate $\alpha_1, \beta_1, \alpha_2, \beta_2$ in the *mixture* of p.d.f.s

$$f(x) = (1 - p) \frac{x^{\alpha_1 - 1} e^{-x/\beta_1}}{\beta_1^{\alpha_1} \Gamma(\alpha_1)} + p \frac{x^{\alpha_2 - 1} e^{x/\beta_2}}{\beta_2^{\alpha_2} \Gamma(\alpha_2)},$$
$$\alpha_1, \beta_1, \alpha_2, \beta_2 > 0, \qquad x \geq 0,$$

which is the p.d.f. of a randomly selected customer. Since we cannot identify service times by type, the likelihood function for the service-time data $X_1, \ldots, X_n$ becomes

$$L = \prod_{i=1}^{n} \left[(1 - p) \frac{X_i^{\alpha_1 - 1} e^{-X_i/\beta_1}}{\beta_1^{\alpha_1} \Gamma(\alpha_1)} + p \frac{X_i^{\alpha_2 - 1} e^{-X_i/\beta_2}}{\beta_2^{\alpha_2} \Gamma(\alpha_2)} \right].$$

Finding $\alpha_1, \beta_1, \alpha_2$, and β_2 that maximize L is a considerably more challenging nonlinear optimization problem than what Section 10.8 describes. The severity of the problem increases if p is unknown and also has to be estimated as in Scenario 4.

The airline reservation problem with unknown service-time distribution merely illustrates a more general data-fitting problem that arises whenever the data are not available in a partition that reflects how a critical attribute or attributes influence the data values. Inevitably, the challenge is to estimate the values of parameters in a *mixture* of distributions

$$f(x) = p_1 f_1(x) + p_2 f_2(x) + \cdots + p_r f_r(x),$$
$$p_i > 0, \qquad i = 1, \ldots, r,$$

$$p + \cdots + p_r = 1,$$

where p_i denotes the proportion of type i and $f_1, \ldots, f_r$ are p.d.f.s or p.m.f.s. Whenever Scenarios 2 and 4 prevail, convenient estimation via the maximum likelihood method is usually not possible.

10.10.2 ATTRIBUTES DO NOT AFFECT DECISION FLOW

When attributes that affect service times do not influence decision flows, one can, in principle, generate these times either by attribute type from the conditional distributions $f_1, \ldots, f_r$ or directly from the unconditional distribution f. Both options induce sample paths during simulation execution that lead to sample averages that converge to correct limits. However, data availability again limits the choice of options.

If Scenario 1 or 3 prevails, the simulationist is well advised to fit $f_1, \ldots, f_r$ separately. Since they are more likely to be unimodal than the unconditional distribution f, the method of Section 10.8 for entries in Tables 10.2 and 10.3 applies to directly fitting $f_1, \ldots, f_r$ more often than to f. When Scenario 2 or 4 prevails, fitting $f_1, \ldots, f_r$ in this way is not possible. Moreover, using the maximum likelihood method to fit f presents a formidable problem.

An alternative fits the data to a possibly multimodal distribution from which samples can be drawn relatively directly during simulation execution. In principle, this approach puts the mixture issue aside and concentrates on curve fitting. Wagner and Wilson (1996) offer Windows-based software, PRIME (Probabilistic Input Modeling Environment), to fit the data to a continuously differentiable f formulated by exploiting the properties of Bézier curves. These are a special class of spline curves whose mathematical and numerical properties are well suited to simulation input modeling.

PRIME offers a user the choice optimizing criteria for fitting the resulting Bézier distribution. These include the methods of least squares, minimum L_1 norm, minimum L_∞ norm, moment matching, percentile matching, and maximum likelihood. A limitation of PRIME is that it assumes bounded support. PRIME is available on the Internet at http://www.ie.ncsu/jwilson by clicking on "software."

10.11 DEPENDENT HISTORICAL DATA

When $X_1, \ldots, X_n$, the historial data to be used to fit a distribution, are not i.i.d. the method of Section 10.8 does not apply. A departure from the i.i.d. assumption occurs if either

$$X_1, \ldots, X_n \text{ are not strictly stationary}$$

or

$$X_1, \ldots, X_n \text{ are strictly stationary but dependent (Section 6.5).}$$

Since nonstrictly stationary input phenomena are inconsistent with a simulation method-ology aimed at deriving estimates of long-run averages, we ignore this case and concentrate on stationary but dependent input phenomena. In what follows, we describe dependence as it relates to an arrival process. However, the applicability of the concepts to a more general setting should be apparent to the reader.

If dependence exists among the observations in a historical record of interarrival times that one plans to use to create an interarrival-time generator in a simulation model, neither does the method of Section 10.8 apply for fitting the data to a distribution nor does the sample-generating procedure for i.i.d. interarrival times apply. To overcome these impediments, we partition our discussion into two parts. Sections 10.11.1 and 10.14 address problems in which the data are all functions of some exogenous phenomenon such as time. Section 10.11.2 addresses the problem in which dependence in the data is tied to considerably more localized considerations.

10.11.1   TIME DEPENDENCE

Problem setting often prompts a simulationist to suspect dependence in historical data. For example, the simulation of the photocopying problem in Section 1.4 focuses on customer arrivals between 10 a.m. and 5 p.m. on weekdays. Elementary considerations about library traffic encourage the suspicion that the arrival rate may vary by time of day and possibly by day of week. Therefore, data on arrivals need to be tested to confirm or reject this suspicion. Section 10.12.1 does this by resorting to the standard statistical tools of the *analysis of variance* for testing.

Dependence related to time of day, day of week, week of month, and month of year frequently exhibits a periodic pattern. Section 10.14 describes how the concept of a non-homogeneous Poisson process can capture the essential features of this time dependence in fitting data and then generating interarrival times. When studying the testing procedures in Section 10.12.1 and modeling concepts in Section 10.14, the reader is encouraged to recognize that these ideas apply to a considerably wider range of problems in which time dependence is suspected.

10.11.2   LOCAL DEPENDENCE

The commonly encountered concept of arrivals to a delay system is that they arise from a large number of independent sources each of which contributes at a relatively small rate. However, other arrival patterns do occur. For example, it may be that short interarrival times occur in clusters, merely reflecting an institutional feature of the external world that provides input to the system being simulated.

How does dependence among interarrival times effect queueing behavior? If depen-dence takes the form of positive correlation and a procedure ignores this property but nevertheless generates interarrival times from the correct marginal distribution, the result-ing sample paths for queue length and waiting time tend to understate the level of congestion in the system. When suspected, the first task is to test for dependence. Testing the data for a flat spectrum using the periodogram and Kolmogorov–Smirnov test provides a comprehensive approach (e.g., Brockwell and Davis 1991, p. 339). Rejection implies the presence of dependence. Whenever applications software is available to perform this test, the simulationist is well advised to take advantage of it. When not available, other less powerful procedures are occasionally used.

One approach relies on the von Neumann ratio (6.65)

$$C_n = 1 - \frac{\sum_{i=1}^{n-1} (X_i - X_{i+1})^2}{2 \sum_{i=1}^{n} (X_i - \bar{X}_n)^2}$$

with a two-sided test. That is, one rejects the hypothesis of independence H at the $100 \times \delta$ percent significance level if

$$C_n < -h(\delta, n) \qquad \text{or} \qquad C_n > h(\delta, n),$$

where

$$h(\delta, n) := \Phi^{-1}(1 - \delta/2)\sqrt{(n - 2)/(n^2 - 1)}, \qquad n > 2.$$

As Section 6.6.8 indicates, this procedure tests the hypothesis $\rho_1 = 0$, where ρ_1 denotes the autocorrelation of lag 1 in the sequence (Section 6.5). If, in reality, the data are dependent with a non-monotone-decreasing autocorrelation function (6.6), then it is possible that ρ_1 is close to zero but subsequent autocorrelations, $\rho_j, |j| > 1$, are substantially different from zero. This consequence limits the value of the procedure.

To overcome this limitation, a third procedure uses the sample autocorrelations

$$\hat{\rho}_j = \frac{\frac{1}{n-j} \sum_{i=1}^{n-j} (X_i - \bar{X}_n)(X_{i+j} - \bar{X}_n)}{\frac{1}{n} \sum_{i=1}^{n} (X_i - \bar{X}_n)^2}, \qquad j = 1, \dots, j_*,$$

for some user-selected lag $j_* \ll n$. If the data are independent, then under relatively general conditions

$$n^{1/2}\hat{\rho}_j \xrightarrow{\text{d}} \mathcal{N}(0, 1) \qquad \text{as} \qquad n \to \infty.$$

We describe one of several ways of exploiting this limiting result. For some $a \in (0, 1)$, suppose a simulationist regards a lagged correlation $\hat{\rho}_j \in [-a, a]$ as of negligible importance. If $A_j(a)$ denotes this event, then for large n

$$\text{pr}\left[A_j(a) \mid \text{H}\right] \approx 2\Phi(an^{1/2}) - 1.$$

As an example, $a = .05$ and $n = 1000$ gives

$$\text{pr}\left[A_j(a) \mid \text{H}\right] \approx 2\Phi(.05\sqrt{1000}) - 1 = .8862,$$

implying that under H we expect to encounter $\mid \hat{\rho}_j \mid \leq a$ about 88.62 percent of the time. If $j_* = 100$ then the mean number of exceedances for $\hat{\rho}_1, \ldots, \hat{\rho}_{100}$ is $100 \times (1 - .8862) = 11.38$. If $a = .05$, $n = 2000$, and $j^* = 100$, the mean number of exceedances under H is $100(1 - .9746) = 2.54$. Observing no more than the average number of exceedances gives credibility to H.

Sometimes a significance level α is specified so that $a = \Phi^{-1}(1 - \alpha/2)$. Then the mean number of exceedances under H is $j_*\alpha$.

10.12 LIBRARY PHOTOCOPYING (SECTION 1.4 EXAMPLE)

The photocopying problem described in Section 1.4 required the simulationist to specify many sampling distributions and to assign numerical values to the parameters of each to facilitate sampling. Principal among these distributions are those for customer interarrival time T, number of pages to be copied N, and the corresponding photocopying time S.

To estimate the parameters, data were collected and analyzed for each floor of the library. The present account describes the analysis for the third floor. For each customer i, the data consisted of

$$T_i := \text{interarrival time},$$
$$N_i := \text{number of pages to be copied},$$
$$S_i := \text{photocopying time}.$$

Times came to the simulationist rounded to the nearest minute. That is, if T_i' and S_i' were customer i's true interarrival and photocopying times, respectively, in minutes, then the data were the integers

$$T_i = \lfloor T_i' + .5 \rfloor$$

and

$$S_i = \lfloor S_i' + .5 \rfloor.$$

Since both interarrival and photocopying times are continuous, the implications of the recording method need to be understood. In particular, $T_i = 0$ for $0 < T_i' < .5$ and $S_i = 0$ for $0 < S_i' < .5$, which substantially misrepresented the true behavior of small interarrival and photocopying times. To reduce the severity of this error, T_i and S_i were redefined as

$$T_i = \begin{cases} .25 & \text{if } T_i' < .5, \\ \lfloor T_1' + .5 \rfloor & \text{if } T_1' \geq 1, \end{cases}$$

and

$$S_i = \begin{cases} .25 & \text{if } S_i' < .5, \\ \lfloor S_i' + .5 \rfloor & \text{if } S_i' \geq .5, \end{cases}$$

respectively. This enables us to regard $\{T_i\}$ and $\{S_i\}$ as sequences of midpoints of the discretization intervals $[0, .5), [.5, 1.5), [1.5, 2.5), \ldots$.

10.12.1 INTERARRIVAL TIMES

Table 10.4 shows the number of arrivals by time of day and day of week. If arrivals were independent of these two phenomena, then a single sampling distribution would suffice for all interarrival times, thereby simplifying the estimation procedure and the subsequent sampling procedure. Accordingly, a two-way analysis of variance (ANOVA) was performed on these data to test the hypotheses

$$H_1 : \text{Arrivals are independent of day of week}$$

and

$$H_2 : \text{Arrivals are independent of hour of day.}$$

Table 10.5 lists the results.

Table 10.4 Number of hourly arrivals

Hour	Mon.	Tues.	Wed.	Thurs.	Fri.	Hourly total
10 a.m.	17	14	18	18	13	80
11	23	32	20	11	17	103
12 p.m.	15	19	19	17	17	87
1	27	21	22	19	25	114
2	23	24	6	14	27	94
3	15	18	13	32	20	98
4	13	18	15	11	8	65
Daily total	133	146	113	122	127	641

Table 10.5 ANOVA for H_1 and H_2

Test	d.f.	F	p-value := pr(exceeding F)
$H_1 \cup H_2$	(10, 24)	1.18	.3520
H_1	(4, 30)	.65	.6327
H_2	(6, 28)	1.53	.2110

For each test, the p-value estimates the probability of observing a test statistic in excess of the corresponding computed F. Hence, the p-value decreases with increasing F. Common practice rejects a hypothesis if its p-value is less than .05.

In the present case, the ANOVA provides little support for rejecting H_1, H_2, or $H_1 \cup H_2$. However, we must be cognizant of two factors. First the statistical power of a test to detect differences is often low when the alternative hypothesis consists of all possible types of differences, as is the case here. Second, the pattern of library usage was known to have fewer arrivals in the early morning (10–11) and in the late afternoon (4–5) than during the remainder of the day (11–4). These factors encouraged testing H_2 against the hypothesis

H_3 : Arrival rates in hours 10–11 and 4–5 differ from those in hours 11–4.

The data in Table 10.4 also arouse suspicion that the large number of arrivals on Tuesday indicate that this day differs from the remaining four. This observation encouraged testing of H_1 against the hypothesis

H_4 : Arrival rate on Tuesday differs from that on the remaining days.

Testing one hypothesis against a specific alternative, as in the case of H_2 versus H_3 and H_1 versus H_4, increases the statistical power of ANOVA to detect differences, when they exist.

Table 10.6 displays results of these ANOVAs. They reject H_2 in favor of H_3 at the .05 level but provide no support for H_4 over H_1. Accordingly, interarrival time data were partitioned into two groups. The analysis that follows focuses on the arrivals in the peak interval 11–4.

The aggregated sample consisted of $n = 493$ interarrival times with sample mean, variance, and standard errors

$$\hat{\mu}_T = 3.030, \qquad \hat{\sigma}_T^2 = 11.12, \qquad (10.4)$$

$$\text{s.e.}(\hat{\mu}_T) = .1502, \qquad \text{r.e.}\,(\hat{\mu}_T) = .04956, \qquad (10.5)$$

the last of which attests to the relative stability of $\hat{\mu}_T$ as an estimate of the true mean μ_T. Figure 10.6a shows the empirical histogram. Together with the discretization, described earlier, the absence of mass at 11, 17, 18, 20, 23, 25, 27, and 28 minutes would severely limit the values of interarrival times, were the histogram used as the basis for sampling in a simulation. Accordingly, we fit a continuous distribution. The shape of the histogram

Table 10.6 ANOVA for H_2 vs. H_3 and H_1 vs. H_4

Test	d.f.	F	p-value
H_2 vs. H_3	$(1, 33)$	6.74	.0140
H_1 vs. H_4	$(1, 33)$	1.63	.2105

encouraged the fitting of a Gamma distribution $\mathcal{G}(\alpha, \beta)$ (Section 8.13). However, the sample coefficient of variation $\hat{\sigma}_T / \hat{\mu}_T = 1.100$ is relatively close to unity, which is the coefficient of variation of the exponential distribution, a special case, $\mathcal{G}(1, \beta)$, of the Gamma. Both options were considered. First, we describe the effect of discretization on the parameter estimation procedure.

Because of discretization, the likelihood function is

$$L(\alpha, \beta, n_0, n_1, \ldots, n_k) = [F(.5)]^{n_0} \ \Pi_{j=1}^{k} [F(j + .5) - F(j - .5)]^{n_j}, \qquad (10.6)$$

where

$$n_0 := \text{\# of observations with } T_i = .25,$$
$$n_j := \text{\# of observations with } T_i = j,$$
$$k - \text{\# of distinct integer categories,}$$

and $F(t) = \int_0^t f(x) dx$, where f is the Gamma p.d.f. in Section 8.13.

Because of the exceptional demands in computing the m.l.e. for β in the exponential case and for α and β in the Gamma case for the likelihood function (10.3), they were computed for the approximate likelihood function

$$L^*(\alpha, \beta, n_0, n_1, \ldots, n_k) = [.5f(.25)]^{n_0} \Pi_{j=1}^{k} [1 \times f(j)]^{n_j} = \Pi_{i=1}^{n} f(T_i). \quad (10.7)$$

In effect, the approximations $.5f(.25) \doteq F(.5)$ and $f(j) \doteq F(j + .5) - F(j - .5)$ for $j = 1, \ldots, k$ were used.

Maximizing with respect to α and β led to the m.l.e.

$$\hat{\alpha} = 1.047, \qquad \hat{\beta} = 2.893,$$
$$\text{s.e.}(\hat{\alpha}) = .05897, \qquad \text{s.e.}(\hat{\beta}) = .2068,$$
$$\text{r.e.}(\hat{\alpha}) = .05631, \qquad \text{r.e.}(\hat{\beta}) = .07147,$$
$$\text{corr}(\hat{\alpha}, \hat{\beta}) = -.7819,$$

where the relative errors attest to the stability of $\hat{\alpha}$ and $\hat{\beta}$ as estimates of the unknown α and β. The corresponding fitted distribution, which would serve as the basis for sampling in the

(a) Empirical histogram

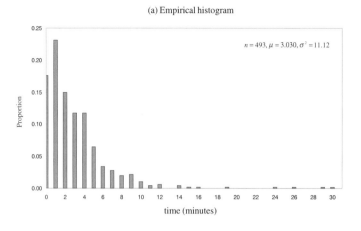

(b) Gamma fit

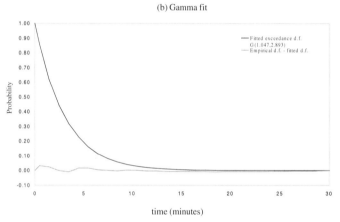

(c) Comparison of difference curves

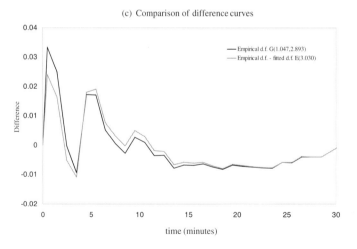

Figure 10.6 Interarrival times

simulation, is

$$\hat{F}(t) = \frac{1}{\hat{\beta}^{\hat{\alpha}} \Gamma(\hat{\alpha})} \int_0^t z^{\hat{\alpha}-1} e^{-z/\hat{\beta}} dz, \quad t \geq 0.$$

Note that $\hat{\alpha} < 1$ implies a monotone decreasing p.d.f. unbounded at $t = 0$. Statistical tests of how well $\{\hat{F}(t)\}$ approximates $\{F(t) = [\beta^{\alpha} \Gamma(\alpha)]^{-1} \int_0^t z^{\alpha-1} e^{-z/\beta} dz, t \geq 0\}$ customarily rely on a comparison of $\hat{F}$ and the empirical d.f.

$$\bar{F}(t) = \frac{1}{n} \sum_{i=1}^n I_{[T_i, \infty)}(t), \qquad t \geq 0. \tag{10.8}$$

Figure 10.6b displays the fitted exceedance d.f. $\{1 - \hat{F}(t), t \geq 0\}$ in black and the differences $\{\bar{F}(t) - \hat{F}(t), t \geq 0\}$ in gray. As discussed in Section 10.3.2, visual assessment of the difference graph provides one statistical evaluation, albeit a subjective one. The appearance of the difference curve suggests a relatively good fit.

Now to the special case of the exponential distribution $\mathcal{G}(1, \beta)$ with p.d.f.

$$f_e(t) = \beta^{-1} e^{t/\beta}, \qquad \beta > 0, \qquad t \geq 0.$$

Clearly, $\hat{\alpha} = 1.047$ adds considerable plausibility to the hypothesis $\Omega: \alpha = 1$. Moreover, the aggregate arrival process to the library's photocopiers fits our description in Section 10.4.1 of a superposition of many independent renewal processes each with comparatively small arrival rate. Thus the distribution of the interarrival times for the aggregate process are well approximated by the exponential where the error of approximation tends to decrease as the number of renewal processes increases.

Maximizing the likelihood function (10.7) with respect to β yields the m.l.e. $\hat{\beta}_e = \hat{\mu}_T = 3.030$, which has the same standard errors as $\hat{\mu}_T$ in expression (10.1). The corresponding fitted d.f. is

$$\hat{F}_e(t) = 1 - e^{-t/\hat{\beta}_e}, \qquad t \geq 0. \tag{10.9}$$

Figure 10.6c shows a marginal superiority for the Gamma fit, which is to be expected.

How are we to choose between the two fits? Let us summarize the facts:

- The interarrival time data estimate the mean as $\hat{\mu}_T = 3.030$ and variance as $\hat{\sigma}_t^2 = 11.12$, implying a coefficient of variation 1.10.

- The Gamma fit estimates the mean as $\hat{\alpha}\hat{\beta} = 3.030$ and variance as $\hat{\alpha}\hat{\beta}^2 = 8.766$, implying a coefficient of variation .9771.

- The exponential fit estimates the mean as $\hat{\beta}_e = 3.030$ and variance as $\hat{\beta}_e^2 = 9.181$, implying coefficient of variation unity.

- Sampling from $\mathcal{E}(\hat{\beta}_e)$ is less time consuming than sampling from $\mathcal{G}(\hat{\alpha}, \hat{\beta})$.

Interestingly, the Gamma fit has the lowest coefficient of variation. As discussed earlier, system congestion tends to grow with increasing coefficient of variation. Therefore, sampling from either $\mathcal{G}(\hat{\alpha}, \hat{\beta})$ or $\mathcal{E}(\hat{\beta}_e)$ induces less congestion than sampling from the empirical d.f. $\{\bar{F}(t)\}$ would. However, this reduction is less for $\mathcal{E}(\hat{\beta}_e)$. This, together with less costly sampling, favors the exponential choice.

In the present case, the proximity of $\hat{\alpha}$ to unity encourages us to choose the exponential fit. However, what would we have done if $\hat{\alpha}$ were, say, 1.1 or 1.2? A perusal of difference curves, as in Figure 10.6c, would provide some help. However, a more objective statistical criterion of acceptability is desirable. The likelihood-ratio test provides a basis for increased objectivity.

10.12.2   LIKELIHOOD RATIO TEST

Recall that $L^*(1, \beta)$ is merely the likelihood function $L^*(\alpha, \beta)$ in expression (10.7) under the hypothesis $\Omega : \alpha = 1$. Because of this constrained optimization in the exponential case, the *likelihood ratio*

$$R := \frac{L^*(1, \hat{\beta}_e, n_0, n_1, \ldots, n_k)}{L^*(\hat{\alpha}, \hat{\beta}, n_0, n_1, \ldots, n_k)}$$

lies in $(0, 1)$. Moreover, it is known that

$$-2 \ln R \xrightarrow{\mathrm{d}} \chi^2_{n-1} \text{ as } n \to \infty, \tag{10.10}$$

where χ^2_{n-1} denotes a chi-squared random variable with $n - 1$ degrees of freedom. In the present case

$$-2 \ln R = 17.15.$$

For the distributional limit (10.10),

$$\text{p-value} = 1 - 8.681 \times 10^{-258},$$

which is an estimate of the probability of observing a χ^2_{492} random variable greater than $-2 \ln R = 17.15$. In the present case the p-value gives no support to rejecting Ω.

10.12.3   OFF-PEAK HOURS

An analysis of the data for the off-peaks hours $(10, 11] \cup (4, 5]$ gave sample mean $\hat{\mu}_T = 3.528$ and variance $\hat{\sigma}_T^2 = 12.80$ for interarrival time. Moreover, maximum likelihood estimation and likelihood ratio testing led to the conclusion that off-peak interarrival times were from $\mathcal{E}(3.528)$. Section 10.14.1 addresses the issue of how to merge the peak and off-peak arrival processes for sampling.

10.12.4 NUMBER OF PAGES

The number-of-pages or page-quantity sequence $N_1, \ldots, N_n$ is a sequence of positive integers, and it is reasonable to assume that they are i.i.d. The Poisson and negative binomial distributions are the most commonly employed for fitting nonnegative integer data with infinite support. If X has the Poisson distribution $P(\lambda)$, then

$$\mathrm{pr}(X = i) = \frac{e^{-\lambda}\lambda^i}{i!}, \qquad \lambda > 0, \qquad i = 0, 1, \ldots.$$

If Y has the negative binomial distribution $\mathcal{N}\mathcal{B}(r, p)$, then

$$\mathrm{pr}(Y = i) = \frac{\Gamma(r + i)}{\Gamma(r)\Gamma(i + 1)}(1 - p)^r p^i, \quad r > 0, \quad 0 < p < 1, \quad i = 0, 1, \ldots.$$

Chapter 8 describes bounded-computing-time algorithms for sampling from each. A distinguishing feature of $P(\lambda)$ is that $\mathrm{E}X = \lambda$, and $\mathrm{var}X = \lambda$, so that

$$\text{variance-to-mean ratio} = \frac{\mathrm{var}X}{\mathrm{E}X} = 1.$$

By contrast, $\mathcal{N}\mathcal{B}(r, p)$ has $\mathrm{E}Y = rp/(1 - p)$ and $\mathrm{var}\,Y = rp/(1 - p)^2$, so that

$$\text{variance-to-mean ratio} = \frac{\mathrm{var}\,Y}{\mathrm{E}Y} = \frac{1}{1 - p} > 1.$$

Since the data yielded the sample mean $\hat{\mu}_N = 22.94$ and sample variance $\hat{\sigma}_N^2 = 1661$ for a sample variance-to-mean ratio 72.40, and since the empirical histogram in Figure 10.7a revealed substantial right skewness uncharacteristic of the Poisson, the negative binomial distribution was chosen for fitting. The resulting m.l.e.s are

$$\hat{r} = .9740, \qquad \hat{p} = .9594,$$
$$\text{s.e.}(\hat{r}) = .0520, \qquad \text{s.e.}(\hat{p}) = .002626,$$
$$\text{r.e.}(\hat{r}) = .05360, \qquad \text{r.e.}(\hat{p}) = .00274,$$
$$\text{corr}(\hat{r}, \hat{p}) = .7947.$$

Figure 10.7b shows the fitted exceedance d.f. in black and the difference between the empirical and fitted d.f.s (in gray). The difference curve reveals no absolute difference greater than .06 between these d.f.s. It also shows that each difference is relatively small when compared with its corresponding ordinate on the fitted d.f.

(a) Empirical histogram

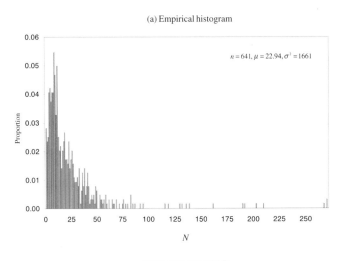

(b) Negative binomial fit

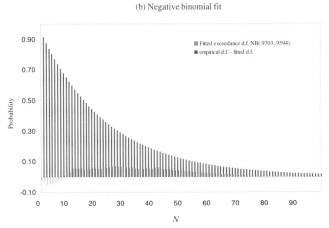

(c) Poisson fit

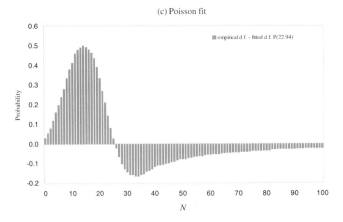

Figure 10.7   Page quantity N

While the sample variance-to-mean ratio supported the decision to fit the negative binomial rather than the Poisson distribution, it is nevertheless instructive to show the results of fitting the latter with $\hat{\lambda} = \hat{\mu}_N = 22.94$. Figure 10.7c shows the differences with the empirical d.f. The superior fit of the negative binomial is unequivocal.

10.12.5 PHOTOCOPYING TIME

Since photocopying time is clearly a function of number of pages to be copied, it is necessary to take this dependence into account when sampling this time. To do this, we need to quantify this relationship and then turn it into a mechanism for sampling.

The plot of $\{(N_i, S_i);\ i = 1, \ldots, I\}$ in Figure 10.8a reveals a linear relationship but with increased variability in S_i for large N_i. To model this behavior, let N denote number of copies, S the corresponding setup plus photocopying time, and let

$$V_j := \text{copying time for page } j$$

with

$$b := EV_j \quad \text{and} \quad \sigma^2 := \text{var } V_j, \quad j = 1, \ldots, N.$$

Multiple copies of the same page or automatic page feeding customarily induces a copying time S directly proportional to the number of pages N and with virtually no time variation per page. However, copying in a library tends to be from journals and books, one page at a time, and thus copying times $V_1, \ldots, V_N$ are reasonably assumed to be i.i.d. Collectively, the graph and the i.i.d. assumption imply the representation

$$S = a + \sum_{j=1}^{N} V_j = a + bN + N^{1/2}\epsilon, \tag{10.11}$$

where

$$a := \text{setup time,}$$

$$\epsilon := N^{1/2} \sum_{j=1}^{N} (V_j - b),$$

$$E\epsilon = 0,$$

$$\text{var } \epsilon = \sigma^2.$$

In terms of the data, we have

$$S_i = a + bN_i + N_i^{1/2}\epsilon_i, \quad i = 1, \ldots, I,$$

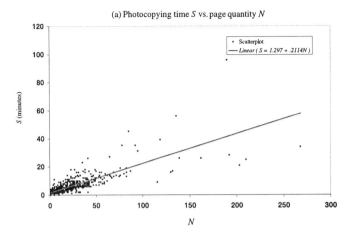

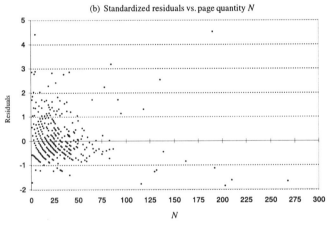

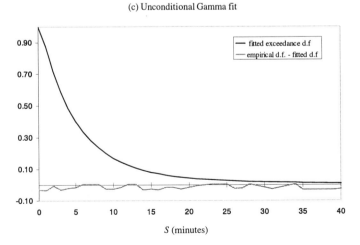

Figure 10.8   Photocopying

with three parameters a, b, and σ^2 to be estimated. This was done by choosing a and b to minimize the sum of squared errors

$$\sum_{i=1}^{I} \epsilon_i^2 = \sum_{i=1}^{I} (S_i - a - bN_i)^2,$$

giving

$$\hat{a} = \frac{I \sum_{i=1}^{I} S_i - \sum_{i=1}^{I} N_i \sum_{j=1}^{I} (j/N_j)}{I^2 - \sum_{i=1}^{I} N_i \sum_{j=1}^{I} N_j^{-1}} = 1.297 \text{ minutes}$$

and

$$\hat{b} = \frac{I \sum_{i=1}^{I} (S_i/N_i) - \sum_{i=1}^{I} S_i \sum_{j=1}^{I} N_j^{-1}}{I^2 - \sum_{i=1}^{I} N_i \sum_{j=1}^{I} N_j^{-1}} = .2114 \text{ minutes} = 12.68 \text{ seconds.}$$

With regard to the residual variance σ^2, we have

$$\hat{\sigma}^2 = \frac{1}{I} \sum_{i=1}^{I} N_i^{-1} (S_i - \hat{a} - \hat{b} N_i)^2$$

$$= \frac{1}{I} \sum_{i=1}^{I} N_i^{-1} \{(S_i - a - b N_i) - [\hat{a} - a + (\hat{b} - b)N_i]\}^2$$

$$= .7685,$$

so that for $\mathbf{N}_I := (N_1, \ldots, N_I)$,

$$E(\hat{\sigma}^2 \mid \mathbf{N}_I) = \sigma^2 (1 - 2/I).$$

Therefore,

$$\tilde{\sigma}^2 = \hat{\sigma}^2 \, I/(I - 2) = .7709$$

is an unbiased estimate of σ^2.

Since $E(\hat{a} \mid \mathbf{N}_I) = a$ and $E(\hat{b} \mid \mathbf{N}_I) = b$, it is easily seen that

$$\text{var}(\hat{a} \mid \mathbf{N}_I) = \frac{\sigma^2 \sum_{i=1}^{I} N_i (I - N_i^{-1} \sum_{j=1}^{I} N_j)^2}{(I^2 - \sum_{i=1}^{I} N_i \sum_{j=1}^{I} N_j^{-1})^2}$$

and

$$\text{var}\,(\hat{b} \mid \mathbf{N}_I) = \frac{\sigma^2 \sum_{i=1}^{I} N_i (I\,N_j^{-1} - \sum_{j=1}^{N} N_j^{-1})^2}{(I^2 - \sum_{i=1}^{I} N_i \sum_{j=1}^{I} N_j^{-1})^2}.$$

Substituting $\tilde{\sigma}^2$ for $\hat{\sigma}^2$, we estimate the standard errors of $\hat{a}$ and $\hat{b}$ as

$$\text{s.e.}(\hat{a}) = .1217 \quad \text{and} \quad \text{s.e.}(\hat{b}) = .008724,$$
$$\text{r.e.}(\hat{a}) = .09383 \quad \text{and} \quad \text{r.e.}(\hat{b}) = .04128.$$

In principle, $(S - a - b\,N)/\sqrt{N\sigma^2} \xrightarrow{\text{d}} \mathcal{N}\,(0, 1)$ as $N \to \infty$. This property suggests that a plot of the standardized residuals $\Delta_i := (S_i - \hat{a} - \hat{b}\,N_i)/\sqrt{N_i \tilde{\sigma}^2}$ versus N_i for $i = 1, \ldots, I$ should reveal a tendency toward symmetry around the zero axis as N_i increases. Although Figure 10.8b confirms this tendency, it also reveals a marked asymmetry and thus nonnormality of the residuals around zero for small N_i.

Since our principal objective is to devise a means of sampling S given N, we need to choose a sampling distribution that accounts for asymmetry for small number of copies N and for the tendency toward normality as N increases. One solution to this problem relies on the Gamma distribution (Section 8.13).

A random variable X from $\mathcal{G}(\alpha, \beta)$ has mean $\alpha\beta$ and variance $\alpha\beta^2$. If we regard S given N as a Gamma distributed random variable with mean $\hat{a} + \hat{b}\,N$ and variance $N\tilde{\sigma}^2$, then it is from $\mathcal{G}(\alpha_N, \beta_N)$, where

$$\alpha_N := \frac{(\hat{a} + \hat{b}N)^2}{N\tilde{\sigma}^2}$$

and

$$\beta_N := \frac{N\tilde{\sigma}^2}{\hat{a} + \hat{b}N},$$

so that $\alpha_1 = 2.951$, $\beta_1 = .5111$, and

$$\frac{\alpha_N}{N} \to .05797 \quad \text{and} \quad \beta_N \to 3.647 \quad \text{as } N \to \infty.$$

One limitation of this approach is that α_N decreases from $\alpha_N = 2.951$ at N to a minimum of $\alpha_N = \hat{a}/\hat{b} = 1.423$ at $N = 6.135$ and increases thereafter. This behavior for small N suggests that convergence to normality for S is not monotone. This fact encourages us to check the adequacy of the fitted distributions.

The Gamma d.f. is

$$F(s; \alpha, \beta) = \frac{1}{\Gamma(\alpha)\beta^\alpha} \int_0^s x^{\alpha-1} e^{-x/\beta}\,dx, \qquad s \geq 0,$$

so that S given N has d.f. $\{F(s; \alpha_N, \beta_N), \ s \geq 0\}$. Then

$$\hat{F}_G(s) := \frac{1}{n} \sum_{i=1}^{n} F(s; \ \alpha_{N_i}, \ \beta_{N_i}), \qquad s \geq 0,$$

provides an estimator of the unconditional d.f. of copying time S independent of the number of copies N. For the data $S_1, \ldots, S_I$, the empirical d.f. is

$$\bar{F}(s) = \frac{1}{I} \sum_{i=1}^{I} I_{(S_i, \infty)}(s), \qquad s \geq 0.$$

Figure 10.8c displays the fitted unconditional exceedance d.f. $\{1 - \hat{F}_G(s)\}$ and the difference $\{\bar{F}(s) - \hat{F}_G(s)\}$. The relatively small absolute differences encourage us to accept this method of approximating the conditional d.f.s of photocopying time.

10.12.6   RELATIONSHIP WITH STANDARD LINEAR REGRESSION

It is of interest to cast this estimation procedure in the context of general linear regression and to contrast the model (10.11) with the more commonly encountered form of linear regression. Let

$$\mathbf{X}^{I \times 2} := \begin{pmatrix} N_1^{-1/2} & N_1^{1/2} \\ \vdots & \vdots \\ N_I^{-1/2} & N_I^{1/2} \end{pmatrix} \quad \text{and} \quad \mathbf{Y}^{I \times 1} := \begin{pmatrix} S_1/N_1^{1/2} \\ \vdots \\ S_I/N_I^{1/2} \end{pmatrix}.$$

Then $\hat{a}$, $\hat{b}$, and $\tilde{\sigma}^2$ have the representations

$$\begin{pmatrix} \hat{a} \\ \hat{b} \end{pmatrix} = \hat{\beta} := \left(\mathbf{X}^{\mathsf{T}}\mathbf{X}\right)^{-1} \mathbf{X}^{\mathsf{T}}\mathbf{Y}$$

and

$$\begin{aligned}
\tilde{\sigma}^2 &= \frac{1}{I-2} \left(\mathbf{Y} - \mathbf{X}\hat{\beta}\right)^{\mathsf{T}} \ \left(\mathbf{Y} - \mathbf{X}\hat{\beta}\right) \\
&= \frac{1}{I-2}\mathbf{Y}^{\mathsf{T}} \left[\mathbf{I} - \mathbf{X}(\mathbf{X}^{\mathsf{T}}\mathbf{X})^{-1}\mathbf{X}^{\mathsf{T}}\right] \mathbf{Y}.
\end{aligned}$$

If the data in Figure 10.8a had revealed deviations $\{S_i - a - bN_i; i = 1, \ldots, I\}$ that fluctuated independently of the N_i, then we would have been tempted to use the more conventional regression model with

$$\mathbf{X} := \begin{pmatrix} 1 & N_1 \\ \vdots & \vdots \\ 1 & N_I \end{pmatrix} \quad \text{and} \quad \mathbf{Y} := \begin{pmatrix} S_1 \\ \vdots \\ S_I \end{pmatrix}.$$

10.13   WASTE GENERATION (SECTION 1.5 EXAMPLE)

An analysis of daily waste-generation data for the incinerator problem in Section 1.5 illustrates one way in which empirical distributions can be the basis for sampling workload. Recall that each of the four sites or campuses produces liquid waste, which it stores in 55-gallon drums until they can be transferred to the incinerator for disposal. Let

$$N_i := \text{number of days of data collected at campus } i$$

and

$$N_{ij} := \text{number of days on which campus } i \text{ generated}$$
$$j \ 55 - \text{gallon drums of liquid waste.}$$

Table 10.7 shows these frequencies. Since no waste generation occurs on weekends and holidays, these days have been omitted. The generating processes at the four campuses proceed independently of day of week, week of month, and month of year. These properties suggest that daily waste generation at each campus be sampled from its own empirical distribution.

Table 10.7   Daily waste generation, $j :=$ number of drums

| Campus i | N_{ij} | | | | | N_i |
	0	1	2	3	4	Total
1	193	32	1	1	0	227
2	145	60	28	13	4	250
3	204	25	13	3	1	246
4	176	6	1	0	0	183

For campus i, let

p_{ij} := probability that on an arbitrarily selected day the campus generates j drums,

$$\mu_i := \sum_{j=0}^{4} j p_{ij} = \text{mean daily waste generation,} \tag{10.12}$$

$$\sigma_i^2 := \sum_{j=0}^{4} (j - \mu_i)^2 p_{ij} = \sum_{j=0}^{4} i^2 p_{ij} - \mu_i^2 = \text{variance of daily waste generation.} \tag{10.13}$$

Given the marginal sums $N_1, \ldots, N_4$, the maximum likelihood estimates of the p_{ij} are

$$\hat{p}_{ij} = \frac{N_{ij}}{N_i} \tag{10.14}$$

with standard errors

$$\text{s.e.}(\hat{p}_{ij}) = \sqrt{p_{ij}(1 - p_{ij})/N_i}$$

and correlations

$$\text{corr}(\hat{p}_{ij}, \hat{p}_{ik}) = -\sqrt{\frac{p_{ij}\, p_{ik}}{(1 - p_{ij})(1 - p_{ik})}}, \qquad j \neq k.$$

Table 10.8 displays the $\hat{p}_{ij}$.

Let $\hat{\mu}_i$ and $\hat{\sigma}_i^2$ denote the maximum likelihood estimates of μ_i and σ_i^2 obtained by substituting $\hat{p}_{ij}$ for p_{ij} in expressions (10.12) and (10.13). For each μ_i and $\delta = .01$, Table 10.9 displays the point estimate $\hat{\mu}_i$ and the approximating $100 \times (1-\delta)$ percent confidence interval $\left[\hat{\mu}_i \pm \Phi^{-1}(1 - \delta/2)\hat{\sigma}_i/N_i^{1/2}\right]$ for μ_i and the relative half-length $\Phi^{-1}(1 - \delta/2)\hat{\sigma}_i/\hat{\mu}_i N_i^{1/2}$. For $\delta = .05$, the relative half-lengths would be .7609 as long. The relatively large half-lengths imply that we would be shortsighted in accepting $\hat{\mu}_i$ as more than a rough approximation to μ_i, especially for Campus 4.

Since waste generation is the driving force in this system, attention needs to be paid to how sensitive system congestion is to the sampling variation in the $\hat{\mu}_i$. We address this topic

Table 10.8 Maximum likelihood estimates of the p_{ij} for daily waste generation

Campus			$\hat{p}_{ij}$		
i	$j = 0$	$j = 1$	$j = 2$	$j = 3$	$j = 4$
1	.8502	.1410	.004405	.004105	0.
2	.5800	.2400	.1120	.05200	.0160
3	.8293	.1016	.05285	.01220	.004065
4	.9617	.03279	.005464	0.	0.

through $T_i :=$ the number of days that elapse between successive positive waste generation at campus i. Since $1 - p_{i0}$ is the probability of waste generation on a given day,

$$\mathrm{pr}(T_i = t) = (1 - p_{i0})p_{i0}^{t-1}, \qquad t = 1, 2, \ldots,$$

with mean

$$\mathrm{E}T_i = \tau_i := \frac{1}{1 - p_{i0}}.$$

Since

$$\mathrm{pr}\left[\frac{|\hat{p}_{ij} - p_{ij}|}{\sqrt{p_{ij}(1 - p_{ij})/N_i}} \leq \Phi^{-1}(1 - \delta/2)\right] \to 1 - \delta \ \text{ as } \ N_i \to \infty,$$

we have the approximating $100 \times (1 - \delta)$ percent confidence interval $[\omega_1(N_{ij}, N_i, \Phi^{-1}(1 - \delta/2)), \omega_2(N_{ij}, N_i, \Phi^{-1}(1 - \delta/2))]$ for p_{ij}, where

$$\omega_l(z, n, \beta) := \frac{z + \beta^2/2 + \delta(-1)^l[\beta^2/4 + z(n - z)/n]^{1/2}}{(n + \beta)^2}.$$

This follows from solving the inequality

$$(\hat{p}_{ij} - p_{ij})^2 \leq \Phi^{-1}(1 - \delta/2)p_{ij}(1 - p_{ij})/N_i$$

for p_{ij}.

For $j = 0$, these results yield the point estimate $1/(1 - \hat{p}_{i0})$ and interval estimate $\left[1/[1 - \omega_1(N_{i0}, N_i, \Phi^{-1}(1 - \delta/2))], 1/[1 - \omega_2(N_{i0}, N_i, \Phi^{-1}(1 - \delta/2))]\right]$ for τ_i. Table 10.10 displays these for $\delta = .01$. The variation inherent in the mean estimate is consistent with that seen in the $\hat{\mu}_i$ of Table 10.9. However, in this form we can more easily offer a sensitivity analysis that exploits the method of sampling.

The most direct way to generate waste daily in the simulation of Section 1.5 is to sample the number of drums D_i for Campus i from the empirical distribution $\{\hat{p}_{i0}, \ldots, \hat{p}_{i4}\}$ via

Table 10.9   Mean daily waste generation

| Campus | | 99% Confidence interval | | | |
| | | | | Relative | |
i	$\hat{\mu}_i$	Lower	Upper	half-length	s.e.$(\hat{\mu}_i)$
1	.1630	.09216	.2338	.4346	.02750
2	.6840	.5257	.8423	.2314	.0615
3	.2602	.1537	.3666	.4092	.04132
4	.04372	0.	.0874	1.000	.01698

Table 10.10   Mean number of days between successive waste generations

Campus	Sample mean	99% Confidence interval	
		Lower	Upper
1	6.676	4.531	10.12
2	2.381	1.994	2.918
3	5.857	4.150	8.490
4	26.14	10.67	66.40

Algorithm ITR in Section 8.1.3. For five ordinates, this algorithm compares favorably in computing time with the cutpoint method in Algorithm CM in Section 8.2 and the alias method in Algorithm ALIAS in Section 8.4, whose superiority materializes for larger numbers of ordinates. However, this direct generating approach has a limitation. For example, if one were to generate waste daily in this way for Campus 1, then the mean number of days between successive positive waste generations in the simulation would be $1/(1 - \hat{p}_{10}) = 227/34 = 6.676$. That is, this approach generates zero drums on 5.676 successive days on average before generating a positive number. This result follows from the observation that the number of days between successive positive waste generations is $1 + Z$, where Z has the geometric distribution $\mathcal{G}e(\hat{p}_{10})$, as in Section 8.19.

To improve computational efficiency, we resort to a procedure that skips the zeros. For $j = 1, \dots, 4$ and $i = 1, \dots, 4$, let

$$p'_{ij} = \frac{p_{ij}}{1 - p_{i0}}$$
$$= \text{probability of generating } j \text{ drums}$$
$$\text{given that at least one drum is generated}$$

with maximum likelihood estimate

$$\hat{p}'_{ij} = \hat{p}_{ij}/(1 - \hat{p}_{ij}) = N_{ij}/(N - N_{i0}).$$

Let

$$\Delta \tau_i := \text{number of days between positive waste generations on Campus } i.$$

Then on the day of the most recent positive waste generation on Campus i, Algorithm WASTE samples $\Delta \tau_i$, schedules the next positive waste generation, and samples the number of drums D_i that are to be generated on that day.

ALGORITHM WASTE

Purpose: To schedule the next positive waste generation for Campus i and to determine its number of drums.

Given: $\hat{p}'_{i0}$ and $\{\hat{q}'_{ij} := \hat{q}'_{i,j-1} + \hat{p}_{ij}; q'_{i0} := 0; j = 1, \ldots, 4\}$.

Method:

Sample Z from $\mathcal{G}e(\hat{p}_{i0})$.
$\Delta\tau_i \leftarrow 1 + Z$.
Schedule the next positive waste generation on day $\tau + \Delta\tau_i$.
Sample U for $\mathcal{U}(0, 1)$.
$j \leftarrow 1$
While $U > q_{ij}$, $j \leftarrow j + 1$.
$D_i \leftarrow j$.

Table 10.11 displays the relevant distributions along with $\hat{p}_{i0}$, $\hat{p}_{i0}/(1 - \hat{p}_{i0})$, the mean number of zero waste generations between successive positive generations, and the mean number of comparisions in sampling $D_i := \hat{p}_{i1} + 2\hat{p}_{i2} + 3\hat{p}_{i3} + 4\hat{p}_{i4}$. Note the substantial mean numbers of days skipped with regard to zero-waste generation. Also note that the relatively small mean numbers of comparisons are consequences of $\hat{p}'_{ij} \geq \hat{p}'_{i,j+1}$ for $j = 1, \ldots, 3$. When these orderings do not arise naturally in a problem, reordering the probabilities so that comparisons proceed from largest to smallest always reduces the mean number of comparisons.

10.14 NONHOMOGENEOUS POISSON PROCESSES

Recall from Section 10.12.1 that peak period interarrival times were from $\mathcal{E}(3.030)$, whereas during off-peak $(10, 11] \cup (16, 17]$ hours interarrival times were from $\mathcal{E}(3.528)$. If one intends to simulate a complete seven-hour day, how does one merge these two processes to reflect the distinct values of their parameters? We begin with the well-known relationship between the exponential and Poisson distributions.

If $T_1, T_2, \ldots$ are i.i.d. from $\mathcal{E}(1/\lambda)$, where $\lambda > 0$, and

$$N(t) := \text{number of arrivals in } (0, t], \tag{10.15}$$

Table 10.11 Waste generation probabilities ($\hat{p}'_{ij} := \hat{p}_{ij}/(1 - \hat{p}_{i0})$; $j = 1, \ldots, 4$)

Campus i	$\hat{p}_{i0}$	$\hat{p}'_{i1}$	$\hat{p}'_{i2}$	$\hat{p}'_{i3}$	$\hat{p}'_{i4}$	$\frac{\hat{p}_{i0}}{1-\hat{p}_{i0}}$	Mean no. of comparisons
1	.8502	.9412	.0294	.0294	0	5.676	1.088
2	.5800	.5714	.2667	.1238	.0381	1.381	1.627
3	.8293	.5952	.3096	.0714	.0238	4.858	1.524
4	.9617	.8571	.1429	0	0	25.11	1.143

then $N(t)$ has the Poisson distribution $\mathcal{P}(\lambda t)$. Suppose we simulate the times of arrivals for a day starting at hour 10 (a.m.) and ending at hour 17 (5 p.m.). Let

$$t_0 := 10,$$
$$t_1 := 11,$$
$$t_2 := 16,$$
$$t_3 := 17,$$
$$\lambda_1 := 60/3.050 = \text{hourly arrival rate in peak interval}(t_1, t_2],$$
$$\lambda_2 := 60/3.528 = \text{hourly arrival rate in off-peak intervals}(t_0, t_1] \cup (t_2, t_3].$$

Then

$$N(t_1 - t_0) \sim \mathcal{P}(\lambda_1(t_1 - t_0)),$$
$$N(t_2 - t_1) \sim \mathcal{P}(\lambda_2(t_2 - t_1)),$$
$$N(t_3 - t_2) \sim \mathcal{P}(\lambda_1(t_3 - t_2)),$$

and by the reproductive property of the Poisson distribution

$$N(t_3 - t_0) = N(t_1 - t_0) + N(t_2 - t_1) + N(t_3 - t_2) \sim \mathcal{P}(\Lambda(t_3)),$$

where

$$\Lambda(t) := \int_{t_0}^{t} \lambda(s)ds, \qquad t_0 < t \le t_3, \tag{10.16}$$

and

$$\lambda(t) := \begin{cases} \lambda_1, & t_0 < t \le t_1, \\ \lambda_2, & t_1 < t \le t_2, \\ \lambda_1, & t_2 < t \le t_3. \end{cases} \tag{10.17}$$

Here interarrival times form a *nonhomogeneous Poisson process* with *instantaneous intensity function* $\{\lambda(t)\}$ that varies with time of day.

A special case of this nonhomogeneous Poisson process is the *homogeneous Poisson process*, wherein $\{\lambda(t)\}$ is invariant with respect to t so that $\Lambda(t) \propto (t - t_0)$. More generally, $\{N(t), t \ge 0\}$ is said to be a nonhomogeneous Poisson process with instantaneous intensity function $\{\lambda(t), t \ge 0\}$ if for each $h \ge 0$ (e.g., Ross, 1980 p. 187)

   i. $N(0) = 0,$

ii. $\{N(t), t \geq 0\}$ has independent increments,

iii. $\mathrm{pr}[N(t+h) - N(t) \geq 2] = \mathrm{o}(h)$,

iv. $\mathrm{pr}[N(t+h) - N(t) = 1] = \lambda(t)h + \mathrm{o}(h)$.

Since $\{\lambda(t)\}$ in (10.17) is right continuous, property iv can be satisfied in the present setting.

The relationship between homogeneous and nonhomogeneous Poisson processes provides a means for sampling interarrival times for the latter, provided that $\{\lambda(t)\}$ is finite. Algorithm NHPP generates samples from the homogeneous Poisson process $\mathcal{P}(\lambda_{max}\, t)$, where $\lambda_{max} \geq \lambda(t)\ \forall t$, and *thins* them by an acceptance–rejection test. The acceptance probability in an arbitrarily selected trial is no less than $\lambda_2/\lambda_1 = .9354$. More generally, if λ_{max} is considerably larger than most values of $\{\lambda(t)\}$, then thinning is not a particularly efficient sampling procedure.

ALGORITHM NHPP

Purpose: To sample an interarrival time from a nonhomogeneous Poisson process with instantaneous intensity function (10.17).

Source: Lewis and Shedler (1979).

Input: $\{\lambda(t),\ t_0 < t \leq t_3\}$, $\lambda_{max} := \max_{t_0 < t \leq t_3} \lambda(t)$, and $\tau := $ current simulated time in hours ($t_0 < \tau \leq t_3$).

Method:

> $Y \leftarrow \tau$.
> Repeat:
>> sample Z from $\mathcal{E}(1/\lambda_{max})$ and U from $\mathcal{U}(0, 1)$
>> $Y \leftarrow Y + Z$
>> if $Y > t_3$, return indicating no additional arrival in $(\tau, t_3]$
> Until $U \leq \lambda(Y)/\lambda_{max}$.
> Return Y as the next arrival time.

10.14.1 CLEARING BOOKS

In addition to the photocopying problem in Section 1.4, the library requested a simulation study to evaluate the benefits of alternative reshelving policies for books and journals that patrons leave on study tables and reshelving carts. Several times daily a student employee makes tours of these areas and collects the materials for eventual reshelving. Hereafter, we use the generic term "book" to mean journal as well as book.

The simulation called for a means of sampling the number of books that are cleared from tables and reshelving carts each time a clearing tour occurs. For this study, a day began at hour 7.75 (7:45 a.m.) and ended at hour 22 (10 p.m.). Table 10.12 shows data collected on 31 tours. It reveals that tours did not occur on a fixed time schedule. This variation in clearing times created a challenge for the simulationist to provide a generating procedure.

Table 10.12 Library book clearing ($k = 31$ observations)

Obs. i	Clearing time interval $s_i < t \le t_i$		No. of books B_i	Books/ hour $B_i/(t_i - s_i)$	Obs. i	Clearing time interval $s_i < t \le t_i$		No. of books B_i	Books/ hour $B_i/(t_i - s_i)$
1	7.75	15	160	22.07	17	18	22	63	15.75
2	15	21	146	24.33	18	7.75	11	60	18.46
3	21	22	10	10	19	11	17	285	47.5
4	7.75	15	133	18.34	20	17	19	63	31.5
5	15	22	92	13.14	21	7.75	10	75	33.33
6	7.75	11	67	20.62	22	10	13	49	16.67
7	11	15	116	29	23	13	16	66	22
8	15	20	66	13.20	24	16	20	15	3.75
9	7.75	11	79	24.31	25	7.75	9	106	84.8
10	11	14	75	25	26	9	13	60	15
11	14	18	208	52	27	13	15.25	55	24.44
12	7.75	14	150	24	28	15.25	17	166	94.86
13	14	16	107	53.50	29	7.75	9	76	60.80
14	7.75	11	51	15.69	30	9	11	25	12.5
15	11	15	216	54	31	11	19	159	19.88
16	15	18	88	29.33					

One approach to this problem again relies on the concept of a nonhomogeneous Poisson process with instantaneous intensity function estimator

$$\hat{\lambda}(t) = \frac{\sum_{i=1}^{k} [B_i/(t_i - s_i)] \, I_{(s_i, t_i]}(t)}{\sum_{i=1}^{k} I_{(s_i, t_i]}(t)}, \qquad 7.75 < t \le 22, \qquad (10.18)$$

where B_i, s_i, and t_i are as defined in Table 10.12 for $i = 1, \ldots, k$. Figure 10.9 shows $\{\lambda(t), 7.75 < t \le 22\}$, which shows lower intensity in the early morning prior to hour 9 and in the evening after hour 17.

Let $\tau_0 := 7.75$ and $\tau_1 := 22$. To devise a procedure for sampling B, the number of books cleared on a tour, from $\mathcal{P}(\Lambda(\tau))$ efficiently for

$$\Lambda(\tau) := \int_{\tau_0}^{\tau} \lambda(t) dt, \qquad \tau_0 < \tau \le \tau_1,$$

where $\{\lambda(t)\}$ is defined in expression (10.18), we first partition the hours $(7.75, 22]$ into 57 quarter-hour intervals of which the jth corresponds to $(\tau_0 + .25(j - 1), \tau_0 + .25j]$. Then

$$\lambda_j := \text{intensity in interval}(\tau_0 + .25(j - 1), \tau_0 + .25j], \qquad (10.19)$$
$$\Lambda_0 := 0,$$
$$\Lambda_j := \Lambda_{j-1} + .25\lambda_j, \qquad j = 1, \ldots, 57.$$

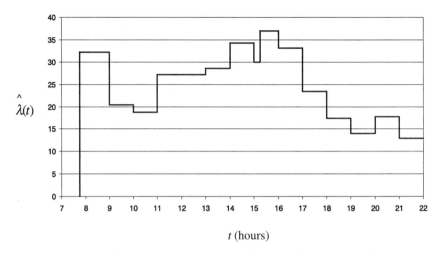

Figure 10.9   Sample instantaneous intensity function for book clearings

In terms of definitions (10.16) and (10.17),

$$\Lambda_j = \Lambda(\tau_0 + .25\,j) \tag{10.20}$$

$$= \int_{\tau_0}^{\tau_0 + .25\,j} \lambda(t)\mathrm{d}t = \sum_{i=1}^{j} \lambda_j, \qquad j = 1, \ldots, 57.$$

By contrast with the need to generate arrival times in the photocopying problem in Section 10.14, the clearing problem provides a daily sequence of clearing times and requires the simulationist to sample B, the number of books cleared, at time τ from $\mathcal{P}(\Lambda(\tau) - \Lambda(t))$, where $t < \tau$ denote the times of two successive clearings on a single day. Algorithm CLEAR does this by retaining $\Lambda_{\mathrm{old}} := \Lambda(t)$ from the previous sampling.

ALGORITHM CLEAR

Purpose: To sample B, the number of cleared books.
Input: Clearing time $\tau \in (\tau_0, \tau_1]$, $\{\lambda_j, j = 1, \ldots, 57\}$, and $\{\Lambda_j; \ j = 1, \ldots, 57\}$.
Comment: Λ_{old} is retained on successive calls within a day.
Method:

  If first tour of day, $\Lambda_{\mathrm{old}} \leftarrow 0$.
  $l \leftarrow 1 + \lceil 4(\tau - \tau_0) \rceil$.
  $\Lambda_{\mathrm{new}} \leftarrow \Lambda_{l-1} + \lambda_l[\tau - 4(l - 1)]$.
  Sample B from $\mathcal{P}(\Lambda_{\mathrm{new}} - \Lambda_{\mathrm{old}})$.
  $\Lambda_{\mathrm{old}} \leftarrow \Lambda_{\mathrm{new}}$.
  Return B.

Definition (10.18) provides a relatively convenient basis for computing a piecewise constant approximation to the instantaneous intensity function $\{\lambda(t)\}$ when the system under study is stationary. In the present case, a day consists of 14.25 hours starting at hour 7.75 and ending at hour 22. If simulated time consists of a sequence of days of 14.25 hours, then for all $t \geq 7.75$ (in hours) stationarity implies that

$$\lambda(t + 14.25j) = \lambda(t), \qquad j = 0, \pm1, \pm2, \ldots. \tag{10.21}$$

Likewise, for the example in Section 10.12.1, if simulated time consists of a sequence of seven-hour days beginning at hour 10 and ending at hour 17, then for all $t \geq 10$ its instantaneous intensity function satisfies

$$\lambda(t + 7j) = \lambda(t), \qquad j = 0, \pm1, \pm2, \ldots. \tag{10.22}$$

Note that the assignment (10.17) satisfies expression (10.22), whereas the assignment (10.18) satisfies expression (10.21).

The forms of these intensity functions suggest another more formal method of estimating $\{\lambda(t)\}$ that exploits their inherent periodicities. In this alternative, $\{\lambda(t)\}$ assumes the form

$$\lambda(t) = \exp[h(t, p, \boldsymbol{\theta})], \qquad t \geq \tau_0, \tag{10.23}$$

where τ_0 denotes the simulation's starting time,

$$h(t, p, \boldsymbol{\theta}) := a_0 + \sum_{k=1}^{p} a_k \sin(b_k t + c_k), \tag{10.24}$$

and

$$\boldsymbol{\theta} := (a_1, \ldots, a_p; b_1, \ldots, b_p; c_1, \ldots, c_p).$$

Here $\{\lambda(t)\}$ is an everywhere continuous and differentiable function, allowing us to regard the phenomenon being modeled as one whose intensity changes relatively smoothly rather than abruptly with time, as is the case in assignments (10.17) and (10.18). Then the challenge is to derive estimates of p and $\boldsymbol{\theta}$, which, when substituted into $\{h(t, p, \boldsymbol{\theta})\}$, make operational the sampling from a nonhomogeneous Poisson process.

Some of the parameters in $\boldsymbol{\theta}$ may be known in advance of estimation. This occurs in the book-clearing problem and in the arrival process for photocopying in Section 10.12.1. For the clearing problem, we have

$$\tau_0 = 7.75,$$
$$b_j = 2\pi j/14.25, \qquad j = 1, \ldots, p,$$

where b_1 is the daily fundamental frequency of book clearing intensity and $b_2, \ldots, b_p$ are its first $p - 1$ harmonics. For the photocopying problem,

$$\tau_0 = 10,$$
$$b_j = 2\pi j/7, \qquad j = 1, \ldots, p.$$

In both cases the relevant p is unknown and needs to be estimated along with $a_1, \ldots, a_p$ and $c_1, \ldots, c_p$.

The form (10.23) guarantees that $\lambda(t)$ is nonnegative for all $t \geq \tau_0$. Moreover, it allows us to cast parameter estimation in terms of the likelihood function. Recall that $N(t)$ denotes the number of arrivals or more generically, the number of events in $(0, t]$. If $N(t)$ is from $\mathcal{P}(\Lambda(t))$, then

$$\mathrm{pr}[N(t_2) - N(t_1) = j] = \frac{[\Lambda(t_1, t_2)]^j e^{-\Lambda(t_1, t_2)}}{j!}, \qquad j = 0, 1, \ldots,$$

where

$$\Lambda(t_1, t_2) := \Lambda(t_2) - \Lambda(t_1), \qquad t_1 \leq t_2.$$

Therefore,

$$\mathrm{pr}[N(t_2) - N(t_1) = 1] = \Lambda(t_1, t_2)e^{-\Lambda(t_i, t_2)} \tag{10.25}$$
$$= \text{probability that exactly one event occurs in}(t_1, t_2).$$

Suppose $\{\lambda(t)\}$ has the form (10.23) and that $\mathbf{Z}_n = \{Z_0 := t_0 < Z_1 < \cdots < Z_n\}$ denotes a sample sequence of times at which n successive events occur. Then expression (10.25) implies the likelihood function

$$L(p, \boldsymbol{\theta}, \mathbf{Z}_n) = \Pi_{j=1}^n \, \Lambda(Z_{j-1}, Z_j) \exp[-\Lambda(Z_{j-1}, Z_j)]$$

and therefore, the loglikelihood function

$$\ln L(p, \boldsymbol{\theta}, \mathbf{Z}_n) = na_0 + \sum_{j=1}^n \sum_{k=1}^p a_k \, \sin(b_k Z_j + c_k) - \int_{Z_0}^{Z_n} e^{h(z, p, \theta)} \, dz.$$

Then the m.l.e. are the solution to

$$\frac{\partial \ln L(p, \boldsymbol{\theta}, \mathbf{Z}_n)}{\partial a_j} = 0, \tag{10.26}$$

$$\frac{\partial \ln L(p, \boldsymbol{\theta}, \mathbf{Z}_n)}{\partial c_j} = 0, \qquad j = 1, \ldots, p,$$

when $b_1, \ldots, b_p$ are known. If they are not, then the additional equations

$$\frac{\partial \ln L(p, \boldsymbol{\theta}, \mathbf{Z}_n)}{\partial b_j} = 0, \qquad j = 1, \ldots, p,$$

apply as well.

The form (10.26) severely impedes a direct approach to computing the m.l.e. Kuhl et al. (1997) describe an iterative technique that overcomes the impediment by exploiting approximations that lead to useful initial values for iterative estimation. Public-domain software that implements their procedures is available at http://www.ie.ncsu.edu/jwilson/page3.html. It also allows for estimation of $\{\lambda(t)\}$ when it contains an exponential polynomial trend. For an alternative approach to estimating $\{\lambda(t)\}$, see Kao and Chang (1988).

10.15 LESSONS LEARNED

- Every discrete-event simulation requires a distribution function (d.f.) to generate samples for each source of stochastic variation it models.

- If a historical data record is available and is sufficiently large, then its empirical distribution function (e.d.f.) may provide an adequate basis for sampling. When this is not so, fitting a theoretical distribution is the next best alternative.

- Every discrete-event simulation requires the assignment of numerical values to all parameters of the theoretical distributions from which stochastic behavior is to be generated. When available, historical data provide the basis for estimating these values.

- The method of maximum likelihood provides a means for estimating the parameters of many distributions. These estimates have optimal properties. Iterative numerical methods need to be used in most cases. In some special cases other methods of estimation may prove more convenient without sacrificing much of this optimality.

- For a specified historical data record, a graphical comparison of its e.d.f. with the fitted d.f. for a particular distribution provides an assessment of how good the fit is. It is especially important for detecting discrepancies in the tails.

- Some data suggest distributions with tails that diminish exponentially. Other data suggest tails that diminish polynomially. The ability to distinguish between tail behaviors increases as the size of the historical data record increases.

- Expert opinion often needs to be used in conjunction with historical data in assigning values to parameters. In the absence of historical data, expert opinion provides the most useful guidance for setting parameter values.

- Regression methods provide a convenient way to formulate generators for conditional sampling.

- Interarrival times are frequently modeled as i.i.d. random variables. When time of day, week, month, or year affects the arrival rate, this assumption is inappropriate. The concept of a nonhomogeneous Poisson process provides a means of accounting for this dependence in generating samples.

10.16 REFERENCES

AbouRizk, S.M., D.W. Halpin, and J.R. Wilson (1994). Fitting Beta distributions based on sample data, *Journal of Construction Engineering and Management*, **120**, 288–305.

Ahrens, J.II., and K.D. Kohut (1981). Computer methods for efficient sampling from largely arbitrary statistical distributions, *Computing*, **26**, 19–31.

Beckman, R.J., and G.L. Tietjen (1978). Maximum likelihood estimation for the Beta distribution, *J. Statist. Comput. Simul.*, **7**, 253–258.

Brockwell, P.J., and R.A. Davis (1991). *Time Series: Theory and Methods*, second edition, Springer-Verlag, New York.

Conover, W.J. (1999). *Practical Nonparametric Statistics*, third edition, Wiley, New York.

Fielitz, B.D., and B.L. Myers (1975). Estimation of parameters in the Beta distribution, *Decision Sciences*, **6**, 1–13.

Fishman, G.S. (1973). *Concepts and Methods in Discrete Event Digital Simulation*, Wiley, New York.

Gnanadesikan, R., R.S. Pinkham, and L.P. Hughes (1967). Maximum likelihood estimation of the parameters of the Beta distribution from smallest order Statistics, *Technometrics*, **9**, 607–620.

Gnedenko, B.V., and I.N. Kovalenko (1989). *Introduction to Queueing Theory*, Birkhäuser, Boston.

Grigelionis, B.I. (1962). Accuracy of approximation of a superposition of renewal processes by a Poisson process, *Litovskiĭ Mathemalicheskiĭ Sbornik*, **II**, (2), 135–143.

Johnson, N.L., S. Kotz, and N. Balakrishnan (1994a). *Continuous Univariate Distributions, Volume I*, second edition, Wiley, New York.

Johnson, N.L., S. Kotz, and N. Balakrishnan (1994b). *Continuous Univariate Distributions, Volume II*, second edition, Wiley, New York.

Kao, E.P.C., and S.-L. Chang (1988). Modeling time-dependent arrivals to service systems: a case in using a piecewise-polynomial rate function in a nonhomogeneous Poisson process, *Man. Sci.*, **34**, 1367–1379.

Köllerstrom, J. (1974). Heavy traffic theory for queues with several servers. I, *J. Appl. Prob.*, **11**, 544–552.

Kuhl, M.E., J.R. Wilson, and M.A. Johnson (1997). Estimating and simulating Poisson processes having trends or multiple periodicities, *IIE Transactions*, **29**, 201–211.

Kulkarni, V.G. (1995). *Modeling and Analysis of Stochastic Systems*, Chapman and Hall, London.

Law, A.M., and W.D. Kelton (1991). *Simulation Modeling and Analysis*, McGraw-Hill, New York.

Lewis, P.A.W., and J. Shedler (1979). Simulation of nonhomogeneous Poisson process by thinning, *Nav. Res. Logist. Quart.*, **26**, 403–413.

Nelson, B.L., and M. Yamnitsky (1998). Input modeling tools for complex problems, *1998 Winter Simulation Conference Proceedings*, D.J. Medeiros, E.F. Watson, J.S. Carson, and M.S. Manivannan, eds., Association for Computing Machinery, New York.AU:see editors note

Ross, S.M. (1980). *Introduction to Probability Models*, Academic Press, New York.

Wagner, M.A., and J.R. Wilson (1996). Using univariate Bézier distributions to model simulation input processes, *IIE Transactions*, **28**, 699–711.

Zacks, S. (1971). *The Theory of Statistical Inference*, Wiley, New York.

APPENDIX

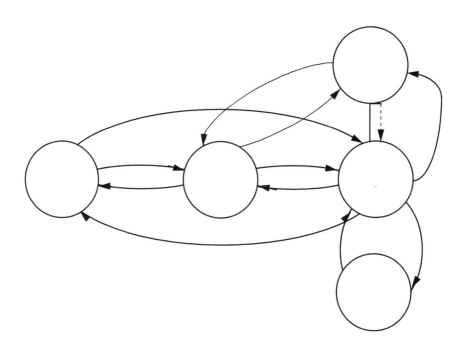

More on the Batch-Means Method

A.1 FNB RULE (SECTION 6.6.4)

Procedure FNB describes the principal steps for implementing the FNB rule in Section 6.6.4.

Procedure A.1 (FNB (LABATCH.2)).

Given: l_1, b_1, and t $(> l_1 b_1)$

$l \leftarrow l_1$ (fix # of batches)

$J(t) \leftarrow 1 + \lfloor \log(t/l_1 b_1)/\log 2 \rfloor$ (# of reviews)

$t'(t) \leftarrow 2^{J(t)-1} l_1 b_1$ (# of observations used on review $J(t)$)

$j \leftarrow 1$, $t_0 \leftarrow 0$, $t_1 \leftarrow l_1 b_1$,

While $t_j \leq t'(t)$:

    Collect $X_{t_{j-1}+1}, \ldots, X_{t_j}$ additional observations

    Compute $\bar{X}_{t_j}$

    Form batches of size b_j based on $X_1, \ldots, X_{t_j}$

    Compute W_{lb_j}

    $b_{j+1} \leftarrow 2b_j$

    $j \leftarrow j + 1$

$$t_j \leftarrow 2t_{j-1} \quad (= l_1 b_j)$$

Collect observations $X_{t'(t)+1}, \ldots, X_t$

Compute $\bar{X}_t$

Output:

$\bar{X}_t$ is the point estimate of μ.

$\mathcal{C}_l(\delta) \leftarrow [\bar{X}_t \pm \tau_{l_{-1}}(1 - \delta/2)\sqrt{b_{J(t)}W_{lb_{J(t)}}/t}\,]$ is the approximating $100 \times (1 - \delta)$ percent confidence interval for μ.

$\{b_j W_{lb_j}; \; j = 1, \ldots, J(t)\}$ is the sequence of successive estimates of σ_∞^2.

As an example, consider again the M/M/1 simulation queueing model with $\nu = .90$ interarrival rate and unit service rate. The simulation began in the steady state, used $l_1 = 7$ and $b_1 = 5$, and terminated when customer $t = 10^7$ entered service. Figure A.1 shows the LABATCH.2-generated sequences $\{\sqrt{V_{t_j}} = \sqrt{b_j W_{l_1 b_j}}; \; j = 1, \ldots, J(t)\}$. The graphs reveal a tendency for V_{t_j} initially to increase with j and eventually to fluctuate around their respective true σ_∞'s.

Figures A.1a and A.2b reveal an additional property. After dissipating systematic variance error, the size of fluctuations remains relatively constant in both graphs, in contrast to the behavior in Figures 6.7a and 6.7b. The origins of this constancy are not hard to find. For $\{l(t_j) = l_1, \; j \geq 1\}$, $\mathrm{E}V_{t_{J(t)}} \to \sigma_\infty^2$ as $t \to \infty$, but

$$\mathrm{var}\, V_{t_{J(t)}} \to \frac{2\sigma_\infty^4 (l_1 + 1)}{(l_1 - 2)^2} + O(1/l_1^2) \quad \text{as } t \to \infty, \tag{A.1}$$

implying that V_t does not converge in mean square and thus is not a consistent estimator of σ_∞^2. However, under relatively weak conditions on $\{X_i\}$, the limit (6.53) holds, implying that $\mathcal{C}_{l_1}(\delta)$ remains an asymptotically valid confidence interval for μ.

Although this settles the issue of validity, it does not address the issue of statistical efficiency, which, in the present case, can be measured by $2\tau_{l_{1-1}}(1 - \delta/2)\sqrt{V_{t_{J(t)}}/t}$, the width of the confidence interval for μ. Expanding $V_t^{1/2}$ in Taylor series about σ_∞ leads to

$$V_t^{1/2} = \sigma_\infty + \frac{V_t - \sigma_\infty^2}{2\sigma_\infty} - \frac{(V_t - \sigma_\infty^2)^2}{4\sigma_\infty^3} + \cdots, \tag{A.2}$$

which, together with expressions (6.50) and (6.51), leads to

$$\mathrm{E}[\tau_{l_{1-1}}(1 - \delta/2)V_{t_{J(t)}}^{1/2}] = \tau_{l_{1-1}}(1 - \delta/2)[1 - \frac{l_1 + 1}{2(l_1 - 1)^2}]\sigma_\infty \quad \text{as } t \to \infty. \tag{A.3}$$

However, if V_t were a strongly consistent estimator of σ_∞^2, then necessarily $l_{J(t)} \to \infty$ as $t \to \infty$, implying that

$$\mathrm{E}[\tau_{l_{J(t)}-1}(1 - \delta/2)V_{t_{J(t)}}^{1/2}] \to \Phi^{-1}(1 - \delta/2)\sigma_\infty \quad \text{as } t \to \infty.$$

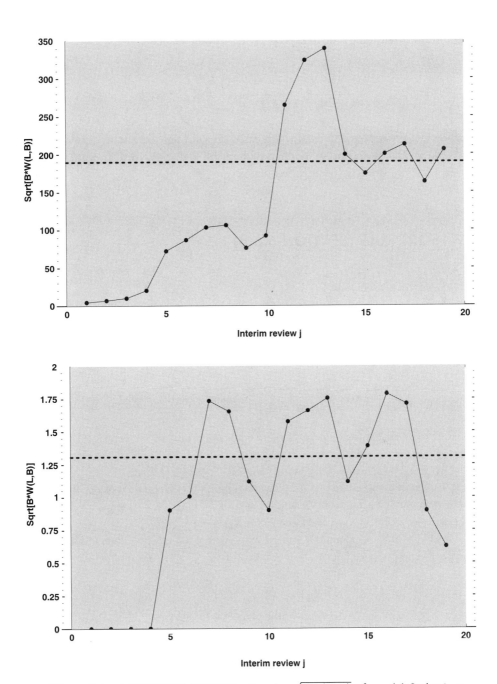

Figure A.1 LABATCH.2 ABATCH estimates $\sqrt{\mathrm{BW(L, B)}}$ of σ_∞, (a) Series 1, $\sigma_\infty = 189.5$ (b) Series 2, $\sigma_\infty = 1.308$

Therefore, for fixed δ, the ratio of expected widths would be

$$\frac{\tau_{l_1-1}(1-\delta/2)\left[1-\frac{l_1+1}{2(l_1-1)^2}\right]}{\Phi^{-1}(1-\delta/2)} \geq 1, \quad \forall l_1 \geq 2, \tag{A.4}$$

revealing that the expected interval width based on a strongly consistent estimator of σ_∞^2 never exceeds the mean interval width based on the FNB rule. In this regard, statistical efficiency favors a strongly consistent estimator.

A.2   SQRT RULE (SECTION 6.6.5)

Let

$$Z_{lb} := \frac{\bar{X}_t - \mu}{\sqrt{W_{lb}/l}} \tag{A.5}$$

and let F_{lb} denote the distribution function (d.f.) of Z_{lb} with Edgeworth expansion

$$F_{lb}(z) = \Phi(z) + \left\{ \frac{\kappa_3(Z_{lb})}{6} - \kappa_1(Z_{lb}) + \left[\frac{\kappa_4(Z_{lb})}{8} - \frac{\kappa_2(Z_{lb}) - 1}{2}\right] z - \frac{\kappa_3(Z_{lb})}{6} z^2 \right.$$
$$\left. - \frac{\kappa_4(Z_{lb})}{24} z^3 + O(1/l) \right\} \frac{1}{\sqrt{2\pi}} e^{-z^2/2}, \quad -\infty < z < \infty,$$

where Φ denotes the standard normal d.f. and $\kappa_i(Z_{lb})$, the ith cumulant of Z_{lb}. If κ_i denotes the ith cumulant of a random variable Y, say for $i = 1, \dots, 4$, then $\kappa_1 = EY$, $\kappa_2 = E(Y - \kappa_1)^2$, $\kappa_3 = E(Y - \kappa_1)^3$, and $\kappa_4 = E(Y - \kappa_1)^4 - 3\kappa_2^2$.

If $E|X_1 - \mu|^{20} < \infty$ and $\{X_i\}$ is ϕ-mixing with $\phi_i = O(i^{-13})$, then (Chien 1989, p. 46)

$$\kappa_1(Z_{lb}) = O(1/l^{1/2})O(1/b^{1/2}),$$
$$\kappa_2(Z_{lb}) = 1 + O(1/l) + O(1/b),$$
$$\kappa_3(Z_{lb}) = O(1/l^{1/2})O(1/b^{1/2}),$$
$$\kappa_4(Z_{lb}) = O(1/l).$$

Actually, Chien gives slightly weaker conditions for the first three cumulants.

As an immediate consequence,

$$F_{lb}(z) - \Phi(z) = O(1/l^{1/2})O(1/b^{1/2}) + O(1/l) + O(1/b), \tag{A.6}$$

revealing that $l(t) \propto t^{1/2}$ and $b(t) \propto t^{1/2}$ induce the fastest convergence of the true coverage rate to the specified theoretical coverage rate $1 - \delta$. In this case, V_t has $O(1/t^{1/2})$ systematic error and $O(1/t^{1/2})$ random error. To exploit these convergence properties, the SQRT rule defines $\{(l_j, b_j); \; j \geq 1\}$ as in expressions (6.55) and (6.56).

Procedure SQRT describes the principal steps in employing the SQRT rule.

Procedure A.2 (SQRT (LABATCH.2)).

Given: l_1, b_1, and $t(> l_1 b_1)$

$\tilde{l}_1 \leftarrow \lfloor \sqrt{2}\, l_1 + .5 \rfloor$

$\tilde{b}_1 \leftarrow 3$; if $b_1 > 1$, $\tilde{b}_1 \leftarrow \lfloor \sqrt{2}\, b_1 + .5 \rfloor$

$J(t) \leftarrow 1 + \lfloor \log(t/l_1 b_1)/\log 2 \rfloor$ (# of reviews)

$t'(t) \leftarrow 2^{J(t)-1} l_1 b_1$ (# of observations used on review $J(t)$)

$j \leftarrow 1, \; t_0 \leftarrow 0, \; t_1 \leftarrow l_1 b_1, \; t_2 \leftarrow \tilde{l}_1, \tilde{b}_1$

While $t_j \leq t'(t)$:

    Collect $X_{t_{j-1}+1}, \ldots, X_{t_j}$ additional observations

    Compute $\bar{X}_{t_j}$

    Compute $W_{l_j b_j}$

    If $j = 1$:

        $l_{j+1} \leftarrow \tilde{l}_1$ and $b_{j+1} \leftarrow \tilde{b}_1$

    Otherwise:

        $l_{j+1} \leftarrow l_{j-1}$ and $b_{j+1} \leftarrow b_{j-1}$

    $j \leftarrow j + 1$

    $t_j \leftarrow l_j b_j$

Collect observations $X_{t'(t)+1}, \ldots, X_t$

Compute $\bar{X}_t$

Output:

    $\bar{X}_t$ is the point estimate of μ.

    $\mathcal{C}_{l_{J(t)}}(\delta) \leftarrow [\bar{X}_t \pm \tau_{l_{J(t)}-1}(1 - \delta/2)\sqrt{b_{J(t)} W_{l_{J(t)} b_{J(t)}}/t}\,]$ is the approximating

        $100 \times (1 - \delta)$ percent confidence interval for μ.

    $\{b_j W_{l_j b_j}; \; j = 1, \ldots, J(t)\}$ is the sequence of successive estimates of σ_∞^2.

Choosing (l_1, b_1) from $\mathcal{B}$ in Table 6.3 ensures that $2l_1 b_1 = \tilde{l}_1 \tilde{b}_1$, so that $t_j = l_j b_j = 2^{j-1} l_1 b_1$, and therefore $t_{j+1}/t_j = 2$, as desired. This constraint proves valuable in Section 6.6.6, which describes batch-size rules that combine the FNB and the SQRT rules.

As illustration, we again use the M/M/1 simulation example with a steady-state sample record $X_1, \ldots, X_t$ of waiting times in queue with $t = 10^7$, $l_1 = 7$, and $b_1 = 5$, so that $\tilde{l}_1 = 10$ and $\tilde{b}_1 = 7$. Thus $2l_1 b_1 = \tilde{l}_1 \tilde{b}_1$, LABATCH.2 generates $J(t) = 19$ interim reviews, and the last review uses $t'(t) = 9, 175, 040$ observations.

Figure A.2 compares $\{\sqrt{b_j W_{l_j b_j}}; \; j = 1, \ldots, 19\}$ for the FNB and SQRT rules. After review 12, the FNB sequence fluctuates around the true σ_∞, whereas the SQRT sequence

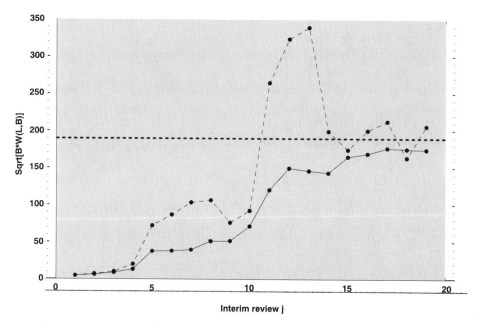

Figure A.2 LABATCH.2 estimation of σ_∞ for waiting time in queue (FNB rule: light dashed line, SQRT rule: solid line, $\sigma_\infty = 189.3$: heavy dashed line)

comes close to σ_∞ after review 15. This contrast in behavior reflects the $O(1/t_j)$ systematic variance error dissipation rate for the FNB rule as compared to the $O(1/t_j^{1/2})$ rate for the SQRT rule. However, sampling fluctuations for the FNB sequence are $O(1)$, in contrast to $O(1/t_j^{1/2})$ for the SQRT rule. We conclude that the FNB rule offers a benefit initially, but the SQRT rule ultimately offers an advantage. The next section describes batching rules that exploit the benefits of both rules.

A.3   CHOOSING t, l_1, AND b_1 (SECTION 6.6.6)

For given $(l_1, b_1) \in \mathcal{B}$, choosing t such that $\log(t/l_1 b_1)/\log 2$ is an integer results in $t'(t) = t$. As a consequence, the LBATCH and ABATCH rules use all t observations to estimate $W_{L_{J(t)} B_{J(t)}}$ and $\bar{X}_{t'(t)} = \bar{X}_t$. Since choosing l_1, b_1, and t subject to the constraint may be too burdensome for some users, LABATCH.2 merely requires that t be specified and then chooses l_1 and b_1 to maximize $t'(t)$.

Let $\mathcal{B}(t)$ denote the subset of $\mathcal{B}$ that maximizes $t'(t)$. LABATCH.2 chooses (l_1, b_1) to be the element of $\mathcal{B}(t)$ that maximizes $J(t)$, the number of interim reviews. As the following example shows, this algorithm reduces the number of elements in $\mathcal{B}$ that need to be considered in maximizing $t'(t)$. Suppose that (a_1, a_2) and $(a_3, a_4) \in \mathcal{B}$ maximize $t'(t)$ and that for some integer $\alpha \geq 1$, $a_1 a_2 = 2^\alpha a_3 a_4$. Then clearly, the choice (a_1, a_2) induces

Table A.1 Data utilization for final estimate of σ^2_∞ $\min_t \left[\frac{1}{t} \max_{(l_1, b_1) \in \mathcal{B}} t'(t) \right]$

	$10 \leq t \leq 24$	$25 \leq t \leq 49$	$50 \leq t \leq 99$	$100 \leq t \leq 499$	$500 \leq t \leq 10^7$
$l_1 \leq 10$	.522	.706	.696	.688	.686
$l_1 \leq 20$	.522	.706	.696	.798	.795
$l_1 \leq 30$	.522	.706	.696	.805	.889
$l_1 \leq 100$	.522	.706	.696	.805	.898

more reviews than (a_3, a_4). As a consequence of this example, LABATCH.2 considers only the underlined 2-tuples in $\mathcal{B}$ in Table 6.3 in maximizing $t'(t)$. As illustration, $t'(10^7) = 9,175,040$ is maximal for each 2-tuple in

$$\mathcal{B}(10^7) = \{(7, 5), (14, 5), (25, 5), (10, 7), (20, 7), (14, 10), (28, 10), (20, 14), (28, 20)\},$$

but $(7, 5)$ allows for the maximal number of reviews, $J(10^7) = 19$.

For $10 \leq t \leq 10^7$, Table A.1 shows the smallest proportion of observations used for $W_{L_{J(t)} B_{J(t)}}$ when $t'(t)$ is maximized subject to alternative upper bounds on l_1. Recall that $l_1 \leq 100$ implies optimization over all entries in $\mathcal{B}$. In this case, choosing $t \geq 500$ implies that no less than 89.8 percent of the data is used for computing $W_{L_{j(t)} B_{J(t)}}$. Alternatively, imposing the constraint $l_1 \leq 30$ in choosing (l_1, b_1) to maximize $t'(t)$ reduces this percentage to 88.9. We recommend either of these options for $t \geq 500$.

We now reconcile

$$A_t := \frac{\bar{X}_t - \mu}{\sqrt{B_{J(t)} W_{L_{J(t)} B_{J(t)}} / t}}, \tag{A.7}$$

whose properties establish the basis for the confidence interval (6.48), with $l(t) = L_{J(t)}$, $b(t) = B_{J(t)}$ and $W_{l(t)b(t)} = V_{t'(t)} / B_{J(t)}$ and

$$G_t := \frac{\bar{X}_{t'(t)} - \mu}{\sqrt{B_{J(t)} W_{L_{J(t)} B_{J(t)}} / t'(t)}}, \tag{A.8}$$

whose properties are described in Theorem 6.6 for $j = J(t)$. Recall that $t'(t) = t_{J(t)} = L_{J(t)} B_{J(t)}$. Whereas A_t uses all t observations to estimate μ, G_t uses only the first $t'(t)$ observations. However, both use $B_{J(t)} W_{L_{J(t)} B_{J(t)}}$ as an estimate of σ^2_∞. Since $(\bar{X}_{t'(t)} - \mu) / \sqrt{\sigma^2_\infty / t'(t)}$ and $(\bar{X}_t - \mu) / \sqrt{\sigma^2_\infty / t}$ are both asymptotically N $(0,1)$, by ASA, and since $B_{L_{J(t)}} W_{L J(t) B_{J(t)}}$ is strongly consistent, it appears that a modification of the proof of Theorem 1 in Yarberry (1993) can yield asymptotic normality for A_t as well. With regard to statistical efficiency, A_t reduces interval width by a factor of $1 - \sqrt{t'(t)/t} \leq 1 - \sqrt{.889} = .0571$ when $(l_1 b_1) \in \mathcal{B}(t)$ for $t \geq 500$.

A.4 REFERENCES

Chien, C.H. (1989). Small-sample theory for steady state confidence intervals, Technical Report-37, Department of Operations Research, Stanford University.

Yarberry, L.S. (1993). Incorporating a dynamic batch size selection mechanism in a fixed-sample-size batch means procedure, unpublished Ph.D. dissertation, Department of Operations Research, University of North Carolina, Chapel Hill.

Author Index

Subject Index